RNA Structure and Folding

Also of Interest

Methods in Protein Biochemistry
Harald Tschesche (Ed.), 2011
ISBN 978-3-11-025233-0, e-ISBN 978-3-11-025236-1

Advances in Algal Cell Biology
Kirsten Heimann, Christos Katsaros (Eds.), 2012
ISBN 978-3-11-022960-8, e-ISBN 978-3-11-022961-5

Kallikrein-Related Peptidases: Characterization, Regulation, and Interactions within the Protease Web
Viktor Magdolen, Christian P. Sommerhoff, Hans Fritz, Manfred Schmitt (Eds.), 2012
ISBN 978-3-11-026036-6, e-ISBN 978-3-11-026037-3

Kallikrein-Related Peptidases: Novel Cancer-Related Biomarkers
Viktor Magdolen, Christian P. Sommerhoff, Hans Fritz, Manfred Schmitt (Eds.), 2012
ISBN 978-3-11-030358-2, e-ISBN 978-3-11-030366-7

Handbook of Flavoproteins
Volume 1, Oxidases, Dehydrogenases and Related Systems
Russ Hille, Susan Miller, Bruce Palfey (Eds.), 2012
ISBN 978-3-11-026842-3, e-ISBN 978-3-11-026891-1

Handbook of Flavoproteins
Volume 2, Complex Flavoproteins, Dehydrogenase and Physical Methods
Russ Hille, Susan Miller, Bruce Palfey (Eds.), 2013
ISBN 978-3-11-029828-4, e-ISBN 978-3-11-029834-5

Biomolecular Concepts
Editor-in-Chief: Pierre Jollès, Isabelle Mansuy
ISSN 1868–503X

Biological Chemistry
Editor-in-Chief: Bernhard Brüne
ISSN 1437–4315

RNA Structure and Folding

Biophysical Techniques and Prediction Methods

Edited by
Dagmar Klostermeier and
Christian Hammann

DE GRUYTER

Editors
Professor
Dr. Dagmar Klostermeier
Institute for Physical Chemistry
University of Münster
Corrensstraße 30
48149 Münster
Germany
dagmar.klostermeier@uni-muenster.de

Professor
Dr. Christian Hammann
School of Engineering and Science
Jacobs University Bremen
Campus Ring 1
28759 Bremen
Germany
c.hammann@jacobs-university.de

ISBN 978-3-11-028459-1
e-ISBN 978-3-11-028495-9

Library of Congress Cataloging-in-Publication Data
A CIP catalog record for this book has been applied for at the Library of Congress.

Bibliographic information published by the Deutsche Nationalbibliothek
The Deutsche Nationalbibliothek lists this publication in the Deutsche Nationalbibliografie; detailed bibliographic data are available in the Internet at http://dnb.dnb.de.

© 2013 Walter de Gruyter GmbH, Berlin/Boston
The publisher, together with the authors and editors, has taken great pains to ensure that all information presented in this work (programs, applications, amounts, dosages, etc.) reflects the standard of knowledge at the time of publication. Despite careful manuscript preparation and proof correction, errors can nevertheless occur. Authors, editors and publisher disclaim all responsibility and for any errors or omissions or liability for the results obtained from use of the information, or parts thereof, contained in this work.

The citation of registered names, trade names, trademarks, etc. in this work does not imply, even in the absence of a specific statement, that such names are exempt from laws and regulations protecting trademarks etc. and therefore free for general use.

Typesetting: Apex CoVantage
Printing and binding: Hubert & Co. GmbH & Co. KG, Göttingen
Cover image: iStockphoto/Thinkstock

∞ Printed on acid-free paper
Printed in Germany
www.degruyter.com

Preface

In the past decade, we have witnessed unprecedented advances in RNA biochemistry. This is exemplified by the groundbreaking research by Fire and Mello, who discovered RNA interference as a central mechanism for gene regulation at the posttranscriptional level. It is now clear that small (and long) noncoding RNA molecules are involved in regulatory pathways on virtually all levels of the flow of genetic information. With the continuous discovery of novel RNA molecules with key cellular functions and of novel regulatory pathways and networks, there is an increasing need for structural information on RNA.

While methods to determine protein structures are now routinely used, studying structure-function relationships of RNA poses additional challenges. This volume provides an introduction to the variety of techniques available to assess the structure and folding of RNA. Each chapter explains the theoretical background of the technique and illustrates its possibilities and limitations in selected application examples.

Chapter 1 provides an overview of optical spectroscopy and calorimetric approaches as tools for the analysis of RNA structure and stability, structural changes, and the interaction with other molecules. The following chapters, Chapter 2 on RNA footprinting by Mélodie Duval et al. and Chapter 3 on RNA probing by Mark Helm, address the use of enzymatic and chemical approaches to obtain information on RNA structure at nucleotide level. In Chapter 4, Claudia Höbartner et al. review the opportunities that the field of chemical biology offers for selective labeling of RNA and how reactive handles can be introduced into RNA for *in vitro* and *in vivo* analyses. Many large RNAs are built up by branched structures, so-called helical junctions. David M. J. Lilley highlights in Chapter 5 how comparative gel electrophoresis can be employed to analyze the conformation of these structural elements and the global conformation of larger RNAs. Chapter 6 by Alexander McPherson introduces atomic force microscopy as a tool in the structural analysis of large RNA and gives a historical account of structural analyses of viruses and viral RNA. Optical tweezers allow measuring forces on the single molecule level. Mathilde Bercy et al. illustrate in Chapter 7 how these approaches can delineate RNA folding pathways and determine the stability of RNA structures. In Chapter 8, the theoretical background of fluorescence resonance energy transfer is presented, and its application to the analysis of RNA conformation and conformational changes from the ensemble to the single molecule level is illustrated. The recent advances of small angle X-ray scattering in the analysis of RNA structure and shape are detailed in the contribution by Alexey G. Kikhney et al. in Chapter 9, which also includes a summary of novel results obtained with this solution method. Cryo–electron microscopy (cryo-EM) allows for the global analysis of large RNA molecules and RNA-protein complexes. This technique is introduced in Chapter 10 by Jean-François Ménétret et al. along with a comparison of cryo-EM with other high-resolution structural approaches. Electron paramagnetic resonance allows for the determination of distances between spin labels attached to the RNA. Chapter 11 by Ivan Krstic et al. describes pulsed electron-electron double resonance spectroscopy and its potential for the structural analysis of RNA molecules. Structural information on RNA at atomic resolution can be obtained by nuclear magnetic resonance (NMR) techniques and by X-ray crystallography. In Chapter 12, Christina R. Mozes and Mirko Hennig summarize the different NMR approaches in the determination of RNA structure and dynamics in solution. Chapter 13 by Adrian R. Ferré-D'Amaré focuses on strategies for crystallization of RNA molecules and RNA-protein complexes as a prerequisite for structure determination by X-ray crystallography. The final two chapters address RNA structure prediction. Chapter 14, by Gerhard Steger and Robert Giegerich, compares different programs for RNA structure predictions, with an emphasis on different energetic constraints. In Chapter 15, Anton I. Petrov et al. summarize and compare prediction methods to identify structural motifs in RNA.

Altogether, this compilation of reviews represents a snapshot of the currently employed methods in the analysis of structure and folding of RNA. We thank all the authors of this volume for their contributions. We would also like to thank Stephanie Dawson and Julia Reindlmeier from De Gruyter for their encouragement and enthusiastic editorial support.

July 2013
Dagmar Klostermeier
Christian Hammann

List of contributing authors

Josette Banroques
CNRS UPR9073 associated with
Université Paris Diderot
Sorbonne Paris Cité
Institut de Biologie Physico-chimique
13 rue Pierre et Marie Curie
75005 Paris, France
Chapter 7

Mathilde Bercy
Laboratoire Nanobiophysique, ESPCI
10 rue Vauquelin
75005 Paris, France
Chapter 7

Thierry Bizebard
CNRS UPR9073 associated with
Université Paris Diderot
Sorbonne Paris Cité
Institut de Biologie Physico-chimique
13 rue Pierre et Marie Curie
75005 Paris, France
Chapter 7

Ulrich Bockelmann
Laboratoire Nanobiophysique, ESPCI
10 rue Vauquelin
75005 Paris, France
e-mail: ulrich.bockelmann@espci.fr
Chapter 7

Lea Büttner
Max Planck Institute for Biophysical Chemistry
Research Group Nucleic Acid
Chemistry
Am Fassberg 11
37077 Göttingen, Germany
Chapter 4

Sebastian Doniach
Applied Physics Department
Stanford University
Stanford, CA 94305, USA
Chapter 9

Mélodie Duval
Architecture et Réactivité de l'ARN, Université de
Strasbourg, IBMC-CNRS
15 rue René Descartes
67084 Strasbourg, France
Chapter 2

Burkhard Endeward
Institute of Physical and Theoretical Chemistry
Faculty of Biochemistry, Chemistry and
Pharmacy
Goethe University Frankfurt am Main
Max-von-Laue-Straße 7
60438 Frankfurt am Main, Germany
Chapter 11

Adrian Ferré-D'Amaré
Laboratory of RNA Biophysics and Cellular
Physiology
National Heart, Lung and Blood Institute
50 South Drive, MSC 8012
Bethesda, MD 20892–8012, USA
e-mail: adrian.ferre@nih.gov
Chapter 13

Olivier Fuchsbauer
Architecture et Réactivité de l'ARN, Université de
Strasbourg, IBMC-CNRS
15 rue René Descartes
67084 Strasbourg, France
Chapter 2

Robert Giegerich
Faculty of Technology and Center for
Biotechnology
Bielefeld University
Universitätsstraße 25
33615 Bielefeld, Germany
e-mail: robert@techfak.uni-bielefeld.de
Chapter 14

Claudia M. Grytz
Institute of Physical and Theoretical Chemistry
Faculty of Biochemistry, Chemistry and
Pharmacy
Goethe University Frankfurt am Main
Max-von-Laue-Straße 7
60438 Frankfurt am Main, Germany
Chapter 11

Airat Gubaev
Institute for Physical Chemistry
University of Münster
Corrensstraße 30
48149 Münster, Germany
e-mail: Ayrat.Gubaev@uni-muenster.de
Chapter 8

Christian Hammann
School of Engineering and Science
Jacobs University Bremen
Campus Ring 1
28759 Bremen, Germany
e-mail: c.hammann@jacobs-university.de
Chapter 1

Isabelle Hazemann
Centre for Integrative Biology (CBI)
Department of Integrated Structural Biology
IGBMC (Institute of Genetics and of Molecular
and Cellular Biology)
Centre National de la Recherche Scientifique
(CNRS) UMR 7104 / Institut National de la Santé
de la Recherche Médicale (INSERM) U964 /
Université de Strasbourg
1 rue Laurent Fries
67404 Illkirch, France
Chapter 10

Anne-Catherine Helfer
Architecture et Réactivité de l'ARN, Université de
Strasbourg, IBMC-CNRS
15 rue René Descartes
67084 Strasbourg, France
Chapter 2

Mark Helm
Institute of Pharmacy und Biochemistry
Johannes Gutenberg University Mainz
Staudinger Weg 5
55128 Mainz, Germany
e-mail: mhelm@uni-mainz.de
Chapter 3

Mirko Hennig
Department of Biochemistry and
Molecular Biology
Medical University of South Carolina
70 President Street
Charleston, SC 29425, USA
e-mail: hennig@musc.edu
Chapter 12

Claudia Höbartner
Max Planck Institute for Biophysical Chemistry
Research Group Nucleic Acid Chemistry
Am Fassberg 11
37077 Göttingen, Germany
e-mail: claudia.hoebartner@mpibpc.mpg.de
Chapter 4

Anne-Sophie Humm
Centre for Integrative Biology (CBI)
Department of Integrated Structural Biology
IGBMC (Institute of Genetics and of Molecular
and Cellular Biology)
Centre National de la Recherche Scientifique
(CNRS) UMR 7104 / Institut National de la Santé
de la Recherche Médicale (INSERM) U964 /
Université de Strasbourg
1 rue Laurent Fries
67404 Illkirch, France
Chapter 10

Fatemeh Javadi-Zarnaghi
Max Planck Institute for Biophysical Chemistry
Research Group Nucleic Acid Chemistry
Am Fassberg 11
37077 Göttingen, Germany
Chapter 4

List of contributing authors

Heena Khatter
Centre for Integrative Biology (CBI)
Department of Integrated Structural Biology
IGBMC (Institute of Genetics and of Molecular and Cellular Biology)
Centre National de la Recherche Scientifique (CNRS) UMR 7104 / Institut National de la Santé de la Recherche Médicale (INSERM)
U964 / Université de Strasbourg
1 rue Laurent Fries
67404 Illkirch, France
Chapter 10

Alexey G. Kikhney
European Molecular Biology Laboratory
Hamburg Outstation
Notkestraße 85, Geb. 25a
22603 Hamburg, Germany
Chapter 9

Bruno P. Klaholz
Centre for Integrative Biology (CBI)
Department of Integrated Structural Biology
IGBMC (Institute of Genetics and of Molecular and Cellular Biology)
Centre National de la Recherche Scientifique (CNRS) UMR 7104 / Institut National de la Santé de la Recherche Médicale (INSERM)
U964 / Université de Strasbourg
1 rue Laurent Fries
67404 Illkirch, France
e-mail: bruno.klaholz@igbmc.fr
Chapter 10

Dagmar Klostermeier
Institute for Physical Chemistry
University of Münster
Correnssstraße 30
48149 Münster, Germany
e-mail: Dagmar.klostermeier@uni-muenster.de
Chapter 1, 8

Ivan Krstic
Institute of Physical and Theoretical Chemistry
Faculty of Biochemistry, Chemistry and Pharmacy
Goethe University Frankfurt am Main
Max-von-Laue-Straße 7
60438 Frankfurt am Main, Germany
Chapter 11

Neocles B. Leontis
Chemistry Department
Bowling Green State University
Bowling Green, OH 43403–0001, USA
419-372-2031
e-mail: leontis@bgsu.edu
Chapter 15

David M. J. Lilley
Cancer Research UK Nucleic Acid Structure Research Group
MSI/WTB Complex
The University of Dundee
Dundee DD1 5EH, UK
e-mail: d.m.j.lilley@dundee.ac.uk
Chapter 5

Pierre Mangeol
Laboratoire Nanobiophysique, ESPCI
10 rue Vauquelin
75005 Paris, France
Chapter 7

Sankar Manicka
Centre for Integrative Biology (CBI)
Department of Integrated Structural Biology
IGBMC (Institute of Genetics and of Molecular and Cellular Biology)
Centre National de la Recherche Scientifique (CNRS) UMR 7104 / Institut National de la Santé de la Recherche Médicale (INSERM)
U964 / Université de Strasbourg
1 rue Laurent Fries
67404 Illkirch, France
Chapter 10

Andriy Marko
Institute of Physical and Theoretical Chemistry
Faculty of Biochemistry, Chemistry and Pharmacy
Goethe University Frankfurt am Main
Max-von-Laue-Straße 7
60438 Frankfurt am Main, Germany
Chapter 11

Stefano Marzi
Architecture et Réactivité de l'ARN, Université de
Strasbourg, IBMC-CNRS
15 rue René Descartes
67084 Strasbourg, France
Chapter 2

Alexander McPherson
Department of Molecular Biology and
Biochemistry
University of California Irvine
Irvine, CA 92697–3900, USA
e-mail: amcphers@uci.edu
Chapter 6

Jean-François Ménétret
Centre for Integrative Biology (CBI)
Department of Integrated Structural Biology
IGBMC (Institute of Genetics and of Molecular
and Cellular Biology)
Centre National de la Recherche Scientifique
(CNRS) UMR 7104 / Institut National de la Santé
de la Recherche Médicale (INSERM)
U964 / Université de Strasbourg
1 rue Laurent Fries
67404 Illkirch, France
Chapter 10

Kareem Mohideen
Centre for Integrative Biology (CBI)
Department of Integrated Structural Biology
IGBMC (Institute of Genetics and of Molecular
and Cellular Biology)
Centre National de la Recherche Scientifique
(CNRS) UMR 7104 / Institut National de la Santé
de la Recherche Médicale (INSERM)
U964 / Université de Strasbourg
1 rue Laurent Fries
67404 Illkirch, France
Chapter 10

Christina R. Mozes
Department of Structural Biology
University of Pittsburgh
Pittsburgh, PA 15213, USA
Chapter 12

Alexander G. Myasnikov
Centre for Integrative Biology (CBI)
Department of Integrated Structural Biology
IGBMC (Institute of Genetics and of Molecular
and Cellular Biology)
Centre National de la Recherche Scientifique
(CNRS) UMR 7104 / Institut National de la Santé
de la Recherche Médicale (INSERM)
U964 / Université de Strasbourg
1 rue Laurent Fries
67404 Illkirch, France
Chapter 10

Igor Orlov
Centre for Integrative Biology (CBI)
Department of Integrated Structural Biology
IGBMC (Institute of Genetics and of Molecular
and Cellular Biology)
Centre National de la Recherche Scientifique
(CNRS) UMR 7104 / Institut National de la Santé
de la Recherche Médicale (INSERM) U964 /
Université de Strasbourg
1 rue Laurent Fries
67404 Illkirch, France
Chapter 10

Anton I. Petrov
Bowling Green State University
Department of Chemistry
Bowling Green, OH 43403–0001, USA
e-mail: apetrov@bgsu.edu
Chapter 15

Thomas F. Prisner
Institute of Physical and Theoretical Chemistry
Faculty of Biochemistry, Chemistry and
Pharmacy
Goethe University Frankfurt am Main
Max-von-Laue-Straße 7
60438 Frankfurt am Main, Germany
e-mail: prisner@chemie.uni-frankfurt.de
Chapter 11

Pascale Romby
Architecture et Réactivité de l'ARN, Université de Strasbourg, IBMC-CNRS
15 rue René Descartes
67084 Strasbourg, France
e-mail: p.romby@ibmc.u-strasbg.fr
Chapter 2

Cedric Romilly
Architecture et Réactivité de l'ARN, Université de Strasbourg, IBMC-CNRS
15 rue René Descartes
67084 Strasbourg, France
Chapter 2

Angelita Simonetti
Centre for Integrative Biology (CBI)
Department of Integrated Structural Biology
IGBMC (Institute of Genetics and of Molecular and Cellular Biology)
Centre National de la Recherche Scientifique (CNRS) UMR 7104 / Institut National de la Santé de la Recherche Médicale (INSERM)
U964 / Université de Strasbourg
1 rue Laurent Fries
67404 Illkirch, France
Chapter 10

Gerhard Steger
Institute of Physical Biology
Heinrich Heine University Düsseldorf
Universitätsstraße 1, Building 26.12.U1
40225 Düsseldorf, Germany
e-mail: steger@biophys.uni-duesseldorf.de
Chapter 14

Dmitri I. Svergun
European Molecular Biology Laboratory
Hamburg Outstation
Notkestraße 85, Geb. 25a
22603 Hamburg, Germany
e-mail: Svergun@EMBL-Hamburg.de
Chapter 9

Blake A. Sweeney
Bowling Green State University
Department of Biological Sciences,
Bowling Green, OH 43403–0001, USA
e-mail: bsweene@bgsu.edu
Chapter 15

N. Kyle Tanner
CNRS UPR9073 associated with
Université Paris Diderot
Sorbonne Paris Cité
Institut de Biologie Physico-chimique
13 rue Pierre et Marie Curie
75005 Paris, France
Chapter 7

Morgan Torchy
Centre for Integrative Biology (CBI)
Department of Integrated Structural Biology
IGBMC (Institute of Genetics and of Molecular and Cellular Biology)
Centre National de la Recherche Scientifique (CNRS) UMR 7104 / Institut National de la Santé de la Recherche Médicale (INSERM)
U964 / Université de Strasbourg
1 rue Laurent Fries
67404 Illkirch, France
Chapter 10

Alexandre Urzhumtsev
Centre for Integrative Biology (CBI)
Department of Integrated Structural Biology
IGBMC (Institute of Genetics and of Molecular and Cellular Biology)
Centre National de la Recherche Scientifique (CNRS) UMR 7104 / Institut National de la Santé de la Recherche Médicale (INSERM)
U964 / Université de Strasbourg
1 rue Laurent Fries
67404 Illkirch, France
Chapter 10

Srividhya Krishnagiri Venkatasubramanian
Centre for Integrative Biology (CBI)
Department of Integrated Structural Biology
IGBMC (Institute of Genetics and of Molecular
and Cellular Biology)
Centre National de la Recherche Scientifique
(CNRS) UMR 7104 / Institut National de la Santé
de la Recherche Médicale (INSERM) U964 /
Université de Strasbourg
1 rue Laurent Fries
67404 Illkirch, France
Chapter 10

Contents

Preface — v

List of contributing authors — vii

Contents — xiii

1 Optical spectroscopy and calorimetry — 1
- 1.1 Introduction — 1
- 1.2 Absorption spectroscopy — 1
- 1.3 Fluorescence — 8
- 1.4 Circular dichroism — 8
- 1.5 Transient electric birefringence — 13
- 1.6 Calorimetry — 17
- 1.6.1 Isothermal titration calorimetry — 18
- 1.6.2 Differential scanning calorimetry — 21
- 1.7 Acknowledgments — 24
- References — 24

2 Footprinting methods for mapping RNA-protein and RNA-RNA interactions — 29
- 2.1 Introduction — 29
- 2.2 Principles and applications of footprinting — 30
- 2.3 Tools for footprinting: what should we know about probes? — 32
- 2.3.1 RNases — 32
- 2.3.2 Chemicals — 35
- 2.3.2.1 Base-specific reagents — 35
- 2.3.2.2 Ribose-phosphate backbone–specific reagents — 35
- 2.4 Examples of RNP or RNA-RNA complexes analyzed by footprinting — 37
- 2.4.1 Determination of the mRNA-binding site of Crc by SHAPE footprinting — 37
- 2.4.2 Footprinting mapping of sRNA-mRNA interaction — 39
- 2.4.3 Footprinting reveals mimicry of mRNA and tRNA for regulation — 41
- 2.4.4 Difficulties in probing transient interactions by footprinting: the case of ribosomal protein S1-RNA complex — 43
- 2.5 Concluding remarks — 45
- 2.6 Acknowledgments — 46
- References — 46

3	Chemical approaches to the structural investigation of RNA in solution —— 51
3.1	Introduction —— 51
3.2	Similar chemistry in different concepts: sequencing, probing, and interference —— 52
3.3	Sequencing and probing by Maxam and Gilbert chemistry —— 53
3.4	Application of Sanger sequencing to probing —— 57
3.5	Further electrophilic small molecule probes —— 58
3.6	Probing agents with nuclease activity —— 59
3.7	Probing agents involving radical chemistry —— 61
3.8	Matching suitable probes to structural features —— 61
3.9	Chemical modification interference —— 62
3.10	Nucleotide analog interference mapping —— 63
3.11	Combination and interplay with other methods —— 65
3.12	Application to an artificial ribozyme —— 66
3.13	Conclusion and outlook —— 69
	References —— 69

4	Bioorthogonal modifications and cycloaddition reactions for RNA chemical biology —— 75
4.1	Introduction —— 75
4.2	Bioorthogonal conjugation strategies —— 76
4.2.1	1,3-dipolar cycloaddition reactions ([3+2] cycloaddition) —— 76
4.2.1.1	Copper-catalyzed azide-alkyne cycloaddition —— 76
4.2.1.2	Strain-promoted azide-alkyne cycloaddition —— 77
4.2.1.3	Nitrile oxides as 1,3-dipoles for metal-free cycloadditions —— 78
4.2.1.4	Photoactivated 1,3-dipolar cycloadditions —— 80
4.2.2	Inverse electron demand Diels-Alder reaction ([4+2] cycloaddition) —— 80
4.2.3	Staudinger reaction of azides and phosphines —— 81
4.3	Synthetic strategies for RNA functionalization: installation of reactive groups for cycloadditions —— 81
4.3.1	Chemical synthesis of modified RNA —— 82
4.3.1.1	Alkyne-containing phosphoramidites for solid-phase synthesis —— 82
4.3.1.2	Solid-phase synthesis of azide-containing RNA —— 83
4.3.1.3	Postsynthetic modification of RNA with azides and alkynes —— 83
4.3.1.4	Functionality transfer reaction using s⁶G-modified DNA —— 84
4.3.2	Enzymatic incorporation of functional groups for click chemistry —— 85
4.3.2.1	In vitro transcription using modified nucleotides —— 86
4.3.2.2	Enzymatic posttranscriptional modification —— 86
4.4	Case studies for applications of click chemistry in RNA chemical biology —— 88

4.4.1	Synthesis of chemically modified ribozymes —— 88	
4.4.2	Monitoring RNA synthesis and turnover by metabolic labeling and click chemistry —— 90	
4.4.3	Bioorthogonal modification of siRNAs for detection, improved stability, and delivery —— 91	
4.5	Summary and conclusions —— 93	
4.6	Acknowledgments —— 93	
	References —— 94	
5	**Analysis of RNA conformation using comparative gel electrophoresis —— 101**	
5.1	The principle behind the analysis of the structure of branched nucleic acids by gel electrophoresis —— 101	
5.2	Helical discontinuities in duplex RNA —— 103	
5.3	The direction of a helical bend —— 104	
5.4	Comparative gel electrophoresis of branched nucleic acids —— 104	
5.5	Comparative gel electrophoresis of four-way DNA junctions —— 108	
5.6	Analysis of the structure of four-way RNA junctions —— 111	
5.7	The 4H junctions of the U1 snRNA and the hairpin ribozyme —— 112	
5.8	A more complex junction found in the HCV IRES —— 112	
5.9	Analysis of the structure of three-way RNA junctions —— 115	
5.9.1	A three-way junction of the HCV IRES element —— 115	
5.9.2	Three-way junctions are the key architectural elements of the VS ribozyme —— 116	
5.9.3	The hammerhead ribozyme is a complex three-way helical junction —— 118	
5.10	Some final thoughts —— 118	
5.11	Acknowledgments —— 120	
	References —— 120	
6	**Virus RNA structure deduced by combining X-ray diffraction and atomic force microscopy —— 125**	
6.1	Introduction —— 125	
6.2	Why don't we learn more about RNA from X-ray crystallography? —— 125	
6.3	X-ray studies revealing RNA —— 126	
6.4	Secondary structure prediction —— 128	
6.5	Generalized ssRNA secondary structural motifs —— 129	
6.6	The folding of RNA in STMV —— 131	
6.7	Atomic force microscopy —— 133	
6.8	Preparation of viral RNA samples for AFM —— 136	
6.9	Atomic force microscopy of viral ssRNAs —— 137	

6.10	AFM results for extended STMV RNA —— 140
6.11	ssRNA in T = 3 icosahedral viruses —— 144
6.12	A model for assembly of STMV inspired by crystallography and AFM —— 147
6.13	AFM of large ssRNA viruses —— 150
	References —— 152
7	**Investigating RNA structure and folding with optical tweezers —— 157**
7.1	Introduction —— 157
7.2	Single-RNA force measurements with optical tweezers —— 158
7.3	Probing RNA and RNA-protein interactions: selected examples —— 160
7.3.1	Probing the structure and the folding dynamics of RNA hairpins —— 160
7.3.2	Exploring the folding dynamics of complex RNA structures in presence of proteins —— 172
7.4	Conclusion —— 178
7.5	Acknowledgments —— 178
	References —— 178
8	**Fluorescence resonance energy transfer as a tool to investigate RNA structure and folding —— 181**
8.1	An introduction to fluorescence resonance energy transfer —— 181
8.2	Introduction of donor and acceptor fluorophores into RNAs and RNA/protein complexes —— 183
8.3	Ensemble FRET —— 184
8.3.1	Steady-state FRET —— 184
8.3.2	Time-resolved FRET —— 186
8.4	Single-molecule FRET —— 189
8.4.1	Instrumentation and experimental procedure —— 191
8.4.2	Data analysis —— 193
8.4.2.1	Identifying single-molecule events —— 193
8.4.2.2	Correction for instrument nonnonideality —— 193
8.4.2.3	The Förster distance R_0 —— 195
8.4.2.4	The orientation factor κ^2 —— 196
8.4.2.5	Analysis of FRET histograms —— 197
8.4.3	FRET data and RNA folding —— 198
8.4.4	From FRET data to structural models of RNA and RNA/protein complexes —— 198
8.5	Selected examples —— 199
8.5.1	Steady-state FRET: ribozymes, rRNA, and RNA polymerase transcription complexes —— 199
8.5.2	Time-resolved FRET: the hairpin ribozyme —— 204

8.5.3	Single-molecule FRET: folding of large ribozymes and transcription by RNA polymerases —— 206	
8.5.4	Single-molecule FRET and modeling of complex structures —— 207	
8.6	Perspectives —— 208	
8.7	Acknowledgments —— 209	
	References —— 209	
9	**RNA studies by small angle X-ray scattering in solution —— 215**	
9.1	Introduction to SAXS —— 215	
9.2	SAXS experiment —— 216	
9.2.1	Sample preparation —— 216	
9.2.2	Form and structure factor: particle interactions —— 217	
9.3	Methods —— 219	
9.3.1	Distance distribution function —— 219	
9.3.2	Overall parameters: radius of gyration, molecular mass, and volume —— 219	
9.4	Modeling —— 221	
9.4.1	Ab initio modeling —— 221	
9.4.1.1	Bead models —— 221	
9.4.1.2	Dummy residue models —— 222	
9.4.1.3	Multiphase models —— 222	
9.4.1.4	Comparison of multiple models —— 223	
9.4.2	SAXS and complementary methods —— 223	
9.4.2.1	High-resolution models —— 223	
9.4.2.2	Rigid body modeling —— 224	
9.4.3	Flexible systems —— 225	
9.4.4	Mixtures —— 225	
9.5	Resolution and ambiguity of SAXS data interpretation —— 226	
9.6	Practical applications —— 227	
9.6.1	Ab initio shape determination —— 227	
9.6.2	Analysis of RNA flexibility —— 229	
9.6.3	Nonstochiometric RNA-protein mixtures and complex formation —— 231	
9.6.4	Structural studies of spliceosome function assisted by SAXS measurements —— 232	
9.6.5	How SAXS helps elucidate riboswitch structure-function relationships —— 233	
9.6.6	Use of SAXS and ASAXS to study the influences of counterions on RNA folding —— 236	
9.6.7	Quantitation of free-energy changes estimated from SAXS 3-D reconstructions —— 236	
9.7	Conclusions and outlook —— 237	

9.8	Acknowledgments —— 238
	References —— 238

10	Integrative structure-function analysis of large nucleoprotein complexes —— 243
10.1	Summary —— 243
10.2	Integrative structure-function analysis of nucleoprotein complexes, example 1: translation complexes —— 250
10.3	Integrative structure-function analysis of nucleoprotein complexes, example 2: transcription complexes —— 253
10.4	Outlook —— 255
10.5	Acknowledgments —— 256
	References —— 257

11	Structure and conformational dynamics of RNA determined by pulsed EPR —— 261
11.1	Introduction —— 261
11.2	Pulse EPR spectroscopy on RNA —— 264
11.2.1	Spin labeling of nucleic acids —— 264
11.2.2	Theoretical description of the PELDOR experiment —— 266
11.2.3	Practical aspects of the PELDOR experiment —— 270
11.2.4	PELDOR experiments with rigid spin labels —— 271
11.2.5	Data analysis and interpretation —— 274
11.3	Application examples —— 276
11.3.1	Applications on dsRNA and DNA —— 277
11.3.2	Application on RNA with more complex structure —— 277
11.3.3	Applications on DNA with rigid spin labels —— 279
11.4	Outlook and summary —— 280
11.5	Acknowledgments —— 281
	References —— 282

12	NMR-based characterization of RNA structure and dynamics —— 287
12.1	Introduction —— 287
12.2	Part I: RNA structure —— 288
12.2.1	Primary structure —— 288
12.2.1.1	RNA sequence determinants on structure —— 288
12.2.1.2	Unusual nucleotides —— 288
12.2.1.3	Torsion angles in the polynucleotide sequence —— 288
12.2.2	Secondary structure: base pairing and helices —— 289
12.2.2.1	Regular structure and base pairing —— 289
12.2.2.2	Helical secondary structure —— 291

12.3	Part II: NMR studies of RNA —— 291	
12.3.1	NMR sample preparation and labeling —— 291	
12.3.1.1	Sample preparation —— 291	
12.3.1.2	Labeling schemes —— 292	
12.3.1.3	RNA purification —— 292	
12.3.2	NMR parameters to characterize RNA structure —— 293	
12.3.2.1	Sequence-specific assignment of NMR resonances —— 293	
12.3.2.2	NMR measurements for torsion angle restraints —— 297	
12.3.2.3	NMR measurements for distance restraints —— 298	
12.3.2.4	Scalar couplings across hydrogen bonds —— 298	
12.3.2.5	Residual dipolar couplings —— 299	
12.3.2.6	NMR-based structure calculation —— 301	
12.3.3	NMR parameters to characterize RNA dynamics —— 301	
12.3.3.1	NMR measurements for RNA dynamics —— 302	
12.3.3.2	Dynamics probed by relaxation parameters —— 303	
12.3.3.3	Dynamics probed by residual dipolar couplings —— 303	
12.4	Part III: examples of RNA tertiary structure —— 304	
12.4.1	Helix-helix interactions —— 304	
12.4.1.1	Coaxial stacking —— 304	
12.4.1.2	A-platform and A-C platform —— 305	
12.4.2	Helix-strand interactions —— 305	
12.4.2.1	Base triples and A-minor motifs —— 305	
12.4.2.2	Tetraloops —— 306	
12.4.3	Loop-loop interactions —— 307	
12.4.3.1	Kissing loop —— 307	
12.4.3.2	Pseudoknot —— 307	
12.5	Conclusion —— 308	
12.6	Acknowledgments —— 308	
	References —— 308	
13	**Crystallization of RNA for structure determination by X-ray crystallography —— 319**	
13.1	Introduction —— 319	
13.2	General strategy for crystallization —— 319	
13.2.1	Oligonucleotides and duplex termini —— 320	
13.2.2	Loop engineering and RNP formation and topological permutation —— 322	
13.2.3	An example of success through construct engineering —— 323	
13.3	Purity and monodispersity —— 325	
13.4	Postcrystallization treatments —— 326	
13.5	Construct design and structure determination —— 328	

13.6	Conclusion —— 329
13.7	Acknowledgments —— 330
	References —— 330

14	**RNA structure prediction —— 335**
14.1	The thermodynamic model of RNA folding —— 336
14.1.1	Free energy and partition function —— 336
14.1.2	Abstract shapes —— 337
14.1.3	Free-energy computation of an RNA structure —— 338
14.1.4	Influence of solvent —— 340
14.2	MFE structure —— 340
14.3	Partition folding —— 341
14.3.1	Suboptimal structures —— 343
14.3.2	Mean and sampled structures —— 344
14.3.3	Shape representative structures and shape probabilities —— 345
14.4	Structure prediction and multiple alignment —— 345
14.5	Beyond secondary structure prediction —— 349
14.5.1	Pseudoknots —— 349
14.5.2	RNA-RNA hybridization —— 353
14.6	Acknowledgments —— 357
	References —— 357

15	**Analyzing, searching, and annotating recurrent RNA three-dimensional motifs —— 363**
15.1	Characteristics of structured RNAs —— 363
15.1.1	RNA molecules are structurally diverse —— 363
15.1.2	"Loops" in RNA secondary structures and RNA 3D motifs —— 364
15.1.3	The 3D motifs and hierarchical organization of RNA —— 366
15.1.4	Linker regions and 3D motifs —— 366
15.2	Structural diversity of RNA 3D motifs —— 368
15.2.1	Contribution of RNA chain flexibility to motif diversity —— 368
15.2.2	Contribution of internucleotide interactions to motif diversity —— 369
15.3	Pairwise nucleotide interactions that stabilize RNA 3D motifs —— 370
15.3.1	Base-pairing interactions and 3D motifs —— 370
15.3.1.1	Occurrence frequencies of base pairs is context dependent —— 370
15.3.1.2	Base-pair isostericity and structure conservation during evolution —— 371
15.3.2	Base-stacking interactions and 3D motifs —— 372
15.3.3	Base-phosphate interactions and 3D motifs —— 372
15.4	Defining RNA 3D motifs —— 373
15.4.1	Role of induced fit in RNA motif structure —— 373

15.4.2	Definition of "classic" RNA 3D motifs —— 374	
15.4.2.1	Definition of modular motifs —— 374	
15.4.2.2	Conservation of motif sequence and structure —— 375	
15.4.3	Recurrent RNA 3D motifs —— 375	
15.5	Tools for searching for RNA 3D motifs in atomic-resolution RNA structures —— 375	
15.5.1	MC-Search —— 376	
15.5.2	NASSAM —— 376	
15.5.3	PRIMOS —— 376	
15.5.4	FR3D and WebFR3D —— 378	
15.5.5	Apostolico et al., 2009 —— 378	
15.5.6	RNAMotifScan —— 378	
15.5.7	FRMF —— 379	
15.5.8	RNA FRABASE 2.0 —— 379	
15.5.9	FASTR3D —— 379	
15.5.10	FRASS —— 380	
15.5.11	R3D-BLAST —— 380	
15.5.12	Comparison of 3D search methods —— 380	
15.6	Classifying RNA 3D motifs —— 381	
15.6.1	Why classify RNA 3D motifs? —— 381	
15.6.2	How to classify RNA 3D motifs? —— 381	
15.6.3	Criteria for grouping motif instances in the same recurrent family —— 382	
15.6.4	Evaluating 3D motif similarity —— 383	
15.6.5	Application of motif classification criteria —— 383	
15.6.6	Automatic classification of RNA 3D motifs —— 384	
15.7	RNA 3D motif collections —— 384	
15.7.1	Motif-oriented collections —— 384	
15.7.1.1	SCOR —— 384	
15.7.1.2	Comparative RNA Web Site —— 386	
15.7.1.3	K-turn database —— 386	
15.7.1.4	RNAMotifScan —— 386	
15.7.1.5	FRMF —— 386	
15.7.1.6	RNA 3D Motif Atlas —— 387	
15.7.2	Loop-oriented collections —— 387	
15.7.2.1	RNAJunction —— 387	
15.7.2.2	RNA STRAND —— 387	
15.7.2.3	RLooM —— 388	
15.7.2.4	RNA CoSSMos —— 388	
15.7.3	Comparing RNA 3D motif collections —— 388	
15.8	RNA 3D motifs that "break the rules" —— 390	

15.8.1	The 3D motifs that contain isolated cWW base pairs —— 390
15.8.2	Composite 3D motifs: 3D motifs composed of more than one loop —— 392
15.8.3	Motifs comprising linker strands —— 393
15.8.4	Motifs interacting with adjacent helices —— 394
15.9	Conclusions —— 395
15.10	Acknowledgments —— 395
	References —— 396

Index —— 399

Dagmar Klostermeier and Christian Hammann
1 Optical spectroscopy and calorimetry

1.1 Introduction

RNA molecules adopt complex three-dimensional structures. Similar to proteins, RNA folds in a hierarchical manner, with the initial formation of secondary structures, duplex regions generated by base pairing. The three-dimensional arrangement of secondary structure elements with respect to each other is mediated by complex interaction networks between individual elements that jointly stabilize the tertiary structures of RNA molecules. As exemplified in the structure of the ribosome [1–3], tertiary interactions frequently involve non-Watson-Crick base pairs [4]. In this chapter, we summarize optical spectroscopy and calorimetric techniques that are widely used in the analysis of RNA structure and structure formation. While probes used in spectroscopy provide a readout for changes in their environment due to structural transitions, calorimetric techniques directly measure the heat exchange that takes place during these transitions.

1.2 Absorption spectroscopy

Absorption spectroscopy is based on the absorption of light by chromophores that undergo a transition from an electronic ground state to an electronically excited state. Absorption occurs when the energy of the incident light corresponds to the energy difference between these states. The nucleobases in RNA absorb light in the ultraviolet spectral region around 260 nm. The absorbance at 260 nm (A_{260}) is commonly used as a measure for RNA concentration. Calculating the concentration from the absorbance via the Lambert-Beer law requires knowledge of the extinction coefficient. Due to nearest-neighbor effects within one RNA strand, and hypochromicity upon duplex formation, the calculation of the extinction coefficient from the sequence is not straightforward for larger, structured RNAs. Extinction coefficients for these RNAs can be determined by hydrolyzing the RNA and measuring the A_{260} of the resulting mixture of nucleotides, for which the extinction coefficient can be calculated according to the nucleobase composition. Comparison of the absorbance of native RNA and nucleotides then yields the extinction coefficient for the structured RNA. For estimates of the concentration for less well-defined samples, such as mixtures of different RNA molecules, or RNAs of unknown sequence, an average value of 50 μg/mL for an absorbance of 1 is often used.

The formation of RNA duplexes causes a decrease in UV absorbance (hypochromic effect). The hypochromicity h is defined as

$$h = \frac{A_{ss} - A_{ds}}{A_{ss}} \tag{1.1}$$

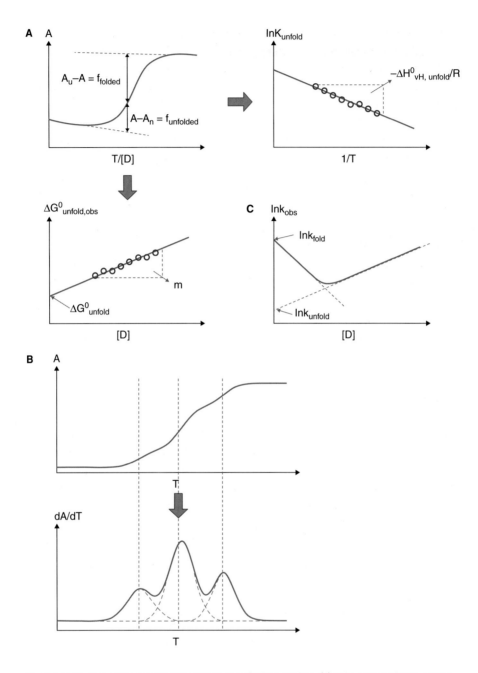

Fig. 1.1 Extraction of thermodynamic data from unfolding studies. (A) Schematic unfolding curve, measured via a spectroscopic probe X, here A for absorbance, as a function of temperature T or denaturant concentration $[D]$. The fraction of folded (f_{folded}) and unfolded molecules ($f_{unfolded}$) can be extracted from the data points in the transition region. The broken lines indicate the baselines for the T or $[D]$ dependence of the observed signal for the folded and unfolded states. The ratio

A_{ss} is the absorbance of the single strands, and A_{ds} is the absorbance of the duplex. The corresponding hyperchromic effect upon duplex dissociation thus allows detection of the disruption of secondary structure elements as an increase in absorbance at 260 nm, either as a function of temperature (melting curve) or as a function of the concentration of a chemical denaturant, such as urea. The resulting curves (Fig. 1.1A) can be analyzed in terms of a two-state model, taking into account the molecularity of the underlying process. For a unimolecular reaction, such as the unfolding of a hairpin, following a two-state folding model, the equilibrium constant K_{unfold} is defined as

$$K_{unfold} = \frac{f_{unfolded}}{f_{folded}} \quad (1.2)$$

K_{unfold} can be related to the measured absorbance at each temperature, and the extrapolated absorbance of folded and unfolded species as

$$K_{unfold} = \frac{A - A_{folded}}{A_{unfolded} - A} \quad (1.3)$$

The temperature dependence of the equilibrium constant K_{unfold} defines the van't Hoff enthalpy of folding, $\Delta H_{vH,unfold}$, which is obtained from the van't Hoff equation (Fig. 1.1A):

$$\frac{d(\ln K_{unfold})}{d(1/T)} = -\frac{\Delta H^0_{vH,unfold}}{R} \quad (1.4)$$

The van't Hoff enthalpy is model dependent and only valid for two-state transitions. In contrast, enthalpies determined in calorimetric experiments are model independent (see Section 1.6).

From the equilibrium constant K_{unfold}, the thermodynamic stability ΔG^0_{unfold} can be calculated as

$$\Delta G^0_{unfold} = -RT \ln K_{unfold} \quad (1.5)$$

Fig. 1.1 (Continued)
$f_{unfolded}/f_{folded}$ defines the equilibrium constant K_{unfold}. The slope of a van't Hoff plot ($\ln K_{unfold}$ as a function of the inverse of the temperature, top right) gives the enthalpy change $\Delta H^0_{vH,unfold}$. From K_{unfold}, the free energy of unfolding, $\Delta G^0_{unfold,obs}$ can be calculated. For denaturant-induced transitions, $\Delta G^0_{unfold,obs}$ can be plotted as a function of the denaturant concentration (bottom left). Extrapolation to 0 M denaturant yields the free energy change ΔG^0_{unfold} in the absence of denaturant. (B) For multiple transitions, analysis of the first derivative of the observed property (here A and dA/dT for absorbance and the first derivative) affords the extraction of thermodynamic data for each transition (see text for details). (C) A plot of the logarithm of apparent folding/unfolding rate constants, $\ln k_{obs}$, as a function of the denaturant concentration [D] is V-shaped. From the linear arms, the (logarithm of the) microscopic rates of folding and unfolding reactions, k_{fold} and k_{unfold}, can be extrapolated.

Together with the Gibbs-Helmholtz equation,

$$\Delta G^0_{unfold} = \Delta H^0_{unfold} - T\Delta S^0_{unfold} \tag{1.6}$$

the folding entropy ΔS^0_{unfold} can be calculated.

Instead of a point-by-point analysis, an expression for the absorbance as a function of the thermodynamic parameters can be derived. The absorbance A measured at each temperature in the transition region can be expressed as the sum of the contributions from folded and unfolded species:

$$A = f_{folded} \cdot A_{folded} + f_{unfolded} \cdot A_{unfolded} \tag{1.7}$$

with

$$f_{folded} + f_{unfolded} = 1 \tag{1.8}$$

The linear baseline preceding the structural transition is the result of the temperature dependence of the absorbance of the hairpin (folded, Fig. 1.1A) and can be described as

$$A_{folded} = A^0_{folded} + m_{folded} \cdot T \tag{1.9}$$

Similarly, the posttransition baseline reflects the temperature dependence of the absorbance of the linear RNA (unfolded, Fig. 1.1A):

$$A_{unfolded} = A^0_{unfolded} + m_{unfolded} \cdot T \tag{1.10}$$

Substituting equations 1.2, 1.5, 1.8, 1.9, and 1.10 into equation 1.7 leads to an expression for the absorbance A as a function of the stabilization energy ΔG^0_{unfold}:

$$A = \frac{A^0_{folded} + m_{folded}T - A^0_{unfolded} - m_{unfolded}T}{1+e^{-\frac{\Delta G^0_{unfold}}{RT}}} + A^0_{unfolded} + m_{unfolded}T \tag{1.11}$$

For measurements of unfolding induced by denaturants, such as urea, $\Delta G^0_{unfold,obs}$ can be calculated from the K_{unfold} at each denaturant concentration, and ΔG^0_{unfold} in the absence of denaturant can be obtained from extrapolation (Fig. 1.1A). Alternatively, an expression similar to equation 1.11 can be derived assuming a linear relationship of the observed $\Delta G^0_{unfold,obs}$ on the denaturant concentration $[D]$:

$$\Delta G^0_{unfold,obs} = \Delta G^0_{unfold} + m[D] \tag{1.12}$$

leading to

$$A = \frac{A^0_{folded} + m_{folded}[D] - A^0_{unfolded} - m_{unfolded}[D]}{1+e^{-(\Delta G^0_{unfold} + m[D])/RT}} + A^0_{unfolded} + m_{unfolded}[D] \tag{1.13}$$

for a unimolecular two-state folding process.

Usually, $\Delta H°_{unfold}$ and $\Delta S°_{unfold}$ are temperature dependent. The temperature dependence of $\Delta H°_{unfold}$ depends on the heat capacity change $\Delta C°_{p,unfold}$:

$$\Delta C°_{p,unfold} = \frac{d\Delta H°_{unfold}}{dT} \quad (1.14)$$

When $\Delta C°_{p,unfold}$ is known, $\Delta H°_{unfold}$ and $\Delta S°_{unfold}$ can be calculated for any given temperature (e.g. from the value at the melting temperature T_m as a reference point), according to equations 1.15 and 1.16:

$$\Delta H°_{unfold}(T) = \Delta H°_{unfold}(T_m) + \Delta C°_{p,unfold}(T - T_m) \quad (1.15)$$

$$\Delta S°_{unfold}(T) = \Delta S°_{unfold}(T_m) + \Delta C°_{p,unfold} \ln\frac{T}{T_m} \quad (1.16)$$

At the melting temperature T_m, the entropy $\Delta S°_{unfold}$ is

$$\Delta S°_{unfold}(T_m) = \frac{\Delta H°_{unfold}(T_m)}{T_m} \quad (1.17)$$

By substituting equations 1.15, 1.16, and 1.17 into equation 1.11, equation 1.18 is obtained, which relates the measured absorbance A to $\Delta H°_{unfold}$, $\Delta S°_{unfold}$, and $\Delta C°_{p,unfold}$:

$$A = \frac{A°_{folded} + m_{folded}T - A°_{unfolded} - m_{unfolded}T}{1 + e^{-\frac{\Delta H°_{vH,unfold}}{R}\left(\frac{1}{T} - \frac{1}{T_m}\right) - \frac{\Delta C°_{p,unfold}}{R}\left(1 - \frac{T_m}{T} - \ln\frac{T}{T_m}\right)}} + A°_{unfolded} + m_{unfolded}T \quad (1.18)$$

Analysis of melting curves representing two-state transitions with equation 1.18 thus yields the thermodynamic parameters for unfolding of the RNA. While it is difficult to obtain $\Delta C°_{p,unfold}$ from absorbance melting curves, its value can be determined reliably in calorimetric experiments (see Section 1.6).

Analogous expressions to equation 1.18 for higher-order reactions can be derived by using the corresponding definitions of the equilibrium constants. With the general definition of α as the fraction folded (i.e. the ratio of concentrations for folded and total RNA, c_{folded} and c_{total}),

$$\alpha = \frac{c_{folded}}{c_{total}} = \frac{c_{folded}}{c_{folded} + c_{unfolded}} \quad (1.19)$$

K_{unfold} for a unimolecular reaction can then be expressed as

$$K_{unfold} = \frac{1 - \alpha}{\alpha} \quad (1.20)$$

For non-unimolecular processes, self-complementarity (e.g. self-association of two identical strands) and non-self-complementarity (e.g. dissociation of a duplex, fork, or helical junction into single strands) has to be distinguished [5]. For a bimolecular reaction of identical strands (self-complementarity), K_{unfold} is

$$K_{unfold} = \frac{2 \cdot c_{total}(1-\alpha)^2}{\alpha} \tag{1.21}$$

For a bimolecular reaction of different strands (non-self-complementarity), K_{unfold} can be expressed as

$$K_{unfold} = \frac{c_{total}(1-\alpha s)^2}{2\alpha} \tag{1.22}$$

Higher-order processes, such as the dissociation of a fork/three-way helical junction formed by three (non-self-complementary) RNA strands or the dissociation of a four-way helical junction into four (non-self-complementary) strands, can be treated accordingly. The general forms of the equilibrium constant as a function of the fraction α have been derived [6] for n^{th}-order processes with non-self-complementarity (equation 1.23) and with self-complementarity (equation 1.24):

$$K_{unfold} = \frac{\left((1-\alpha)\frac{c_{total}}{n}\right)^n}{\alpha \cdot \frac{c_{total}}{n}} = \frac{\left(\frac{c_{total}}{n}\right)^{n-1}(1-\alpha)^n}{\alpha} \tag{1.23}$$

$$K_{unfold} = \frac{((1-\alpha) \cdot c_{total})^n}{\alpha \cdot \frac{c_{total}}{n}} = \frac{n \cdot c_{total}^{n-1}(1-\alpha)^n}{\alpha} \tag{1.24}$$

With these definitions of the equilibrium constant, an expression for the absorbance A as a function of the thermodynamic parameters equivalent to equations 1.11, 1.13, and 1.18 can be obtained, and data analysis can be performed as illustrated previously for unimolecular, two-state transitions.

While analytical expressions for multiple transitions can be derived, it is more convenient to plot the first derivative of the measured absorbance (dA/dT) as a function of temperature (Fig. 1.1B). Structural transitions then appear as peaks (reminiscent of differential scanning calorimetry [DSC] curves, see Section 1.6.2) and can be analyzed by describing each transition i with its fractional signal change ΔA_i, the corresponding enthalpy change $\Delta H^0_{i,unfold}$, and the transition midpoint $T_{m,i}$ [7]:

$$\frac{dA}{dT} = \sum_i \Delta A_i \frac{\Delta H^0_{i,unfold} K_{i,unfold}}{RT^2(1+K_{i,unfold})^2} \tag{1.25}$$

$K_{i,\text{unfold}}$ for each transition is defined as [8]

$$K_{i,\text{unfold}} = e^{\frac{\Delta H^0_{i,\text{unfold}}}{R}\left(\frac{1}{T_{m,i}} - \frac{1}{T}\right)} \tag{1.26}$$

$\Delta G^0_{i,\text{unfold}}$ can be calculated from these equilibrium constants, and $\Delta S^0_{i,\text{unfold}}$ is obtained from the Gibbs-Helmholtz equation (equation 1.6).

Although the largest change in absorbance is caused by formation or loss of secondary structure, absorbance measurements are sometimes also sensitive to tertiary structure transitions. In these cases, individual secondary structure transitions of a large RNA molecule can be assigned by performing control experiments with deletion constructs. Once the formation of all secondary structure elements has been accounted for, the remaining transitions then must reflect tertiary structure formation [8]. Hypochromicity can also be used as a probe to monitor the kinetics of secondary structure formation when measurements are performed as a function of time. The observed rate constants k_{obs} for folding or unfolding reactions are the sum of the two rate constants for the forward and reverse reactions. To extract individual rate constants, the observed rate constants have to be determined over a wide range of denaturant concentrations. A plot of the logarithm of these rate constants as a function of denaturant concentration is V-shaped, with a minimum around the denaturant concentration where the transition occurs [9]. From extrapolations of the two linear regions to the y-axis, the (logarithm of the) rate constants for folding and unfolding in the absence of denaturant can be extracted (Fig. 1.1C). For folding pathways beyond two-state folding, individual transitions can be separated experimentally using single or double jump approaches.

UV melting curves can be measured in standard absorbance spectrometers with a thermostable cuvette holder, preferentially controlled by a Peltier element for automated temperature ramps. A multiple cuvette holder is convenient for parallel measurements and facilitates comparison of melting curves for different RNAs.

Absorbance has been widely used to characterize the thermodynamic stability of secondary structure elements that are building blocks of larger RNAs. The extensive thermodynamic characterization of secondary structure formation in model RNAs has delivered a library of reference values, and it has led to programs that predict secondary structures of RNA from sequence and calculate T_m values accurate to within a few degrees [10–20]. Chemical modifications can also be taken into account in these approaches [21].

To dissect RNA folding pathways, and to distinguish between secondary and tertiary structure formation, absorbance is often used in combination with circular dichroism (CD; see Section 1.4). The folding of various RNA molecules has been studied using absorbance methods, among them transfer RNAs (tRNAs) [22, 23], ribosomal RNAs (rRNAs) [24, 25], ribozymes [26, 27], pseudoknots [7, 28], and kissing loops [29]. The effect of posttranslational modifications on the stability of tRNA[Phe] has been determined by comparing melting curves of *in vitro* transcribed and isolated, modified tRNA[Phe] [22]. A study of mitochondrial tRNAs also found a stabilizing effect of modifications [23]. The folding pathway of pseudoknots that constitute regulatory elements in bacteriophage messenger RNA (mRNA) leader sequences has been analyzed using absorbance spectroscopy. Thermodynamic parameters for multiple transitions were extracted from the first derivative of melting curves and assigned to secondary and tertiary structure transitions [7]. The Mg^{2+} dependence of these transitions has been dissected using the same approach [28].

1.3 Fluorescence

Fluorescence is the emission of light upon return from the excited state to the ground state. As RNA does not contain fluorescent groups, extrinsic fluorophores have to be used for fluorescence experiments with RNA. Organic fluorophores can be introduced during RNA chemical synthesis. Alternatively, fluorophores can be attached to functional groups introduced during synthesis. A number of fluorescent base analogs have been developed in the past to facilitate fluorescence studies with RNA. A detailed overview about suitable fluorophores for studying RNA structure and folding has been given by Wilhelmsson [30].

Pyrene, covalently attached to the RNA of interest, is a fluorescent probe for secondary and tertiary structure transitions (reviewed by Smalley and Silverman [31]). Pyrene fluorescence has been used to monitor local and global folding of the *Tetrahymena* ribozyme [32–34]. 2-Aminopurine (2-AP) is a fluorescent base that base pairs with uridine and can replace A in A-U base pairs. Its fluorescence depends on base stacking, and it therefore serves as a local probe for structural transitions (reviewed by Hall [35]). 2-AP fluorescence has been monitored to dissect contributions from local folding to the formation of the global structure of model RNA hairpins [36] and to follow the dynamics of stem and loop during unfolding [37]. Similarly, local rearrangements of hairpin and hepatitis delta virus ribozymes have been followed by 2-AP fluorescence [38]. 2-AP fluorescence has been used to dissect local events in the folding of riboswitches [39–42]. Fluorescence resonance energy transfer (FRET) reports on the distance between two fluorescent dyes and has been widely used as a probe for global RNA folding (see Chapter 8). Chemical biology approaches useful for introducing fluorescent labels into RNA are described in Chapter 4.

1.4 Circular dichroism

The *E*-field vector of linearly polarized light periodically changes the amplitude but maintains its direction in one plane, the plane of polarization. In contrast, the *E*-field vector of circularly polarized light has a constant amplitude but rotates around the propagation axis (Fig. 1.2A). As a result, its tip describes a helix. Linearly polarized light can be viewed as a superposition of right-handed and left-handed circularly polarized light (Fig. 1.2A). Optical activity is caused by different interaction of matter

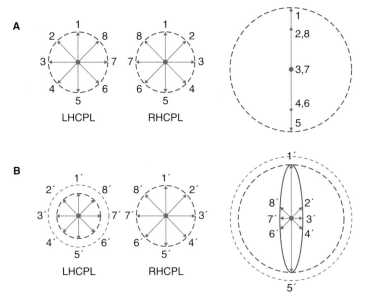

Fig. 1.2 Circular dichroism. (A) Linearly polarized light (right) can be constructed as a superposition of left- and right-handed circularly polarized components (LHCPL and RHCPL, left and center). The numbers from 1 to 8 depict the projection of the electric field vector at different time points. By vectorial addition, the resulting electric field vector can be calculated (right). (B) A sample that exhibits circular dichroism shows different extinction coefficients for left- and right-handed circularly polarized light (left). The left- and right-handed circularly polarized components (bottom left and center) of the transmitted light have different amplitudes. The resulting elliptically polarized light (right) can be constructed by vectorial addition of both components for each time point (1′ to 8′). The tips of the vectors lie on an ellipse (blue line) whose half-axes define the tangents of the ellipticity (see text). Optically active samples also exhibit different refractive indices n for left- and right-handed circularly polarized light, leading to different traveling velocities of the two components in the sample, and a phase shift in the transmitted light. As a result, the ellipse is tilted relative to the polarization plane of the incident light (not shown).

with the right- and left-handed components. CD results from the different absorption of right-handed and left-handed circularly polarized light.

$$\Delta\varepsilon = \varepsilon_l - \varepsilon_r \tag{1.27}$$

Optical activity and CD are linked to chirality. Chiral chromophores and intrinsically symmetric chromophores in an asymmetric environment can give rise to CD. Periodic coupling of multiple chromophores, such as in helices, also creates optically active structures. The chromophores in RNA are the nucleobases, and their CD is related to sequence and stacking geometry. CD spectra allow distinction between A-, B-, and Z-form helices, and between duplex and unstacked single strands. CD provides a sensitive qualitative measure of conformation, and both secondary and tertiary structure transitions can be followed by changes in CD.

The different absorption of right- and left-handed components by the sample leads to elliptically polarized transmitted light (Fig. 1.2B). The tangent of the ellipticity θ is defined as the ratio of the small and large (half-)axes of the ellipse. Its value (in degrees) can be calculated from the difference ΔA in absorbance of left- and right-handed circularly polarized light:

$$\theta = 32.98 \cdot \Delta A \tag{1.28}$$

The absorption differences between right- and left-handed circularly polarized light are small. Therefore, it is not possible to measure them separately and calculate the CD from their difference. Instead, CD instruments modulate the linearly polarized incident light, such that the illumination of the sample alternates between an excess of the right- and left-handed polarized components with a modulation frequency ω. If the sample shows CD, the intensity of the transmitted light varies with the same modulation frequency ω and the amplitude that reflects the different absorption of the two components. The signal generated by the photomultiplier thus is a superposition of a large direct current, proportional to the average overall absorption, and a small alternating current caused by the CD. The circular dichroitic absorbance ΔA is calculated from the ratio of the alternating and direct current signals and converted into the ellipticity θ (equation 1.28).

To compare CD spectra for different RNAs, it is useful to normalize the measured ellipticity θ by the number of nucleotides. Instead of the measured ellipticity (in degrees), the CD is often reported in $\Delta \varepsilon$ (in cm² dmol⁻¹):

$$\Delta \varepsilon = \frac{\theta}{32.98 \cdot c \cdot d \cdot n} \tag{1.29}$$

Here, c is the concentration in M, d the optical path length in cm, and n the number of nucleotides of the RNA.

Secondary structure transitions give rise to changes in the CD signal at 260 nm. The analysis of transition curves of the CD signal as a function of temperature or urea concentration can be analyzed to derive the thermodynamic stability of the folded state similar to the procedures described for absorbance spectroscopy (see Section 1.2). The expression for the dependence of the measured ellipticity θ on the denaturant concentration $[D]$, corresponding to equation 1.13, is

$$\theta = \frac{\theta^0_{folded} + m_{folded}[D] - \left(\theta^0_{unfolded} + m_{unfolded}[D]\right)}{1 + e^{-\left(\Delta G^0_{fold} + m[D]\right)/RT}} + \theta^0_{unfolded} + m_{unfolded}[D] \tag{1.30}$$

The equation for a temperature-induced unfolding transition, corresponding to equation 1.18, is

$$\theta = \frac{\theta^0_{folded} + m_{folded}T - \theta^0_{unfolded} + m_{unfolded}T}{1+e^{\frac{\Delta H_{vH,unfold}}{R}\left(\frac{1}{T}-\frac{1}{T_m}\right) - \frac{\Delta C_{p,unfold}}{R}\left(1-\frac{T_m}{T}-\ln\frac{T}{T_m}\right)}} + \theta^0_{unfolded} + m_{unfolded}T \quad (1.31)$$

Tertiary structure transitions lead to a change in the CD signal between 275 and 290 nm. The optimal parameters have to be determined from difference spectra of folded and unfolded states of the RNA molecule studied. Tertiary structure of RNA is often stabilized by the specific binding of Mg^{2+} ions, and tertiary structure transitions can be induced and followed by measuring the CD signal as a function of the Mg^{2+} concentration. Mg^{2+} binding can be described quantitatively by the Hill equation, using as parameters the dissociation constant K_d and the Hill coefficient n. The Hill coefficient is a measure of cooperativity, and sets a limit for the minimum number of binding sites. For a single transition $A \to B$, equation 1.32 describes the measured ellipticity θ as a function of the Mg^{2+} concentration:

$$\theta(Mg^{2+}) = \theta_A + \left[\frac{[Mg^{2+}]^n}{[Mg^{2+}]^n + K^n_{1/2}}\right](\theta_B - \theta_A) \quad (1.32)$$

where θ_A and θ_B are the ellipticities of states A and B [43]. For two sequential transitions $A \to B \to C$, the corresponding equation is

$$\theta(Mg^{2+}) = \frac{\theta_A + \theta_B([Mg^{2+}]/K_{1,1/2})^{n_1} + \theta_C([Mg^{2+}]/K_{1,1/2})^{n_1}([Mg^{2+}]/K_{2,1/2})^{n_2}}{1 + ([Mg^{2+}]/K_{1,1/2})^{n_1} + ([Mg^{2+}/K_{1,1/2}])^{n_1}([Mg^{2+}]/K_{2,1/2})^{n_2}} \quad (1.33)$$

with θ_A, θ_B, and θ_C as the ellipticities of states A, B, and C [43]. Equation 1.33 reduces to

$$\theta(Mg^{2+}) = \theta_A + \left[\frac{[Mg^{2+}]^{n_1}}{[Mg^{2+}]^{n_1} + K^{n_1}_{1,1/2}}\right](\theta_B - \theta_A) + \left[\frac{[Mg^{2+}]^{n_2}}{[Mg^{2+}]^{n_2} + K^{n_2}_{2,1/2}}\right](\theta_C - \theta_B) \quad (1.34)$$

if the individual transitions are well separated [43].

CD spectroscopy has been used to characterize secondary and tertiary structures of a number of different RNAs, often in combination with absorbance measurements. The secondary structure of *Escherichia coli* 5S rRNA has been characterized by a combined approach using absorbance and CD spectroscopy [24, 25]. A large set of reference spectra was measured for a representative set of nucleotides, dinucleotides, and various oligonucleotides, whose combination reflects any nearest-neighbor property possible in the 5S rRNA. The measured CD spectra for the 5S rRNA were then analyzed in terms of these basis spectra to estimate the secondary structure content; the number of A-U, G-C, and G-U bases pairs; and the nearest-neighbor base-pair content. The results ruled out two of the existing structural models [24, 25]. In a study with RNA that forms a pseudoknot at low temperatures and a hairpin structure at higher temperatures [44], the approach was validated and its general applicability underlined. The Mg^{2+}-induced folding of yeast 5S rRNA has been dissected kinetically and thermodynamically, using CD and absorbance, providing evidence for the rearrangement of base pairs in tertiary structure formation [45].

Urea-induced unfolding and Mg^{2+}-dependent folding of tRNA^{Phe} followed by absorbance and CD spectroscopy, combined with chemical probing (see Chapter 3), have delineated the folding pathway of this tRNA [46]. RNA unfolding was monitored by the change in CD at 260 nm, either as a function of urea concentration at different Mg^{2+} concentrations or as a function of Mg^{2+} in the absence of urea. The data showed that tRNA^{Phe} folds into the native state via two Mg^{2+}-dependent transitions. Hydroxyl radical and nuclease protection experiments suggested that the second transition corresponds to tertiary structure formation [46]. The effect of mono- and divalent ions on the folding pathway has been addressed in a subsequent study [47]. CD measurements as a function of the Mg^{2+}, Na^+, and urea concentrations revealed that the tRNA folds into the native state via an intermediate both in the presence of Mg^{2+} or Na^+. However, different dependencies of the ΔG on the urea concentration (m-value) for the second transition suggested that these intermediates have different structures. Chemical probing (see Chapter 3) showed that the intermediate in the presence of Na^+ contains all four helices, and the second transition reflects the formation of tertiary structure. In contrast, the intermediate in the presence of Mg^{2+} contains only three helices, and the acceptor stem is formed in the second step together with the tertiary structure [47]. CD measurements were also employed to investigate differences in stability between unmodified and mature, posttranscriptionally modified *E. coli* tRNA^{Phe} [22]. The results demonstrated that modified nucleotides stabilize the tertiary structure without affecting the folding mechanism.

The folding of ribonuclease (RNase) P RNA has been studied extensively with CD. The pathway for Mg^{2+}-dependent folding of the *Bacillus subtilis* RNase P catalytic domain has been dissected by measuring the CD signal as a function of the Mg^{2+} concentration at 260 nm (secondary structure formation) and 287 nm (tertiary structure formation) [26, 48]. The overall stability for the catalytic domain was found to be similar to the stabilities of small globular proteins [48]. Combined with fluorescence, hydroxyl radical protection, and activity measurements, a three-step folding mechanism for RNase P RNA with two Mg^{2+}-dependent transitions was formulated. Analysis of the lifetime of the intermediate as a function of urea revealed that this rate-limiting step is accelerated by adding urea, suggesting that formation of the native state requires the disruption of previously formed interactions [26]. An intermediate in the folding pathway of the isolated specificity domain was later characterized structurally using a combination of CD data with chemical and nuclease probing (see Chapters 2 and 3), small angle X-ray scattering (see Chapter 9), and molecular modeling [49]. CD studies on RNase P have also contributed to unraveling the strategies for stabilization of RNAs from thermophilic organisms in comparison to their mesophilic counterparts [50, 51]. Comparative studies with the catalytic domain of RNase P RNA from *B. subtilis* and

Bacillus stearothermophilus demonstrated that the last intermediate in the folding pathway is less structured for the RNase P RNA with thermophilic origin. As a consequence, the degree of structure formation and the cooperativity of this step are increased, resulting in higher functional stability [50]. The same stabilization strategy was identified in a second study with the specificity domains of RNase P RNA from *E.coli* and *Thermus thermophilus* [51]. Interestingly, using the hammerhead ribozyme as a model system, cold denaturation of RNAs has been demonstrated for the first time with CD spectroscopy [52]. CD has also been applied to investigate structural changes in RNA and protein upon complex formation for human immunodeficiency virus (HIV) Rev and the Rev-responsive element [53], and HIV Tat and the Tar RNA [54].

1.5 Transient electric birefringence

In birefringent media, the refractive index n, and thus the traveling velocity of light, depends on the propagation direction. Birefringence Δn is defined as the difference in the refractive indices between two rays of light polarized in orthogonal directions x and y:

$$\Delta n = n_x - n_y \tag{1.35}$$

The anisotropy of the refraction index results from different polarizabilities in the different directions. Homogenous solutions normally do not show birefringence due to the random distribution of the molecules. Birefringence can be induced by orienting the molecules of interest in an external electric field (TEB), in a magnetic field, or with a gradient of shear forces. Nucleic acids can be oriented in an electric field, leading to optical anisotropy. A solution of oriented nucleic acid duplexes exhibits negative birefringence due to the high polarizability in the plane of the nucleobases. The decay of TEB after switching off the electric field depends on the rotational motion of the molecules and can therefore be used to distinguish molecules of different sizes or shapes (reviewed in several studies [55–57]). The analysis of RNA conformations with TEB is based on the different rotational mobility of compact, bent molecules in comparison with linear control molecules [56] (Fig. 1.3A). To investigate the effect of a structural element on bending, long duplex arms are placed on either side of the central element (e.g. a bulge, loop, or protein-binding site) (Fig. 1.3A,B). The required micromolar concentrations of RNA preclude the analysis of protein binding for systems with K_d values below micromolar, but stoichiometric effects can still be studied. The TEB decay is measured for this molecule and the linear control (Fig. 1.3A,B). From these experiments, a measure for the interhelical angle can be obtained. To obtain reliable results, the two molecules should be as similar as possible, both in length and in sequence. Hence, the duplex arms should be long in comparison to the central element.

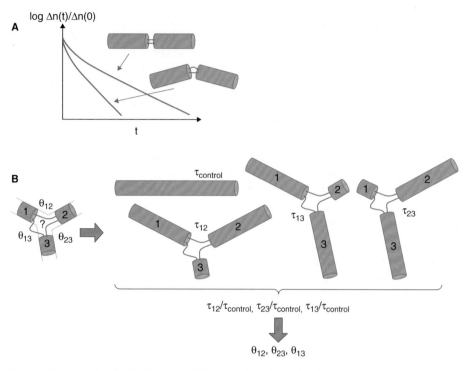

Fig. 1.3 Transient electric birefringence. (A) For transient electric birefringence measurements, molecules are oriented in an electric field. After the field is switched off, the difference Δn in the refractive index in two orthogonal directions is measured as a function of time. The TEB decays more rapidly for a linear RNA than for a bent or kinked RNA. Helical regions are depicted as cylinders. (B) The geometry of a central RNA element (lines) can be deduced from a set of TEB experiments where two of the three arms are extended by long helices (cylinders). From the decay times of the TEB signal for the different constructs, relative to the steady-state birefringence, $n(0)$, the interhelical angles θ_{12}, θ_{13}, and θ_{23} can be calculated (see text for details).

In a typical TEB experiment [55], the sample is illuminated with linearly polarized laser light. In the detection path, a polarizer is placed at a 90° angle with respect to the polarization of the incident light, such that no transmission is observed for an optically isotropic sample. The electric field for orientation of the molecules is generated by a high-voltage pulse generator. After the RNA has been oriented by a microsecond electric pulse, the polarizability in the plane of the bases leads to the preferential retardation of light in one direction. The resulting transmitted light is elliptically polarized. A corresponding fraction of light can now pass the polarizer in the detection path and will be detected by a photodiode. Detectable signals require micromolar RNA concentrations. When the orienting field is switched off, the TEB will decay to zero upon randomization of the RNA orientation. The decay times τ depend strongly on the length and the shape of the RNA (Fig. 1.3A). In the range from 100 to 200 bp, τ increases with the duplex length to the power of 2.5 [55]. A 90° bend centered on a 150 bp RNA leads to a two-fold reduction of τ [58, 59]. To optimize the sensitivity of TEB measurements, the non-helix element should be placed at the center of the RNA helix (Fig. 1.3A,B). In this case, the shortening of the molecule in comparison to the linear control is maximized. The signal will decrease substantially if the element is off center, even more so when a protein-binding site is studied [55]. The duplexes flanking the element should be short enough not to affect the structure. Duplexes of <200–250 bp, the persistence length of RNA, ensure a monoexponential TEB decay in the absence of a bend [60]. The sequence of the duplex should be optimized such that no alternative secondary structures are formed. Altogether, the helical arms should be four to five times longer than the element studied to minimize other hydrodynamic contributions. While such effects can be excluded by performing experiments with molecules carrying arms of different length, applications have frequently used arms of 70 bp (e.g. [59, 61, 62]).

TEB decays are measured repeatedly, averaged, and converted to the fractional change. Depending on the experimental setup, this response can then be converted into the birefringence response $\Delta n(t)$, which is normalized by the steady-state birefringence $n(0)$. The birefringence response is analyzed by a double exponential function [63].

$$\frac{\Delta n(t)}{\Delta n(0)} = \alpha_1 e^{-\frac{t}{\tau_1}} + \alpha_2 e^{-\frac{t}{\tau_2}} \tag{1.36}$$

α_1 and α_2 reflect the amplitudes of the fast and slow components of the decay with the decay times τ_1 and τ_2. Ultimately, birefringence decays need to be interpreted in terms of a hydrodynamic model that depends on the shape and flexibility of the molecule. The ratio of the decay times for the linear and the bent RNA shows little sensitivity to structural parameters and the hydrodynamic model [63]. The ratio of the decay times for the molecule of interest and the linear control can be converted into a bend angle according to the empirically derived equation 1.37 [58]:

$$e = 1.46\cos^{-1}\left(\frac{\tau}{\tau_c}\right) + 0.005\left(\sin^{-1}\left(1 - \frac{\tau}{\tau_c}\right)\right)^{2.3} \tag{1.37}$$

Alternative equations have been derived for specific cases [58].

The resulting bend angle is an apparent value that does not specify the flexibility of the molecule. To analyze the flexibility, two bends have to be introduced into a single molecule at various distances. For rigid conformers, the τ ratios for the different molecules vary with their spacing, depending on the individual bend angles. Flexibility of the individual bends reduces the effects of the distance variation on the τ ratios [55].

TEB has been applied to study DNA and RNA conformations. In addition to bend angles, determination of interhelical angles for different constructs of three- or four-way junctions, in which two of the helical arms have been extended, allows for the determination of the global conformation (see the study by Hagerman [57] for references). Toward establishing a framework for understanding effects of bulges and loops on folds of larger RNA, a systematic study has investigated the effect of bulges [59] and loops [61] within poly-A- or poly-U-duplexes. The interhelical angles of 148 bp poly-A- or poly-U duplex RNA with bulges of 1–6 nucleotides increased from 7° to 93° with n [59]. When symmetric internal loops (n = 2, 4, and 6) were introduced into the center of 150- to 154-bp RNA, the apparent bend angles only varied between 20° and 40° [61], demonstrating that loops affect the helix geometry less than bulges.

The HIV TAR RNA contains a bulge that induces a 50° bend [62]. TEB studies showed that binding of a peptide derived from the HIV Tat protein straightens the molecule. tRNA conformation has also been studied using TEB [64, 65]. Determination of the angle between the anticodon and the acceptor stem in tRNA^{Phe} at different Mg^{2+} concentrations revealed a conformational change with a change of the apparent angle from ~150° to ~70° at 200 µM Mg^{2+}. At higher Mg^{2+} concentrations, the angle increases to ~80°–90° in a further rearrangement [66]. Mitochondrial (m)tRNAs often lack conserved elements of the canonical cloverleaf that confer rigidity to the folded tRNA. Bovine mtRNA^{Ser} lacks the dihydrouridine arm [67]. Two structural models have been proposed that differ with respect to the angle between acceptor and anticodon stems. With TEB it was shown that the inter-stem angle for mtRNA is 120°, and thus larger than the corresponding angle for yeast tRNA^{Phe} (70°–80°), demonstrating that truncated mtRNAs adopt structures deviating from the canonical L-form geometry [67]. In a subsequent study, a set of 11 mitochondrial and nonmitochondrial tRNAs, comprising canonical and noncanonical species, confirmed such a variation in global structure [68] and showed a tendency toward more obtuse angles between acceptor and anticodon arms for truncated mtRNAs than for the canonical cloverleaf tRNAs. The dispersion of the anticodon-acceptor angle for bovine mtRNA^{Ser} was investigated by placing a core mtRNA bend and a second bend of known bend angle and rigidity at different distances [69]. The core of the mtRNA was more flexible than the canonical yeast tRNA^{Phe}. Possibly, the increased flexibility, combined with a larger separation of the two helical arms, explains why non-canonical and canonical mtRNAs can function within the same protein synthesis machinery [69]. tmRNAs are RNAs that consist of a tRNA domain and a second domain that serves as an mRNA. tmRNA is involved in liberating ribosomes that are stalled on mRNA fragments without stop codons or when tRNAs are scarce. TEB experiments revealed an angle of 110° between the anticodon and acceptor stems in tmRNA, similar to canonical tRNAs [70].

The conformations of a number of three-helix junctions have been mapped by TEB, among them the central three-helix junction of the 5S rRNA from *Sulfolobus acidocaldarius* [71], in which two helices are colinear and the third shows flexibility with respect to their axis, and a 16S rRNA helical junction [72] that exhibits equal interhelical angles in the absence of the ribosomal protein S15 but adopts a more compact conformation with two colinear helices and a third helix that forms a 40° angle with their axis. Similarly, the hammerhead ribozyme has been investigated in its pre-cleavage and post-cleavage forms [73, 74]. The angle between the two helices flanking the catalytic center is similar in both forms, explaining the high reversibility of the hammerhead cleavage-ligation reaction [75]. Altogether, these studies have demonstrated the suitability of TEB to investigate the geometry of RNA molecules and the effect of protein binding on their conformation.

1.6 Calorimetry

In general, chemical reactions produce or absorb heat. Changes in heat can be analyzed in calorimetry experiments. Two different formats have been established: isothermal titration calorimetry (ITC) and DSC. With ITC, thermodynamic parameters of

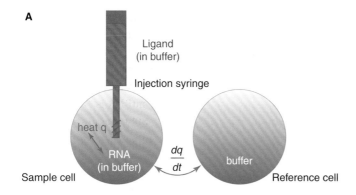

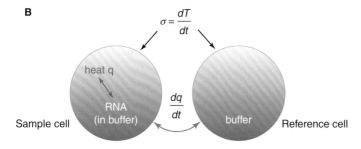

Fig. 1.4 Principles of calorimeter instruments. (A) In an isothermal titration calorimeter, a ligand contained in an injection syringe is titrated to RNA molecules in the sample cell. The heat q that is absorbed or released due to this interaction is determined as the power dq/dt that is required to maintain identical temperatures in the sample and the reference cell. Both cells are situated in a thermally insulated jacket (not shown). (B) In a differential scanning calorimeter, the sample cell contains RNA molecules and the reference cell buffer only. Both cells are heated with a constant rate σ. The heat q that is absorbed or released during RNA unfolding is again determined as the power to maintain identical temperatures in both cells. As in ITC, both cells are situated in a thermally insulated jacket (not shown).

binding events can be determined, resulting in information on the stoichiometry of the interaction, its enthalpy ΔH, and its equilibrium binding constant. Knowledge of these terms allows calculating the entropic contributions (ΔS) to the free energy ΔG. If ITC experiments are performed at several temperatures, then changes in the heat capacity ΔC_p can also be determined for the interaction under investigation. DSC, on the other hand, predominantly addresses (un)folding of macromolecules and allows the determination of heat capacities ΔC_p, melting temperatures T_m, and enthalpies ΔH of the transitions. ITC and DSC experiments yield a complete set of thermodynamic parameters for the underlying process.

Both methods require sophisticated instrumentation, in which the heat change of a sample cell is determined relative to a reference cell (Fig. 1.4). A great asset is that native and label-free RNA molecules can be used for the determination of thermodynamic parameters, which, however, often requires high RNA concentrations. An in-depth treatment of the theoretical background to both methods is given by Ladbury and Chowdhry [76], and detailed experimental protocols, including valuable measures of precautions, are summarized by Sokoloski and Bevilacqua [77].

1.6.1 Isothermal titration calorimetry

ITC is a very convenient method for the direct determination of thermodynamic parameters of interactions of a biological macromolecule with a ligand. Possible ligands of RNA molecules include proteins, small organic molecules, or metal ions, but also interacting RNA molecules. ITC allows for the model-independent analysis of the enthalpy ΔH and the determination of stoichiometry n and binding constants K_{eq} of the interaction. From these data, the entropic contribution ($-T\Delta S$) to the free energy (ΔG) of the interaction can be calculated with the Gibbs-Helmholtz equation (equation 1.6).

In a conventional ITC experiment, the RNA is placed in the sample cell and the ligand in the syringe. The sample cell is housed together with the reference cell in a thermally insulated jacket, which is set to the experimental temperature with an accuracy of $10^{-4}\,°C$. Upon reaching the constant reaction temperature, a series of small volume injections is performed. The heat change that takes place after each injection is measured in a feedback control circuit as the power (dq/dt, μcal s^{-1}; Fig. 1.5), which is necessary to maintain equal temperatures in the sample and the reference cell [78]. The thereby obtained raw data can be further analyzed by integrating over time, and thus the area under the ITC peaks (Fig. 1.5) represents the heat $\Delta Q_{cal}(i)$ (kcal/mol or kJ/mol of injected ligand), which had to be added to maintain the thermal equilibrium between the sample and the reference cell after the ith injection. What ITC determines, however, is the total heat change of the system, of which the RNA-ligand interaction is only a part, albeit usually the larger fraction. Additional sources of heat changes are the rearrangement of solvation states or the drastic (around 100-fold) dilution of ligand upon injection. To correct for these

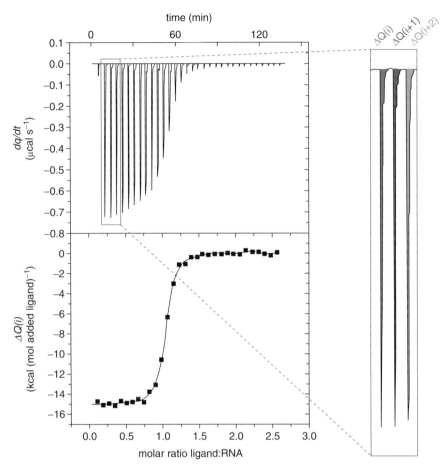

Fig. 1.5 Isothermal titration calorimetry data analysis. ITC raw data are plotted as the power dq/dt that is required to maintain identical temperatures in the sample and the reference cell against time (TOP). The area under each peak represents the total heat exchange $\Delta Q(i)$ of injection i, as shown for the three injections I, $i + 1$, and $i + 2$ in the inset. To allow for the establishment of the new equilibrium after the injection, it is important that the signal reaches the baseline again before the next injection starts (see inset). For determination of thermodynamic parameters, the $\Delta Q(i)$ values are plotted against the molar ratio n of ligand to RNA concentration (bottom). Simulations of binding curves are calculated to describe the experimental data. The result of this analysis is the fitted curve in gray. Its intersection with the y-axis at a molar ratio of 0 represents the molar enthalpy ΔH^0 of the interaction, and its slope at the inflection point corresponds to K_A. The value of n for the interaction can be inferred from the point of inflection of the fitted curve.

effects, a second series of injections is performed as a control experiment, in which the ligand is injected in buffer. The resulting heat values $\Delta Q_{buff}(i)$ for ligand dilution are subtracted from the $\Delta Q_{cal}(i)$ terms, such that the heat values for the ligand binding to RNA $\Delta Q(i)$ can be obtained (equation 1.38):

$$\Delta Q(i) = \Delta Q_{cal}(i) - \Delta Q_{buff}(i) \tag{1.38}$$

Each $\Delta Q(i)$ value corresponds to the establishment of a new equilibrium between ligand-free and ligand-bound RNA at the prevalent concentrations of ligand and RNA after injection i. Thus, for the determination of thermodynamic parameters from these measurements, it is essential that the individual injections are sufficiently separated to allow reaching equilibrium [79] before a new injection starts (see Fig. 1.5).

Because the experimental setup in a titration calorimeter is isobaric, the total heat Q is equivalent to the enthalpy change ΔH. For the determination of the enthalpy, the individual $\Delta Q(i)$ values, representing the first derivative of the total heat Q, are plotted against relative ligand concentration (Fig. 1.5). In the case of a bimolecular interaction, where n ligand molecules interact with one type of binding site in the RNA, the total heat content Q can be expressed as

$$Q = nf[RNA]_{tot} \Delta H V_0 \tag{1.39}$$

in which f represents the fraction of sites occupied by the ligand, $[RNA]_{tot}$ the total RNA concentration, and V_0 the sample cell volume. The association constant K_A of this interaction can be described as

$$K_A = \frac{f}{(1-f)[L]_{free}} \tag{1.40}$$

in which $[L]_{free}$ is the free concentration of ligand. The latter relates to the total ligand concentration $[L]_{tot}$ by

$$[L]_{tot} = [L]_{free} + nf[RNA]_{tot} \tag{1.41}$$

Combining equations 1.40 and 1.41 results in the quadratic equation

$$f^2 - f\left(1 + \frac{[L]_{tot}}{n[RNA]_{tot}} + \frac{1}{nK_A[RNA]_{tot}}\right) + \frac{[L]_{tot}}{n[RNA]_{tot}} = 0 \tag{1.42}$$

The solution of equation 1.42 inserted in equation 1.39 yields

$$Q = \frac{n[RNA]_{tot} \Delta H V_0}{2}\left(1 + \frac{[L]_{tot}}{n[RNA]_{tot}} + \frac{1}{nK_A[RNA]} - \sqrt{\left(1 + \frac{[L]_{tot}}{n[RNA]_{tot}} + \frac{1}{nK_A[RNA]_{tot}}\right)^2 - \frac{4[L]_{tot}}{n[RNA]_{tot}}}\right) \tag{1.43}$$

Thus, for the determination of n, K_A, and ΔH, the accurate concentrations of RNA and ligand need to be known (see, for example Section 1.2). The individual

$\Delta Q(i)$ values that are measured upon completion of injection i in an ITC experiment can be

$$\Delta Q(i) = Q(i) + \frac{dV_i}{V_0}\left[\frac{Q(i)+Q(i-1)}{2}\right] - Q(i-1) \tag{1.44}$$

expressed as in which dV_i is the injection volume, and $Q(i)$ and $Q(i-1)$ are the total heat contents after the ith and the $(i-1)$st injection.

Simulations of binding curves to the experimentally determined $\Delta Q(i)$ values as a function of the relative concentrations of ligand to RNA are required to determine values for n, K_A, and ΔH, as exemplified in Fig. 1.5. Highly concentrated solutions of RNA and ligand show nonideal behavior. Therefore, the effective concentration (activity) of these compounds might deviate from the total concentrations, as, for example, determined by UV spectroscopy, which can lead to noninteger numbers of n. In these cases, concentrations are commonly adjusted during fitting simulations, which strictly requires knowledge of the stoichiometry from alternative methods. On the other hand, for the investigation of metal ion binding to RNA, noninteger numbers of n are not unexpected given the polyanionic character of the nucleic acid, as exemplified in an ITC analysis of the metal-ion-induced folding of the hammerhead ribozyme [78].

Upon determination of the n, K_A, and ΔH values from ITC data, this information can be used to determine the free energy change of the interaction by equation 1.5. From this, the entropic contributions to the binding event with the Gibbs-Helmholtz equation (equation 1.6).

In addition, changes in the molar heat capacities of the RNA-ligand interaction, ΔC_p, can be determined by ITC experiments carried out at different temperatures, according to

$$\Delta C_p = \frac{\Delta H_{T2} - \Delta H_{T1}}{T2 - T1} \tag{1.45}$$

Experiments at multiple temperatures allow for the direct distinction of a temperature-independent ΔC_p (linear relationship between ΔH and T) or temperature-dependent ΔC_p (nonlinear dependence of ΔH on T) and thus yield reliable values for ΔC_p.

1.6.2 Differential scanning calorimetry

In DSC, the two-chamber principle of the ITC is also employed; however, the sample cell containing the RNA (in the presence or absence of ligands) and the reference cell are closed (Fig. 1.4B). Practically, a DSC instrument measures the difference in power that is required to maintain the temperature constant between reference and sample cell. Both cells are heated at a constant rate σ,

$$\sigma = \frac{dT}{dt} \tag{1.46}$$

and the differential power that is required due to unfolding processes in the sample cell is measure. This raw data can be converted to the excess molar heat capacity $C_{p,excess}$ using equation 1.47:

$$\frac{dq}{dt}\frac{1}{\sigma n} - C_{p,excess} \tag{1.47}$$

in which q is the heat absorbed and n is the number of moles of RNA in the cell. $C_{p,excess}$ can be plotted against the temperature range covered in the DSC experiment. From this curve, the melting temperature T_m and the change in the molar heat capacity $\Delta C_{p\,unfold}$ between the folded and unfolded state can be determined (Fig. 1.6).

Upon determination of $\Delta C_{p,\,unfold}$ and the melting temperature T_m, the entropy and enthalpy of unfolding, $\Delta H°_{unfold}$ and $\Delta S°_{unfold}$, can be calculated according to equations 1.14–1.16. Particularly for complex RNA structures, the DSC signal likely represents the sum of several individual unfolding events that superimpose. These multiple events might reflect secondary and/or tertiary RNA structure changes. The analysis of such DSC curves can be carried out in analogy to the procedure described previously for higher-order processes analyzed by absorption spectroscopy (see Section 1.2, equations 1.25 and 1.26).

Both ITC and DSC have been used widely in the analysis of thermodynamic parameters of RNA-ligand interactions and the analysis of RNA secondary and tertiary structure. The two methods have individual advantages and disadvantages [80] and can be used complementarily, ideally in combination with other structural or biochemical methods, such as NMR (see Chapter 12), X-ray crystallography (see Chapter 13), or FRET (see Chapter 8). An early calorimetric investigation of the helix-coil transition of tRNA[Phe] showed the temperature dependence of cooperativity in RNA folding, which manifested in distinct van't Hoff enthalpies and calorimetrically measured enthalpies [81]. The unfolding of tRNA[Phe] was further detailed by calorimetry, resulting in the observation of five distinct unfolding events, due to melting of different tRNA domains [82]. While in that study the RNA unfolding reaction appeared to exhibit no strong change in heat capacity, a recent summary of published ΔC_p values associated with RNA structure changes revealed a surprising variability in heat capacity changes for folding of individual RNA molecules [83]. Since calorimetry measures the heat changes of the entire system, the obtained data strongly depends on the design of the molecules under investigation, and the experimental setup. This can be seen for example when comparing thermodynamic values of tRNA unfolding, as summarized by Mikulecky and Feig [83]. Similarly, two studies of the thermodynamics of hammerhead ribozyme folding revealed dramatic differences due to the different experimental setups [78, 84]. Next to the design of the investigated RNA molecules, the experimental

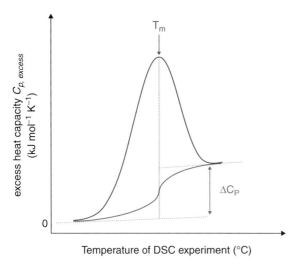

Fig. 1.6 Differential scanning calorimetry data analysis. The excess heat capacity that is determined between the sample and the reference cell in a DSC experiment is plotted against the temperature range, in which the experiment was carried out. The molar heat capacity change between folded and unfolded state ΔC_p corresponds to the distance of the baselines at the beginning and the end of the experiment. The melting temperature T_m can be determined as the temperature of the maximum. The enthalpy of unfolding ΔH^0_{unfold} can be calculated by integrating the area under the DSC curve using the gray baseline.

conditions also strongly influence the interaction. For example, a detailed analysis of the interaction of several aminoglycosides with A-site model RNAs revealed that the protonation state of the antibiotics was critical for the specific interaction with this rRNA mimic [85]. This observation clearly can serve to optimize drug-target interactions of aminoglycoside antibiotics [86]. In addition, monovalent ions strongly influenced the heat capacity changes observed in the course of RNA duplex formation [87].

As indicated earlier, ligands used in ITC experiments can be other, (partially) complementary RNA molecules, proteins, or small molecules. Calorimetric studies on RNA-RNA interactions, such as tetraloop-receptor binding, revealed a predominantly enthalpy-driven formation of the binary complexes [88] that was shown to be realized as a multistep and hierarchical process. ITC studies also revealed the folding pathway of the hepatitis delta virus ribozyme to be hierarchical and enthalpically driven, with a large unfavorable entropic contribution [89]. Conversely, the Feig lab has shown that the folding of the three-way junction of the hammerhead ribozyme can be entropically driven [84].

The discovery of riboswitches has led to the widespread use of ITC for the thermodynamic characterization of the specific interaction between the riboswitch and its cognate small molecule ligand. Throughout, these interactions were characterized as enthalpically favored and entropically disfavored, in line with an increase in order upon ligand binding [90–93]. Interestingly, enthalpy-entropy compensation, a phenomenon frequently observed in the analysis of proteins [94], is also observed for near cognate interactions of RNA with nonprotein ligands [95]. Indeed, enthalpy-entropy compensation appears to be a recurring theme for protein-RNA or protein-nucleotide interactions (e.g. [96, 97]).

Several RNA-protein interactions have been analyzed by ITC. A particularly intriguing example is the analysis of double-stranded RNA-binding domains (dsRBDs) of the antiviral PKR protein: unlike the name suggests, the dsRBDs required bulges or loops in the RNA target for high-affinity binding, which was observed by ITC to take place in a 1:1 stoichiometry [98].

As noted previously for small molecule ligand-riboswitch interactions, the complex formation between proteins and RNAs also appears to be generally enthalpy driven, with a large offset by unfavorable entropy contributions. This was, for example, observed in an ITC study of the bacterial Hfq protein that can bind many cellular RNAs. This study also revealed different stoichiometries for Hfq binding to individual RNA molecules [99].

Overall, spectroscopic and calorimetric techniques thus are invaluable approaches to analyze thermodynamic parameters underlying RNA folding and stability. In combination with structural methods, these approaches provide a basis to interpret the energetics of RNA folding and stability, and of the interaction of RNA with ligands on a molecular level.

1.7 Acknowledgments

This work was supported by a fellowship in the Heisenberg programme of the Deutsche Forschungsgemeinschaft to CH (HA3459/5).

References

1. Ban N, Nissen P, Hansen J, Moore PB, Steitz TA. The complete atomic structure of the large ribosomal subunit at 2.4 A resolution. Science. 2000;289:905–20.
2. Schluenzen F, Tocilj A, Zarivach R, et al. Structure of functionally activated small ribosomal subunit at 3.3 angstroms resolution. Cell. 2000;102:615–23.
3. Wimberly BT, Brodersen DE, Clemons WM Jr, et al. Structure of the 30S ribosomal subunit. Nature. 2000;407:327–39.
4. Leontis NB, Stombaugh J, Westhof E. The non-Watson-Crick base pairs and their associated isostericity matrices. Nucleic Acids Res. 2002;30:3497–531.
5. Puglisi JD, Tinoco I Jr. Absorbance melting curves of RNA. Methods Enzymol. 1989;180:304–25.
6. Marky LA, Breslauer KJ. Calculating thermodynamic data for transitions of any molecularity from equilibrium melting curves. Biopolymers. 1987;26:1601–20.
7. Qiu H, Kaluarachchi K, Du Z, Hoffman DW, Giedroc DP. Thermodynamics of folding of the RNA pseudoknot of the T4 gene 32 autoregulatory messenger RNA. Biochemistry. 1996;35:4176–86.

8. Draper DE, Gluick TC. Melting studies of RNA unfolding and RNA-ligand interactions. Methods Enzymol. 1995;259:281–305.
9. Matthews CR. Effect of point mutations on the folding of globular proteins. Methods Enzymol. 1987;154:498–511.
10. Freier SM, Kierzek R, Jaeger JA, et al. Improved free-energy parameters for predictions of RNA duplex stability. Proc Natl Acad Sci U S A. 1986;83:9373–7.
11. Turner DH, Sugimoto N, Freier SM. RNA structure prediction. Annu Rev Biophys Biophys Chem. 1988;17:167–92.
12. Jaeger JA, Turner DH, Zuker M. Improved predictions of secondary structures for RNA. Proc Natl Acad Sci U S A. 1989;86:7706–10.
13. Zuker M. Computer prediction of RNA structure. Methods Enzymol. 1989;180:262–88.
14. Zuker M. On finding all suboptimal foldings of an RNA molecule. Science. 1989;244:48–52.
15. SantaLucia J Jr, Turner DH. Measuring the thermodynamics of RNA secondary structure formation. Biopolymers. 1997;44:309–19.
16. Zuker M. Mfold web server for nucleic acid folding and hybridization prediction. Nucleic Acids Res. 2003;31:3406–15.
17. Markham NR, Zuker M. DINAMelt web server for nucleic acid melting prediction. Nucleic Acids Res. 2005;33:W577–81.
18. Mathews DH, Turner DH. Prediction of RNA secondary structure by free energy minimization. Curr Opin Struct Biol. 2006;16:270–8.
19. Mathews DH, Turner DH, Zuker M. RNA secondary structure prediction. Curr Protoc Nucleic Acid Chem. 2007;Chapter 11:Unit 11.2.
20. Markham NR, Zuker M. UNAFold: software for nucleic acid folding and hybridization. Methods Mol Biol. 2008;453:3–31.
21. Mathews DH, Disney MD, Childs JL, Schroeder SJ, Zuker M, Turner DH. Incorporating chemical modification constraints into a dynamic programming algorithm for prediction of RNA secondary structure. Proc Natl Acad Sci U S A. 2004;101:7287–92.
22. Serebrov V, Vassilenko K, Kholod N, Gross HJ, Kisselev L. Mg^{2+} binding and structural stability of mature and in vitro synthesized unmodified *Escherichia coli* tRNAPhe. Nucleic Acids Res. 1998;26:2723–8.
23. Jones CI, Spencer AC, Hsu JL, et al. A counterintuitive Mg^{2+}-dependent and modification-assisted functional folding of mitochondrial tRNAs. J Mol Biol. 2006;362:771–86.
24. Johnson KH, Gray DM. A method for estimating the nearest neighbor base-pair content of RNAs using CD and absorption spectroscopy. Biopolymers. 1991;31:373–84.
25. Johnson KH, Gray DM. An estimate of the nearest neighbor base-pair content of 5S RNA using CD and absorption spectroscopy. Biopolymers. 1991;31:385–95.
26. Pan T, Sosnick TR. Intermediates and kinetic traps in the folding of a large ribozyme revealed by circular dichroism and UV absorbance spectroscopies and catalytic activity. Nat Struct Biol. 1997;4:931–8.
27. Horton TE, DeRose VJ. Cobalt hexammine inhibition of the hammerhead ribozyme. Biochemistry. 2000;39:11408–16.
28. Nixon PL, Giedroc DP. Equilibrium unfolding (folding) pathway of a model H-type pseudoknotted RNA: the role of magnesium ions in stability. Biochemistry. 1998;37:16116–29.
29. Chang KY, Tinoco I Jr. The structure of an RNA "kissing" hairpin complex of the HIV TAR hairpin loop and its complement. J Mol Biol. 1997;269:52–66.
30. Wilhelmsson LM. Fluorescent nucleic acid base analogues. Q Rev Biophys. 2010;43:159–83.
31. Smalley MK, Silverman SK. Fluorescence of covalently attached pyrene as a general RNA folding probe. Nucleic Acids Res. 2006;34:152–66.
32. Bevilacqua PC, Johnson KA, Turner DH. Cooperative and anticooperative binding to a ribozyme. Proc Natl Acad Sci U S A. 1993;90:8357–61.

33. Li Y, Turner DH. Effects of Mg^{2+} and the 2' OH of guanosine on steps required for substrate binding and reactivity with the *Tetrahymena* ribozyme reveal several local folding transitions. Biochemistry. 1997;36:11131–9.
34. Silverman SK, Cech TR. RNA tertiary folding monitored by fluorescence of covalently attached pyrene. Biochemistry. 1999;38:14224–37.
35. Hall KB. 2-Aminopurine as a probe of RNA conformational transitions. Methods Enzymol. 2009;469:269–85.
36. Ballin JD, Bharill S, Fialcowitz-White EJ, Gryczynski I, Gryczynski Z, Wilson GM. Site-specific variations in RNA folding thermodynamics visualized by 2-aminopurine fluorescence. Biochemistry. 2007;46:13948–60.
37. Sarkar K, Nguyen DA, Gruebele M. Loop and stem dynamics during RNA hairpin folding and unfolding. RNA. 2010;16:2427–34.
38. Walter NG, Harris DA, Pereira MJ, Rueda D. In the fluorescent spotlight: global and local conformational changes of small catalytic RNAs. Biopolymers. 2001;61:224–42.
39. Rieder R, Lang K, Graber D, Micura R. Ligand-induced folding of the adenosine deaminase A-riboswitch and implications on riboswitch translational control. ChemBioChem. 2007;8:896–902.
40. Lang K, Rieder R, Micura R. Ligand-induced folding of the thiM TPP riboswitch investigated by a structure-based fluorescence spectroscopic approach. Nucleic Acids Res. 2007;35:5370–8.
41. Rieder U, Kreutz C, Micura R. Folding of a transcriptionally acting preQ1 riboswitch. Proc Natl Acad Sci U S A. 2010;107:10804–9.
42. Haller A, Rieder U, Aigner M, Blanchard SC, Micura R. Conformational capture of the SAM-II riboswitch. Nat Chem Biol. 2011;7:393–400.
43. Sosnick TR, Fang X, Shelton VM. Application of circular dichroism to study RNA folding transitions. Methods Enzymol. 2000;317:393–409.
44. Johnson KH, Gray DM. Analysis of an RNA pseudoknot structure by CD spectroscopy. J Biomol Struct Dyn. 1992;9:733–45.
45. Maruyama S, Sugai S. Folding of yeast 5S ribosomal RNA induced by magnesium binding. J Biochem. 1980;88:151–8.
46. Shelton VM, Sosnick TR, Pan T. Applicability of urea in the thermodynamic analysis of secondary and tertiary RNA folding. Biochemistry. 1999;38:16831–9.
47. Shelton VM, Sosnick TR, Pan T. Altering the intermediate in the equilibrium folding of unmodified yeast tRNAPhe with monovalent and divalent cations. Biochemistry. 2001;40:3629–38.
48. Fang X, Pan T, Sosnick TR. A thermodynamic framework and cooperativity in the tertiary folding of a Mg^{2+}-dependent ribozyme. Biochemistry. 1999;38:16840–6.
49. Baird NJ, Westhof E, Qin H, Pan T, Sosnick TR. Structure of a folding intermediate reveals the interplay between core and peripheral elements in RNA folding. J Mol Biol. 2005;352:712–22.
50. Fang XW, Golden BL, Littrell K, et al. The thermodynamic origin of the stability of a thermophilic ribozyme. Proc Natl Acad Sci U S A. 2001;98:4355–60.
51. Baird NJ, Srividya N, Krasilnikov AS, Mondragon A, Sosnick TR, Pan T. Structural basis for altering the stability of homologous RNAs from a mesophilic and a thermophilic bacterium. RNA. 2006;12:598–606.
52. Mikulecky PJ, Feig AL. Heat capacity changes in RNA folding: application of perturbation theory to hammerhead ribozyme cold denaturation. Nucleic Acids Res. 2004;32:3967–76.
53. Daly TJ, Rusche JR, Maione TE, Frankel AD. Circular dichroism studies of the HIV-1 Rev protein and its specific RNA binding site. Biochemistry. 1990;29:9791–5.
54. Metzger AU, Schindler T, Willbold D, et al. Structural rearrangements on HIV-1 Tat (32–72) TAR complex formation. FEBS Lett. 1996;384:255–9.
55. Hagerman PJ. Transient electric birefringence for determining global conformations of nonhelix elements and protein-induced bends in RNA. Methods Enzymol. 2000;317:440–53.

56. Hagerman PJ, Amiri KMA. "Hammering away" at RNA global structure. Curr Opin Struct Biol. 1996;6:317–21.
57. Hagerman PJ. Sometimes a great motion: the application of transient electric birefringence to the study of macromolecular structure. Curr Opin Struct Biol. 1996;6:643–9.
58. Vacano E, Hagerman PJ. Analysis of birefringence decay profiles for nucleic acid helices possessing bends: the r-ratio approach. Biophys J. 1997;73:306–17.
59. Zacharias M, Hagerman PJ. Bulge-induced bends in RNA: quantification by transient electric birefringence. J Mol Biol. 1995;247:486–500.
60. Kebbekus P, Draper DE, Hagerman P. Persistence length of RNA. Biochemistry. 1995;34:4354–7.
61. Zacharias M, Hagerman PJ. The influence of symmetric internal loops on the flexibility of RNA. J Mol Biol. 1996;257:276–89.
62. Zacharias M, Hagerman PJ. The bend in RNA created by the trans-activation response element bulge of human immunodeficiency virus is straightened by arginine and by Tat-derived peptide. Proc Natl Acad Sci U S A. 1995;92:6052–6.
63. Zacharias M, Hagerman PJ. Influence of static and dynamic bends on the birefringence decay profile of RNA helices: Brownian dynamics simulations. Biophys J. 1997;73:318–32.
64. Thompson MR, Williams RC, O'Neal CH. Studies on proteins and tRNA with transient electric birefringence. Biophys J. 1978;24:264–6.
65. Friederich MW, Gast FU, Vacano E, Hagerman PJ. Determination of the angle between the anticodon and aminoacyl acceptor stems of yeast phenylalanyl tRNA in solution. Proc Natl Acad Sci U S A. 1995;92:4803–7.
66. Friederich MW, Hagerman PJ. The angle between the anticodon and aminoacyl acceptor stems of yeast tRNA(Phe) is strongly modulated by magnesium ions. Biochemistry. 1997;36:6090–9.
67. Frazer-Abel AA, Hagerman PJ. Determination of the angle between the acceptor and anticodon stems of a truncated mitochondrial tRNA. J Mol Biol. 1999;285:581–93.
68. Frazer-Abel AA, Hagerman PJ. Variation of the acceptor-anticodon interstem angles among mitochondrial and non-mitochondrial tRNAs. J Mol Biol. 2004;343:313–25.
69. Frazer-Abel AA, Hagerman PJ. Core flexibility of a truncated metazoan mitochondrial tRNA. Nucleic Acids Res. 2008;36:5472–81.
70. Stagg SM, Frazer-Abel AA, Hagerman PJ, Harvey SC. Structural studies of the tRNA domain of tmRNA. J Mol Biol. 2001;309:727–35.
71. Shen Z, Hagerman PJ. Conformation of the central, three-helix junction of the 5 S ribosomal RNA of *Sulfolobus acidocaldarius*. J Mol Biol. 1994;241:415–30.
72. Orr JW, Hagerman PJ, Williamson JR. Protein and Mg^{2+}-induced conformational changes in the S15 binding site of 16 S ribosomal RNA. J Mol Biol. 1998;275:453–64.
73. Amiri KM, Hagerman PJ. Global conformation of a self-cleaving hammerhead RNA. Biochemistry. 1994;33:13172–7.
74. Amiri KMA, Hagerman PJ. The global conformation of an active hammerhead RNA during the process of self-cleavage. J Mol Biol. 1996;261:125–34.
75. Amiri KMA, Hagerman PJ. The global conformation of an active hammerhead RNA during the process of self-cleavage. J Mol Biol. 1996;261:125–34.
76. Ladbury JE, Chowdhry BZ, eds. Biocalorimetry. Chichester, Wiley; 1998.
77. Sokoloski JE, Bevilacqua PC. Analysis of RNA folding and ligand binding by conventional and high-throughput calorimetry. Methods Mol Biol. 2012;905:145–74.
78. Hammann C, Cooper A, Lilley DMJ. Thermodynamics of ion-induced RNA folding in the hammerhead ribozyme: an isothermal titration calorimetric study. Biochemistry. 2001;40:1423–9.
79. Blandamer MJ. Thermodynamic background to isothermal titration calorimetry. In: Ladbury JE, Chowdhry BZ, eds. Biocalorimetry. Chichester, Wiley; 1998: 5–25.

80. SantaLucia J Jr, Turner DH. Measuring the thermodynamics of RNA secondary structure formation. Biopolymers. 1997;44:309–19.
81. Bode D, Schernau U, Ackermann T. Calorimetric investigations of the helix-coil conversion of phenylalaninespecific transfer ribonucleic acid. Biophys Chem. 1974;1:214–21.
82. Hinz HJ, Filimonov VV, Privalov PL. Calorimetric studies on melting of tRNA Phe (yeast). Eur J Biochem/FEBS. 1977;72:79–86.
83. Mikulecky PJ, Feig AL. Heat capacity changes associated with nucleic acid folding. Biopolymers. 2006;82:38–58.
84. Mikulecky PJ, Takach JC, Feig AL. Entropy-driven folding of an RNA helical junction: an isothermal titration calorimetric analysis of the hammerhead ribozyme. Biochemistry. 2004;43:5870–81.
85. Pilch DS, Kaul M, Barbieri CM, Kerrigan JE. Thermodynamics of aminoglycoside-rRNA recognition. Biopolymers. 2003;70:58–79.
86. Salim NN, Feig AL. Isothermal titration calorimetry of RNA. Methods. 2009;47:198–205.
87. Takach JC, Mikulecky PJ, Feig AL. Salt-dependent heat capacity changes for RNA duplex formation. J Am Chem Soc. 2004;126:6530–1.
88. Vander Meulen KA, Davis JH, Foster TR, Record MT Jr, Butcher SE. Thermodynamics and folding pathway of tetraloop receptor-mediated RNA helical packing. J Mol Biol. 2008;384:702–17.
89. Reymond C, Bisaillon M, Perreault JP. Monitoring of an RNA multistep folding pathway by isothermal titration calorimetry. Biophys J. 2009;96:132–40.
90. Kulshina N, Edwards TE, Ferre-D'Amare AR. Thermodynamic analysis of ligand binding and ligand binding-induced tertiary structure formation by the thiamine pyrophosphate riboswitch. RNA. 2010;16:186–96.
91. Montange RK, Mondragon E, van Tyne D, Garst AD, Ceres P, Batey RT. Discrimination between closely related cellular metabolites by the SAM-I riboswitch. J Mol Biol. 2010;396:761–72.
92. Trausch JJ, Ceres P, Reyes FE, Batey RT. The structure of a tetrahydrofolate-sensing riboswitch reveals two ligand binding sites in a single aptamer. Structure. 2011;19:1413–23.
93. Weigand JE, Schmidtke SR, Will TJ, et al. Mechanistic insights into an engineered riboswitch: a switching element which confers riboswitch activity. Nucleic Acids Res. 2011;39:3363–72.
94. Cooper A, Johnson CM, Lakey JH, Nollmann M. Heat does not come in different colours: entropy-enthalpy compensation, free energy windows, quantum confinement, pressure perturbation calorimetry, solvation and the multiple causes of heat capacity effects in biomolecular interactions. Biophys Chem. 2001;93:215–30.
95. Muller M, Weigand JE, Weichenrieder O, Suess B. Thermodynamic characterization of an engineered tetracycline-binding riboswitch. Nucleic Acids Res. 2006;34:2607–17.
96. Niedzwiecka A, Darzynkiewicz E, Stolarski R. Thermodynamics of mRNA 5' cap binding by eukaryotic translation initiation factor eIF4E. Biochemistry. 2004;43:13305–17.
97. Moll D, Schweinsberg S, Hammann C, Herberg FW. Comparative thermodynamic analysis of cyclic nucleotide binding to protein kinase A. Biol Chem. 2007;388:163–72.
98. Kim I, Liu CW, Puglisi JD. Specific recognition of HIV TAR RNA by the dsRNA binding domains (dsRBD1-dsRBD2) of PKR. J Mol Biol. 2006;358:430–42.
99. Mikulecky PJ, Kaw MK, Brescia CC, Takach JC, Sledjeski DD, Feig AL. *Escherichia coli* Hfq has distinct interaction surfaces for DsrA, rpoS and poly(A) RNAs. Nat Struct Mol Biol. 2004;11:1206–14.

Mélodie Duval, Cedric Romilly, Anne-Catherine Helfer,
Olivier Fuchsbauer, Pascale Romby, and Stefano Marzi

2 Footprinting methods for mapping RNA-protein and RNA-RNA interactions

2.1 Introduction

Ribonucleoprotein (RNP) complexes have multiple roles in a wide range of processes including RNA editing, processing, transport, translation, and degradation. Indeed, from its transcription to its decay, an RNA molecule is constantly making interactions with proteins to govern its fate. RNP complexes involving messenger RNAs (mRNAs) are responsible for a large variety of regulatory mechanisms of gene expression. It is now accepted that mRNAs often harbor *cis*-acting regulatory and structured elements mostly present in their 5' and 3' untranslated regions (UTRs) [1,2]. In bacteria, these *cis*-acting elements can respond to temperature variations, metabolic changes, and represented specific signatures for the recognition of a variety of *trans*-acting ligands such as proteins, RNAs, and metabolites. The vast majority of translational regulatory proteins are repressors, although phage studies have identified activator proteins [3,4]. Besides proteins, another important class of bacterial regulators targeting mRNAs comprises the small noncoding RNAs, the so-called sRNAs. These sRNAs include antisense RNAs (asRNAs), which are encoded at the same locus of their target mRNA leading to perfect base-pairing complementarity [5,6], and *trans*-encoded sRNAs that can regulate several mRNAs, tolerating some mismatches [7–10]. In all these systems, the regulatory activities of RNAs are linked to the dynamics and versatility of their structures.

Although RNA-RNA recognition is mainly based on the formation of intermolecular Watson-Crick base pairings, RNA-binding proteins (RBPs) interact with RNA with different stabilities and recognition mechanisms [11]. Several classes of RNA-binding domains have been identified (e.g. the RBD [also known as RRM] [12], double-stranded RBD [dsRBD] [13], KH [14], La [15], zinc finger [16], CAT [17], and S1 [18] domains) and their structures have been determined by X-ray crystallography or nuclear magnetic resonance (NMR). Each RBP recognizes specific RNA sequences, structures, or both, and in many cases, an induced-fit mechanism was also described. A special class of RBP includes proteins that facilitate RNA folding, the so-called RNA chaperone proteins [19–21]. Among these proteins, some help the RNA to get its active structure by disrupting misfolded or inactive RNA structures. Unlike RNA helicases, these proteins do not require adenosine triphosphate but interact weakly and transiently with the RNA. It was demonstrated that the weak stability and transient binding are two important features for RNA chaperone activity. Indeed, stronger RNA binding is connected to decreased RNA chaperone activity and is detrimental to RNA folding [22]. The characterization of RNA chaperone protein–RNA complexes is one major challenge, which requires the development of appropriate tools in order to get access to the dynamics of the complex formation. Given the importance of RNPs in all aspects of life, our information about their composition, structure, and dynamics of assembly and disassembly is still limited. Therefore, getting the rules of recognition and understanding the impact of the RNA structure on binding are major challenges ahead to understand gene regulation.

Information regarding the final structure of sRNA-mRNA or RNP complexes can be obtained by high-resolution structure methods (cryo–electron microscopy, NMR, or crystallography, see Chapters 10, 12, and 13). Besides these powerful approaches, other biochemical and genetic methods have been developed to analyze complex regulatory systems and to provide complementary data to the determination of high-resolution structures. These approaches include cross-linking, fluorescent-based approaches (see Chapter 8), chemical interference (see Chapter 3), and footprinting. Combined with genetic approaches, it is thus possible to gain some knowledge about the RNA-binding site, the RNA conformational changes, and the key sites of interaction on RNA molecules of any size. Among them, footprinting, which is based on the accessibility of each nucleotide of the RNA toward chemicals and enzymes, follows the RNA structure interacting with various ligands, provides information of RNA structure within purified RNP complexes, and allows probing the RNA structure in vivo.

This chapter describes the different probes and their specificities that can be used for footprinting. Using specific examples, we will illustrate the advantages and the limitations of the approaches to study both RNP and RNA-RNA regulatory complexes.

2.2 Principles and applications of footprinting

The footprinting approach was first developed on DNA to map protein-binding sites on the chromatin, and of the RNA polymerase on promoter sequences using dimethyl sulfate (DMS) or deoxyribonuclease I [23,24]. In parallel, enzymes and chemicals were adapted for RNA sequencing [25–27] and later on for probing the structure of end-labeled RNAs [28]. Another major technical improvement was the introduction of primer extension using 5'-end-labeled DNA probes to detect the modification or cleavage events allowing the analysis of large RNA molecules [29,30]. A derived approach based on DNA sequencing technology took advantage of capillary electrophoresis combined with fluorescent DNA probes for primer extension to analyze very long RNA molecules in one single step [31]. The knowledge gained on RNA molecules, for which the crystal structures were solved, facilitated the interpretation of the probing data. In particular, they showed that the reactivity of nucleotides toward the chemicals was well correlated with their accessibility in the RNA structure [28,32,33]. The footprinting method has been used largely to map protein-, RNA- and metabolite-binding sites.

Footprints are the results of the shielding effect that the ligand exerts when the RNA is being probed. Briefly, the RNA complex is treated with a reagent, small organic molecule, metal ion, or ribonuclease (RNase), which can induce either cleavage in the RNA or a covalent adduct with the RNA. The experiments are performed under limited and statistical conditions so that less than one modification or cleavage per molecule can take place. For the first experiment, the probe:RNA ratio must be adapted so that the majority of full-length RNA is not modified or cleaved. The detection of the modification/cleavages is done using various methodologies depending on the length of the RNA molecule and on the nature of the modification. The direct detection approach, which uses end-labeled RNAs, detects RNA cleavages. Due to the resolution of the sequencing gel, this approach is limited to RNAs containing less than 300 nucleotides. The second approach uses primer extension to detect stops of reverse transcriptase at modified nucleotides or cleavages and can be applied to RNAs of any size. Footprinting experiments can be done to map specific binding sites of one ligand and to follow the assembly of RNPs. These experiments can be coupled with a quench-flow apparatus to investigate the path and dynamics of ligand binding [34]. Moreover, footprinting can be adapted under more complex physiological conditions and can be carried out in living cells. However, comparison with the in vitro footprinting provides complementary data and is of great help in the interpretation of the data. Given the high complexity of the cellular environment, it is advisable to perform the experiment in a comparative manner (i.e. by studying two different culture conditions [35] or by performing a transient induction of a regulatory molecule [36]). The use of probes is, however, limited by their inability to penetrate the cell wall and the membrane due to their size, structure, and/or charge [37,38]. The approaches that have gained widespread use for in vivo RNA probing are induced by DMS [35], to a lesser extent β-ethoxy-α-ketobutyraldehyde (kethoxal) [39], and lead(II) [36], next to selective 2'-hydroxyl acylation analyzed by primer extension (SHAPE) (see Section 2.4.1) [40,41] and footprinting using synchrotron-generated hydroxyl radicals [42,43].

Chemical interference (see Chapter 3) is derived from footprinting experiments; it defines a set of nucleotides that have lost the capability to interact with a ligand when they are modified by a chemical probe (e.g. [44,45]). In this respect, nucleotide analog interference mapping is a powerful method to investigate the importance of specific nucleotide functional groups on the RNA structure or on ligand binding [46,47]. Although chemical interference and footprinting methods have been successfully used for the investigation of stable RNA-ligand complexes, they are less appropriate for the study of transient RNA-ligand complexes. Site-directed chemical probing (SDCP) is a technique of choice. A chemical probe (usually Fe–ethylenediaminetetraacetic acid [EDTA]) is tethered to a cysteine residue of protein or to RNA, and upon binding, the modified ligand induces RNA cleavage. Thus, this method visualizes a positive signature of the ligand during binding and provides topographical information (e.g. [48,49]).

2.3 Tools for footprinting: what should we know about probes?

The choice of the probe is very critical, and its size impacts the resolution of the data that can be achieved (Table 2.1). Because enzymes and RNases are the largest probes, they are sensitive to steric hindrance and therefore cannot be used to precisely determine the binding site of a protein. Conversely, chemicals or radicals, which attack specific positions of nucleotides, generate data that can be used to delimitate a minimal binding site. For in vitro probing, it is worth spending time to design renaturation protocols to obtain a homogeneous RNA population and to test the biological relevance of the preparation (enzymatic activity and ligand binding). The conformational homogeneity can be analyzed by native polyacrylamide gel electrophoresis, as different conformers show different electrophoretic mobilities. Probing experiments have to be conducted for all the probes under strictly defined conditions of buffers, pH, temperature, and ionic conditions. Incubation controls also have to be performed in the absence of the probes to take into account the unspecific cleavages induced by the incubation treatment. For both in vitro and in vivo footprinting, it is important to define the best concentration of the probe and the time of the reaction in order to work under statistical conditions. Binding of the ligand is expected to induce protection of RNA toward the chemical and the enzymatic probes. However, protection may also result from conformational rearrangement or stabilization of an alternative structure of the RNA. On the other hand, sites of enhanced reactivity or cleavage unambiguously reveal RNA conformational rearrangements. The use of diverse probes can help to distinguish between direct interaction and conformational changes.

2.3.1 RNases

Although their size renders them quite sensitive to steric hindrance, RNases are easy to use and should certainly not be neglected. It is noteworthy that if the hydrolysis is too strong, RNases can generate secondary cleavage events, which do not reflect the native RNA structure. These cleavage events are often not reproducibly found and are weak. Most RNases induce cleavage at specific positions within single-stranded regions of the RNA (Table 2.1). Commonly used RNases are RNase T1 (specific for unpaired guanines), RNase T2 (with preference for unpaired adenines), and RNase V1 (specific for double-stranded regions). RNase V1, extracted from cobra venom, is the only probe that gives a positive signature for the existence of helical or stacked regions. It is thus an appropriate enzyme for mapping RNA-RNA interactions. The ubiquitous RNase III specifically recognizes and cleaves double-stranded RNA (dsRNA) in bacteria and in eukaryotes [50]. The enzyme was shown to bind and cleave a variety of structures such as coaxially stacked short helices, dsRNA interrupted by internal loops, or structures formed by loop-loop interactions that are often found in sRNA-mRNA

Table 2.1 Structure-specific probes for RNA. DMS, dimethyl sulfate; DEPC, diethylpyrocarbonate; CMCT, 1-cyclohexyl-3-(2-morpholinoethyl) carbodiimide metho-p-toluene sulfonate. SHAPE reagents: NMIA (N-methylisatoic anhydride), 1M7 (1-methyl-7-nitroisatoic anhydride) and BzCN (benzoyl cyanide). Detection method: direct, detection of cleavage events on end-labeled RNA molecule; indirect, detection by primer extension with reverse transcriptase using either ^{32}P-labeled primer or primer labeled at the 5' end with a fluorophore. Direct*/indirect*: a chemical treatment is necessary to cleave the ribose-phosphate chain prior to the detection. In vivo mapping: probes that diffuse efficiently across membranes and walls (+), other probes that can be used only after permeabilization of the cell (+*), or probes that have never been used in vivo so far. Molecular weight, specificity, and products generated by the probe action are indicated. Most of the probes provide information useful to build a secondary structure model (II) and elements of the tertiary fold (III), and to map the binding sites of proteins, RNA, and metabolites. NMIA, in-line: due to the long reaction time, this method might not be appropriate for mapping interaction sites.

Probes		Target and modification	Detection technique	RNA structure	In vivo use	Molecular weight (Da)	Conditions to be used
Ribonucleases	RNase T1	Unpaired (...Gp3')	Direct/indirect	II, III	+*	11,000	Wide range of conditions (T °C between 4°C and 55°C, works with or without magnesium ion and salt, urea).
	RNase T2	Unpaired A>>U, C,G (...Ap3')	Direct/indirect	II, III	+*	36,000	Various conditions (T °C between 4°C and 55°C, works with or without magnesium ion).
	RNase V1	Paired or stacked nt (5'pN...)	Direct/indirect	II, III	+*	15,900	Requires magnesium ion. Active between 4°C and 50°C.

Category	Reagent	Target/Effect	Direct/Indirect			Ref	Comments
Chemicals	DMS	A(N1) N1-CH3	Indirect	II, III	+	126	Reactive at pH ranging from 4.5 to 9 and $T°C$ from 4°C to 90°C. Tris buffer should be avoided since DMS reacts with amine groups.
		C(N3) N3-CH3	Direct*/indirect	II, III	+		
		G(N7) N7-CH3	Direct*/indirect*	II, III	+		
	Kethoxal	G (N1, N2), links N1 and N2	Indirect	II, III	+		pH 8; stabilized by borate ions.
	DEPC	A(N7) N7-CO_2H_2	Direct*/indirect	II, III	+*	174	See DMS.
	CMCT	G(N1) N1-adduct	Indirect	II, III	+*	424	Optimal reactivity at pH 8, $T°C$ from 4°C to 90°C. CMCT is soluble until 300 mg/mL in water.
		U(N3) N3-adduct	Indirect	II, III	+*		
	NMIA	2'OH ribose O-acylation (unpaired nt)	Indirect	II, III		177	Active under a wide range of conditions; rapidly hydrolyzed in water with a half-life of 8.5 min at 37°C.
	1M7	2'OH ribose O-acylation (unpaired nt)	Indirect	II, III	+	222	Active under a wide range of conditions; rapidly hydrolyzed in water with a half-life of 14 s at 37°C.
	BzCN	2'OH ribose O-acylation (unpaired nt)	Indirect	II, III		117	Active under a wide range of conditions; rapidly hydrolyzed in water with a half-life of 0.2 s at 37°C.
Divalent ions and hydrolytic cleavages	Fe-EDTA (radical hydroxyl)	Cleavage at ribose (C1', C4')	Direct, indirect	II, III	+		Reactivity relatively insensitive to buffers, pH, $T°C$, Mg^{2+}, and cations. Only sodium phosphate should be avoided as well as glycerol (lower than 0.5%) and Tris buffers.
	Pb(II)-acetate	Unpaired nt and unstable helices; phosphate linkage cleavage	Direct, indirect	II, III	+		Formation of precipitate if chlorure; use acetate instead; avoid acidic pH (optimal pH 7.5–8); EDTA stops the reaction.
	In-line	Unpaired nt; phosphate linkage cleavage	Direct, indirect	II, III	?		Works under a wide range of conditions: salt and Mg, $T°C$, and optimal pH 7.5–8. Long incubation time.

regulatory complexes. Although not yet commercially available, it is a valuable tool to gain insight into the recognition mechanism between sRNA and its target mRNAs [51].

2.3.2 Chemicals

Chemicals modify either bases or the ribose-phosphate backbone (see Table 2.1). Detailed protocols for probing RNA with chemicals have been previously reported [37,52]. However, precautions need to be taken, as chemicals may modify the protein moiety. Thus, controls have to be performed to test the stability of the complexes in the presence of the chemical reagent. For example, diethyl pyrocarbonate (DEPC) reacts faster with histidine and cysteine residues than with adenines (N7) and is usually avoided for the analysis of RNA-protein complexes.

2.3.2.1 Base-specific reagents

Base-specific reagents modify specific position of bases and are sensitive to secondary structure and base stacking. They are commonly used to probe the structure of RNA molecules and of RNA-ligand complexes. The combined use of DMS, 1-cyclohexyl-3-(2-morpholinoethyl) carbodiimide metho-p-toluene sulfonate (CMCT), and β-ethoxy-α-ketobutyraldehyde (kethoxal) allows testing the four nucleotides at positions that are involved in Watson-Crick interactions. DMS methylates position N1 of adenines and to a lesser extent position N3 of cytosines, and it also methylates position N7 of guanines [28,53]. CMCT modifies position N3 of uridines and to a lesser degree N1 of guanines. Kethoxal reacts with guanine, giving a cyclic adduct between positions N1 and N2 and its two carbonyls [54]. Cleavage by these base-specific reagents provides information on unpaired regions in the RNA secondary structure. In contrast, nonreactivity points to the formation of Watson-Crick or noncanonical base pairs. From comparisons in the presence or in the absence of a ligand (RNA, protein, or metabolite), ligand-induced protections indicate either direct contacts or changes in RNA secondary or tertiary structure.

2.3.2.2 Ribose-phosphate backbone–specific reagents

These reagents attack the ribose-phosphate backbone (Table 2.1). Due to the fact that they are less sensitive to the secondary structure of RNAs, these probes are appropriate to map a ligand-binding site (see Section 2.4.1). However, the choice of the probe will depend on the ligand analyzed in the footprinting assays. Because the SHAPE and lead(II)-induced cleavage events are sensitive to the flexibility of nucleotides, they can be used to map protein- or RNA-binding sites. Hydroxyl radicals have been shown to map the accessible surface of compact tertiary RNA structures. They are thus appropriate for following the folding pathways of an RNA molecule, such as the metabolite-sensing riboswitches, but less suitable for mapping RNA-RNA interactions.

Cleavage of RNAs by coordinated metal ions such as Pb^{2+} is a simple and sensitive test for footprinting both in vivo and in vitro (e.g. [36] and references therein). The prevalent mechanism for cleavage involves a nucleophilic attack of the 2'-hydroxyl group of the adjacent ribose resulting in products containing 2',3'-cyclic phosphate and 5' hydroxyl termini. Lead(II)-induced cleavage has been largely used to probe RNA, RNA-protein, and RNA-RNA interactions. The metal induces two types of cleavage events: strong cuts at or near the site of coordination of the ion that often corresponded to a magnesium-binding site (e.g. [55]) and low-intensity cleavage at multiple sites in flexible regions comprising helices of weak stability and unpaired nucleotides. It is a highly sensitive probe allowing detection of subtle conformational changes upon ligand binding.

SHAPE reagents are a new generation of chemicals that modify 2'OH groups of riboses and are highly sensitive to the local flexibility of each nucleotide [56]. Although purine residues are slightly more reactive than pyrimidines, it was shown that these differences are subtle in contrast to the reactivity differences between paired and unpaired nucleotides [57]. SHAPE-induced ribose modifications can only be detected as stops of reverse transcription. As in footprinting experiments, ligand-induced protection can be the result of either direct shielding or a conformational change. Suitable reagents include N-methylisatoic anhydride [58], 1-methyl-7-nitroisatoic anhydride [57,59], and benzoyl cyanide [31,56].

The hydroxyl radical (•OH) is the smallest chemical species that has been used for footprinting at higher resolution. These highly reactive and oxidizing species cleave the nucleic acid backbone by abstracting hydrogens of positions C1' and C4' of the ribose or deoxyribose sugar rings [60]. Because the sugars are ubiquitously exposed at the surface of nucleic acid helices, the reaction is not highly sensitive to the RNA secondary structure, and thus ligand-induced protection usually reflects direct contacts [61]. However, this reagent has been used to probe the inside and the outside of compact tertiary RNA structure and the nonreactivity of a specific residue in RNAs might imply its involvement in a molecular interaction (hydrogen bonding or ion coordination) [32]. An additional advantage is that •OH radicals interact nonspecifically with nucleotides, thereby allowing us to monitor the individual behavior of all residues that are involved in tertiary interactions or ligand binding. Hydroxyl radicals are generated by the Fenton reaction [62,63], which involves oxidizing Fe^{2+} (EDTA) complexes with H_2O_2 (hydrogen peroxide). The resulting Fe^{3+} is then reduced to Fe^{2+} by a reducing agent [ascorbate, dithiothreitol, dithioerythritol, or tris(2-carboxyethyl)phosphine] to amplify the hydroxyl radical production. The use of the chelating agent EDTA to form the Fe^{2+}(EDTA) complex prevents the iron from interacting electrostatically with the RNA, so that the only reactants are the hydroxyl radicals generated in solution [64]. Hydroxyl radicals can also be obtained through the ionization of intracellular water by X-rays to perform footprinting experiments on a shorter timescale [65,66].

In-line probing has exploited the capacity of RNA to undergo spontaneously cleavage owing to a nucleophilic attack of the 2'OH group [67]. This procedure involves a long time of incubation (>40 h) of the RNA. Although this method has been successfully used to probe conformational changes of the riboswitch-sensing elements, the long incubation time is not favorable for monitoring the binding site of proteins.

2.4 Examples of RNP or RNA-RNA complexes analyzed by footprinting

In the following section, we present four examples of bacterial RNP or RNA-RNA complexes studied by footprinting using different probes (SHAPE, enzymes, or chemicals). The examples highlight the power of the technique when stable complexes are the object of the investigation (see Sections 2.4.1-3), but also show the limits encountered when unstable/transitory complexes are studied (see Section 2.4.4).

2.4.1 Determination of the mRNA-binding site of Crc by SHAPE footprinting

Crc is the master regulator of catabolite repression in *Pseudomonas* [68]. By binding to its target mRNAs encoding transcriptional regulatory proteins (e.g. *alkS*, *benR*, and *xylR*) and/or proteins involved in the transport and/or catabolism, it affects the expression of more than 135 different genes. Because the RNA-binding site was not known, in vitro footprinting was performed on *alkS* mRNA using RNase T2, RNase V1, SHAPE, and DMS (Fig. 2.1; [69]). For the SHAPE experiments, data were quantified using the SAFA software [70]. This procedure was necessary to interpret the data and to fully appreciate the flexibility of each nucleotide as reflected by their reactivity. The combination of probes demonstrates the existence of a hairpin structure, which sequesters the Shine-Dalgarno (SD) sequence and the AUG codon. Indeed, the presence of the helix was supported by weak reactivity of nucleotides toward SHAPE and of adenines at N1 toward DMS, and by the concomitant presence of RNase V1 cleavage events, whereas most of the nucleotides of the apical loop were cleaved by RNase T2 and were reactive toward SHAPE and DMS. Adding increasing concentrations of Crc induced strong protections in a particular sequence motif UAAUAAUAA (Fig. 2.1). Mutations were then introduced at this site to monitor the specificity of the interaction using gel retardation assays. Together with sequence comparison of the known mRNA targets, a novel RNA sequence motif NAANAANAA (where N is C=U>A) was identified as the recognition motif of Crc [69,71]. This sequence motif was always located upstream or downstream of the AUG codon of target mRNAs. Further experiments showed that Crc binding to the apical loop of *alkS* mRNA prevented the formation of the active initiation complexes [69]. In this study, the footprinting experiment was essential to highlighting the

38 — Mélodie Duval et al.

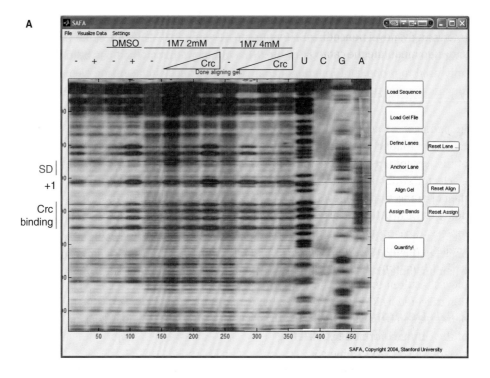

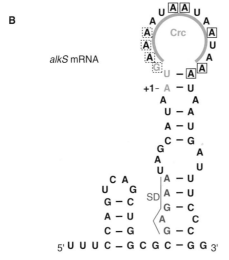

Fig. 2.1 Determination of the *Pseudomonas putida* Crc protein-binding site on *alkS* mRNA. (A) Gel fractionation of ribose modification of *alkS* mRNA obtained after primer extension with reverse transcriptase. This picture shows the quantification of the reactivity of each nucleotide toward 1M7 (SHAPE) modification using the SAFA software (A). The *alkS* mRNA was incubated in the absence (−) or in the presence of Crc (300 nM, 600 nM, or 1.2 μM). Incubation controls (no 1M7) were performed in the absence (−) or in the presence (+) of Crc. Dimethyl sulfoxide (DMSO; the solvent used for 1M7) was added in the incubation controls presented in the third and fourth lanes. Two different amounts

signature of Crc, and later on it led to the discovery of additional mRNA targets [72,73]. Interestingly, novel noncoding RNAs carrying several NAANAANAA repeat motifs were shown to modulate the catabolite repression by sequestering the Crc protein [71,73].

2.4.2 Footprinting mapping of sRNA-mRNA interaction

In pathogenic bacteria, sRNAs are key regulators of virulence gene expression [51,74,75]. Among these sRNAs, *Staphylococcus aureus* RNAIII was shown to be one of the key intracellular effectors of the quorum-sensing system [76]. RNAIII acts as a selector switch at high cell density to regulate temporally the virulence factors at the posttranscriptional level [51]. Detailed mechanistic studies revealed that specific domains of RNAIII activate the translation of *hla* mRNA encoding hemolysin α and repress the expression of various mRNAs encoding early expressed virulence factors (protein A and coagulase) and the transcriptional regulator Rot. For repression, RNAIII functions as an RNA that anneals to target mRNAs, and the formed complexes result in the repression of translation initiation and in rapid mRNA degradation triggered by RNase III. Footprinting was essential to gain insights into the regions of interaction between RNAIII and the mRNAs. We show here the particular example of the complex formed by RNAIII and *coa* mRNA encoding a major virulence factor coagulase [77]. The stable complex was probed using RNases T1 and V1 and several chemicals, which modified bases at specific positions (DEPC, DMS, and CMCT). The combination of the probes was essential to propose a secondary structure model of *coa* mRNA, which is characterized by two main stem-loop structures. The first one presented the SD sequence in the apical loop, and a second stable hairpin structure was occurring within the coding sequence of the mRNA (Fig. 2.2). Unexpectedly, addition of increasing concentrations of RNAIII revealed two distant regions of interactions. Strong protections were observed against RNase T1 and RNase V1 in a region of 35 nucleotides within the ribosome-binding site (RBS). Interestingly, these protections were accompanied by enhanced RNase V1 cleavage indicative of the formation of a long duplex (Fig. 2.2). More surprisingly, a second region of interaction was highlighted by strong protection of guanines located in the apical loop of HII against RNase T1 and by the concomitant appearance of RNase V1 cleavage sites in the same area (Fig. 2.2A). The same experiment performed on RNAIII revealed that two hairpin domains (H13 and H7) of RNAIII were protected by *coa* mRNA. Site-directed mutational analysis revealed that the 3' end domain of RNAIII (H13) binds to the RBS of *coa* mRNA, while the central domain of RNAIII (H7) formed a loop-loop interaction

---

Fig. 2.1 (*Continued*)
of 1M7 (2 mM and 4 mM) have been used. Lanes U, C, G, and A are sequencing ladders. (B) Secondary structure of the ribosome-binding site of *alkS* mRNA showing the residues protected from 1M7 modification by Crc binding. Dashed squares highlight residues that are weakly protected while lined squares represent those strongly protected. The Crc-binding site is underlined [69].

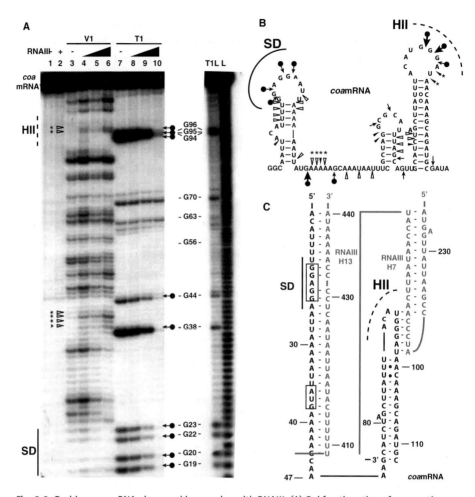

Fig. 2.2 Probing *coa* mRNA alone and in complex with RNAIII. (A) Gel fractionation of enzymatic cleavages of 5'-end-labeled *coa* mRNA. Lanes 1, 2: incubation controls of mRNA free or bound to RNAIII, respectively; lane 3: RNase V1 hydrolysis on free *coa* mRNA; lanes 4–6: RNase V1 hydrolysis on *coa* mRNA in the presence of increasing concentrations RNAIII (lane 4, 50 nM; lane 5, 100 nM; lane 6, 500 nM), respectively; lane 7: RNase T1 hydrolysis on free *coa* mRNA; lanes 8–10: RNase T1 hydrolysis on *coa* mRNA in the presence of increasing concentrations of RNAIII (lane 8, 50 nM; lane 9, 100 nM; lane 10, 500 nM), lanes T1L, L: RNase T1 under denaturing conditions and alkaline ladders, respectively. (B) Enzymatic cleavage sites are reported on the schematic representation of the secondary structure of *coa* mRNA. Black-and-white triangles are for strong and moderate RNase V1 cleavage, respectively. Plain and dashed arrows are for strong and moderate RNase T1 cuts. Effect of RNAIII binding: black-and-white circles are for strong and moderate protection, respectively; stars are for enhancement and new cleavage events. (C) The secondary structure model of RNAIII-*coa* mRNA duplex based on data from Chevalier et al. [77] and Romilly et al. [51].

with the hairpin loop HII of the mRNA (Fig. 2.2B). The two domains contributed to the regulation. Binding of RNAIII to the RBS prevented ribosome binding. The loop-loop interaction further enhanced the stability of the complex and promoted binding of RNase III, leading to rapid degradation of the mRNA [51,77]. Thus, in this study, footprinting assays highlight an unexpected bipartite binding site that is essential for efficient control. Another probe that has been used largely to monitor RNA-RNA pairing is lead(II) (e.g. [36,78]).

2.4.3 Footprinting reveals mimicry of mRNA and tRNA for regulation

Unlike Crc, which regulates the level of expression of many genes, other RBPs regulate only a few genes or act solely on their own mRNA transcript to repress their synthesis by a feedback mechanism. This is the case for the *Escherichia coli thrS* gene encoding threonyl-tRNA synthetase (ThrRS). The enzyme is negatively autoregulated at the translational level [79]. Detailed structure-probing experiments using RNases and base-specific chemicals revealed that the operator region is folded into four distinct domains ([80]; Fig. 2.3A). The first domain contains an unstructured RBS encompassing the SD sequence and the AUG codon. Two stem-loop structures (domains 2 and 4) are connected via a 24 nucleotide-long loop (domain 3). The binding of ThrRS was investigated by footprinting experiments using chemical and enzymatic probes. Again, footprinting was essential to determine the regions of interaction between the operator region of the mRNA and ThrRS. ThrRS-induced strong protections were found in the apical loops of domains 2 and 4 ([81] and references therein). Mutations that were introduced in domains 2 and 4, which represent structural analogies with the anticodon stem and loop of tRNA^{Thr}, affected regulation [79,82]. Later on, the crystal structure of ThrRS complexed with domain 2 indeed showed that the hairpin loop occupied the same binding site as the anticodon arm of tRNA^{Thr} (Fig. 2.3A; [83]). Hence, the combination of footprinting with genetic and structural approaches determined precisely the translational operator and identified the structural elements essential for ThrRS binding.

Further experiments highlighted how ThrRS represses initiation of translation. The deletion of the most N-terminal domain of ThrRS (ThrRSΔN) abolishes its regulatory ability, although the binding affinity to *thrS* mRNA is maintained [84]. The crystal structure of *thrS* mRNA bound to the 70S ribosome suggested that the regulatory mechanism involves a steric clash between the N-terminal domain of ThrRS and the ribosome [85,86]. A ternary complex of ThrRSΔN, *thrS* mRNA, and 70S ribosomes could possibly form. This was illustrated by footprinting experiments using RNase T2 as the probe and a truncated version of *thrS* mRNA (Fig. 2.3B). In the free RNA, RNase T2 cleaves the apical loop of domain 2 and the RBS. Adding either the ribosome or ThrRSΔN revealed distinct protection patterns because the ribosome protected

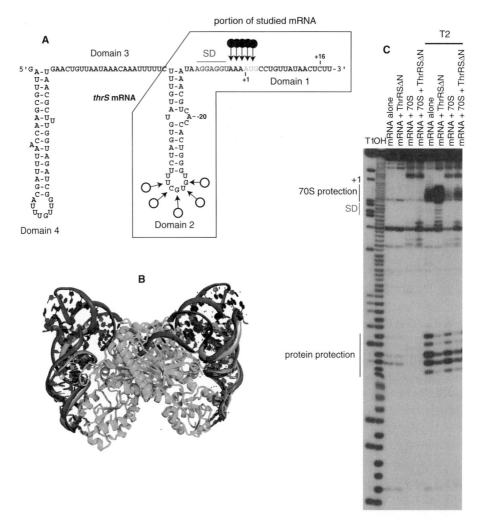

Fig. 2.3 Analysis of the binding sites of the ribosome and ThrRSΔN on *E. coli thrS* mRNA. (A) Secondary structure of *thrS* mRNA operator. The Shine-Dalgarno (SD) sequence is annotated in red and the AUG codon in blue. The RNase T2 cleavage sites are shown by arrows. The protections induced by 70S binding are denoted by white dots, and the protections induced by ThrRSΔN are represented by black dots. (B) Tertiary structure of ThrRSΔN (green) bound to tRNAs[Thr] (red) with the superposition of the hairpin domain 2 of *thrS* mRNA. The model was obtained by aligning the individual structures of ThrRS complexed with domain 2 of the *thrS* mRNA operator from *E. coli* (pdb 1KOG [83]) and of ThrRSΔN complexed to tRNA[Thr] from *E. coli* (pdb 1QF6 [86]). (C) Enzymatic footprint using RNase T2 on *thrS* mRNA. The RNase T2 assay was performed in parallel on the free RNA and in the presence of the ribosome (70S), ThrRSΔN protein, and both of them. Incubation controls were carried out in the absence of RNase T2. T1 and alkaline ladder [54] are respectively RNase T1 and alkaline ladders performed on a denatured RNA. The +1 position (A of AUG codon) and SD sequences are shown using landmarks.

primarily the RBS and ThrSΔN protected the apical loop of domain 2 (Fig. 2.3). Interestingly, performing the assays in the presence of the mutant ThrRSΔN and the ribosome showed protections in both domains. This data suggested the concomitant presence of ThrRSΔN and the ribosome on the same mRNA molecule (Fig. 2.3C). However, to fully ascertain the simultaneous binding, additional experiments would be required such as purification of the complex using gel filtration or ultracentrifugation on a glycerol gradient. Taken together, footprinting helps to determine the regions of *thrS* mRNA that are essential for control and highlights the molecular mimicry between the mRNA and the tRNA, which is the basis of the feedback regulation.

2.4.4 Difficulties in probing transient interactions by footprinting: the case of ribosomal protein S1-RNA complex

Some RNA-protein interactions are transient and very dynamic. This is the case for proteins, which have an RNA chaperone activity. *E. coli* ribosomal protein S1 belongs to this class of proteins [87,88]. S1 is essential for the translation of many mRNAs and for viability [89]. It binds preferentially to unstructured regions harboring AU stretches assisting mRNA-ribosome interaction [90]. Moreover, S1 has an RNA unwinding activity that is crucial for the accommodation of structured mRNAs in the decoding channel of the 30S subunit [87,88,91]. For example, we have recently shown that S1 promotes the active initiation of translation of *rpsO* mRNA encoding ribosomal protein S15 [88]. The 5' UTR region of *rpsO* mRNA is folded into a metastable pseudoknot structure that is known to be stabilized by S15 to repress its own translation [92]. It is noteworthy that the enzymatic probing revealed the concomitant presence of single-strand-specific RNase T2 and double-strand-specific RNase V1 cleavage events, illustrating the presence of two alternative conformations in vitro [92]. Only the pseudoknot structure is the active form of the mRNA that is recognized by the 30S subunit because the SD is present in an unpaired region (Fig. 2.4). Using a variety of approaches, including gel retardation assays and surface plasmon resonance, a specific interaction has been monitored between S1 and the pseudoknot structure of *rpsO* mRNA, with a K_d value of around 100 nM [88]. We have probed the structure of the 5' UTR of *rpsO* mRNA in the presence of S1 using RNases (Fig. 2.4A) and various base-specific reagents (results not shown). However, the addition of increasing concentrations of S1 failed to induce protection or enhanced cleavages in *rpsO* mRNA even if the concentration of S1 was largely above the K_d value (up to 1 µM) (Fig. 2.4). This strongly suggested that the kinetic properties of S1-RNA interaction do not prevent the accessibility of the RNA toward the probes. Therefore, alternative strategies are required to identify the binding site, such as mutational analysis, cross-linking experiments, or SDCP. This latter approach, which employs Fe(EDTA) tethered to a specific cysteine residue of the protein, produces local hydroxyl radicals upon binding

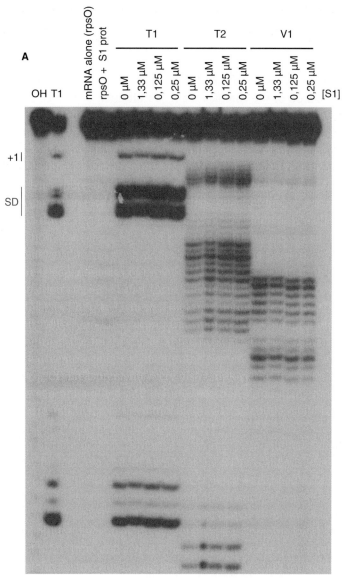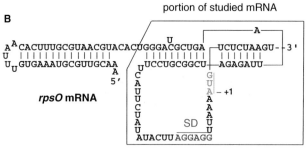

that can cleave the proximal RNA. This approach was successfully used for S1-*rpsO* complex characterization [88].

2.5 Concluding remarks

RNA functions rely on the ability of RNA to adopt unique three-dimensional structures that are specifically recognized by ligands. The structural rules that govern RNA-ligand interactions are based on the identification of the nucleotides and the atoms involved in the specific recognition, and on the mechanism by which the RNA and its ligand adapt to each other (possible induced-fit mechanism). In the absence of high-resolution structures, biochemical and genetic approaches have been useful for mapping the ligand-binding site on RNA (footprinting), for providing topographical information on ligand-RNA complexes (SDCP and cross-linking), and for identifying specific contacts and their relative contribution to binding (chemical interference and site-directed mutagenesis). It is certainly the combination of these approaches that can give insights into the structure-function relationships. Because new regulatory roles continue to emerge for RNAs, there is a growing need to develop technologies that are amenable to studying any classes of RNAs and RNA-ligand complexes in vivo. A recent approach has been described to monitor the kinetic analysis of the preribosome structure in living *Saccharomyces cerevisiae* cells [94]. This study is based on in vivo structure probing by DMS modification combined with affinity purification of newly synthesized 20S preribosomal RNA over a time course of metabolic labeling with 4-thiouracil. Furthermore, with the development of high-throughput sequencing technologies, it is now possible to gain simultaneously quantitative, single nucleotide-resolution secondary and tertiary structural information for many RNA molecules. This methodology has been adapted on a pool of RNAs in vitro using SHAPE [95] or using various RNases [96,97]. Given the fact that DMS, lead(II)-induced cleavages [36], and SHAPE [40,41] are convenient tools for probing RNA/ligand complexes in vivo, it now becomes possible to obtain an overall picture of the RNA structure and dynamics in response to stress, environmental changes, or induction of a repressor/activator (for reviews, see Westhof and Romby [98] and Wan et al. [99]).

Fig. 2.4 Footprinting analysis of transient mRNA-protein interaction. (A) Gel fractionation of enzymatic cleavages of 5'-end-labeled *rpsO* mRNA in the absence or in the presence of increasing concentrations of ribosomal protein S1. As enzymes RNase T1 (unpaired guanines), RNase T2 (unpaired adenines), and RNase V1 (paired nucleotides) were employed. T1 and alkaline ladders OH were used to identify the positions of cleavage sites in the mRNA sequence. Incubation controls in the absence of RNase were carried out in parallel. (B) Secondary structure of *rpsO* mRNA according to Philippe et al. [93]. The Shine-Dalgarno (SD) sequence and AUG codon are underlined.

2.6 Acknowledgments

We are thankful to all the team's members for helpful discussions. This work was supported by the Centre National de la Recherche Scientifique (CNRS), and the "Laboratoires d'excellence" (LABEX) NetRNA grant ANR-10-LABX-36 (PR) in the frame of « Programme d'Investissements d'Avenir ».

References

1. Gebauer F, Preiss T, Hentze MW. From *cis*-regulatory elements to complex RNPs and back. Cold Spring Harb Perspect Biol. 2012;4:a012245.
2. Sonenberg N, Hinnebusch AG. Regulation of translation initiation in eukaryotes: mechanisms and biological targets. Cell. 2009;136:731–45.
3. Romby P, Springer M. Translational Control in Prokaryotes. In: Matthews MBSN, Hershey JWB, eds. Translational Control in Biology and Medicine. Cold Spring Harbor: Cold Spring Harbor Laboratory Press; 2007:803–27.
4. Marzi S, Fechter P, Chevalier C, Romby P, Geissmann T. RNA switches regulate initiation of translation in bacteria. Biol Chem. 2008;389:585–98.
5. Wagner EG, Altuvia S, Romby P. Antisense RNAs in bacteria and their genetic elements. Adv Genetics. 2002;46:361–98.
6. Brantl S. Bacterial chromosome-encoded small regulatory RNAs. Future Microbiol. 2009; 4:85–103.
7. Massé E, Gottesman S. A small RNA regulates the expression of genes involved in iron metabolism in *Escherichia coli*. Proc Natl Acad Sci U S A. 2002;99:4620–5.
8. Guillier M, Gottesman S, Storz G. Modulating the outer membrane with small RNAs. Genes Dev. 2006;20:2338–48.
9. Papenfort K, Vogel J. Multiple target regulation by small noncoding RNAs rewires gene expression at the post-transcriptional level. Res Microbiol. 2009;160:278–87.
10. Wassarman KM. Native gel electrophoresis to study the binding and release of RNA polymerase by 6S RNA. Methods Mol Biol. 2012;905:259–71.
11. Masliah G, Barraud P, Allain FH. RNA recognition by double-stranded RNA binding domains: a matter of shape and sequence. Cell Mol Life Sci. 2012.
12. Birney E, Kumar S, Krainer AR. Analysis of the RNA-recognition motif and RS and RGG domains: conservation in metazoan pre-mRNA splicing factors. Nucleic Acids Res. 1993;21:5803–16.
13. Saunders LR, Barber GN. The dsRNA binding protein family: critical roles, diverse cellular functions. FASEB J. 2003;17:961–83.
14. Nagai K. RNA-protein complexes. Curr Opin Struct Biol. 1996;6:53–61.
15. Alfano C, Sanfelice D, Babon J, et al. Structural analysis of cooperative RNA binding by the La motif and central RRM domain of human La protein. Nat Struct Mol Biol. 2004;11:323–9.
16. Berg JM. Fingering nucleic acids: the RNA did it. Nat Struct Biol. 2003;10:986–7.
17. Yang Y, Declerck N, Manival X, Aymerich S, Kochoyan M. Solution structure of the LicT-RNA antitermination complex: CAT clamping RAT. EMBO J. 2002;21:1987–97.
18. Bycroft M, Hubbard TJ, Proctor M, Freund SM, Murzin AG. The solution structure of the S1 RNA binding domain: a member of an ancient nucleic acid-binding fold. Cell. 1997;88:235–42.
19. Herschlag D. RNA chaperones and the RNA folding problem. J Biol Chem. 1995;270:20871–4.
20. Weeks KM. Protein-facilitated RNA folding. Curr Opin Struct Biol. 1997;7:336–42.

21. Rajkowitsch L, Chen D, Stampfl S, et al. RNA chaperones, RNA annealers and RNA helicases. RNA Biol. 2007;4:118–30.
22. Mayer O, Rajkowitsch L, Lorenz C, Konrat R, Schroeder R. RNA chaperone activity and RNA-binding properties of the *E. coli* protein StpA. Nucleic Acids Res. 2007;35:1257–69.
23. Mirzabekov AD, Melnikova AF. Localization of chromatin proteins within DNA grooves by methylation of chromatin with dimethyl sulphate. Mol Biol Rep. 1974;1:379–84.
24. Pribnow D. Nucleotide sequence of an RNA polymerase binding site at an early T7 promoter. Proc Natl Acad Sci U S A. 1975;72:784–8.
25. Donis-Keller H, Maxam AM, Gilbert W. Mapping adenines, guanines, and pyrimidines in RNA. Nucleic Acids Res. 1977;4:2527–38.
26. Peattie DA. Direct chemical method for sequencing RNA. Proc Natl Acad Sci U S A. 1979;76:1760–4.
27. Maxam AM, Gilbert W. A new method for sequencing DNA. Proc Natl Acad Sci U S A. 1977;74:560–4.
28. Peattie DA, Gilbert W. Chemical probes for higher-order structure in RNA. Proc Natl Acad Sci U S A. 1980;77:4679–82.
29. Lempereur L, Nicoloso M, Riehl N, Ehresmann C, Ehresmann B, Bachellerie JP. Conformation of yeast 18S rRNA. Direct chemical probing of the 5' domain in ribosomal subunits and in deproteinized RNA by reverse transcriptase mapping of dimethyl sulfate-accessible. Nucleic Acids Res. 1985;13:8339–57.
30. Moazed D, Stern S, Noller HF. Rapid chemical probing of conformation in 16 S ribosomal RNA and 30 S ribosomal subunits using primer extension. J Mol Biol. 1986;187:399–416.
31. Mortimer SA, Weeks KM. Time-resolved RNA SHAPE chemistry: quantitative RNA structure analysis in one-second snapshots and at single-nucleotide resolution. Nat Protoc. 2009;4:1413–21.
32. Latham JA, Cech TR. Defining the inside and outside of a catalytic RNA molecule. Science. 1989;245:276–82.
33. Romby P, Moras D, Dumas P, Ebel JP, Giege R. Comparison of the tertiary structure of yeast tRNA(Asp) and tRNA(Phe) in. J Mol Biol. 1987;195:193–204.
34. Fabbretti A, Milon P, Giuliodori AM, Gualerzi CO, Pon CL. Real-time dynamics of ribosome-ligand interaction by time-resolved chemical probing methods. Methods Enzymol. 2007;430:45–58.
35. Giuliodori AM, Di Pietro F, Marzi S, et al. The cspA mRNA is a thermosensor that modulates translation of the cold-shock protein CspA. Mol Cell. 2010;37:21–33.
36. Lindell M, Romby P, Wagner EG. Lead(II) as a probe for investigating RNA structure in vivo. RNA. 2002;8:534–41.
37. Helfer A-C, Romilly C, Chevalier C, Lioliou E, Marzi S, Romby P. Probing RNA structure in vitro with enzymes and chemicals. In: Hartmann RK, Bindereif A, Schön A, Westhof E, eds. Handbook of RNA Biochemistry. 2nd ed. Wiley-VCH Verlag GmbH & Co. KGaA.
38. Wells SE, Hughes JM, Igel AH, Ares M. Use of dimethyl sulfate to probe RNA structure in vivo. Methods Enzymol. 2000;318:479–93.
39. Balzer M, Wagner R. A chemical modification method for the structural analysis of RNA and RNA-protein complexes within living cells. Anal Biochem. 1998;256:240–2.
40. Gherghe C, Lombo T, Leonard CW, et al. Definition of a high-affinity Gag recognition structure mediating packaging of a retroviral RNA genome. Proc Natl Acad Sci U S A. 2010;107:19248–53.
41. Spitale RC, Crisalli P, Flynn RA, Torre EA, Kool ET, Chang HY. RNA SHAPE analysis in living cells. Nat Chem Biol. 2013;9:18–20.
42. Adilakshmi T, Soper SF, Woodson SA. Structural analysis of RNA in living cells by in vivo synchrotron X-ray footprinting. Methods Enzymol. 2009;468:239–58.
43. Sclavi B. Time-resolved footprinting for the study of the structural dynamics of DNA-protein interactions. Biochem Soc Trans. 2008;36:745–8.

44. Skripkin E, Paillart JC, Marquet R, Ehresmann B, Ehresmann C. Identification of the primary site of the human immunodeficiency virus type 1 RNA dimerization in vitro. Proc Natl Acad Sci U S A. 1994;91:4945–9.
45. Schlegl J, Gegout V, Schläger B, et al. Probing the structure of the regulatory region of human transferrin receptor messenger RNA and its interaction with iron regulatory protein-1. RNA. 1997;3:1159–72.
46. Ryder SP, Ortoleva-Donnelly L, Kosek AB, Strobel SA. Chemical probing of RNA by nucleotide analog interference mapping. Methods Enzymol. 2000;317:92–109.
47. Waldsich C. Dissecting RNA folding by nucleotide analog interference mapping (NAIM). Nat Protoc. 2008;3:811–23.
48. Joseph S, Noller HF. Directed hydroxyl radical probing using iron(II) tethered to RNA. Methods Enzymol. 2000;318:175–90.
49. Culver GM, Noller HF. Directed hydroxyl radical probing of RNA from iron(II) tethered to proteins in ribonucleoprotein complexes. Methods Enzymol. 2000;318:461–75.
50. Drider D, Condon C. The continuing story of endoribonuclease III. J Mol Microbiol Biotechnol. 2004;8:195–200.
51. Romilly C, Chevalier C, Marzi S, et al. Loop-loop interactions involved in antisense regulation are processed by the endoribonuclease III in *Staphylococcus aureus*. RNA Biol. 2012;9:1461–72.
52. Fechter P, Parmentier D, Fuchsbauer O, Romby P, Marzi S. Traditional chemical mapping of RNA structure in vitro and in vivo. In: Turner DHM, Douglas H, eds. RNA Structure Determination: Methods and Protocols. Humana Press; 2013.
53. Gilham PT. An addition reaction specific for uridine and guanosine nucleotides and its application to the modification of ribonuclease action. J Am Chem Soc. 1962;84:687–9.
54. Shapiro R, Cohen BI, Shiuey SJ, Maurer H. On the reaction of guanine with glyoxal, pyruvaldehyde, and kethoxal, and the structure of the acylguanines. A new synthesis of N2-alkylguanines. Biochemistry. 1969;8:238–45.
55. Brown RS, Dewan JC, Klug A. Crystallographic and biochemical investigation of the lead(II)-catalyzed. Biochemistry. 1985;24:4785–801.
56. Mortimer SA, Weeks KM. Time-resolved RNA SHAPE chemistry. J Am Chem Soc. 2008;130: 16178–80.
57. Wilkinson KA, Vasa SM, Deigan KE, Mortimer SA, Giddings MC, Weeks KM. Influence of nucleotide identity on ribose 2'-hydroxyl reactivity in RNA. RNA. 2009;15:1314–21.
58. Wilkinson KA, Merino EJ, Weeks KM. Selective 2'-hydroxyl acylation analyzed by primer extension (SHAPE): quantitative RNA structure analysis at single nucleotide resolution. Nat Protoc. 2006;1:1610–6.
59. Watts JM, Dang KK, Gorelick RJ, et al. Architecture and secondary structure of an entire HIV-1 RNA genome. Nature. 2009;460:711–6.
60. Hertzberg RP, Dervan PB. Cleavage of DNA with methidiumpropyl-EDTA-iron(II): reaction conditions and product analyses. Biochemistry. 1984;23:3934–45.
61. Powers T, Noller HF. Hydroxyl radical footprinting of ribosomal proteins on 16S rRNA. RNA. 1995;1:194–209.
62. Imlay JA, Chin SM, Linn S. Toxic DNA damage by hydrogen peroxide through the Fenton reaction in vivo and in vitro. Science. 1988;240:640–2.
63. Dixon WJ, Hayes JJ, Levin JR, Weidner MF, Dombroski BA, Tullius TD. Hydroxyl radical footprinting. Methods Enzymol. 1991;208:380–413.
64. Tullius TD, Dombroski BA, Churchill ME, Kam L. Hydroxyl radical footprinting: a high-resolution method for mapping protein-DNA contacts. Methods Enzymol. 1987;155:537–58.

65. Hayes JJ, Kam L, Tullius TD. Footprinting protein-DNA complexes with gamma-rays. Methods Enzymol. 1990;186:545–9.
66. Ottinger LM, Tullius TD. High-resolution in vivo footprinting of a protein-DNA complex using gamma-radiation. J Am Chem Soc. 2000;122:5901–2.
67. Soukup GA, Breaker RR. Relationship between internucleotide linkage geometry and the stability of RNA. RNA. 1999;5:1308–25.
68. Rojo F. Carbon catabolite repression in *Pseudomonas*: optimizing metabolic versatility. FEMS Microbiol Rev. 2010;34:658–84.
69. Moreno R, Marzi S, Romby P, Rojo F. The Crc global regulator binds to an unpaired A-rich motif at the *Pseudomonas putida* alkS mRNA coding sequence and inhibits translation initiation. Nucleic Acids Res. 2009;37:7678–90.
70. Das R, Laederach A, Pearlman SM, Herschlag D, Altman RB. SAFA: semi-automated footprinting analysis software for high-throughput quantification of nucleic acid footprinting experiments. RNA. 2005;11:344–54.
71. Sonnleitner E, Abdou L, Haas D. Small RNA as global regulator of carbon catabolite repression in *Pseudomonas*. Proc Natl Acad Sci U S A. 2009;106:21866–71.
72. Moreno R, Fonseca P, Rojo F. The Crc global regulator inhibits the *Pseudomonas putida* pWW0 toluene/xylene assimilation pathway by repressing the translation of regulatory and structural genes. J Biol Chem. 2010;285:24412–9.
73. Moreno R, Fonseca P, Rojo F. Two small RNAs, CrcY and CrcZ, act in concert to sequester the Crc global. Mol Microbiol. 2012;83:24–40.
74. Papenfort K, Vogel J. Regulatory RNA in bacterial pathogens. Cell Host Microbe. 2010;8:116–27.
75. Felden B, Vandenesch F, Bouloc P, Romby P. The *Staphylococcus aureus* RNome and its commitment to virulence. PLoS Pathog. 2011;7:e1002006.
76. Novick RP, Ross HF, Projan SJ, Kornblum J, Kreiswirth B, Moghazeh S. Synthesis of staphylococcal virulence factors is controlled by a regulatory RNA molecule. EMBO J. 1993;12:3967–75.
77. Chevalier C, Boisset S, Romilly C, et al. *Staphylococcus aureus* RNAIII binds to two distant regions of coa mRNA to arrest translation and promote mRNA degradation. PLoS Pathog. 2010;6:e1000809.
78. Darfeuille F, Unoson C, Vogel J, Wagner EG. An antisense RNA inhibits translation by competing with standby ribosomes. Mol Cell. 2007;26:381–92.
79. Springer M, Graffe M, Butler JS, Grunberg-Manago M. Genetic definition of the translational operator of the threonine-tRNA ligase gene in *Escherichia coli*. Proc Natl Acad Sci U S A. 1986;83:4384–8.
80. Moine H, Romby P, Springer M, et al. Messenger RNA structure and gene regulation at the translational level in *Escherichia coli*: the case of threonine:tRNAThr ligase. Proc Natl Acad Sci U S A. 1988;85:7892–6.
81. Romby P, Springer M. Bacterial translational control at atomic resolution. Trends Genetics. 2003;19:155–61.
82. Brunel C, Caillet J, Lesage P, et al. Domains of the *Escherichia coli* threonyl-tRNA synthetase translational operator and their relation to threonine tRNA isoacceptors. J Mol Biol. 1992;227:621–34.
83. Torres-Larios A, Dock-Bregeon AC, Romby P, et al. Structural basis of translational control by *Escherichia coli* threonyl tRNA synthetase. Nat Struct Biol. 2002;9:343–7.
84. Caillet J, Nogueira T, Masquida B, et al. The modular structure of *Escherichia coli* threonyl-tRNA synthetase as both an enzyme and a regulator of gene expression. Mol Microbiol. 2003;47:961–74.

85. Jenner L, Romby P, Rees B, et al. Translational operator of mRNA on the ribosome: how repressor proteins exclude ribosome binding. Science. 2005;308:120–3.
86. Sankaranarayanan R, Dock-Bregeon AC, Romby P, et al. The structure of threonyl-tRNA synthetase-tRNA(Thr) complex enlightens its repressor activity and reveals an essential zinc ion in the active site. Cell. 1999;97:371–81.
87. Rajkowitsch L, Schroeder R. Dissecting RNA chaperone activity. RNA. 2007;13:2053–60.
88. Duval M, Korepanov A, Fuchsbauer O, et al. *Escherichia coli* ribosomal protein S1 unfolds structured mRNAs onto the ribosome for active translation initiation. Submitted 2013.
89. Sørensen MA, Fricke J, Pedersen S. Ribosomal protein S1 is required for translation of most, if not all, natural mRNAs in *Escherichia coli* in vivo. J Mol Biol. 1998;280:561–9.
90. Hajnsdorf E, Boni IV. Multiple activities of RNA-binding proteins S1 and Hfq. Biochimie. 2012;94:1544–53.
91. Qu X, Lancaster L, Noller HF, Bustamante C, Tinoco I. Ribosomal protein S1 unwinds double-stranded RNA in multiple steps. Proc Natl Acad Sci U S A. 2012;109:14458–63.
92. Marzi S, Myasnikov AG, Serganov A, et al. Structured mRNAs regulate translation initiation by binding to the platform of the ribosome. Cell. 2007;130:1019–31.
93. Philippe C, Portier C, Mougel M, et al. Target site of *Escherichia coli* ribosomal protein S15 on its messenger RNA. J Mol Biol. 1990;211:415–26.
94. Swiatkowska A, Wlotzka W, Tuck A, Barrass JD, Beggs JD, Tollervey D. Kinetic analysis of pre-ribosome structure in vivo. RNA. 2012;18:2187–200.
95. Lucks JB, Mortimer SA, Trapnell C, et al. Multiplexed RNA structure characterization with selective 2'-hydroxyl acylation analyzed by primer extension sequencing (SHAPE-Seq). Proc Natl Acad Sci U S A. 2011;108:11063–8.
96. Underwood JG, Uzilov AV, Katzman S, et al. FragSeq: transcriptome-wide RNA structure probing using high-throughput sequencing. Nat Methods. 2010;7:995–1001.
97. Kertesz M, Wan Y, Mazor E, et al. Genome-wide measurement of RNA secondary structure in yeast. Nature. 2010;467:103–7.
98. Westhof E, Romby P. The RNA structurome: high-throughput probing. Nat Methods. 2010;7:965–7.
99. Wan Y, Kertesz M, Spitale RC, Segal E, Chang HY. Understanding the transcriptome through RNA structure. Nat Rev Genetics. 2011;12:641–55.

Mark Helm
3 Chemical approaches to the structural investigation of RNA in solution

3.1 Introduction

The structural biology of RNA relies on numerous techniques that furnish structural information of a very diverse nature. Assuming that a scientist wants a maximum of structural resolution in a minimum of time, the considerations guiding the choice of method are typically dominated by weighing the amount of material available versus the turnaround time. On one end of spectrum of available methods are classical high-resolution techniques such as nuclear magnetic resonance (NMR) (see Chapter 12) and crystallography (see Chapter 13), which typically require large amounts of RNA and a significant amount of time. On the other end, low-resolution data can be obtained by bioinformatics in a comparatively short time and requiring extremely little material.

Structural probing of RNA in solution and related techniques hold a middle ground: requiring small amounts of RNA, these approaches produce experimental data reflecting relevant details of secondary and tertiary structure in a reasonable amount of time. This compromise makes it the method of choice for many applications, as it is experimentally accessible for the typical RNA biochemist who has limited experience in, and equipment for, crystallography and bioinformatics. Several further advantages have led to a widespread application of probing, including in particular the investigation of dynamic features not accessible to, for example, crystallography. The combination of structural probing with data from cross-linking experiments, phylogenetic analysis, and computer-based modeling can predict secondary structure with considerable accuracy, and the computation of three-dimensional models that come within a few angstroms of crystal structures resolution is an interesting competition in the field [1].

The basic concept of structural probing is to measure the accessibility of atoms or functional groups in RNA toward a chemical agent or an enzyme. Highest accessibility typically correlates with an exposed position in the RNA structure, as well as with a certain structural flexibility of the probed residue. The concept as such dates from the very early days of RNA biochemistry and was developed along with separation and sequencing techniques. Because the latter are an integral part of structural probing, significant progress in the field was regularly induced by leaps in nucleic acid–sequencing technique. Thus, the invention of Maxam and Gilbert and Sanger sequencing has impacted the literature of structural probing to an extreme extent, and the recent transition to high-throughput ("next generation") sequencing may have similar effects. A more steady progress comes from the development of new probes, especially chemical reagents targeting previously inaccessible functional groups in RNA (see also Chapter 2).

A technically related approach that has received much less attention is chemical modification interference. This technique starts with a chemical modification step of an unfolded RNA population and uses a physical separation step to segregate RNA populations into those that fold properly and those that cannot. Comparative analysis of the chemical modifications in both groups allows determining which residues are important for segregation. The same principle is also used in nucleotide analog interference mapping (NAIM), where the chemical modifications are introduced cosynthetically into the RNA by use of nucleotide analogs.

This book chapter will cover conceptual aspects of RNA structural probing, chemical modification interference, and nucleotide analog interference mapping and outline their interplay with neighboring techniques in RNA structural biology. For practical advice and protocols, the reader is referred to method and protocol papers covering these topics [1–13].

3.2 Similar chemistry in different concepts: sequencing, probing, and interference

A simple visualization may compare biopolymer structures to knots of spaghetti. Thus, if an RNA resembles a pile of spaghetti, then structural probing is a technique suitable to reveal which parts of which piece of pasta are on the surface of the pile. The trick is to spray tomato sauce onto the pile, and the only remaining technical problem is then to catalogue all the red parts of the pasta.

The corresponding idea of structural probing is to apply a chemical reagent to a properly folded RNA and then to analyze which residues of the RNA have had contact with the reagent. Therefore, the reagent is required to leave some kind of molecular trace (i.e. a chemical modification on the RNA), and there has to be a method to detect such traces. Later on, probing experiments will be classified by the type of reagent as well as by the detection method used. Progress in these two aspects has been the main driving force in the development of structural probing. Many of the early probing experiments were performed on transfer RNAs (tRNAs) well before a crystal structure was available. In a typical experiment, RNA was modified with a chemical agent under so-called native conditions (i.e. buffer, salt, and temperature conditions that are presumed to allow RNA folding similar to that in the living cell). For detection of the modified residues, the RNA was degraded to fragments of defined size by digestion with nucleases such as, for example, ribonuclease (RNase) T1, which faithfully cleaves at each guanosine residue. The resulting fragments were then reisolated by chromatography, and the fragments were further analyzed for the presence of modified groups to map the modification onto the RNA sequence [14]. The methods back then consumed RNA quantities that are huge by today's standards and relied on time-consuming chromatography techniques that are hardly in use any more. The information output was relatively low: publications reporting on the accessibility of a few or even a single residue of a tRNA to a given reagent were not unusual [15]. For comparison, sequencing techniques of the latest generation now allow mapping almost every single nucleotide in an entire transcriptome [16–18] or viral RNA genome [19]. This represents an increase in information gain by six to eight orders of magnitude, which is intrinsically related to the several revolutions of nucleic acid sequencing. The detection of nucleic acid modifications by mass spectrometry for structural probing has also made significant inroads [20], although high-throughput capacity is still lagging behind the conventional separation techniques.

3.3 Sequencing and probing by Maxam and Gilbert chemistry

A boost to structural probing came with the invention by Maxam and Gilbert of sequencing of end-labeled material by chemical scission. This technique is a direct application of the organic chemistry of RNA modification, in particular of RNA alkylation. For example, dimethyl sulfate (DMS), a strong methylating agent, was developed for sequencing back then and is still a widely used reagent for structural probing. In this method, exposure of the RNA (or DNA) to a chemical agent that reacts with specific nitrogens of the RNA's nucleobases is followed by a chemical treatment that induces chain scission only at those residues that have been affected by the previous exposure. In the case of DMS, alkylation occurs at several different nucleobases; chemical treatments for chain scission have been developed to reveal the alkylation of *N3* in cytidines and of *N7* in guanosines (Fig. 3.1A,B). Because the method employs homogeneous end-labeled samples, the length of the resulting labeled fragment (Fig. 3.2A) specifies the position of the modified residue and can be analyzed by gel electrophoresis and autoradiography in connection with appropriate controls (Fig. 3.2B).

The similarity of concepts and procedures in sequencing, probing, and chemical modification interference is readily illustrated by limited modification of RNA in various states of folding as outlined in Fig. 3.2C. The presented example is limited to the alkylation of the *N3* of cytidines, a reaction followed by two subsequent chemical treatments, first with hydrazine and then with aniline at low pH. The latter induces chain scission at the backbone of the modified residue as detailed in Fig. 3.1D. For sequencing, the methylation reaction is conducted under denaturing conditions, typically high temperatures, to unfold the RNA structure and promote equal accessibility of all residues to the modifying reagent (Fig. 3.2C, right panel). Occasionally, in very strongly structured nucleic acids, some residual structure remains despite heating. As a result, the corresponding residues are not fully accessible and give

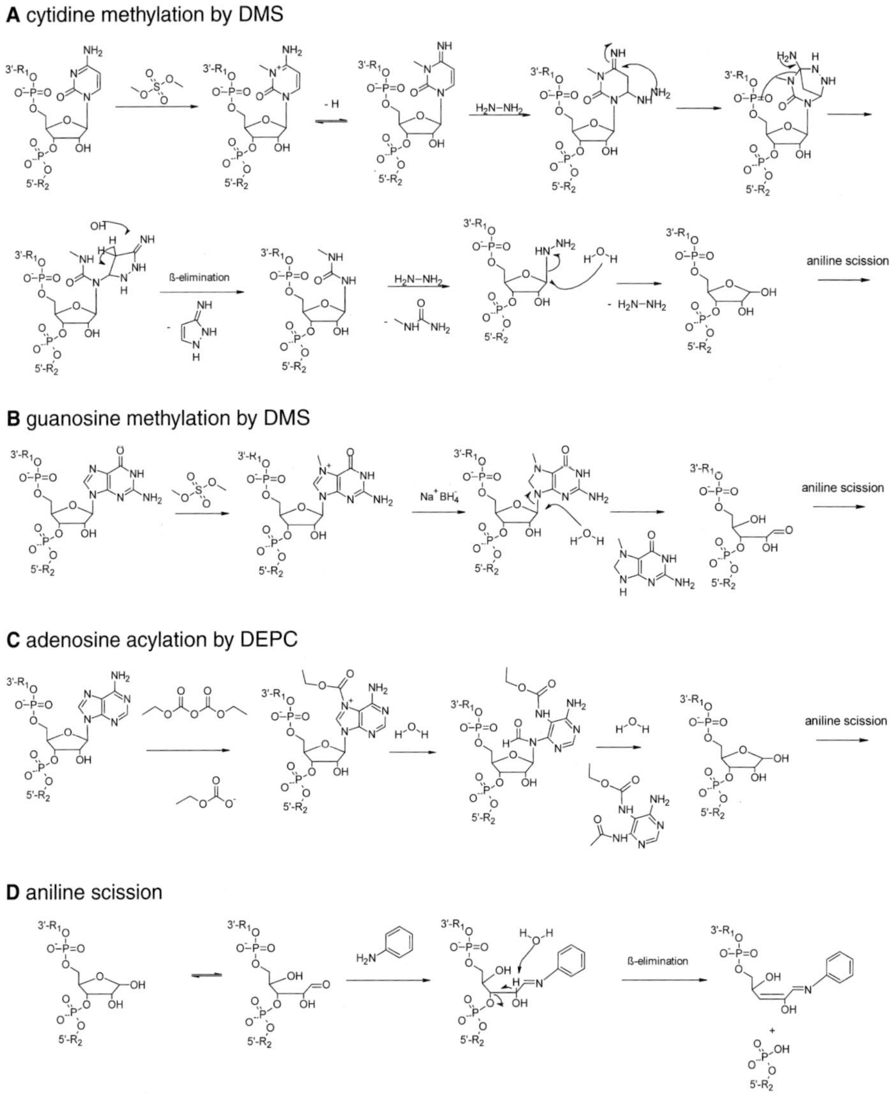

A cytidine methylation by DMS

B guanosine methylation by DMS

C adenosine acylation by DEPC

D aniline scission

Fig. 3.1 The classical reactions of chemical probing. (A) Methylation of the *N3* of cytidine by dimethyl sulfate is illustrated along with the mechanism by which the subsequent treatment with hydrazine renders the ribose susceptible to aniline scission. (B) Methylation of the *N7* of guanosine by dimethyl sulfate is illustrated along with the mechanism by which the subsequent treatment with sodium borohydride renders the ribose susceptible to aniline scission. (C) Acylation of the *N7* of adenosines directly leads to ring opening and renders the ribose susceptible to aniline scission (D) without further treatment.

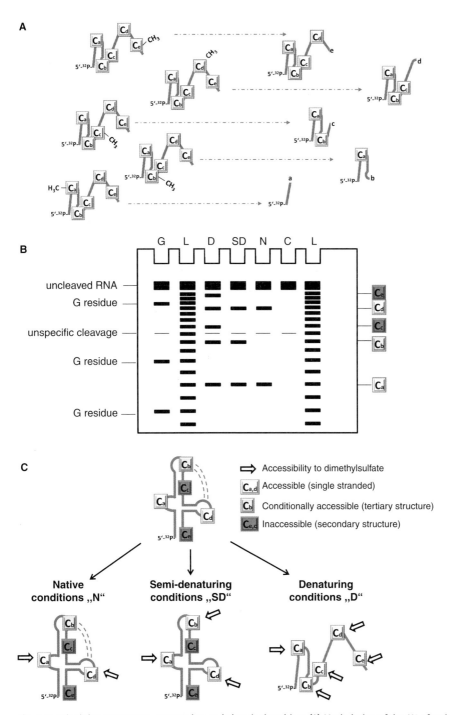

Fig. 3.2 Principles common to sequencing and chemical probing. (A) Methylation of the *N3* of cytidine by dimethyl sulfate under denaturing conditions produces random distribution of methylated

weak or no signals in the sequencing experiment. Probing makes a rule of this exception: in order to identify all adenosines involved in RNA structure, conditions are chosen to allow RNA structure formation to different degrees. In addition to native conditions (Fig. 3.2C left panel), which have already been mentioned, omission of magnesium ions under otherwise native conditions leads to so-called semidenaturing conditions (Fig. 3.2A,B middle panel). With few exceptions, the absence of magnesium ions abolishes or at least weakens tertiary interactions in RNA to a point that the residues involved become at least partially accessible to modifying agents. Outside the Watson-Crick interactions of canonical secondary structure, the $N3$ of cytidines are typically accessible under semidenaturing conditions. After DMS treatment, the RNA is subjected to the strand scission protocol and then to denaturing polyacrylamide gel electrophoresis (PAGE). An autoradiography of the gel slab has the typical aspect of old-fashioned sequencing gels. Indeed, assignment of bands in probing experiments is typically based on concomitant partial sequencing. In our example, partial sequencing is embodied by the lane labeled "D," which contained the sample treated under denaturing conditions – it displays a band for every cytidine residue present in the RNA (Fig. 3.2A). This is frequently supplemented by a so-called ladder "L," an RNA sample subjected to limited, nonspecific hydrolysis (e.g. by extended heating at high temperatures in water or imidazole). G-ladders, created by limited digestion with RNase T1 to faithfully produce signals for every guanosine in the sequence ("G"), are also very popular for partial RNA sequencing [21]. In addition to lanes devoted to sequencing and band assignment, our virtual probing gel in Fig. 3.2B includes the crucial lanes with samples treated under semidenaturing ("SD") and native ("N") conditions [22], which contain the information of interest, plus a lane with an untreated control sample to identify sites of spontaneous cleavage and degradation products not related to the action of the probing agent. Of note, reaction conditions must be established for "limited" modification (i.e. they

---

Fig. 3.2 (*Continued*)
residues. Cleavage by hydrazine/aniline treatment of end-labeled RNA gives rise to one population of characteristic length per cytidine residue in the RNA sequence, which can be analyzed by PAGE. (B) Virtual example of a PAGE analysis of a probing experiment of N3 in cytidines. The lane annotation "G" stands for an RNase T1 based partial sequencing reaction of G residues; "L" is an alkaline ladder used as a ruler to identify the migration behavior of all possible RNA fragments; "D" identifies the sequencing reaction under denaturing conditions shown in panel (A); "SD" stands for the same reaction under semidenaturing conditions (i.e. at room temperature but absent any divalent cations); "N" is the actual probing experiments in which the methylation was conducted at room temperature and in the presence of divalent ions; "C" denotes a control suitable to identify unspecific cleavages occurring during a mock incubation. (C) Symbolic representation of a structured RNA containing five C residues (annotated a through e) accessible to DMS methylation under native conditions (boxed in white), semidenaturing conditions (white and light gray), and denaturing conditions (all boxed). Note that the omission of magnesium ions under semidenaturing conditions leads to the breakdown of tertiary interactions (dotted lines) and thus to accessibility of Cb.

are chosen such that each RNA molecule carries one modification only). In practice, this means that about 10% of the RNA molecules may be modified before the statistical occurrence of multiple events on the same molecule becomes significant. This is necessary to avoid mapping any structural influence of the methyl groups, besides the fact that multiple methylations on a single RNA molecule cannot be detected with this technique. For the previously discussed reasons, the strongest signal in each lane of a probing gel typically originates from uncleaved material (Fig. 3.2B). With chemistry and protocol being essentially identical, the difference between sequencing and probing is limited to the accessibility of residues: all give a signal in sequencing, but only those in weakly structured (or unstructured) parts of a native RNA give a signal in probing. Of note, the so-called direct method of Maxam and Gilbert type probing has been accurately validated by comparing experimental reactivities to the accessible surface in solution [23–28].

3.4 Application of Sanger sequencing to probing

As mentioned, the previous concept requires purified samples of RNA with a homogenous and labeled extremity at either the 5' or the 3' end because inhomogeneous ends produce multiple signals [14]. Also, since it is based on Maxam and Gilbert type sequencing, it is severely restricted to chemical modifications for which chain scission chemistry can be devised. These drawbacks are overcome by the Sanger sequencing technique, which is based on primer extension. The primer length clearly specifies a homogenous 5' end of the resulting labeled complementary DNA and therefore obviates the need for purified, labeled RNA of homogenous length. The sequence information contained in the primer allows working in a mixture of RNA species (e.g. in preparations of total RNA from biological samples). While the original procedure was developed for the extension of radiolabeled primers, subsequent developments enabled the use of fluorescently labeled primers or even labeled triphosphate nucleotides. Again, advances in sequencing have impacted probing: the read length of primer extension typical of sequencing gels may be somewhat improved by using capillary electrophoresis embodied in automated DNA sequencers in combination with fluorescent labeling [29, 30]. Ultimately, Sanger sequencing is the basis for the recently developed probing of entire transcriptomes by application of the latest generation of high-throughput sequencing techniques [16–18]. The technical advantages of Sanger sequencing have significantly enlarged the arsenal of probes because an arrest of reverse transcription is sufficient for detection, instead of the chemical treatments that lead to chain scission. Although the conditions for efficient arrest of primer extension may vary among the various reverse transcriptases employed, modifications of the Watson-Crick face are generally highly efficient. Modifications on the *N7* of purines may lead to inefficient arrest, as in the case of

N7-methylated guanosines [31]. However, subsequent reaction cascades may lead to further changes in the nucleotide structure, which will be picked up by reverse transcriptase. Thus, diethyl pyrocarbonate (DEPC) treatment of adenosines is efficiently detected by primer extension because the resulting adduct undergoes a cascade of reaction steps that ultimately results in an abasic site. This site is both the basis for chemical scission by aniline (Fig. 3.1D) and a strong stop for primer elongation by reverse transcriptase. Especially for classical reagents of chemical probing, namely, modification by DEPC and DMS, both types of sequencing have been very frequently used. When making a decision on whether to use the "direct" method of chain scission or the "indirect" method of primer extension, one should be aware that the direct method is somewhat more precise because it produces less background signal. The reverse transcriptase is sensitive to structural features of the template RNA itself, which may lead to strong stop signals unrelated to chemical modification. On the other hand, primer extension is the only way for probing RNA regions remote from either extremity by more than ~80–90 nucleotides. This technical limitation, which is owed to the resolution of sequencing gels, effectively restricts probing by the direct approach to RNAs of up to ~150 nucleotides. A highly interesting application of primer extension, which is otherwise not accessible, is structural probing in vivo [10, 32, 33]. Here, the RNA is chemically modified or fragmented inside the living cell, isolated immediately afterward, and analyzed by primer extension.

3.5 Further electrophilic small molecule probes

Tab. 3.1 gives an overview of chemical probes and sequencing methods used for their detection as described in the literature. Conceivably, high-throughput sequencing methods may be applied to all probes that respond to Sanger sequencing, as well as to all those inducing a chain scission. The corresponding atoms, functional groups, or moieties of standard RNA nucleotides that can be probed are displayed in Fig. 3.3. This figure also indicates the target atoms of probes that employ Fenton or related chemistry, as well as in-line probing [6]. Among the classical electrophiles that, in addition to DMS and DEPC, enjoy widespread use, cyclohexylcarbodiimide (CMCT), kethoxal and ethylnitrosourea (ENU) merit explicit mention. While CMCT and kethoxal can only be detected by primer extension, alkylation of backbone phosphates by ENU renders these susceptible to alkaline cleavage, allowing direct detection [25, 34]. The selective 2'-hydroxyl acylation analyzed by primer extension (SHAPE) reagents are one of the most recently developed types of agents; they are derivatives of N-methylisatoic anhydride, which acylate free 2'-OH residues in single-stranded regions [35] (see also Chapter 2).

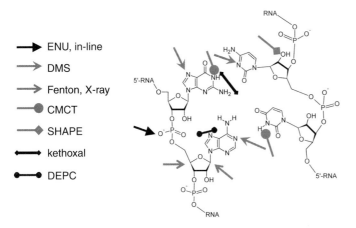

Fig. 3.3 Chemical probes and their targets in RNA on the atomic level.

3.6 Probing agents with nuclease activity

Probing can also be performed with agents that cleave RNA without further chemical treatment. Classically, that is the case for proteinaceous RNases, and the concept has been extended to include RNase mimics and divalent cations, all of which act on the phosphodiester bond of the RNA backbone. Several RNases with specificity for single-stranded residues are known, which will induce cleavage on single-stranded RNA to different degrees as a consequence of their divergent sizes and preferences for certain nucleobases. For example, RNase T1 is well known to be highly specific for unpaired guanosine residues. Another important enzyme is the popular RNase V1, which preferentially cleaves in structured regions of an RNA. The attribute "structured" in this case includes double-stranded RNA, but also elements of tertiary structure and long-range interactions, a fact often overlooked [34]. Because nucleases, once bound to the RNA, are likely to induce several successive cleavage events, structural alteration as a consequence of the primary cleavage may lead to an artificial secondary cleavage, which is not in keeping with the original RNA structure. In contrast, multiple modification events with alkylating agents are much more likely to be mutually independent. Comparison of probing experiments conducted with both, 5'- and 3'-labeled RNA [21] is an established approach to eliminate secondary cleavages [36]. Cleavage of the phosphodiester bond by RNases proceeds by two different mechanisms, which lead to the phosphate ending up on either the 5' ribose of the 3' fragment (nucleases S1, P1, and RNase V1) or on the 3' ribose of the 5' fragment (RNases A,T1,T2,U2, and others) [34]. The latter mechanism also applies to imidazole-based RNase A mimics [3], multivalent cations, and general acid/general base catalysis, and thus to the cleavage products of in-line

Table 3.1 Structural probes for RNA

	Common name and/or abbreviation(s)	Target structure	Chromatography	Strand scission	Primer extension	High-throughput sequencing	Reference
Alkylating agents and electrophiles							
Diethyl pyrocarbonate	DEPC	$N7$-A	+	+			[22]
Dimethyl sulfate	DMS	$N3$-C	+	+	+		[22, 31]
	DMS	$N7$-G	+	+			[22]
	DMS	$N1$-A	+		+		[31]
Chloroacetaldehyde			+				[78]
Alpha-carbonylaldehydes, glyoxal, kethoxal		$N1$-G, $N2$-G	+				[15, 34]
Ethylnitrosourea	ENU	Phosphate					[25]
Cyclohexylcarbodiimide	CMCT	$N3$-U, $N1$-G	+		+		[79]
1-Fluoro-2,4-dinitrobenzene	Sanger's reagent	Adenosine, 2'OH/3'OH	+				[80]
N-methylisatoic anhydride and derivatives	SHAPE reagents	2'OH			+	+	[35, 81]
$BenzN_2^+$, PuN_2^{3+}, $SperN_2^{5+}$ (light activated)	Aryl-diazonium salts	Phosphate		+			[82]
Oxidations, radicals, and photochemistry							
T4MPyP, T2MPyP, TMAP	Porphyrin cations	Apex of helices, hinges		+			[38]
KONOO	Peroxo-nitrite	Ribose CH		+			[37]
Synchroton radiation	X-ray	Ribose CH		+	+		[2, 5]
Fe(II)/EDTA/H_2O_2, various ligands	Fenton reagents	Ribose CH		+	+		[83]
bis(orthophenanthroline)-copper(I)	Copper reagents	Ribose CH		+			[4]
Rh(DIP)33+, Rh(phen)2phi3+/hv	Rhodium reagents	G-U bp, tertiary structure		+			[39, 84]
Ni-TMAPES/KHSO5, Ni-CR/KHSO5	Nickel reagents	$N7$-G, G-U bp		+			[7, 41, 85]
Phosphodiester hydrolysis							
Nuclease S1, P1, Rn I, N. crassa	S1, P1	ssRNA		+			[34, 86, 87]
RNase T2	T2	ssRNA		+			[88]
RNase T1	T1	ssG		+			[21]
RNase U2	U2	ssA		+			[87]
RNase CL3	CL3	ssC		+			[89]
RNase V1	V1	dsRNA		+			[90]
General acid/general base	In-line probing	Flexible phospho-diester		+			[6, 91]
Imidazol and conjugates	RNase A mimics	ssRNA, py-A sites		+			[3]
Pb^{2+}, Zn^{2+}, Mg^{2+}, Ca^{2+}, Sr^{2+}, Ln^{2+}	Multivalent metal ions	Binding pockets		+	+		[3, 32, 92–97]

probing [6]. On a sequencing gel, this is relevant for the interpretation of small RNA fragments (<15 nucleotides) because fragments of otherwise identical length resulting from different cleavage mechanisms show differential migration behavior, thus complicating signal assignment.

3.7 Probing agents involving radical chemistry

Analogous considerations apply to fragments resulting from aniline cleavage (see Section 3.3) or from cleavage chemistry involving radicals or redox pairs of heavy atoms, of which Fenton-type chemistry is probably the most popular. Although the highly reactive species produced by the combination of Fe(II) and hydrogen peroxide is likely not the often cited free hydroxyl radical (HO·), this chemistry yields cleavage patterns similar to those obtained with peroxynitrite [37] or synchrotron radiation [2, 5], both bona fide radical generators. This type of chemistry is thought to produce strand cleavage via hydrogen abstraction from the CH bonds in the ribose, and therefore to map general solvent accessibility of the respective sugars. Because the species involved react very fast, it allows one to conduct experiments with time resolution on a scale relevant for RNA folding [5].

3.8 Matching suitable probes to structural features

The previous classification of the plethora of available structural probes is based on their (bio)chemistry and thus on technical aspects. However, to tackle a scientific problem related to RNA structure, it is more helpful to decide on a set of probes based on their specificity [34]. For an unknown RNA structure, nucleases provide a relatively fast and experimentally simple way to match digestion patterns to potential secondary structures. This is especially relevant for larger RNAs, where contributions of tertiary interactions are less likely to significantly override the parameters used in prediction algorithms of secondary structure. Probing with divalent cations, typically lead ions, is similarly fast and simple and yields data on the flexibility of nucleotides as well as identification of binding pockets for divalent cations. The latter are typically occupied by magnesium ions under physiological conditions and may make a major contribution to the overall structure. More detailed information on the nucleotide or even on the atomic level may be obtained by probing with small molecules, typically the alkylating agents introduced initially. A number of more exotic probes are then available to identify or confirm the existence or particular structural features such as the end of a helix [38] or noncanonical G-U base pairs [39–41]. The differential capability of probes to pick up the structural flexibility of the polymer chain has been pointed out recently [42]. Two of the more recent probing techniques (i.e. in-line probing [6] and SHAPE reagents [35]) are rather potent in the discrimination between rigid and flexible nucleotides. Weeks [42] has proposed grouping small molecule probes into three categories, based on whether they are (1) base selective, (2) mapping solvent accessibility, or (3) assessing local nucleotide dynamics. In addition, crosslinking agents and bifunctional agents can provide three-dimensional constraints for model building by identifying local neighbors through space. While this classification is a helpful concept, not all entries in Tab. 3.1 are readily assigned to these categories.

3.9 Chemical modification interference

As outlined previously, the closely related techniques of probing and Maxam and Gilbert–style sequencing differ only in the folding status of the RNA at the time of chemical modification. The chemical modification interference in its oldest embodiment is yet another technique using that same chemistry and protocols that are closely related. The conceptual novelty of modification interference schemes is the introduction of a selection step before modification detection. This selection occurs after the chemical modification as illustrated in Fig. 3.2C. In our example, this again means methylation of cytidines under denaturing conditions, only after which the RNA is allowed to adopt its native fold under native conditions. As an important consequence of the chemical modification, which occurs in a statistic distribution evenly across all cytidine residues, this RNA sample is not structurally homogenous. Rather, it now contains different populations of RNA molecules, some of which are functionally folded as the unmodified RNA, and some of which cannot adopt the native folding as a consequence of the chemical modification. Like indenter and imprint, these two populations contain complementary information on the importance of the modified residues for RNA folding. In our example, the pool of misfolded RNA contains information on all cytidines whose chemical modification *interferes* with proper folding –hence the name of the technique. The selection step in chemical modification interference must be designed to allow separately interrogating these two populations for the structural information in question. Hence, the selection step usually consists of a physical segregation of these two populations. This can, for example, be achieved by separation on a nondenaturing gel, where well-folded RNAs typically move faster than unfolded or incompletely folded species [43], or through binding to a partner macromolecule [44]. In applications to ribozymes, the selection step is typically based on catalytic activity [45, 46]. After isolation from the gel, application of aniline scission to the folded population, followed by analysis on a sequencing gel, must obviously give signals only for those adenosine residues whose modification (and subsequent depurination) did not interfere with proper folding. Thus, all bands appearing here signal the relative unimportance of the respective residues. Inversely, application of aniline scission to the "nonfunctional" sample should only produce bands for adenosine residues whose *N7* is crucial to folding. However, while information content of both populations is, in principle, complementary, in most practical cases crucial information is drawn from the functional sample (e.g. because other factors such as metastable folding intermediates may enhance the nonfunctional population).

Although chemical modification interference schemes were first developed with Maxam and Gilbert chemistry, primer extension can be applied equally well [47]. As with probing, this technique enlarges the arsenal of available reagents. All chemicals used for structural probing (except for those with nuclease activity) can in principle be applied to chemical modification interference as well, as was illustrated, for example, with DMS, DEPC, kethoxal, CMCT [44, 47–53], and even a nickel-based reagent [43]. In addition, reagents may be used that are not suitable for probing under native conditions but which modify the RNA under more stringent conditions, such as, for example, hydrazine treatment, a popular agent in early chemical modification interference experiments [45, 52, 54–57]. Hydrazine is a potent nucleophile, which preferentially attacks the C6 of pyrimidines in a reaction of the Michael addition type (Fig. 3.1A). Depending on the conditions, it will preferentially attack pyrimidines of low electron density (such as m^3C in probing [22]), or under more stringent conditions, it can be used to generally attack all pyrimidines [58]. However, the concentrations applied range in the double-digit percentile [58] and therefore are not compatible with native conditions.

As opposed to the hitherto treated techniques of postsynthetic chemical RNA modification, incorporation of altered nucleoside structures into RNA for interference experiments can also be carried out cosynthetically. For interrogation of the involvement of phosphate oxygens in RNA structure (e.g. in magnesium ion coordination), phosphorothioate nucleotides can be incorporated into RNA by RNA polymerase. Importantly, the combinatorial distribution of the chemically altered nucleotides, which is achieved in structural probing under conditions of incomplete modification, must be emulated here, too. This is achieved by spiking a relatively small amount of the thiolated NTP into the NTP pool of the transcription mixture. As phosphorothioates are labile toward a treatment with iodine, their position is easily revealed after the separation of functional and nonfunctional populations. Of note, phosphorothioates in RNA are chiral, and thus stereochemical aspects have to be carefully considered in the interpretation [59].

3.10 Nucleotide analog interference mapping

From the discovery of phosphorothioates and their usefulness in structural and stereochemical studies emerged a technique that has found widespread application especially in the ribozyme field. NAIM in its originally developed form is based on the interference scheme using phosphorothioates as described previously [8, 11–13]. In addition to information on the phosphate backbone, NAIM can provide detailed information on the atomic resolution level at positions elsewhere in the phosphorothioate nucleotide. With this technique, nucleotide modifications can be employed for structural studies, which would otherwise be undetectable by either Sanger or Maxam and Gilbert sequencing. This is achieved by the use of special phosphorothioate triphosphates in which the atoms or functional groups of interest are chemically altered, as is shown in a few example cases in Fig. 3.4. As in chemical modification interference, a pool of RNAs containing such modified phosphorothioates in a combinatorial distribution is submitted to a protocol suited to separate functional and nonfunctional molecules. However, interference with proper RNA structure may now originate either from the sulfur in the phosphate, as described previously, or from the additional modification elsewhere in the nucleotide. To distinguish between these two possibilities, an additional pool is treated as a control experiment, which contains unmodified phosphorothioates only, and thus in fact corresponds exactly to the chemical inference assay described in the end of the preceding section [11, 13]. NAIM has become especially popular for applications in ribozyme biochemistry, where the physical separation can typically be made dependent on catalytic activity of the RNA. Because in this way NAIM picks up effects occurring during turnover, it

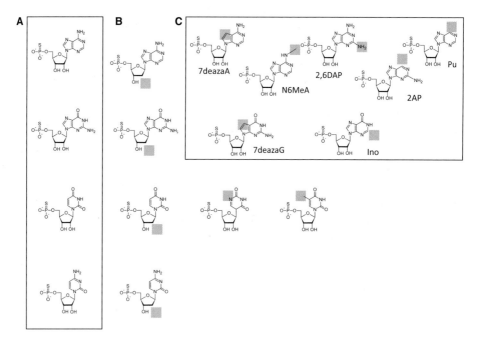

Fig. 3.4 Selected nucleoside analogue phosphorothioates for NAIM studies. (A) Parent phosphorothioates of the four canonical nucleotides (B,C) analogues in which modifications are boxed in gray. (C) Analogues used in the study of the Diels-Alderase ribozyme (Kraut et al. [46]). Abbreviations: 7deazaA, 7-deazaadenosine; 7deazaG, 7-deazaguanosine; Pu, purine; 2AP, 2-aminopurine; 2,6DAP, 2,6, diaminopurine; Ino, inosine; N6MeA, 6-methyladenosine.

can provide information on the importance of individual functional groups or atoms for catalysis, which are not necessarily accessible in static crystal structures.

3.11 Combination and interplay with other methods

Structural probing is a powerful tool capable of determining secondary structures, revealing elements of tertiary structure, identifying candidates for binding pockets of divalent ions, and monitoring structural rearrangements in response to RNA binding to either small molecules or other macromolecules including RNA and proteins. However, probing should not be viewed as a stand-alone approach because information gain is vastly increased in combination with other methods such as secondary structure prediction (see Chapter 14), phylogenetic analysis, modification interference, mutational analysis, thermal melting studies (see Chapter 1), three-dimensional structural modeling, molecular dynamics, crystallography (see Chapter 13), and/or NMR (see Chapter 12) studies [60–63]. A typical exploration of an RNA structure starts with the prediction of a secondary structure from its sequence, guided by Watson-Crick constraints applied by either brain or computing power. For most RNAs, conserved nucleotides can be identified by sequence comparison from phylogenetic analysis, which may be of biological origin, or, in the case of artificial ribozymes and aptamers, originate from sequencing at the end of a Systematic Evolution of Ligands by Exponential Enrichment (SELEX) experiment. Typically, phylogenetic data make a strong case for one particular secondary structure model, and they may hint which other conserved elements could be implicated in tertiary interactions [61, 64]. Mutational analysis implies a certain level of structural understanding of the RNA in question, such that meaningful mutants can be designed and tested either for altered structure (e.g. by probing) or in a functional readout. Thermal melting studies, classically conducted by monitoring the absorption of UV light as a function of temperature, are suited to characterizing the interplay between different structural domains, which may be thermally unfolded either independently or in a cooperative fashion. Such thermal melting processes can then be mapped onto a structural model by temperature-dependent probing experiments [65, 66]. In a typical course of an RNA's extensive structural characterization, three-dimensional models are generated, which integrate all information and are produced in the abovementioned types of experiments. Modern modeling approaches integrate detailed high-resolution knowledge on structural motifs extracted from X-ray structures [61, 64, 67]. While their accuracy has remained at the 4–5 Å resolution level for some time, recent advances are remarkable, which become increasingly independent on experimental data [1]. For RNAs of high interest, X-ray or NMR-based structures typically become available within a few years, obliterating the use of such models in subsequent studies. Despite this, the relationship between structural probing and modeling on one hand, and NMR and crystal structures on the other hand, is not nearly as competitive as may be anticipated. This is because probing and modeling provide essential information in the rate-limiting step of both NMR and crystallography, namely, the design of suitable model constructs that contain the relevant structural characteristics of the parent RNA and are structurally rigid enough to crystallize or yield exploitable signals in NMR studies.

3.12 Application to an artificial ribozyme

The interconnection of probing, interference, and related techniques as recounted previously is very well illustrated by the case of the Diels-Alderase ribozyme, which has been extensively characterized by a variety of techniques. The central structural motif of this small ribozyme emerged from a SELEX experiment, which produced a number of longer sequences from which a consensus sequence was integrated into a 49-nucleotide-long minimal motif. This RNA catalyzes the Diels-Alder reaction of two small organic molecules, namely, an anthracen and a maleimide in a stereospecific pericyclic reaction, and, of note, can do so with multiple turnover [68, 69]. Because this is the only example of such a case so far, studies of the structure as well as of the folding dynamics and the catalytic mechanism are of high interest. In addition to the multiple turnover reaction with free reactants, the Diels-Alder reaction can also be conducted in an "in *cis*" format, where the anthracene is covalently tethered to the 5' end of the RNA, and a maleimide derivative containing a biotin may be used to tag active ribozymes for physical separation (e.g. in interference or NAIM experiments).

The phylogenetic data compiled from several clones of the original SELEX experiment have accurately identified the secondary structure as is displayed in Fig. 3.5. It consists of three double-helical stems (boxed in Fig. 3.5A,B), an asymmetric internal loop composed of a pentanucleotide and a hexanucleotide, and the formally single-stranded 5'-GGAG tetranucleotide, with the first G being the attachment site for the anthracene. Probing of end-labeled RNA with RNases S1, T2, T1, and U2 confirmed the secondary structure by revealing cleavages in the unpaired regions, while RNase V1 identified residues in helical double-stranded structures and also stacked single-stranded residues [70] (Fig. 3.5B). Indeed, Fig. 3.5B shows high correlation of single-stranded regions with the corresponding nuclease cleavage sites, as well as an accurate identification of helices by RNase V1. Of note, only few nucleotides do not show susceptibility to any nuclease, and a few selected are cleaved by nucleases of both types. A typical feature of RNA is that, in many cases, the functionally interesting structural elements are not contained in standard Watson-Crick helices, but rather in loops or bulges that appear single stranded in predications of secondary structure and show appreciable but incomplete accessibility toward probes that recognize single-stranded regions. Many residues crucial to function or structure are involved in tertiary interactions and

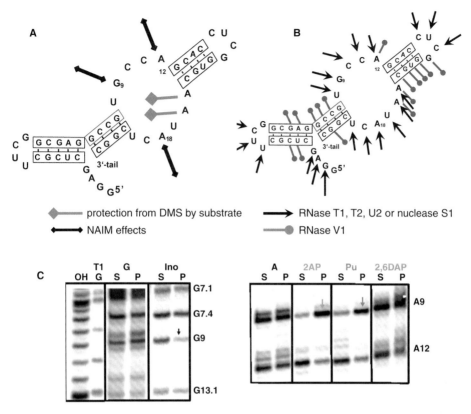

Fig. 3.5 Chemical approaches to the structure of the Diels-Alderase ribozyme. (A) Secondary structure of a minimal motif 49mer ribozyme with Diels-Alderase activity. (B) Nuclease probing confirms the secondary structure prediction. Cleavage by RNase V1 faithfully identifies helices and residues involved in higher-order structure. All single-stranded residues are identified by cleavage of at least one of several nucleases specific for single-stranded RNA. (C) NAIM experiments identify purine residues important for catalysis. Left panel: guanosine analogues; "OH" denotes a sequencing ladder (compare "L" in Fig. 3.2), and "T1-G" partial sequencing by cleavage after G residues by RNase T1. "S" denotes signal lanes of substrate population, and "P" those of product populations. Right panel: adenosine analogues. Abbreviations of analogues as shown in Fig. 3.4. Clear differences between the "S" and "P" populations, as indicated by arrows, are illustrated for positions 9 and 12.

noncanonical base pairs [67, 71, 72], which require detailed investigations on the atomic level. First implications can be drawn from the cleavage data of residues A15 and A16 as indicated in Fig. 3.5B. The fact that these adenosines are cleaved by RNase V1 may be taken as a first indication of involvement in more sophisticated structures, which must be addressed with smaller probes capable of picking up details at higher resolution. One such probe is lead acetate. Probing with lead ions in the presence of increasing concentrations of magnesium ions indicated residues implicated in a putative coordination site for divalent cations. Thus, out of a handful of the cleavages in the hexanucleotide region bulge induced by lead ions, cleavage at C19 and A15 diminished with increasing concentrations of magnesium ions. The notion of this region as a metal ion coordination site was later confirmed by phosphorothioate interference [46] and crystallography [73]. Further small molecule probing of end-labeled RNA with DEPC and DMS, revealed by Maxam and Gilbert chemistry, was used to assess the solvent accessibility of *N3*-C and *N7*-A, while reactivity of *N1*-A residues toward DMS was analyzed by primer extension. The detected reactivity patterns were largely in agreement with the secondary structure, and in addition, a strong influence of magnesium ions was again observed for nearly all residues forming the internal bulge. The latter confirmed yet again the existence of sophisticated structure, making this an obvious candidate for the catalytic center. This notion was further enhanced by the fact that accessibility of A15 and A16 to methylation by DMS was quantitatively suppressed in the presence of anthracene substrate. Of note, the hitherto recounted probing data address the ribozyme structure before catalysis or in the presence of one substrate. Some insight into the role of the 5' stretch of apparently single-stranded nucleotides in catalysis was obtained from "permutational" analysis of this region. Diels-Alderase activity was especially sensitive to exchange of the A3 and G4, U20 and C19. Since compensatory mutations according to Watson-Crick rules restored activity, base pairing of these residues was integrated as a constraint, along with the probing data, into a three-dimensional model [70], whose overall architecture and major structural elements were later confirmed by a crystal structure of the Diels-Alderase ribozyme [73].

Using the large volume of structural data as a starting point, dynamic features of the ribozyme were investigated (e.g. by single molecule spectroscopy [74], NMR [62], and Molecular dynamics (MD) simulations [60, 75], which will not be treated here as they are out of our scope). Of high interest to the topic of this book chapter is, however, that mutational analysis on the atomic level and interference methods, in particular NAIM, continued to make major contributions to the elucidation of the catalytic mechanism of this rather unusual ribozyme. In addition to the phosphorothioate interference that has been mentioned previously as helpful in confirming suspected binding sites for divalent cations, a variety of nucleoside analogue phosphorothioates were employed in NAIM studies (Fig. 3.5C). These identified three particular purine residues, G9, A12, and A18, as relevant for catalysis, all of which are part of the internal bubble [46]. All methods applied so far have outlined that this region contains the catalytic pocket, composed of various interesting noncanonical RNA structure elements. A number of atomic mutants were generated and tested for catalytic activity, carrying, for example, 2'-deoxy, or 2'-O-methyl substituents, 7-deazaguanosine, 8-oxo-2'-deoxyguanosine, isoguanosine, and isocytidine substitutions [46]. The most striking result here was the dependence of catalytic activity on the presence of a single H bond between the cytidine-NH_2 (C10) and the uridine-O2 (U17). Altogether, while the X-ray crystal structures of the Diels-Alderase provided a first detailed glimpse at the structural principles governing the architecture of this remarkable ribozyme, the subsequent NAIM and atomic mutagenesis studies allowed the classification of structural elements and the assignment of function.

3.13 Conclusion and outlook

Structural probing of RNA in solution is a venerable technique that still encounters strongly innovative developments. Instead of being an alternative to, or in competition with, other methods, probing is most powerful when combined with methods such as phylogenetic and mutational analysis, secondary structure prediction, chemical modification interference, NAIM, and atomic mutagenesis. Alone or in combination with any of the latter, probing is well suited to laying the groundwork for NMR, crystallography, and even single molecule studies. Given the latest developments in the field, which include in-line probing [6], SHAPE chemistry [35], and the application of the latest generation of sequencing techniques [16–18, 76], it is easy to imagine widespread use of high-throughput, partially automated pipelines [77] for structural analysis emerging in the near future. It is to be anticipated that high-throughput sequencing will eventually substitute the good, old probing gels and, by virtue of potentiating the analytical performance in primer extension assays, may boost the development of new types of chemical probes as well.

References

1. Cruz JA, Blanchet MF, Boniecki M, et al. RNA-puzzles: a CASP-like evaluation of RNA three-dimensional structure prediction. RNA. 2012;18:610–25.
2. Adilakshmi T, Soper SF, Woodson SA. Structural analysis of RNA in living cells by in vivo synchrotron X-ray footprinting. Methods Enzymol. 2009;468:239–58.
3. Giege R, Felden B, Zenkova MA, Sil'nikov VN, Vlassov VV. Cleavage of RNA with synthetic ribonuclease mimics. Methods Enzymol. 2000;318:147–65.
4. Hermann T, Heumann H. Structure and distance determination in RNA with copper phenanthroline probing. Methods Enzymol. 2000;318:33–43.
5. Ralston CY, Sclavi B, Sullivan M, et al. Time-resolved synchrotron X-ray footprinting and its application to RNA folding. Methods Enzymol. 2000;317:353–68.
6. Regulski EE, Breaker RR. In-line probing analysis of riboswitches. Methods Mol Biol. 2008; 419:53–67.
7. Rokita SE, Burrows CJ. Probing nucleic acid structure with nickel- and cobalt-based reagents. Curr Protoc Nucleic Acid Chem 2001;Chapter 6:Unit 6.4.
8. Ryder SP, Strobel SA. Nucleotide analog interference mapping. Methods. 1999;18:38–50.
9. Low JT, Weeks KM. SHAPE-directed RNA secondary structure prediction. Methods. 2010; 52:150–8.
10. Liebeg A, Waldsich C. Probing RNA structure within living cells. Methods Enzymol. 2009;468:219–38.
11. Waldsich C. Dissecting RNA folding by nucleotide analog interference mapping (NAIM). Nat Protoc. 2008;3:811–23.
12. Suydam IT, Strobel SA. Nucleotide analog interference mapping. Methods Enzymol. 2009; 468:3–30.
13. Cochrane JC, Strobel SA. Probing RNA structure and function by nucleotide analog interference mapping. Curr Protoc Nucleic Acid Chem. 2004;Chapter 6:Unit 6.9.

14. Giegé R, Helm M, Florentz C. Classical and novel chemical tools for RNA structure probing. In: Söll D, Nishimura S, Moore P, eds. RNA. Oxford: Pergamon Press; 2001:71–89.
15. Cramer F. Three-dimensional structure of tRNA. Prog Nucleic Acid Res Mol Biol. 1971;11:391–421.
16. Lucks JB, Mortimer SA, Trapnell C, et al. Multiplexed RNA structure characterization with selective 2'-hydroxyl acylation analyzed by primer extension sequencing (SHAPE-Seq). Proc Natl Acad Sci U S A. 2011;108:11063–8.
17. Underwood JG, Uzilov AV, Katzman S, et al. FragSeq: transcriptome-wide RNA structure probing using high-throughput sequencing. Nat Methods. 2010;7:995–1001.
18. Kertesz M, Wan Y, Mazor E, et al. Genome-wide measurement of RNA secondary structure in yeast. Nature. 2010;467:103–7.
19. Watts JM, Dang KK, Gorelick RJ, et al. Architecture and secondary structure of an entire HIV-1 RNA genome. Nature. 2009;460:711–6.
20. Fabris D. MS analysis of nucleic acids in the post-genomic era. Anal Chem. 2011;83:5810–6.
21. Wrede P, Wurst R, Vournakis J, Rich A. Conformational changes of yeast tRNAPhe and *E. coli* tRNA2Glu as indicated by different nuclease digestion patterns. J Biol Chem. 1979;254:9608–16.
22. Peattie DA, Gilbert W. Chemical probes for higher-order structure in RNA. Proc Natl Acad Sci U S A. 1980;77:4679–82.
23. Lavery R, Pullman A, Pullman B. The electrostatic molecular potential of yeast tRNAPhe. (I). The potential due to the phosphate backbone. Nucleic Acids Res. 1980;8:1061–79.
24. Lavery R, Pullman A, Pullman B, de Oliveira M. The electrostatic molecular potential of tRNAPhe. IV. The potentials and steric accessibilities of sites associated with the bases. Nucleic Acids Res. 1980;8:5095–111.
25. Vlassov VV, Giege R, Ebel JP. The tertiary structure of yeast tRNAPhe in solution studied by phosphodiester bond modification with ethylnitrosourea. FEBS Lett. 1980;120:12–6.
26. Holbrook SR, Kim SH. Correlation between chemical modification and surface accessibility in yeast phenylalanine transfer RNA. Biopolymers. 1983;22:1145–66.
27. Lavery R, Pullman A. A new theoretical index of biochemical reactivity combining steric and electrostatic factors. An application to yeast tRNAPhe. Biophys Chem. 1984;19:171–81.
28. Romby P, Moras D, Dumas P, Ebel JP, Giege R. Comparison of the tertiary structure of yeast tRNA(Asp) and tRNA(Phe) in solution. Chemical modification study of the bases. J Mol Biol. 1987;195:193–204.
29. Glasner W, Merkl R, Schmidt S, Cech D, Fritz HJ. Fast quantitative assay of sequence-specific endonuclease activity based on DNA sequencer technology. Biol Chem Hoppe Seyler. 1992;373:1223–5.
30. Schmidt C, Welz R, Muller S. RNA double cleavage by a hairpin-derived twin ribozyme. Nucleic Acids Res. 2000;28:886–94.
31. Lempereur L, Nicoloso M, Riehl N, Ehresmann C, Ehresmann B, Bachellerie J-P. Conformation of yeast 18S rRNA. Direct chemical probing of the 5' domain in ribosomal subunits and in deproteinized RNA by reverse transcriptase mapping of dimethyl sulfate-accessible sites. Nucleic Acids Res. 1985;13:8339–57.
32. Lindell M, Romby P, Wagner EG. Lead(II) as a probe for investigating RNA structure in vivo. RNA. 2002;8:534–41.
33. Zemora G, Waldsich C. RNA folding in living cells. RNA Biol. 2010;7:634–41.
34. Ehresmann C, Baudin F, Mougel M, Romby P, Ebel JP, Ehresmann B. Probing the structure of RNAs in solution. Nucleic Acids Res. 1987;15:9109–28.
35. Weeks KM, Mauger DM. Exploring RNA structural codes with SHAPE chemistry. Acc Chem Res. 2011;44:1280–91.
36. Favorova OO, Fasiolo F, Keith G, Vassilenko SK, Ebel JP. Partial digestion of tRNA—aminoacyl-tRNA synthetase complexes with cobra venom ribonuclease. Biochemistry. 1981;20:1006–11.

37. Götte M, Marquet R, Isel C, et al. Probing the higher order structure of RNA with peroxonitrous acid. FEBS Lett. 1996;390:226–8.
38. Celander DW, Nussbaum JM. Efficient modification of RNA by porphyrin cation photochemistry: monitoring the folding of coaxially sacked RNA helices in tRNAPhe and the human immunodeficiency virus type 1 rev response element RNA. Biochemistry. 1996;35:12061–9.
39. Chow CS, Barton JK. Recognition of G-U mismatches by tris(4,7-diphenyl-1,10-phenanthroline)rhodium(III). Biochemistry. 1992;31:5423–9.
40. Burgstaller P, Hermann T, Huber C, Westhof E, Famulok M. Isoalloxazine derivatives promote photocleavage of natural RNAs at G.U base pairs embedded within helices. Nucleic Acids Res. 1997;25:4018–27.
41. Hickerson RP, Watkins-Sims CD, Burrows CJ, Atkins JF, Gesteland RF, Felden B. A nickel complex cleaves uridine in folded RNA structures: application to *E. coli* tmRNA and related engineered molecules. J Mol Biol. 1998;279:577–87.
42. Weeks KM. Advances in RNA structure analysis by chemical probing. Curr Opin Struct Biol. 2010;20:295–304.
43. Pan J, Woodson SA. Folding intermediates of a self-splicing RNA: mispairing of the catalytic core. J Mol Biol. 1998;280:597–609.
44. Skripkin E, Paillart JC, Marquet R, Ehresmann B, Ehresmann C. Identification of the primary site of the human immunodeficiency virus type 1 RNA dimerization in vitro. Proc Natl Acad Sci U S A. 1994;91:4945–9.
45. Belinsky MG, Britton E, Dinter-Gottlieb G. Modification interference analysis of a self-cleaving RNA from hepatitis delta virus. FASEB J. 1993;7:130–6.
46. Kraut S, Bebenroth D, Nierth A, Kobitski AY, Nienhaus GU, Jaschke A. Three critical hydrogen bonds determine the catalytic activity of the Diels-Alderase ribozyme. Nucleic Acids Res. 2012;40:1318–30.
47. Wagner C, Ehresmann C, Ehresmann B, Brunel C. Mechanism of dimerization of bicoid mRNA: initiation and stabilization. J Biol Chem. 2004;279:4560–9.
48. Jossinet F, Paillart JC, Westhof E, et al. Dimerization of HIV-1 genomic RNA of subtypes A and B: RNA loop structure and magnesium binding. RNA. 1999;5:1222–34.
49. Mat-Arip Y, Garver K, Chen C, Sheng S, Shao Z, Guo P. Three-dimensional interaction of Phi29 pRNA dimer probed by chemical modification interference, cryo-AFM, and cross-linking. J Biol Chem. 2001;276:32575–84.
50. Majerfeld I, Yarus M. Isoleucine:RNA sites with associated coding sequences. RNA. 1998;4:471–8.
51. Nagel RJ, Lancaster AM, Zahler AM. Specific binding of an exonic splicing enhancer by the pre-mRNA splicing factor SRp55. RNA. 1998;4:11–23.
52. Batey RT, Williamson JR. Interaction of the *Bacillus stearothermophilus* ribosomal protein S15 with 16 S rRNA: II. Specificity determinants of RNA-protein recognition. J Mol Biol. 1996;261:550–67.
53. Green L, Waugh S, Binkley JP, Hostomska Z, Hostomsky Z, Tuerk C. Comprehensive chemical modification interference and nucleotide substitution analysis of an RNA pseudoknot inhibitor to HIV-1 reverse transcriptase. J Mol Biol. 1995;247:60–8.
54. Rymond BC, Rosbash M. A chemical modification/interference study of yeast pre-mRNA spliceosome assembly and splicing. Genes Dev. 1988;2:428–39.
55. Bindereif A, Wolff T, Green MR. Discrete domains of human U6 snRNA required for the assembly of U4/U6 snRNP and splicing complexes. EMBO J. 1990;9:251–5.
56. Yu YT, Nilsen TW. Sequence requirements for maturation of the 5' terminus of human 18 S rRNA in vitro. J Biol Chem. 1992;267:9264–8.

57. Gunzl A, Cross M, Bindereif A. Domain structure of U2 and U4/U6 small nuclear ribonucleoprotein particles from *Trypanosoma brucei*: identification of trans-spliceosomal specific RNA-protein interactions. Mol Cell Biol. 1992;12:468–79.
58. Peattie DA. Direct chemical method for sequencing RNA. Proc Natl Acad Sci U S A. 1979;76:1760–4.
59. Vörtler LC, Eckstein F. Phosphorothioate modification of RNA for stereochemical and interference analyses. Methods Enzymol. 2000;317:74–91.
60. Berezniak T, Jaschke A, Smith JC, Imhof P. Stereoselection in the Diels-Alderase ribozyme: a molecular dynamics study. J Comput Chem. 2012;33:1603–14.
61. Masquida B, Beckert B, Jossinet F. Exploring RNA structure by integrative molecular modelling. N Biotechnol. 2010;27:170–83.
62. Manoharan V, Furtig B, Jaschke A, Schwalbe H. Metal-induced folding of Diels-Alderase ribozymes studied by static and time-resolved NMR spectroscopy. J Am Chem Soc. 2009;131:6261–70.
63. Furtig B, Buck J, Manoharan V, et al. Time-resolved NMR studies of RNA folding. Biopolymers. 2007;86:360–83.
64. Westhof E, Masquida B, Jossinet F. Predicting and modeling RNA architecture. Cold Spring Harb Perspect Biol. 2011 Feb 1;3(2). pii: a003632. doi: 10.1101/cshperspect.a003632.
65. Waldminghaus T, Kortmann J, Gesing S, Narberhaus F. Generation of synthetic RNA-based thermosensors. Biol Chem. 2008;389:1319–26.
66. Waldminghaus T, Gaubig LC, Klinkert B, Narberhaus F. The *Escherichia coli* ibpA thermometer is comprised of stable and unstable structural elements. RNA Biol. 2009;6:455–63.
67. Leontis NB, Westhof E. Analysis of RNA motifs. Curr Opin Struct Biol. 2003;13:300–8.
68. Seelig B, Jaschke A. A small catalytic RNA motif with Diels-Alderase activity. Chem Biol. 1999;6:167–76.
69. Seelig B, Keiper S, Stuhlmann F, Jaschke A. Enantioselective ribozyme catalysis of a bimolecular cycloaddition. Angew Chem Int Ed Engl. 2000;39:4576–9.
70. Keiper S, Bebenroth D, Seelig B, Westhof E, Jaschke A. Architecture of a Diels-Alderase ribozyme with a preformed catalytic pocket. Chem Biol. 2004;11:1217–27.
71. Leontis NB, Westhof E. Geometric nomenclature and classification of RNA base pairs. RNA. 2001;7:499–512.
72. Leontis NB, Stombaugh J, Westhof E. The non-Watson-Crick base pairs and their associated isostericity matrices. Nucleic Acids Res. 2002;30:3497–531.
73. Serganov A, Keiper S, Malinina L, et al. Structural basis for Diels-Alder ribozyme-catalyzed carbon-carbon bond formation. Nat Struct Mol Biol. 2005;12:218–24.
74. Kobitski AY, Nierth A, Helm M, Jaschke A, Nienhaus GU. Mg^{2+}-dependent folding of a Diels-Alderase ribozyme probed by single-molecule FRET analysis. Nucleic Acids Res. 2007;35:2047–59.
75. Berezniak T, Zahran M, Imhof P, Jaschke A, Smith JC. Magnesium-dependent active-site conformational selection in the Diels-Alderase ribozyme. J Am Chem Soc. 2010;132:12587–96.
76. Weeks KM. RNA structure probing dash seq. Proc Natl Acad Sci U S A 2011;108:10933–4.
77. Aviran S, Trapnell C, Lucks JB, et al. Modeling and automation of sequencing-based characterization of RNA structure. Proc Natl Acad Sci U S A. 2011;108:11069–74.
78. Krzyzosiak WJ, Ciesiolka J. Long-range conformational transition in yeast tRNAPhe, induced by the Y-base removal and detected by chloroacetaldehyde modification. Nucleic Acids Res. 1983;11:6913–21.
79. Rhodes D. Accessible and inaccessible bases in yeast phenylalanine transfer RNA as studied by chemical modification. J Mol Biol. 1975;94:449–60.

80. Watanabe K, Cramer F. Reaction of tRNAPhe from yeast with 1-fluoro-2,4-dinitrobenzene. Eur J Biochem. 1978;89:425–32.
81. Mortimer SA, Weeks KM. A fast-acting reagent for accurate analysis of RNA secondary and tertiary structure by SHAPE chemistry. J Am Chem Soc. 2007;129:4144–5.
82. Garcia A, Giege R, Behr JP. New photoactivatable structural and affinity probes of RNAs: specific features and applications for mapping of spermine binding sites in yeast tRNA(Asp) and interaction of this tRNA with yeast aspartyl-tRNA synthetase. Nucleic Acids Res. 1990;18:89–95.
83. Latham JA, Cech TR. Defining the inside and outside of a catalytic RNA molecule. Science. 1989;245:276–82.
84. Chow CS, Behlen LS, Uhlenbeck OC, Barton JK. Recognition of tertiary structure in tRNAs by Rh(phen)2phi3+, a new reagent for RNA structure-function mapping. Biochemistry. 1992;31:972–82.
85. Chen X, Woodson SA, Burrows CJ, Rokita SE. A highly sensitive probe for guanine N7 in folded structures of RNA: application to tRNA(Phe) and *Tetrahymena* group I intron. Biochemistry. 1993;32:7610–6.
86. Donis-Keller H, Maxam AM, Gilbert W. Mapping adenines, guanines, and pyrimidines in RNA. Nucleic Acids Res. 1977;4:2527–38.
87. Knapp G. Enzymatic approaches to probing of RNA secondary and tertiary structure. Methods Enzymol. 1989;180:192–212.
88. Vary CP, Vournakis JN. RNA structure analysis using T2 ribonuclease: detection of pH and metal ion induced conformational changes in yeast tRNAPhe. Nucleic Acids Res. 1984;12:6763–78.
89. Florentz C, Briand JP, Romby P, Hirth L, Ebel JP, Glege R. The tRNA-like structure of turnip yellow mosaic virus RNA: structural organization of the last 159 nucleotides from the 3' OH terminus. EMBO J. 1982;1:269–76.
90. Auron PE, Weber LD, Rich A. Comparison of transfer ribonucleic acid structures using cobra venom and S1 endonucleases. Biochemistry. 1982;21:4700–6.
91. Dock-Bregeon AC, Moras D. Conformational changes and dynamics of tRNAs: evidence from hydrolysis patterns. Cold Spring Harb Symp Quant Biol. 1987;52:113–21.
92. Brown RS, Hingerty BE, Dewan JC, Klug A. Pb(II)-catalysed cleavage of the sugar-phosphate backbone of yeast tRNAPhe – implications for lead toxicity and self-splicing RNA. Nature. 1983;303:543–6.
93. Werner C, Krebs B, Keith G, Dirheimer G. Specific cleavages of pure tRNAs by plumbous ions. Biochim Biophys Acta. 1976;432:161–75.
94. Streicher B, Westhof E, Schroeder R. The environment of two metal ions surrounding the splice site of a group I intron. EMBO J. 1996;15:2556–64.
95. Matsuo M, Yokogawa T, Nishikawa K, Watanabe K, Okada N. Highly specific and efficient cleavage of squid tRNA(Lys) catalyzed by magnesium ions. J Biol Chem. 1995;270:10097–104.
96. Ciesiolka J, Marciniec T, Krzyzosiak W. Probing the environment of lanthanide binding sites in yeast tRNA(Phe) by specific metal-ion-promoted cleavages. Eur J Biochem. 1989;182:445–50.
97. Marciniec T, Ciesiolka J, Wrzesinski J, Krzyzosiak WJ. Identification of the magnesium, europium and lead binding sites in *E. coli* and lupine tRNAPhe by specific metal ion-induced cleavages. FEBS Lett. 1989;243:293–8.

Claudia Höbartner, Lea Büttner, and Fatemeh Javadi-Zarnaghi
4 Bioorthogonal modifications and cycloaddition reactions for RNA chemical biology

4.1 Introduction

Specific bioconjugation and labeling strategies for RNA play important roles for various applications in RNA chemical biology. In vitro investigations of RNA structure and function by spectroscopic methods require site-specific attachment of labels or functional groups. Studies of RNA synthesis, localization, and turnover that make use of advanced imaging technologies demand the selective attachment of fluorophores for visualization. Ideal labeling reactions need to be precise in their selectivity and proceed under mild conditions that are compatible with the delicate nature of RNA. Such bioorthogonal conjugation reactions need to run reliably in aqueous systems and complex biological milieus at near-ambient temperatures (4°C–37°C). To achieve high selectivity, reactive functional groups are employed that are normally absent from biomolecules (i.e. abiotic groups), which provide exclusive mutual reactivity. Bioorthogonal click reactions meet these criteria and are defined as reactions that are fast and highly selective, and give a single reaction product in high yield [1]. A number of diverse bioorthogonal reactions have been recently summarized [2], which have been heavily employed in the context of protein modifications. An early prominent example for bioorthogonal covalent bond formation in chemical biology is the Staudinger ligation of an azide with a phosphine, which was first used for staining of cell-surface carbohydrates [3]. In the past decade, several cycloaddition reactions involving alkenes or alkynes have gained increasing attention for biomolecular labeling and ligation [4–5].

The application of bioorthogonal reactions demands the installation of one of the reaction partners on the biomolecule, by chemical, enzymatic, metabolic, or genetic routes. For RNA, which consists primarily of four types of ribonucleotides, potential attachment sites include the ribose moiety (mostly the 2'-hydroxy group), the phosphate backbone, and the nucleobase (Fig. 4.1). Most prominently, modifications are appended at carbon position 5 of pyrimidine nucleotides, carbon 8 of purine nucleotides, or the exocyclic amino groups of adenosine, guanosine, and cytidine. In addition to internal nucleoside modifications, additional functional groups are often attached at the 5' or 3' terminus of RNA oligonucleotides.

Bioconjugation approaches are important for installation of labels for imaging applications, but click reactions also prove generally useful for conjugation reactions to ligate oligonucleotides, to cross-link nucleic acids with protein, or to attach functional units that improve delivery and cellular stability. In this chapter, we give an overview of the current state of the art of click reactions for RNA chemical biology, mostly focusing

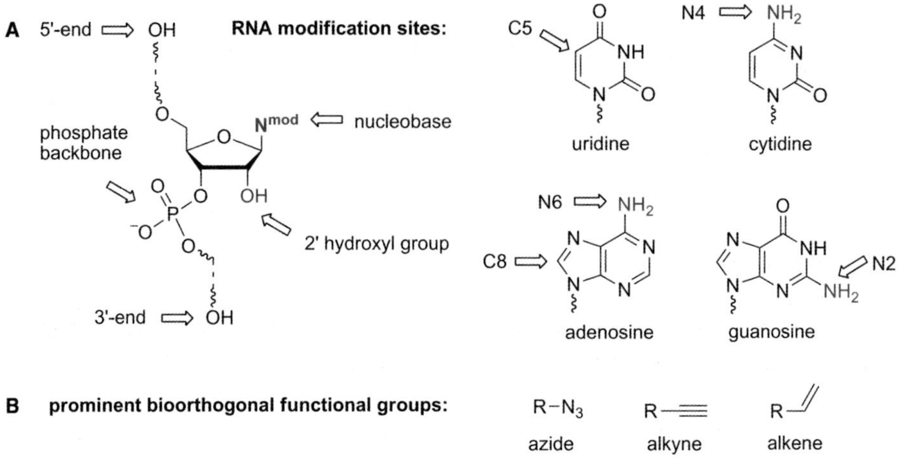

Fig. 4.1 (A) Attachment sites for bioorthogonal functional groups on RNA. (B) Most prominent examples of bioorthogonal functional groups used on RNA.

on cycloaddition reactions. We summarize the most common approaches for installation of the reactive handles by chemical and enzymatic methods, and we discuss three examples of applications of click chemistry in RNA research, including generation of modified ribozymes, monitoring of cellular RNA synthesis, and manipulation of small interfering RNAs (siRNAs).

4.2 Bioorthogonal conjugation strategies

4.2.1 1,3-dipolar cycloaddition reactions ([3+2] cycloaddition)

4.2.1.1 Copper-catalyzed azide-alkyne cycloaddition

The Huisgen 1,3-diploar azide-alkyne cycloaddition (AAC) has been developed into a widely applicable click reaction upon the seminal reports by the groups of Meldal [6] and Sharpless [7] that Cu(I) efficiently catalyzes AAC reactions to produce exclusively 1,4-substituted triazoles under mild conditions (Fig. 4.2). The high selectivity and reliable nature of Cu-catalyzed azide-alkyne cycloaddition (CuAAC) reactions resulted in numerous applications in diverse areas of research [8–11]. For the conjugation of biomolecules, the presence of reactive Cu(I) species posed nontrivial challenges, due to the thermodynamic instability of Cu(I), which is easily oxidized to Cu(II) or can disproportionate to Cu (0) and Cu(II). For nucleic acids, the presence of free Cu(I/II) ions could cause unspecific strand cleavage due to the production of reactive oxygen species leading to disintegration of the phosphodiester

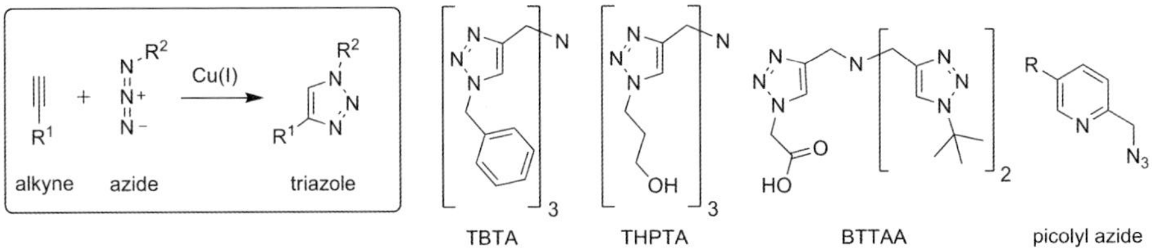

Fig. 4.2 Copper-catalyzed azide-alkyne cycloaddition (CuAAC) and Cu(I)-stabilizing ligands. Copper-chelating azides, such as picolyl azide, show fast reactions at low Cu(I) concentrations.

backbone via radical-induced reactions. Exclusion of oxygen can reduce damaging effects, but maintaining inert reaction conditions is often not an option for applications in chemical biology. The Cu(I) catalyst is usually generated in situ by reduction of readily available Cu(II) salts with ascorbate or ascorbic acid as reducing agent [7]. To circumvent the difficulties associated with oxidative stress, Cu(I)-binding ligands, such as tris(benzyltriazolylmethyl)amine (TBTA) and water-soluble analogs tris(hydroxypropyltriazolylmethyl)amine (THPTA) and bis(tert-butyltriazolylmethyl)-(2-carboxymethyl)amine (BTTAA) (Fig. 4.2), are used to stabilize Cu(I) and to accelerate the reaction [12–14]. For in vitro applications, Cu(I)-stabilizing ligands can also be replaced by using acetonitrile as a cosolvent, instead of dimethyl sulfoxide or other organic solvents that are necessary for solubility of the TBTA ligand [15, 16]. Until recently, only few reports existed on the application of CuAAC in the cellular context due to toxicity issues caused by Cu and the formation of reactive oxygen species [17]. The cell compatibility of CuAAC has been greatly improved by reducing the Cu concentrations down to 30 µM in the presence of BTTAA [14]. However, for efficient labeling at even lower Cu concentrations (10 µM), the application of a new type of azide substrates is crucial [18]. The Cu-chelating property of picolyl azides (Fig. 4.2) has been shown to impressively accelerate the cycloaddition, which was first explored on small fluorogenic molecules [19, 20], and then further improved and adopted for several examples of RNA and protein labeling in the cellular environment [18]. Interestingly, the chelation assistance of CuAAC, which in vitro also proceeds efficiently with Cu(OAc)$_2$ in the absence of added reducing agent, raised new mechanistic questions on catalysis, which are currently addressed by experimental and theoretical investigations [20, 21].

4.2.1.2 Strain-promoted azide-alkyne cycloaddition

As an alternative for Cu-catalyzed click reactions, the strain-promoted azide-alkyne cycloaddition (SPAAC) was developed, taking advantage of the inherent ring strain of cyclooctynes (Fig. 4.3A). This strategy was pioneered by Bertozzi and coworkers [22], who exploited the increased intrinsic reactivity of cyclooctynes, causing a lower activation barrier for the cycloaddition reaction with azides. Although SPAAC shows improved biocompatibility and prevents cytotoxicity due to the absence of Cu(I), it is usually considerably slower than copper-catalyzed click reactions. Depending on substrates and reaction conditions, 10- to 100-fold slower reaction rates have been reported [13, 14, 23]. A number of cyclooctyne derivatives were investigated to improve the reaction rate by lowering the energy of the lowest unoccupied molecular orbital (LUMO) of the strained alkyne by introducing fluoride substituents or nitrogen heteroatoms into the ring system [24]. Reaction rates were also increased by further increasing the strain energy due to fused aryl rings in dibenzocyclootynes (DIBO) or a fused cyclopropane in bicyclononyne (BCN) (Fig. 4.3B) [23, 25–27]. Based on recent developments and theoretical considerations, it is expected that new and even smaller substrates for strain-promoted cylcoaddition reactions will become available in the near future [28]. Since the first reports of SPAAC, this bioconjugation strategy has found widespread applications for labeling of proteins and cell-surface glycans in vitro and in living cells [29–32]. Recently, the advantages of SPACC have also been increasingly harnessed for the modification of RNA and DNA [33–37]. Moreover, SPAAC has proved to be a powerful approach for coupling processes in nanotechnology and in materials and polymer science. Excellent reviews on applications of SPAAC in chemical biology and beyond are recommended for further reading [4, 38–41].

4.2.1.3 Nitrile oxides as 1,3-dipoles for metal-free cycloadditions

The general concept of the Huisgen 1,3-dipolar cycloadditions is not limited to azides and alkynes but has been demonstrated for a wide variety of substrate combinations [42]. Nitrile oxides represent a class of strong electrophiles that react with unsaturated dipolarophiles to form isoxazole derivatives (Fig. 4.3C). For the modification of oligonucleotides, the proof of principle was demonstrated by Singh et al. [43] for derivatization of a terminal alkyne of a DNA oligonucleotide on the solid support. Carell and coworkers [44] described norbornene as an alternative substrate for the cycloaddition with nitrile oxides on DNA oligonucleotides. The nitrile oxides were generated in situ from hydroximoyl chloride precursors. The click reactions proceeded under mild conditions and could be performed on the solid support or on purified deprotected DNA. The nitrile oxide cycloaddition with norbornenes can potentially be used as an orthogonal reaction to CuAAC because terminal alkynes react significantly slower with nitrile oxides than the strained norbornene. Disadvantages of this reaction are found in the production of regioisomers (1,4 and 1,5-exo products) and the potential background reaction of nitrile oxides with unmodified DNA nucleotides (after 6 h, 50% of DNA was modified by the undesired background reaction; however, such long reaction times are not necessary since full conversion of the norbornene was achieved after 10 min) [44].

Recently, the nitrile oxide cycloaddition was extended to the modification of RNA oligonucleotides. Singh, Heaney, and coworkers [35] exploited the ring strain

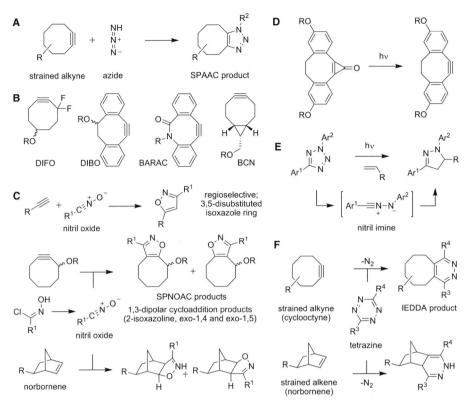

Fig. 4.3 (A) Strain-promoted azide-alkyne cycloaddition (SPAAC). (B) Selected examples of strained cyclooctynes. DIFO, difluorocyclooctyne; DIBO, dibenzocyclooctyne; BARAC, biarylazacyclooctynone; BCN, bicyclo[6.1.0]nonyne. (C) 1,3-dipolar cycloadditions of nitrile oxides with strained dipolarophiles or terminal alkynes; SPNOAC = strain promoted nitrile oxide alkyne cycloaddition. (D) Masked cyclooctyne for phototriggered SPAAC. (E) Photoclick reaction: 1,3-dipolar cycloaddition of nitrile imines (generated upon photolysis of diaryl tetrazoles) with terminal alkenes. (F) Inverse electron demand Diels-Alder (IEDDA) reactions of strained alkyne and alkene dienophiles with (diaryl)tetrazines.

of cyclooctynes for the fast reaction with nitrile oxides to install fluorescent reporter groups at the 5' end of solid-phase-bound RNA oligonucleotides. In situ generation of nitrile oxides from stable oxime precursors enabled the efficient strain-promoted nitrile oxide alkyne cycloaddition. After cleavage and deprotection, modified RNA was isolated and characterized. The compatibility of this reaction with deprotected RNA oligonucleotides has not yet been shown. Sterically demanding nitrile oxides were conjugated efficiently, while analogous azides did not produce any cycloaddition product under comparable conditions, which underlines the potential orthogonality [35]. Similar to the reaction with norbornenes, cyclooctynes yielded regioisomeric isoxazole conjugates. In contrast, terminal alkynes produced 3,5-disubstituted isoxazoles regioselectively [43]. Other strain-promoted cycloadditions that have recently been reported for proteins, e.g. using nitrones as 1,3-dipoles [27, 45–47], are yet to be explored in the context of nucleic acids.

4.2.1.4 Photoactivated 1,3-dipolar cycloadditions

An interesting aspect of 1,3-dipolar cycloadditions pertains to the in situ generation of the 1,3-dipole or the dipolarophile in a light-controlled manner. This attractive approach has been reported for manipulation of proteins, but analogous modifications of nucleic acids have not yet been demonstrated. To spur potential developments in this direction, the general concepts are briefly mentioned and depicted in Fig. 4.3D,E.

A phototriggered version of SPAAC was realized by Boons and coworkers [48], who introduced a dibenzocyclooctyne derivative, in which the triple bond was masked as cyclopropenone (Fig. 4.3D). UV-induced decarbonylation generated the strained alkyne in situ and allowed selective ligation to azides in living cells.

In an alternative photochemical strategy, nitrile imines are used as 1,3-dipoles, which are generated in situ by photolysis of tetrazoles (Fig. 4.3E). The principle of this 1,3-diploar cycloaddition with terminal alkenes also dates back to early investigations by Huisgen and coworkers [49]. Recently, Lin and coworkers [50–52] exploited this transformation for bioconjugation applications and used the fluorogenic photoclick reaction to visualize proteins in vitro and in cells.

4.2.2 Inverse electron demand Diels-Alder reaction ([4+2] cycloaddition)

An attractive addition to the repertoire of bioorthogonal conjugation reactions exploits the selectivity and very fast kinetics of inverse electron demand Diels-Alder reactions (IEDDA or DAinv). For the [4+2] cycloaddition, electron-deficient tetrazines have been described to react efficiently with certain dienophiles in the context of proteins [53–57] and nucleic acids [58–60]. Trans-cyclooctene, norbornene, and bicylco[6.1.0]nonynes (BCN) were site-specifically incorporated into biomolecules and demonstrated high reactivity in the cycloaddition reaction (Fig. 4.3F). The reaction is irreversible due to the elimination of N_2 in the course of the reaction. For the application of this powerful conjugation reaction to RNA, Jäschke and coworkers [60] reported a norbornene-modified transcription initiator (see Section 4.3.2.1). An attractive feature of the metal-free IEDDA chemistry with *trans*-cyclooctenes is based on the almost perfect orthogonality to CuAAC, which has been demonstrated for the site-specific dual labeling of DNA in a one-pot procedure for the installation of fluorescence resonance energy transfer (FRET) probes [59].

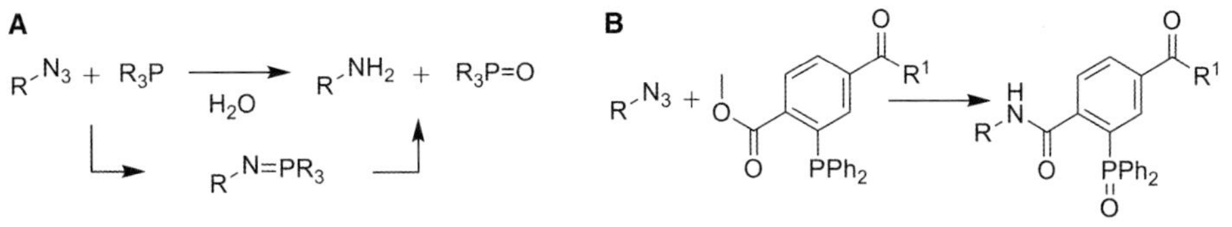

Fig. 4.4 (A) Staudinger reaction of azides with phosphines. Hydrolysis of the iminophosphoran liberates the free amine. (B) Bioorthogonal amide bond formation via Staudinger ligation.

4.2.3 Staudinger reaction of azides and phosphines

Even before the advantageous properties of azides as a 1,3-dipole were explored for bioconjugation reactions, the N_3 group was used as powerful bioorthogonal functional group in the Staudinger reaction. The unique reactivity of azides with phosphines leads to reduction to amines via the intermediate formation of iminophosphoranes. Triarylphosphines equipped with an ester functionality directly lead to formation of an amide bond, thereby promoting the Staudinger ligation (Fig. 4.4). Similar to many of the Huisgen cycloadditions discussed previously, the principles of the Staudinger reaction have long been known in the chemical literature [61] but gained renewed importance for bioconjugation reactions [62]. Due to the rather slow reaction kinetics of the Staudinger ligation, cycloaddition reactions are often preferred for modern applications of azide-modified biomolecules, especially for probing fast biochemical processes. Nevertheless, the Staudinger ligation has found broad applications in vitro for postsynthetic labeling of peptides, carbohydrates, and nucleic acids, and it was the first bioorthogonal reaction used in living systems [63]. Attractive applications of the Staudinger reduction involve the phosphine-mediated unmasking of azide-containing chromophores, which have enabled the detection and visualization of RNA in living cells [64]. For more details, excellent overviews on various aspects and applications of the Staudinger reaction/ligation on nucleic acids and other biomolecules are available in the recent literature [65, 66].

4.3 Synthetic strategies for RNA functionalization: installation of reactive groups for cycloadditions

This chapter gives a brief overview on general strategies available for the installation of alkyne and azide modifications on oligonucleotides and provides some recent examples of synthetic building blocks. With the ever-increasing number of applications of click chemistry approaches as efficient tools in nucleic acid chemical biology, this is not a comprehensive overview of all reported modifications but rather a random selection that tries to cover the diversity of approaches used.

4.3.1 Chemical synthesis of modified RNA

Solid-phase synthesis is the method of choice for site-specific installation of modifications in oligonucleotides. Reactive groups can be attached to the 5' or 3' termini, or at internal strand positions. On nucleosides, three generally possible sites for attachment can be distinguished: the nucleobase, the ribose, or the phosphate backbone. For click chemistry of RNA, these possibilities have been explored in various combinations. This chapter summarizes some common examples.

4.3.1.1 Alkyne-containing phosphoramidites for solid-phase synthesis

Attachment of terminal and strained alkynes to the 5' end of oligonucleotides can be achieved using nonnucleosidic phosphoramidites such as, for example, compounds (1) and (2) (Fig. 4.5A), in which the desired unsaturated functional group is attached via an alkyl chain [35, 67]. Several terminal phosphoramidites are available from commercial sources, including but not limited to Jena Bioscience, Baseclick, Berry & Associates, ChemGenes, and others [68]. For installation of alkynes between nucleotides at an internal position, the nonnucleoside phosphoramidite (3a) and several analogs thereof have been introduced (Fig. 4.5B) [34]. Analogous hydroxyprolinol derivates (3b) were also used for the preparation of solid supports that enabled 3'-terminal modification of oligonucleotides.

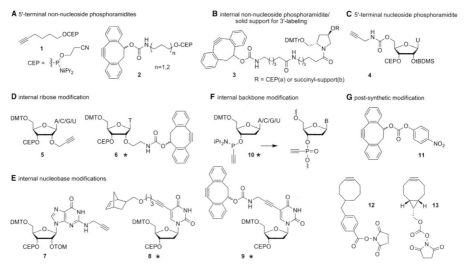

Fig. 4.5 Selected building blocks for chemical synthesis of terminal or internally alkyne-modified oligonucleotides. The examples marked with an asterisk have so far only been used on DNA but are also promising for targeting RNA. See text for description of panels A-G.

Nucleoside phosphoramidites have also been reported for attachment of alkynes at 5'-terminal (e.g. (4), Fig 4.5C [69]) and internal positions (Fig. 4.5D–F). Commonly used ribose modifications (Fig. 4.5D) include commercially available 2'-O-propargyl (5) and analogous 2'-O-butinyl phosphoramidites. More recently, phosphoramidite (6) has been reported containing dibenzocyclooctyne (DIBO) attached via a 2'-O-aminoethyl linker [37]. Examples for the modification of the nucleobase (Fig 4.5E) with terminal alkynes include 5-ethynyluridine (EU, available from Baseclick), N^2-propargyl-2-aminopurine [70], and most recently N^2-propargylguanosine (7) [71]. Position C5 of pyrimidine nucleosides is a popular site for nucleoside modification in the major groove. Examples include the norbornene-modified analog (8) [44] and the DIBO-extended uridine derivative (9) [37], both of which have so far only been used for DNA oligonucleotides. Besides ribose and nucleobase modifications, the phosphate backbone provides a third possibility for attachment of alkyne functional groups (Fig 4.5F), exemplified via phosphoramidite analogues of type (10) [72], which can in principle be used to address every backbone position in a sequence-independent manner.

4.3.1.2 Solid-phase synthesis of azide-containing RNA

Unlike alkynes, azide functional groups within nucleosides are not easily compatible with standard phosphoramidite chemistry due to the general reactivity of phosphor-III species with azides according to the Staudinger reaction. To overcome this limitation, different strategies have been developed for the efficient introduction of azides into oligonucleotides using solid-phase synthesis. Lönnberg and coworkers [73] employed H-phosphonate chemistry to install azide-modified nucleotides into DNA. Micura and coworkers [74, 75] demonstrated reliable solid-phase synthesis of azide-modified RNA oligonucleotides using the phosphodiester building blocks shown in Fig. 4.6A. In both cases, the unmodified nucleotides for chain elongation after coupling of the azido building block (under phosphotriester or H-phosphonate coupling conditions) could be installed as standard phosphoramidites. Although the Staudinger reaction prevents the synthesis of azide-containing phosphoramidites, once N_3-modified nucleotides are incorporated in a growing oligonucleotide chain, the azido group does not interfere further with strand assembly by phosphoramidite chemistry.

4.3.1.3 Postsynthetic modification of RNA with azides and alkynes

Postsynthetic modification strategies are often preferred for the attachment of bulky residues or moieties that are sensitive to the conditions encountered during solid-phase synthesis. A number of building blocks are available or have been reported that enable the introduction of primary amino groups on RNA, either at the 5' or 3' termini, or at internal positions via nucleobase or ribose modification. Popular

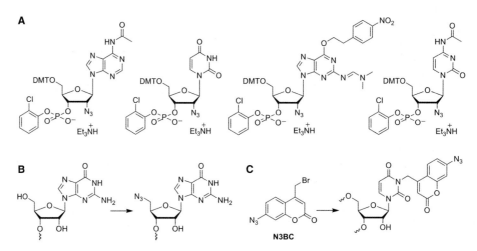

Fig. 4.6 Synthesis of azide-containing RNA. (A) Solid-phase synthesis using 2'-N_3-phosphodiester building blocks. (B) Postsynthetic conversion of 5'-OH into N_3 [76]. (C) Selective labeling of uridine via N^3-alkylation [77].

examples include 5-aminoallyl- and 2'-amino- or 2'-aminoethyl-modified pyrimidine nucleotides (see recent reviews for a more comprehensive summary of functionalization strategies) [66, 78–81]. The attachment of reactive handles for click conjugation to prefunctionalized RNA is achieved via formation of amide or carbamate bonds. Examples of activated carbonates (11) and esters (12, 13) of strained alkynes are depicted in Fig. 4.5G. Analogous strategies apply for the attachment of azides (e.g. via N-hydroxysuccinimid [NHS] ester of azidohexanoic acid) [76, 82].

Fig. 4.6B,C depicts two strategies for postsynthetic installation of azido groups on unmodified RNA. The 5'-terminal hydroxyl group of RNA can be changed into a 5'-N_3 group in a two-step procedure (Fig. 4.6B) that involves intermediate formation of a 5'-iodide upon treatment of the support-bound RNA with methyltriphenylphosphonium iodide [76, 83]. Another approach for functionalization of RNA reports on the new alkylation reagent 7-azido-4-(bromo-methyl)coumarin (Fig. 4.6C) [77], which selectively modifies N_3 of uridine via nucleophilic substitution. The azide moiety can be used for cross-linking studies, further derivatization by CuAAC, or to generate a coumarin fluorophore upon photolysis or reduction.

4.3.1.4 Functionality transfer reaction using s⁶G-modified DNA

A completely different approach to introduce reactive functional groups into RNA for click chemistry was developed by Sasaki and coworkers [84]. In the functionality-transfer reaction (FTR, Fig. 4.7), 2-methyliden-1,3-diketone functional groups are selectively transferred from a 6-thioguanosine donor nucleotide in a DNA onto the exocyclic amino group of a cytidine or guanosine nucleotide in the target RNA [85, 86].

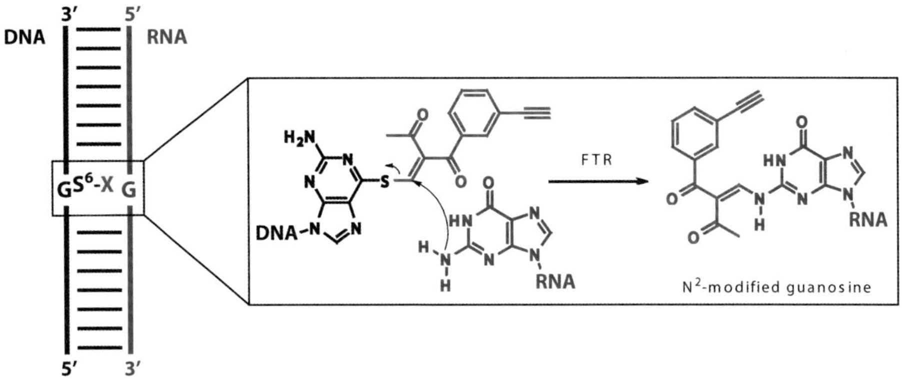

Fig. 4.7 Functionality transfer reaction for internal modification of RNA [84].

The acceptor nucleotide is placed opposite to the modified s^{6}G in a Watson-Crick base-paired duplex, which brings the reactive sites into close proximity. The transfer reaction proceeds at neutral pH for cytidine but requires alkaline conditions for the transfer to guanosine. Interestingly, in the presence of nickel ions, guanosine was also shown to be reactive under neutral conditions [87]. This FTR strategy was expanded for the transfer of an acetylene-linked methyliden-1,3-diketone to the 2-amino group of guanosine (Fig. 4.7) [84]. The transfer of the alkyne scaffold proceeded rapidly, and the CuAAC conjugation with various azide-modified compounds showed high incorporation yields, demonstrating the general applicability for further RNA investigations.

4.3.2 Enzymatic incorporation of functional groups for click chemistry

A number of enzymatic methods have been described for the incorporation of azides as well as terminal alkynes into RNA in vitro and in the cellular context. One of the first examples of CuAAC in RNA chemical biology employed enzymatic incorporation of 5-ethynyluridine (Fig. 4.8, (14)) for metabolic labeling of RNA (see also Section 4.4.2) [88]. Similarly, the nucleotide analogs 5-ethynyl-2'-deoxyuridine [89] and 5-ethynyl-2'-fluoroarabinouridine (15) [90] have been successfully used to detect DNA synthesis in proliferating cells. More recently, N^6-propargyladenosine was added to the list of alkyne-modified nucleotides incorporated by cellular RNA polymerases [91]. For cellular applications, high-density labeling of newly synthesized nucleic acids is desirable for good signal intensity upon conjugation to fluorophores via click conjugation. On the other hand, in vitro applications often prefer site-specific labeling of RNA. Here, we summarize several strategies that have been developed for the installation of azides and alkynes using enzymatic manipulation of RNA.

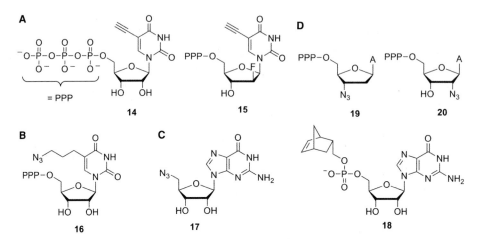

Fig. 4.8 Selected examples of building blocks for enzymatic incorporation of alkynes, azides, and strained alkenes into RNA using polymerase enzymes. (A) Cellular incorporation. (B) In vitro transcription. (C) Initiation by T7 RNA polymerase. (D) Single-nucleotide extension by poly-A-polymerase.

4.3.2.1 In vitro transcription using modified nucleotides

Modifications at the 5 position of pyrimidine nucleotides are well tolerated by polymerase enzymes, such as T7 RNA polymerase [92, 93]. A recent example for RNA is the azido-modified uridine (16) (Fig. 4.8B), which was efficiently incorporated into RNA and was shown to be accessible for derivatization via 1,3-dipolar cycloaddition with alkynes or via Staudinger ligation [94, 95]. Similar modifications have earlier been incorporated into DNA by thermophilic DNA polymerases in PCR reactions [96, 97].

In vitro transcription with T7 RNA polymerase can also be used to install unnatural nucleotides at the 5' end of the RNA transcript, by using chemically modified initiator nucleotides (Fig. 4.8C). In this way, 5'-azido groups were installed using 5'-azidoguanosine (17), which was employed in fourfold excess over GTP in the transcription reaction [15]. An analogous approach was used to attach the norbornene modification at the 5' end of an RNA transcript, using the GMP analog (18) [60].

4.3.2.2 Enzymatic posttranscriptional modification

Enzymatic posttranscriptional modification of RNA was achieved using several different enzymes. For functionalization of the 3' end, Paredes and Das [15] report the application of poly(A)polymerase (PAP) with 3'-azido-2',3'-deoxyadenosine triphosphate (19) (Fig. 4.8D), following the principles of early work by Martin and Keller [98]. The presence of the 3'-terminal azide prevented further extension. A related approach by Jäschke and coworkers used 2'-azido-2'-deoxyadenosine triphosphate (20) [36]. Here, the presence of the 3'-OH group enabled further enzymatic manipulation (e.g. via polyadenylation using nonmodified adenosine triphosphate) or enzymatic ligation to a second RNA fragment. Thereby, the terminal azide was converted into an internal one that could still undergo further modification (e.g. via CuAAC). Using various enzymes, this approach was extended to attaching any 2'-azide-containing nucleotide at the 3' end of RNA. The best results were achieved with yeast and *Escherichia coli* PAP, using short reaction times in the presence of $MnCl_2$, which allowed for efficient single incorporation of 2'-N3-2'-deoxynucleotides.

Site-specific internal modification of RNA can be achieved with enzymes that modify the nucleobases or the ribose unit. Examples par excellence are methyltransferase enzymes, which recognize specific target nucleotides and use *S*-adenosyl methionine (SAM/AdoMet, Fig. 4.9A) as cofactor for transferring a methyl group to exocyclic amino groups of nucleobases, to carbon or nitrogen atoms of the nucleobase heterocycles, but also to the ribose 2'-OH group within RNA [99]. Methyltransferases have been diverted from methylation to alkylation reactions by providing modified SAM cofactors that carry azide- or alkyne-containing transferable groups instead of the natural methyl group substrate. This approach was pioneered in the DNA field by the groups of Weinhold and Klimašauskas that introduced aziridine-based suicide substrates for DNA methyltransferases [100], as well as double-activated SAM analogs in which the transition state of the transfer reaction is stabilized by conjugation to a double bond in the side chain ("double-activated" cofactors) [101]. These endeavors culminated in the development of the sequence-specific methyltransferase-induced labeling and

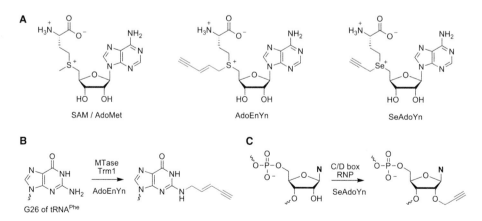

Fig. 4.9 Methyltransferase-catalyzed alkynylation of RNA. (A) Structures of S-adenosyl methionine and alkynylated analogs. (B) Site-specific labeling of tRNA by transfer of the EnYn side chain to the N^2-amino group of guanosine. (C) Site-specific, RNA-guided transfer of the propargyl residue to 2'-OH group by C/D box RNPs.

methyltransferase-directed transfer of activated groups approaches for nucleic acid labeling [102, 103]. Alternatives to reactive aziridine derivatives are nitrogen mustard precursors, reported by Rajski and coworkers [104] for DNA-methyltransferase-moderated click chemistry on DNA. Recently, Helm and coworkers [105] devised a methyltransferase-based strategy for the site-specific modification of RNA for click chemistry. The tagging reaction is catalyzed by the transfer RNA (tRNA):methyltransferase Trm1, which transfers a 5-carbon chain with a terminal alkyne from the double-activated SAM analog AdoEnYn onto the N^2-amino group of guanosine 26 in tRNA^{Phe} (Fig. 4.9B) [105]. The scope of this method resides in the site-specificity of the RNA methyltransferase, which can be considered an asset but also a limitation. Screening a large number of other RNA methyltransferases will be needed to reveal the generality of this approach. However, nature also makes use of an additional class of methyltransferases that are guided to their target sites via small nucleolar guide RNAs. These RNA methyltransferases are part of C/D box snoRNPs (small nucleolar ribonucleoproteins) that target ribose 2'-hydroxy groups in various RNA substrates. Molecular engineering of such RNPs has been successfully used to divert natural messenger RNA (mRNA) splicing in *Xenopus* oocytes, demonstrating the programmable nature of snoRNAs for molecular biology applications [106]. Exchanging the natural cofactor of C/D box snoRNPs against SeAdoYn enabled the development of a site-specific labeling approach for RNA. Klimašauskas and coworkers [107] employed an in vitro reconstituted functional C/D box RNP from a thermophilic archaeon, *Pyrococcus abyssi*, for the site-specific alkynyl-tagging of defined target sites in tRNA and pre-mRNA (Fig. 4.9C). The selenium-containing derivative was chosen as a preferred substrate due the increased stability and longer half-life in neutral buffered media compared to the analogous methionine derivative [108, 109]. The RNP-based chemoenzymatic technique offers a broad range of possibilities for labeling biologically relevant RNA targets for various applications, including imaging and affinity-based experiments. Moreover, the programmable nature provides flexibility in target site selection and at the same time sequence specificity for labeling of RNA with single-nucleotide precision.

4.4 Case studies for applications of click chemistry in RNA chemical biology

4.4.1 Synthesis of chemically modified ribozymes

Studying structure and function of large RNAs, such as ribozymes or riboswitches, by biophysical methods requires the synthesis of long RNA sequences with site-specific modifications. Click chemistry provides an orthogonal alternative to postsynthetic modification of amines with activated esters (mostly NHS esters) for the installation of, for example, fluorescent labels for FRET experiments.

In addition, click chemistry, most prominently CuAAC (Section 4.2.1.1), provides a new strategy for covalent ligation of RNA fragments to access large RNA structures. This offers an alternative to enzyme-mediated RNA ligation using T4 RNA or T4 DNA ligase. Replacement of the natural 3'-5'-phosphodiester bond with unnatural triazole-containing linkages has been described for three examples of RNA-cleaving ribozymes, including the hammerhead, hairpin, and hepatitis delta virus (HDV) ribozymes.

El-Sagheer and Brown [76] prepared a functional hairpin ribozyme from three fragments of chemically synthesized RNA molecules that contain alkyne and/or azide handles attached via the C5 position of 2'-deoxyuridine nucleosides. The alkyne modification (21) was introduced during solid-phase synthesis using a modified thymidine phosphoramidite, and the azide in (22) was installed postsynthetically via derivatization of a precursor nucleotide presenting a primary amine. Positioning these modifications in neighboring base pairs of two complementary RNA/DNA hybrid strands allowed the formation of interstrand cross-links in the major groove of the double helix (Fig. 4.10A). A second strategy employed a splint-mediated CuAAC ligation to connect the 3'-terminal alkyne (23) and the 5'-terminal azide (24) to form a modified hammerhead ribozyme bearing a triazole modification that replaces a phosphodiester linkage in the catalytic core of the ribozyme (Fig 4.10B) [76]. A related approach was reported by Paredes and Das to connect two

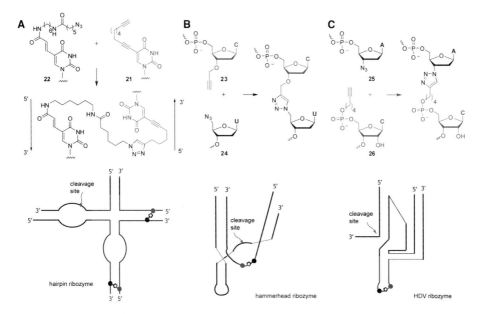

Fig. 4.10 (A) CuAAC ligation of complementary fragments of the hairpin ribozyme via a major groove linkage. (B) Replacement of a phosphodiester linkage in the active site of a hammerhead ribozyme. (C) Connecting two fragments of the HDV ribozyme at a remote location via an extended triazole-containing linkage. In all cases, the ribozymes maintained substantial activity in the presence of the unnatural linkage.

fragments of the HDV ribozyme via a click reaction between the 3'-terminal azide (25) and the 5'-terminal alkyne (26) (Fig 4.10C) [15]. The RNA fragment 25 was generated via PAP-mediated 3'-N3-2', 3'-dideoxyadenosine triphosphate ligation (Section 4.3.2.2), while the 5'-terminal alkyne was introduced via phosphoramidite (1) (see Fig. 4.5A). In this case, the triazole linkage was significantly longer compared to a standard phosphodiester bond, but at a remote location (stem-loop not involved in the active site) this artificial linkage was well tolerated and did not interfere with ribozyme activity. But even in the case of the hammerhead ribozyme, the short triazole linkage from (23)+(24) was reasonably well tolerated. The ribozyme kinetics were only slightly affected, and cleavage site accuracy was not compromised.

4.4.2 Monitoring RNA synthesis and turnover by metabolic labeling and click chemistry

Recent technical developments for RNA imaging allowed molecular biologists to address questions about RNA synthesis, transport, localization, and turnover in vivo. The biosynthetic incorporation of 5-ethynyluridine (EU) into RNA by RNA polymerase I, II, and III in cells enables visualization of newly synthesized RNA by CuAAC conjugation of fluorescent azides for microscopic imaging. Jao and Salic [110] demonstrated that this approach allows monitoring of total RNA turnover in cells. Moreover, these authors also used EU incorporation for metabolic labeling of RNA in whole animals, which enabled the comparison of transcription rates in various tissues. In these experiments, EU was added to the culture medium or directly injected into mice. After cellular uptake, EU enters the ribonucleoside salvage pathway to become phosphorylated to the triphosphate (14) (Fig 4.8A), before being a substrate for RNA polymerases. A number of control experiments demonstrated that EU is specifically incorporated only into RNA but not into DNA. It was shown that ribonucleotide reductase is not able to convert EU into 2'-deoxy-5-ethynyluridine, a known substrate for DNA polymerases that would result in concomitant labeling of DNA [89]. To further confirm incorporation by RNA polymerases, cellular RNA was isolated, digested, and analyzed by high-performance liquid chromatography. Moreover, mass spectrometric analysis confirmed the presence of the modified uridine in isolated mRNAs, ribosomal RNAs, and tRNAs.

In a second application, the EU incorporation and click labeling strategy was used to study localization and turnover of coronavirus RNA synthesis in infected cells. Hagemeijer et al. [111] treated Murine LR7 fibroblast cells that were infected with the mouse hepatitis virus with EU and click conjugated the modified RNA transcripts with an azide-modified AlexaFluor dye. To avoid labeling of host RNA, the experiments were carried out in the presence of actinomycin D, which inhibits the cellular DNA-dependent RNA polymerase but does not affect viral RNA transcription. A similar strategy was applied to examine RNA synthesis activity after bacterial infection with *Salmonella* [112]. Such studies have recently been facilitated by the availability of commercial ethynyluridine-based click chemistry kits for RNA imaging [113]. Using EU in combination with click chemistry allowed visualization of nascent RNA and the study of RNA dynamics in infected cells. The CuAAC-based labeling strategy was shown to be compatible with immunohistochemistry and with detection of green fluorescent protein (GFP) conjugated proteins to address question of cellular localization [111].

Besides EU as an alkyne-modified pyrimidine analog, N^6-propargyladenosine (p^6A) was used for enzymatic incorporation of an alkyne-modified purine analog. Grammel et al. [91] reported efficient incorporation of p^6A into RNA in mammalian (HeLa) cells by all three RNA polymerases I, II and III. In addition, polyadenylate polymerase was capable of incorporating p^6A into RNA. A pulse-chase strategy combining EU and p^6A labeling was developed to study poly(A) tail dynamics in human embryonic kidney (HEK) cells.

Several advantages can be highlighted for the click chemistry approach compared to established procedures based on metabolic incorporation of other modified nucleotide analogs, such as 4-thiouridine (s^4U) or 5-bromouridine (5BrU). While s^4U-modified RNA can be selectively conjugated after isolation [114], it is incompatible with direct labeling in the cell due to the presence of competing thiols. Incorporation of 5BrU (or 5-bromo-2'-deoxyuridine for DNA labeling) can be detected on fixed cells by immunostaining using BrdU antibodies. With typical molecular weights of about 150 kDa, antibodies are rather big labels that are characterized by slow diffusion, limited tissue penetration, and incomplete epitope recognition [115, 116]. In contrast, the alkyne tag and the dyes for click labeling are very small in size, which should allow more precise labeling; this is of high relevance with the increasing importance of superresolution light microscopy techniques [117].

4.4.3 Bioorthogonal modification of siRNAs for detection, improved stability, and delivery

siRNAs are short double-stranded RNAs capable of specific and efficient gene silencing, via the intracellular RNA interference (RNAi) pathway. Synthetic siRNAs have been intensively used as molecular biology tools but are also investigated as potential therapeutic agents. There has been substantial progress in the field, but major challenges yet to be overcome include the poor cell penetration and limited stability in physiological environment. Click chemistry–based bioconjugation strategies have been used on siRNA to address questions of detection, stability against degradation, and cellular uptake.

To monitor delivery and localization of siRNAs, fluorescent labels have been conjugated at various positions via CuAAC. Micura and coworkers synthesized 2'-azido-modified siRNAs [74], demonstrated the tolerance of this modification in the guide strand, attached an alkyne-modified fluorophore, and explored localization of siRNAs in chicken DF1 cells using fluorescence microscopy [75]. Instead of fluorphore-labeling on the ribose, Helm, Domingo, and coworkers [71] reported incorporation of N^2-alkyne-modified guanosine into siRNAs (via phosphoramidite (7), Fig. 4.5E), followed by selective fluorophore labeling via CuAAC. As for the fluorescent siRNA generated via 2'-azido nucleotides, the RNAi activity of the nucleobase-labeled strands was only slightly diminished. The azide-alkyne click chemistry has also been used to attach ^{18}F-containing residues to alkyne-modified siRNAs [118]. This labeling strategy will potentially be useful for in vivo imaging of siRNA via positron emission tomography.

To increase nuclease resistance of siRNAs, several modifications of the nucleobase, the ribose, or the phosphodiester backbone have been investigated [119, 120]. Besides other nonionic backbone modifications [121], CuAAC-generated triazole linkages between peptide nucleic acid derived nucleotide analogs have been explored in the sense and antisense strand of siRNA duplices [122]. A triazole linkage close to the 5' end of the sense strand reduced the siRNA potency, while incorporation toward the 3' end enhanced the silencing effect. In general, the special triazole linkage in place of a phosphodiester backbone was compatible with the RNAi machinery, even at internal positions (within the base-paired region) of the siRNA duplex that is generally more sensitive to modifications than the outer regions.

Several strategies have been explored to increase the cellular uptake and to improve pharmacokinetic and pharmacodynamic properties of siRNA as therapeutic agents [123, 124]. Diverse approaches, including formulations using nanoparticles, liposomes, or polycationic polymers are currently used for delivery of siRNAs, but the direct covalent conjugation of lipophilic molecules or recognition elements to RNA strands is gaining increasing importance. Manoharan, Rajeev, and coworkers [69] reported the site-specific conjugation of alkyne-modified siRNAs via CuAAC with azide-functionalized ligands, including lipophilic alkyl chains, cholesterol, oligoamine, and carbohydrate derivatives. Most efficient click conjugation was achieved by microwave-assisted CuAAC between azido-functionalized ligands and fully protected solid-support-bound alkyne-modified RNAs (carrying the alkynes at terminal or internal ribose residues synthesized, e.g. via phosphoramidites (4) or (5) (Fig. 4.5C,D)). Examination of RNAi activity in HeLa cells revealed high efficiency of lipophilic siRNA conjugates for silencing of a luciferase reporter gene.

Recently, CuAAC has been used to conjugate cell-type-specific ligands to siRNAs, in order to explore receptor-mediated endocytosis as an alternative delivery strategy. Carell and coworkers [125] synthesized azide-modified anandamide, folate, and cholesterol derivatives for conjugation to alkyne-modified siRNAs (via C5 position of a 3'-terminal deoxyuridine residue). Anandamide targets the cannabinoid receptor that is present in human immune B cells and is overexpressed in certain cancer cells. The uptake of folate-modified siRNAs was studied with HeLa cancer cells, which overexpress the folate receptor. Receptor-mediated uptake was monitored by confocal microscopy (using fluorescein-labeled antisense strands), and the potent RNAi effect was demonstrated using a dual luciferase reporter assay. The conjugation of receptor-specific ligands is promising for delivery of siRNAs into difficult-to-transfect cell types, such as immune cells, or to address neuronal uptake.

4.5 Summary and conclusions

In summary, bioorthogonal conjugation reactions have revolutionized chemical biology based on the unprecedented selectivity for reactions on biomolecules in complex systems. For tagging of nucleic acids, click reactions span a broad range of chemistries and provide promising tools for an expanding range of applications. Most important are strategies for labeling of RNA with fluorophores or isotope probes for spectroscopy or imaging. As superresolution microscopy and single-molecule fluorescence spectroscopy are becoming increasingly popular, click chemistry approaches for fast and reliable labeling of RNA are of tremendous importance. Although click chemistry applications on DNA have already gained invaluable weight in nucleic acid chemical biology, applications for RNA are still lagging behind. The majority of click chemistry reports on RNA so far made use of CuAAC, but other fast covalent bond formations under metal-free conditions will soon gain increasing importance, mostly because issues of toxicity and other undesired side effects of copper-catalyzed reactions can be circumvented. However, several challenges are associated with strain-promoted fast click reactions, which often include limited solubility of the conjugation reagents in aqueous media, and/or limited stability of the reactive species. These difficulties are met for example with tetrazines used in IEDDA reactions, or for in situ generated dipoles that react with strained alkenes or alkynes. Other issues concern the regiochemistry and stereoselectivity of the conjugation reactions. While CuAAC is known to selectively produce 1,4-substituted triazoles, the situation may not be as clear with other types of mentioned conjugation chemistries (see Section 4.2). Using mutually orthogonal reactions, bioconjugation strategies may be employed simultaneously in systems of increasing complexity. For these advancements, there remains a need for new reagents with fewer side reactions and increased efficiencies. Nevertheless, bioorthogonal chemistry has added and will continue to add exquisite tools to examine and control biological events in defined and intact cellular systems. As enabling technology, bioorthogonal click reactions on RNA will advance future chemical biology research to precisely measure, detect, and perturb nucleic acid structures and their functions with spatial and temporal resolution.

4.6 Acknowledgments

Research in the nucleic acid chemistry group is generously supported by the Max Planck Society. We gratefully acknowledge support by an IMPRS Molecular Biology fellowship to F.J.-Z., a GGNB fellowship (DFG Grant GSC 226/1) to L.B., and by the

Cluster of Excellence and DFG Research Center Nanoscale Microscopy and Molecular Physiology of the Brain (CNMPB).

References

1. Kolb HC, Finn MG, Sharpless KB. Click chemistry: diverse chemical function from a few good reactions. Angew Chem Int Ed. 2001;40:2004–21.
2. Sletten EM, Bertozzi CR. Bioorthogonal chemistry: fishing for selectivity in a sea of functionality. Angew Chem Int Ed. 2009;48:6974–98.
3. Saxon E, Bertozzi CR. Cell surface engineering by a modified Staudinger reaction. Science. 2000;287:2007–10.
4. Jewett JC, Bertozzi CR. Cu-free click cycloaddition reactions in chemical biology. Chem Soc Rev. 2010;39:1272–9.
5. Debets MF, van Berkel SS, Dommerholt J, Dirks AT, Rutjes FP, van Delft FL. Bioconjugation with strained alkenes and alkynes. Acc Chem Res. 2011;44:805–15.
6. Tornoe CW, Christensen C, Meldal M. Peptidotriazoles on solid phase: [1,2,3]-triazoles by regiospecific copper(I)-catalyzed 1,3-dipolar cycloadditions of terminal alkynes to azides. J Org Chem. 2002;67:3057–64.
7. Rostovtsev VV, Green LG, Fokin VV, Sharpless KB. A stepwise Huisgen cycloaddition process: copper(I)-catalyzed regioselective "ligation" of azides and terminal alkynes. Angew Chem Int Ed. 2002;41:2596–9.
8. El-Sagheer AH, Brown T. Click chemistry with DNA. Chem Soc Rev. 2010;39:1388–405.
9. Amblard F, Cho JH, Schinazi RF. Cu(I)-catalyzed Huisgen azide-alkyne 1, 3-dipolar cycloaddition reaction in nucleoside, nucleotide, and oligonucleotide chemistry. Chem Rev. 2009;109: 4207–20.
10. Liang LY, Astruc D. The copper(I)-catalyzed alkyne-azide cycloaddition (CuAAC) "click" reaction and its applications. An overview. Coord Chem Rev. 2011;255:2933–45.
11. He XP, Xie J, Tang Y, Li J, Chen GR. CuAAC click chemistry accelerates the discovery of novel chemical scaffolds as promising protein tyrosine phosphatases inhibitors. Curr Med Chem. 2012;19:2399–405.
12. Chan TR, Hilgraf R, Sharpless KB, Fokin VV. Polytriazoles as copper(I)-stabilizing ligands in catalysis. Org Lett. 2004;6:2853–5.
13. Presolski SI, Hong V, Cho SH, Finn MG. Tailored ligand acceleration of the Cu-catalyzed azide-alkyne cycloaddition reaction: practical and mechanistic implications. J Am Chem Soc. 2010;132:14570–6.
14. Besanceney-Webler C, Jiang H, Zheng T, et al. Increasing the efficacy of bioorthogonal click reactions for bioconjugation: a comparative study. Angew Chem Int Ed. 2011;50:8051–6.
15. Paredes E, Das SR. Click chemistry for rapid labeling and ligation of RNA. ChemBioChem. 2011;12:125–31.
16. Paredes E, Das SR. Optimization of acetonitrile co-solvent and copper stoichiometry for pseudo-ligandless click chemistry with nucleic acids. Bioorg Med Chem Lett. 2012;22:5313–6.
17. Hong V, Steinmetz NF, Manchester M, Finn MG. Labeling live cells by copper-catalyzed alkyne – azide click chemistry. Bioconjug Chem. 2010;21:1912–6.
18. Uttamapinant C, Tangpeerachaikul A, Grecian S, et al. Fast, cell-compatible click chemistry with copper-chelating azides for biomolecular labeling. Angew Chem Int Ed. 2012;51:5852–6.
19. Brotherton WS, Michaels HA, Simmons JT, Clark RJ, Dalal NS, Zhu L. Apparent copper(II)-accelerated azide-alkyne cycloaddition. Org Lett. 2009;11:4954–7.

20. Kuang GC, Michaels HA, Simmons JT, Clark RJ, Zhu L. Chelation-assisted, copper(II)-acetate-accelerated azide-alkyne cycloaddition. J Org Chem. 2010;75:6540–8.
21. Kuang GC, Guha PM, Brotherton WS, et al. Experimental investigation on the mechanism of chelation-assisted, copper(II) acetate-accelerated azide-alkyne cycloaddition. J Am Chem Soc. 2011;133:13984–4001.
22. Agard NJ, Prescher JA, Bertozzi CR. A strain-promoted [3 + 2] azide-alkyne cycloaddition for covalent modification of biomolecules in living systems. J Am Chem Soc. 2004;126:15046–7.
23. Jewett JC, Sletten EM, Bertozzi CR. Rapid Cu-free click chemistry with readily synthesized biaryl-azacyclooctynones. J Am Chem Soc. 2010;132:3688–90.
24. Agard NJ, Baskin JM, Prescher JA, Lo A, Bertozzi CR. A comparative study of bioorthogonal reactions with azides. ACS Chem Biol. 2006;1:644–8.
25. Ning X, Guo J, Wolfert MA, Boons GJ. Visualizing metabolically labeled glycoconjugates of living cells by copper-free and fast huisgen cycloadditions. Angew Chem Int Ed. 2008;47:2253–5.
26. Debets MF, van Berkel SS, Schoffelen S, Rutjes FP, van Hest JC, van Delft FL. Aza-dibenzocyclooctynes for fast and efficient enzyme PEGylation via copper-free (3+2) cycloaddition. Chem Commun. 2010;46:97–9.
27. Dommerholt J, Schmidt S, Temming R, et al. Readily accessible bicyclononynes for bioorthogonal labeling and three-dimensional imaging of living cells. Angew Chem Int Ed. 2010;49:9422–5.
28. Liang Y, Mackey JL, Lopez SA, Liu F, Houk KN. Control and design of mutual orthogonality in bioorthogonal cycloadditions. J Am Chem Soc. 2012;134:17904–7.
29. Baskin JM, Prescher JA, Laughlin ST, et al. Copper-free click chemistry for dynamic in vivo imaging. Proc Natl Acad Sci U S A. 2007;104:16793–7.
30. Baskin JM, Dehnert KW, Laughlin ST, Amacher SL, Bertozzi CR. Visualizing enveloping layer glycans during zebrafish early embryogenesis. Proc Natl Acad Sci U S A. 2010;107:10360–5.
31. Beatty KE, Fisk JD, Smart BP, et al. Live-cell imaging of cellular proteins by a strain-promoted azide-alkyne cycloaddition. ChemBioChem. 2010;11:2092–5.
32. Chang PV, Prescher JA, Sletten EM, et al. Copper-free click chemistry in living animals. Proc Natl Acad Sci U S A. 2010;107:1821–6.
33. Shelbourne M, Chen X, Brown T, El-Sagheer AH. Fast copper-free click DNA ligation by the ring-strain promoted alkyne-azide cycloaddition reaction. Chem Commun. 2011;47:6257–9.
34. Jayaprakash KN, Peng CG, Butler D, et al. Non-nucleoside building blocks for copper-assisted and copper-free click chemistry for the efficient synthesis of RNA conjugates. Org Lett. 2010;12:5410–3.
35. Singh I, Freeman C, Madder A, Vyle JS, Heaney F. Fast RNA conjugations on solid phase by strain-promoted cycloadditions. Org Biomol Chem. 2012;10:6633–9.
36. Winz ML, Samanta A, Benzinger D, Jäschke A. Site-specific terminal and internal labeling of RNA by poly(A) polymerase tailing and copper-catalyzed or copper-free strain-promoted click chemistry. Nucleic Acids Res. 2012;40:e78.
37. Shelbourne M, Brown T, El-Sagheer AH. Fast and efficient DNA crosslinking and multiple orthogonal labelling by copper-free click chemistry. Chem Commun. 2012;48:11184–6.
38. Lutz JF. Copper-free azide-alkyne cycloadditions: new insights and perspectives. Angew Chem Int Ed. 2008;47:2182–4.
39. Lallana E, Riguera R, Fernandez-Megia E. Reliable and efficient procedures for the conjugation of biomolecules through Huisgen azide-alkyne cycloadditions. Angew Chem Int Ed. 2011;50:8794–804.
40. Best MD. Click chemistry and bioorthogonal reactions: unprecedented selectivity in the labeling of biological molecules. Biochemistry. 2009;48:6571–84.
41. Debets MF, van der Doelen CW, Rutjes FP, van Delft FL. Azide: a unique dipole for metal-free bioorthogonal ligations. ChemBioChem. 2010;11:1168–84.

42. Huisgen R. 1.3-dipolar cycloadditions – past and future. Angew Chem Int Ed. 1963;2:565–632.
43. Singh I, Vyle JS, Heaney F. Fast, copper-free click chemistry: a convenient solid-phase approach to oligonucleotide conjugation. Chem Commun. 2009;45:3276–8.
44. Gutsmiedl K, Wirges CT, Ehmke V, Carell T. Copper-free "click" modification of DNA via nitrile oxide-norbornene 1,3-dipolar cycloaddition. Org Lett. 2009;11:2405–8.
45. McKay CS, Moran J, Pezacki JP. Nitrones as dipoles for rapid strain-promoted 1,3-dipolar cycloadditions with cyclooctynes. Chem Commun. 2010;46:931–3.
46. McKay CS, Blake JA, Cheng J, Danielson DC, Pezacki JP. Strain-promoted cycloadditions of cyclic nitrones with cyclooctynes for labeling human cancer cells. Chem Commun. 2011;47:10040–2.
47. Ning X, Temming RP, Dommerholt J, et al. Protein modification by strain-promoted alkyne-nitrone cycloaddition. Angew Chem Int Ed. 2010;49:3065–8.
48. Poloukhtine AA, Mbua NE, Wolfert MA, Boons GJ, Popik VV. Selective labeling of living cells by a photo-triggered click reaction. J Am Chem Soc. 2009;131:15769–76.
49. Clovis JS, Eckell A, Huisgen R, Sustmann R. 1.3-Dipolare Cycloadditionen. 25. Der Nachweis Des Freien Diphenylnitrilimins Als Zwischenstufe Bei Cycloadditionen. Chem Ber. 1967;100:60–70.
50. Song W, Wang Y, Qu J, Lin Q. Selective functionalization of a genetically encoded alkene-containing protein via "photoclick chemistry" in bacterial cells. J Am Chem Soc. 2008;130:9654–5.
51. Song W, Wang Y, Qu J, Madden MM, Lin Q. A photoinducible 1,3-dipolar cycloaddition reaction for rapid, selective modification of tetrazole-containing proteins. Angew Chem Int Ed. 2008;47:2832–5.
52. Lim RK, Lin Q. Photoinducible bioorthogonal chemistry: a spatiotemporally controllable tool to visualize and perturb proteins in live cells. Acc Chem Res. 2011;44:828–39.
53. Blackman ML, Royzen M, Fox JM. Tetrazine ligation: fast bioconjugation based on inverse-electron-demand Diels-Alder reactivity. J Am Chem Soc. 2008;130:13518–9.
54. Lang K, Davis L, Torres-Kolbus J, Chou C, Deiters A, Chin JW. Genetically encoded norbornene directs site-specific cellular protein labelling via a rapid bioorthogonal reaction. Nat Chem. 2012;4:298–304.
55. Lang K, Davis L, Wallace S, et al. Genetic Encoding of bicyclononynes and trans-cyclooctenes for site-specific protein labeling in vitro and in live mammalian cells via rapid fluorogenic Diels-Alder reactions. J Am Chem Soc. 2012;134:10317–20.
56. Borrmann A, Milles S, Plass T, et al. Genetic encoding of a bicyclo[6.1.0]nonyne-charged amino acid enables fast cellular protein imaging by metal-free ligation. ChemBioChem. 2012;13:2094–9.
57. Plass T, Milles S, Koehler C, et al. Amino acids for Diels-Alder reactions in living cells. Angew Chem Int Ed. 2012;51:4166–70.
58. Schoch J, Wiessler M, Jäschke A. Post-synthetic modification of DNA by inverse-electron-demand Diels-Alder reaction. J Am Chem Soc. 2010;132:8846–7.
59. Schoch J, Staudt M, Samanta A, Wiessler M, Jäschke A. Site-specific one-pot dual labeling of DNA by orthogonal cycloaddition chemistry. Bioconjug Chem. 2012;23:1382–6.
60. Schoch J, Ameta S, Jäschke A. Inverse electron-demand Diels-Alder reactions for the selective and efficient labeling of RNA. Chem Commun. 2011;47:12536–7.
61. Staudinger H, Meyer J. On new organic phosphorus bonding. III. Phosphine methylene derivatives and phosphinimine. Helv Chim Acta. 1919;2:635–46.
62. Kohn M, Breinbauer R. The Staudinger ligation – a gift to chemical biology. Angew Chem Int Ed. 2004;43:3106–16.
63. Prescher JA, Dube DH, Bertozzi CR. Chemical remodelling of cell surfaces in living animals. Nature. 2004;430:873–7.

64. Pianowski Z, Gorska K, Oswald L, Merten CA, Winssinger N. Imaging of mRNA in live cells using nucleic acid-templated reduction of azidorhodamine probes. J Am Chem Soc. 2009;131:6492–7.
65. van Berkel SS, van Eldijk MB, van Hest JC. Staudinger ligation as a method for bioconjugation. Angew Chem Int Ed. 2011;50:8806–27.
66. Weisbrod SH, Marx A. Novel strategies for the site-specific covalent labelling of nucleic acids. Chem Commun. 2008:5675–85.
67. van Delft P, Meeuwenoord NJ, Hoogendoorn S, et al. Synthesis of oligoribonucleic acid conjugates using a cyclooctyne phosphoramidite. Org Lett. 2010;12:5486–9.
68. See, for example, http://www.jenabioscience.com, http://www.baseclick.eu, http://www.berryassoc.com, and http://www.chemgenes.com. Accessed October 2012.
69. Yamada T, Peng CG, Matsuda S, et al. Versatile site-specific conjugation of small molecules to siRNA using click chemistry. J Org Chem. 2011;76:1198–211.
70. Peacock H, Maydanovych O, Beal PA. N(2)-Modified 2-aminopurine ribonucleosides as minor-groove-modulating adenosine replacements in duplex RNA. Org Lett. 2010;12:1044–7.
71. Seidu-Larry S, Krieg B, Hirsch M, Helm M, Domingo O. A modified guanosine phosphoramidite for click functionalization of RNA on the sugar edge. Chem Commun. 2012;48:11014–6.
72. Krishna H, Caruthers MH. Alkynyl phosphonate DNA: a versatile "click"able backbone for DNA-based biological applications. J Am Chem Soc. 2012;134:11618–31.
73. Kiviniemi A, Virta P, Lönnberg H. Utilization of intrachain 4'-C-azidomethylthymidine for preparation of oligodeoxyribonucleotide conjugates by click chemistry in solution and on a solid support. Bioconjug Chem. 2008;19:1726–34.
74. Aigner M, Hartl M, Fauster K, Steger J, Bister K, Micura R. Chemical synthesis of site-specifically 2'-azido-modified RNA and potential applications for bioconjugation and RNA interference. ChemBioChem. 2011;12:47–51.
75. Fauster K, Hartl M, Santner T, et al. 2'-Azido RNA, a versatile tool for chemical biology: synthesis, X-ray structure, siRNA applications, click labeling. ACS Chem Biol. 2012;7:581–9.
76. El-Sagheer AH, Brown T. New strategy for the synthesis of chemically modified RNA constructs exemplified by hairpin and hammerhead ribozymes. Proc Natl Acad Sci U S A. 2010;107:15329–34.
77. Kellner S, Seidu-Larry S, Burhenne J, Motorin Y, Helm M. A multifunctional bioconjugate module for versatile photoaffinity labeling and click chemistry of RNA. Nucleic Acids Res. 2011;39:7348–60.
78. Wachowius F, Höbartner C. Chemical RNA modifications for studies of RNA structure and dynamics. ChemBioChem. 2010;11:469–80.
79. Phelps K, Morris A, Beal PA. Novel modifications in RNA. ACS Chem Biol. 2012;7:100–9.
80. Wachowius F, Höbartner C. Chemical synthesis of modified RNA. In: Mayer G. The Chemical Biology of Nucleic Acids. Chichester John Wiley & Sons; 2010:1–37.
81. Rublack N, Nguyen H, Appel B, Springstubbe D, Strohbach D, Müller S. Synthesis of specifically modified oligonucleotides for application in structural and functional analysis of RNA. J Nucleic Acids. 2011;2011:805253.
82. Kocalka P, El-Sagheer AH, Brown T. Rapid and efficient DNA strand cross-linking by click chemistry. ChemBioChem. 2008;9:1280–5.
83. Miller GP, Kool ET. Versatile 5'-functionalization of oligonucleotides on solid support: amines, azides, thiols, and thioethers via phosphorus chemistry. J Org Chem. 2004;69:2404–10.
84. Onizuka K, Shibata A, Taniguchi Y, Sasaki S. Pin-point chemical modification of RNA with diverse molecules through the functionality transfer reaction and the copper-catalyzed azide-alkyne cycloaddition reaction. Chem Commun. 2011;47:5004–6.
85. Onizuka K, Taniguchi Y, Sasaki S. Site-specific covalent modification of RNA guided by functionality-transfer oligodeoxynucleotides. Bioconjug Chem. 2009;20:799–803.

86. Onizuka K, Taniguchi Y, Sasaki S. A new usage of functionalized oligodeoxynucleotide probe for site-specific modification of a guanine base within RNA. Nucleic Acids Res. 2010;38:1760–6.
87. Onizuka K, Taniguchi Y, Sasaki S. Activation and alteration of base selectivity by metal cations in the functionality-transfer reaction for RNA modification. Bioconjug Chem. 2010;21:1508–12.
88. Jao CY, Salic A. Exploring RNA transcription and turnover in vivo by using click chemistry. Proc Natl Acad Sci U S A. 2008;105:15779–84.
89. Salic A, Mitchison TJ. A chemical method for fast and sensitive detection of DNA synthesis in vivo. Proc Natl Acad Sci U S A. 2008;105:2415–20.
90. Neef AB, Luedtke NW. Dynamic metabolic labeling of DNA in vivo with arabinosyl nucleosides. Proc Natl Acad Sci U S A. 2011;108:20404–9.
91. Grammel M, Hang H, Conrad NK. Chemical reporters for monitoring RNA synthesis and poly(A) tail dynamics. ChemBioChem. 2012;13:1112–5.
92. Hocek M, Fojta M. Cross-coupling reactions of nucleoside triphosphates followed by polymerase incorporation. Construction and applications of base-functionalized nucleic acids. Org Biomol Chem. 2008;6:2233–41.
93. Mayer G. The chemical biology of aptamers. Angew Chem Int Ed. 2009;48:2672–89.
94. Rao H, Sawant AA, Tanpure AA, Srivatsan SG. Posttranscriptional chemical functionalization of azide-modified oligoribonucleotides by bioorthogonal click and Staudinger reactions. Chem Commun. 2012;48:498–500.
95. Rao H, Tanpure AA, Sawant AA, Srivatsan SG. Enzymatic incorporation of an azide-modified UTP analog into oligoribonucleotides for post-transcriptional chemical functionalization. Nat Protoc. 2012;7:1097–112.
96. Weisbrod SH, Marx A. A nucleoside triphosphate for site-specific labelling of DNA by the Staudinger ligation. Chem Commun. 2007;43:1828–30.
97. Gierlich J, Gutsmiedl K, Gramlich PM, Schmidt A, Burley GA, Carell T. Synthesis of highly modified DNA by a combination of PCR with alkyne-bearing triphosphates and click chemistry. Chemistry. 2007;13:9486–94.
98. Martin G, Keller W. Tailing and 3'-end labeling of RNA with yeast poly(A) polymerase and various nucleotides. RNA. 1998;4:226–30.
99. Motorin Y, Helm M. RNA nucleotide methylation. Wiley Interdiscip Rev RNA 2011;2:611–31.
100. Pljevaljcic G, Pignot M, Weinhold E. Design of a new fluorescent cofactor for DNA methyltransferases and sequence-specific labeling of DNA. J Am Chem Soc. 2003;125:3486–92.
101. Dalhoff C, Lukinavicius G, Klimašauskas S, Weinhold E. Direct transfer of extended groups from synthetic cofactors by DNA methyltransferases. Nat Chem Biol. 2006;2:31–2.
102. Pljevaljcic G, Schmidt F, Weinhold E. Sequence-specific methyltransferase-induced labeling of DNA (SMILing DNA). ChemBioChem. 2004;5:265–9.
103. Lukinavicius G, Lapiene V, Stasevskij Z, Dalhoff C, Weinhold E, Klimašauskas S. Targeted labeling of DNA by methyltransferase-directed transfer of activated groups (mTAG). J Am Chem Soc. 2007;129:2758–9.
104. Weller RL, Rajski SR. DNA methyltransferase-moderated click chemistry. Org Lett. 2005;7:2141–4.
105. Motorin Y, Burhenne J, Teimer R, et al. Expanding the chemical scope of RNA: methyltransferases to site-specific alkynylation of RNA for click labeling. Nucleic Acids Res. 2011;39:1943–52.
106. Ge J, Liu H, Yu YT. Regulation of pre-mRNA splicing in *Xenopus* oocytes by targeted 2'-O-methylation. RNA. 2010;16:1078–85.
107. Tomkuviene M, Clouet-d'Orval B, Cerniauskas I, Weinhold E, Klimašauskas S. Programmable sequence-specific click-labeling of RNA using archaeal box C/D RNP methyltransferases. Nucleic Acids Res. 2012;40:6765–73.

108. Willnow S, Martin M, Luscher B, Weinhold E. A selenium-based click AdoMet analogue for versatile substrate labeling with wild-type protein methyltransferases. ChemBioChem. 2012;13:1167–73.
109. Bothwell IR, Islam K, Chen Y, et al. Se-adenosyl-L-selenomethionine cofactor analogue as a reporter of protein methylation. J Am Chem Soc. 2012;134:14905–12.
110. Jao CY, Salic A. Exploring RNA transcription and turnover in vivo by using click chemistry. Proc Natl Acad Sci U S A. 2008;105:15779–84.
111. Hagemeijer MC, Vonk AM, Monastyrska I, Rottier PJ, de Haan CA. Visualizing coronavirus RNA synthesis in time by using click chemistry. J Virol. 2012;86:5808–16.
112. Eulalio A, Frohlich KS, Mano M, Giacca M, Vogel J. A candidate approach implicates the secreted *Salmonella* effector protein SpvB in P-body disassembly. PLoS One. 2011;6:e17296.
113. For example, Click-iT® RNA Alexa Fluor Imaging Kits from Invitrogen.
114. Rabani M, Levin JZ, Fan L, et al. Metabolic labeling of RNA uncovers principles of RNA production and degradation dynamics in mammalian cells. Nat Biotechnol. 2011;29:436–42.
115. Opazo F, Levy M, Byrom M, et al. Aptamers as potential tools for super-resolution microscopy. Nat Methods. 2012;9:938–9.
116. Ries J, Kaplan C, Platonova E, Eghlidi H, Ewers H. A simple, versatile method for GFP-based super-resolution microscopy via nanobodies. Nat Methods. 2012;9:582–4.
117. Weil TT, Parton RM, Davis I. Making the message clear: visualizing mRNA localization. Trends Cell Biol. 2010;20:380–90.
118. Mercier F, Paris J, Kaisin G, et al. General method for labeling siRNA by click chemistry with fluorine-18 for the purpose of PET imaging. Bioconjug Chem. 2011;22:108–14.
119. Amarzguioui M, Holen T, Babaie E, Prydz H. Tolerance for mutations and chemical modifications in a siRNA. Nucleic Acids Res. 2003;31:589–95.
120. Behlke MA. Chemical modification of siRNAs for in vivo use. Oligonucleotides. 2008;18:305–19.
121. Iwase R, Toyama T, Nishimori K. Solid-phase synthesis of modified RNAs containing amide-linked oligoribonucleosides at their 3'-end and their application to siRNA. Nucleosides Nucleotides Nucleic Acids. 2007;26:1451–4.
122. Efthymiou TC, Huynh V, Oentoro J, Peel B, Desaulniers JP. Efficient synthesis and cell-based silencing activity of siRNAS that contain triazole backbone linkages. Bioorg Med Chem Lett. 2012;22:1722–6.
123. Manoharan M. Oligonucleotide conjugates as potential antisense drugs with improved uptake, biodistribution, targeted delivery, and mechanism of action. Antisense Nucleic Acid Drug Dev. 2002;12:103–28.
124. Tiemann K, Rossi JJ. RNAi-based therapeutics-current status, challenges and prospects. EMBO Mol Med. 2009;1:142–51.
125. Willibald J, Harder J, Sparrer K, Conzelmann KK, Carell T. Click-modified anandamide siRNA enables delivery and gene silencing in neuronal and immune cells. J Am Chem Soc. 2012;134:12330–3.

David M. J. Lilley
5 Analysis of RNA conformation using comparative gel electrophoresis

Electrophoresis in a polyacrylamide gel matrix has been exploited to provide relative angular information about the disposition of helical arms about a branch point in nucleic acid junctions. Although originally applied to DNA junctions [1–3], it has been extensively used to study junctions in RNA molecules. The method is very simple and requires inexpensive apparatuses. However, a drawback of the method is that electrophoresis is not underpinned by a good theoretical understanding compared to other structural methods, so that the conclusions that emerge are seldom quantitative. While the symmetry of a given junction may be determined, its interaxial angles will not be measured beyond a relative sense. Nevertheless, comparison of the angles can frequently provide enough information to give a good description of the structure. This comparative gel electrophoresis method is very frequently performed in parallel with other biophysical methods, in particular the use of fluorescence resonance energy transfer (FRET, see Chapter 8), which provides highly complementary information. The electrophoretic method can also sometimes resolve structural ambiguities that other biophysical methods may leave.

5.1 The principle behind the analysis of the structure of branched nucleic acids by gel electrophoresis

Most theoretical treatments of the gel electrophoresis of nucleic acids are based on de Gennes's [4] concept of reptation. The DNA or RNA is considered to migrate rather in the manner of a snake through a "tube" formed by the polymer matrix of the gel. The force of the electric field pulling the nucleic acid through the gel is balanced by the retardation due to friction. Lumpkin and Zimm [5] derived the equation of motion of the nucleic acid along a gel tube as

$$\mu = \frac{Q}{\xi} \left\langle \frac{h_x^2}{L^2} \right\rangle \tag{5.1}$$

where μ is the mobility of the nucleic acid chain, Q is the effective charge, ξ is the frictional coefficient, h_x is the end-to-end length in the direction of the electric field, and L is the contour length. The presence of a kink or bend in a double-stranded DNA or RNA molecule will reduce the end-to-end distance, and thus its electrophoretic mobility will be lowered. The kinked species will be retarded relative to unkinked species. This relatively simple theory has been successfully applied to the analysis of kinked nucleic acid geometry. The method is illustrated by comparing a series of otherwise identical RNA duplexes that are kinked because of a bulge comprising an increasing number of unopposed adenine nucleotides, which has been shown by FRET experiments to give progressive kinking as the bulge size increases [6]. In the electrophoretic analysis, as the number of adenines increase the migration is more and more retarded (Fig. 5.1).

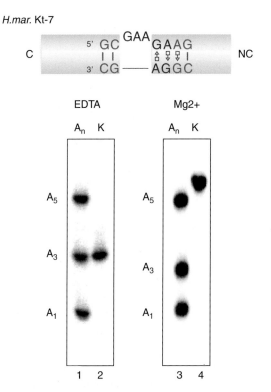

Fig. 5.1 Electrophoretic retardation of oligoadenine bulges and a k-turn motif in duplex RNA. For these experiments, the overall length of the double-stranded species and sequence context are carefully matched in each case. The central sequence of *Haloarcular marismortui* Kt-7 is shown at the top. It comprises a three-base bulge followed by critical G · A and A · G base pairs. The conventionally base paired helix is termed the C (canonical) helix, while that with the G · A pairs is termed the NC (noncanonical) helix. The lower panels show the result of electrophoresis in 15% polyacrylamide gels of a mixture of duplexes with central A_1, A_3, and A_5 bulges (tracks headed A_n) alongside a matched duplex containing the *H. marismortui* Kt-7 sequence (tracks headed K). The left-side gel was electrophoresed in 90 mM Tris-borate (pH 8.3) buffer containing 2 mM ethylenediaminetetraacetic acid (EDTA), while that on the right was run in the same buffer containing 500 µM Mg^{2+} ions in place of EDTA. Note the progressive retarded mobility with the increase in bulge size. The k-turn RNA migrates alongside the A_3 bulge in the absence of metal ions but migrates more slowly than the A_5 bulge in the presence of Mg^{2+} ions.

The theory has been refined [7–12] to include nucleic acid elasticity, for example [13]. There is no well-developed theory at the present time for nucleic acids containing helical junctions, although there have been some attempts to incorporate this [14,15]. However, most electrophoretic data on junctions are analyzed empirically. The general rule that greater kinking results in lower mobility works well in practice, unless bending is so severe that helices clash (an example of this was found in our laboratory some years ago) [16]. At its heart, the conclusions emerge from the interpretation of relative mobilities of carefully matched species.

5.2 Helical discontinuities in duplex RNA

A specific sequence or structural feature located in a specific point in a duplex may introduce a discrete bend or kink. This might be intrinsic, or result upon the binding of a particular protein. In principle, the element might be a point of flexibility, or it could be a precise axial bend that is relatively rigid.

An example of the latter is provided by the kink turn (k-turn) RNA [17], which introduces a sharp kink with an included angle of ~50°. This is an extremely widespread motif in many functional RNA species, including both ribosomal subunits, the small nucleolar RNA (snoRNA) s involved in the site-specific methylation and pseudouridylation of archaeal and eukaryotic RNA [18–21], U4 small nuclear RNA (snRNA) involved in spliceosome assembly [22,23], and a number of riboswitches and ribozymes [24–27].

The structural consequences of a k-turn may be studied by placing the motif at the center of a long RNA duplex (although DNA extensions can be used for synthetic convenience) that is electrophoresed in a polyacrylamide gel under particular conditions of salt composition and so forth. Kinked species exhibit reduced electrophoretic mobility as the end-to-end distance is reduced, as expected from equation 5.1. The species under study could be run alongside other species such as a perfect duplex or a duplex with a central bulge kink for comparison. Sequence or functional group changes that affect the stability of the kinked geometry will be manifest as altered electrophoretic mobility. This is illustrated for the well-studied k-turn Kt-7 in Fig. 5.1, where its mobility is compared with a series of oligo-adenine bulges under two sets of conditions [16]. In the absence of metal ions, the k-turn migrates similarly to an A_3 bulge, as might be anticipated by inspection of its secondary structure. However, in the presence of ≥70 µM Mg^{2+} ions [28], the k-turn folds into a sharply kinked structure, with a mobility that is more retarded than an A_5 bulge under the same conditions. This folding transition has been confirmed by steady-state, time-resolved [16], and single-molecule [29] FRET measurements (see Chapter 8). And many crystal structures of k-turns show that the structure is kinked with an included angle of 50°, stabilized by A-minor interactions involving a number of specific hydrogen bonds in the center of the structure.

5.3 The direction of a helical bend

A given bend or kink in double-stranded RNA is characterized by the included angle and also its direction. The simple electrophoretic experiment may provide some insight into the magnitude of the bend, but no information on its direction. However, if two such bends are introduced into the same duplex molecule, then electrophoresis now becomes sensitive to their relative direction because this will generate a three-dimensional trajectory of the molecule, as first shown with A-tract bends in DNA [30].

The outer helical sections beyond the kinks will be related by a dihedral angle, which will be a function of the local geometry of the kink (i.e. the direction of the bend), the spacing between the two kinks, and the periodicity of the helix. The shape of the molecule, and thus its frictional properties and electrophoretic mobility, will vary systematically with the length of helix between the kinks. In these experiments (often termed *phasing experiments*) the overall length of the duplex can be kept constant by removing base pairs from the arms as the spacer is lengthened. If the kinks have a fixed direction of bending, the end-to-end distance should vary sinusoidally with the length of the helix length between the kinks, generating fastest mobility when the structure is *trans*-planar and slowest for the *cis*-planar form. The variation in mobility will be a progressively damped sine function, with a period equivalent to that of the spacer helix and a phase determined by the structure of the kinks. Such phasing experiments provided the first measurements of the periodicity of double-stranded RNA [31–33]. The sinusoidal modulation of mobility proves that component kinks have a particular direction and provides a sensitive way of revealing axial kinking when the extent of deviation is small; this was employed to reveal kinking at the human immunodeficiency virus trans activation response (TAR) element, for example [34]. Another example is provided by our study of the bending introduced by the highly asymmetric internal loop of the U1A box motif [35] (Fig. 5.2). The phasing method was also used to analyze kinking by the k-turn [16], showing that this motif had a precise direction, and to determine the handedness of the four-way junction from the hairpin ribozyme [36].

5.4 Comparative gel electrophoresis of branched nucleic acids

Helical branch points are created where three or more helices radiate from a center, connected by the covalent continuity of their component strands [37]. In DNA, a perfect four-way junction is the well-known Holliday junction that is the central intermediate of homologous genetic recombination important for the repair of double-strand breaks. In RNA, three-, four-, and five-way and even higher-order junctions abound. The junctions may be perfectly base paired, or more often elaborated by additional

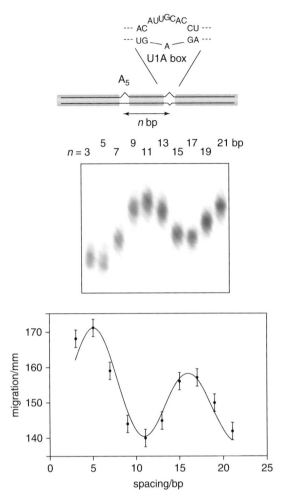

Fig. 5.2 Electrophoretic experiments using test the phase relationship between two bent RNA structures to measure the direction of kinking. Two kinks will generate a dihedral angle between the outer arms that depends on the number of base pairs in the helix between them. In such experiments, one kink is frequently a base bulge, while the other can be a bulge, loop, or junction. Modulation of the electrophoretic mobility as a function of central helix length indicates that the kink examined has a definite direction, which can be inferred from the phase. The asymmetric internal loop of the U1A messenger RNA untranslated region (UTR) box 2 is shown at the top. The box 2 loop comprises a single adenine opposing 7 nucleotides on the opposite strand and exhibits marked electrophoretic retardation when contained as a single motif in duplex RNA [35]. In this experiment, it has been placed in a 67 bp duplex, separated from an A_5 bulge by a spacer sequence 3–21 bp in length. These species were electrophoresed in a 15% polyacrylamide gel (center). The pronounced modulation of mobility with spacing clearly indicates that the UTR box bends with a fixed direction. The mobility has been plotted against spacer length and fitted to a model based on Lumpkin-Zimm theory assuming bends of fixed angle and dihedral angle for both the A_5 bulge and the U1A box 2 (lower).

formally unpaired bases at the point of strand exchange; these can be classified by a standard nomenclature [38], according to which the simple four-way junction is 4H.

The potential of gel electrophoresis for the analysis of the angular distribution of helical arms in a four-way junction was first suggested from studies of a pseudo-cruciform structure [1]. This showed that placing a four-way junction within a linear DNA fragment led to a marked reduction in electrophoretic mobility that was greater the more centrally located the junction. The electrophoretic retardation indicated that the axis of the DNA fragment was kinked by the presence of the junction. This approach was then extended to compare the angles subtended between all the helical pairs of a given junction [2,3] (Fig. 5.3). In a junction with n helical arms, there are $^nC_2$ combinations of the arms taken two at a time, where

$$^nC_2 = \frac{n!}{2!(n-2)!} \tag{5.2}$$

In the basic long-short arm method we compare the electrophoretic mobility of the $^nC_2$ species with two long arms (most typically 40 bp, but longer arms give greater sensitivity to angular disposition) and the remaining arms short (e.g. 15 bp). For a four-way junction (4H, where $n = 4$), there are six species to compare, each of which has two long and two short arms. A three-way junction (3H, $n = 3$) gives three species, each of which has a single short arm. The symmetry and pattern of the relative mobilities of the different species is highly informative on the global shape of the junction.

There are a variety of ways to generate the long-short arm species. Cooper and Hagerman [2] first generated their longer arms by ligation of extensions onto a core of a four-way junction in the six possible combinations. By contrast, Duckett et al. [3] assembled DNA four-way (4H) junctions from four oligonucleotides each of 80 bp (four arms of 40 bp) and shortened the required arms by cleavage with specific restriction enzymes. These methods are only really feasible for DNA and are not easily extended to the study of RNA junctions. However, with the ease of modern chemical synthesis it is now perfectly feasible to synthesize the required 16 oligonucleotides and then simply assemble the 6 long-short arm species by hybridization of the required 4 oligonucleotides for each. For RNA, these 16 species can also be generated by transcription. Where chemical synthesis is used, the outer segments are usually made of DNA for synthetic convenience.

The method may be well illustrated by the four-way junction (Fig. 5.3). Let us first take the structure as an open X-shape, where each of the arms is extended toward the corners of a square. We generate the six long-short arm species and name them according to the long arms, so that species BH has long B and H arms, and short R and X arms, for example. For the square structure, two quite different classes of species have been generated, depending on whether the long arms are adjacent (BH, BX, HR, and RX) or diametrically opposed (BR and HX) around the junction. Our previous

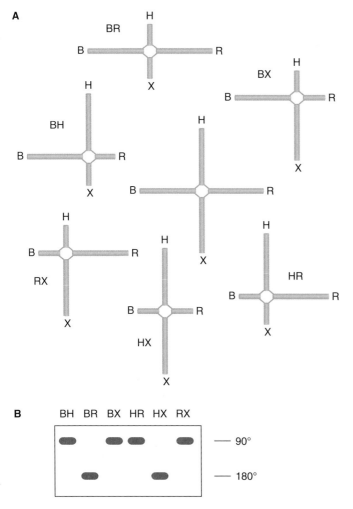

Fig. 5.3 The principle of the long-short arm method for the electrophoretic analysis of a four-way junction. (A) The six species can be considered to be derived from the central junction comprising four long arms. These are named sequentially B, H, R, and X as in our original studies of DNA junctions [3]. The long-short arm species are generated by shortening two helical arms, to give the six species shown, named by their long arms. (B) shows a simulation of the pattern of mobilities expected if the conformation were an open square. In that case, the BH, BX, HR, and RX species are slow (90° between the long arms), and BR and HX are fast (180° between the long arms)

discussion about the effect of kinking on electrophoretic mobility would lead us to expect that the quasi-linear species (BR and HX) would migrate faster than the remaining four species where the angles between the long arms are approximately 90°. So if the species are loaded in the order BH, BR, BX, HR, HX, and RX (i.e. sequentially around the junction), we would anticipate the pattern slow, fast, slow, slow, fast, slow. Note that very few assumptions go into the interpretation. All the conclusions are drawn from the relative mobilities of the multiple long-short arm species, and no calibration against other species is required.

5.5 Comparative gel electrophoresis of four-way DNA junctions

Although the four-way junction, or Holliday junction, is a structure exclusive to DNA, it provided the incentive on which much of the methodology of comparative gel electrophoretic was developed, and as it illustrates the general principles so well it will be considered here. When a four-way DNA junction was analyzed in the absence of divalent metal ions, exactly the pattern described in the previous section was found (Fig. 5.4, upper gel), from which we deduced that under these conditions the junction adopted an open-square conformation [3]. The open center was consistent with an observed reactivity of thymine bases at the point of strand exchange to electrophilic attack. The square structure was subsequently confirmed by ensemble FRET analysis [39–41] (see Chapter 8).

The comparative gel electrophoretic analysis next revealed that the structure of the four-way DNA junction depended strongly on the presence or absence of metal ions. It was found that the pattern of the six long-short arm species changed very significantly upon addition of Mg^{2+} ions [3]. An example for the well-studied sequence Junction 3 is shown in Fig. 5.4 (lower gel). There are now three distinct mobilities, and the pattern (with the samples applied to the gel in the same order as before) can be described at slow, intermediate, fast, fast, intermediate, slow. This indicates that the six interhelical angles fall into three classes, and that the symmetry has been lowered from the square structure. The pattern may be explained by the formation of the stacked X-structure, formed by coaxial helical alignment of helices in pairs (B with X and H with R), and a rotation between the two axes.

The lowering of symmetry to twofold divides the strands into two kinds; continuous strands that run down the length of the axes, and the exchanging strands that change axes at the junction. Earlier dogma had it that the continuous strands should run parallel, but if the assignment of the different species was correct, then the structure was in fact approximately antiparallel. Thus, the exchanging strands do not cross at the center. Analysis of the structure by FRET [39,40] and hydrodynamic methods [42] confirmed the antiparallel stacked X-structure in solution, and stereochemically acceptable models of the structure were constructed [43]. There were numerous attempts to get diffracting crystals of four-way DNA junctions, but this did not happen for many years. Eventually, a number of atomic resolution structures were obtained, although ironically in each case the successful laboratories were not trying to crystallize junctions. The first was a DNA-RNA hybrid in which one of the stacked

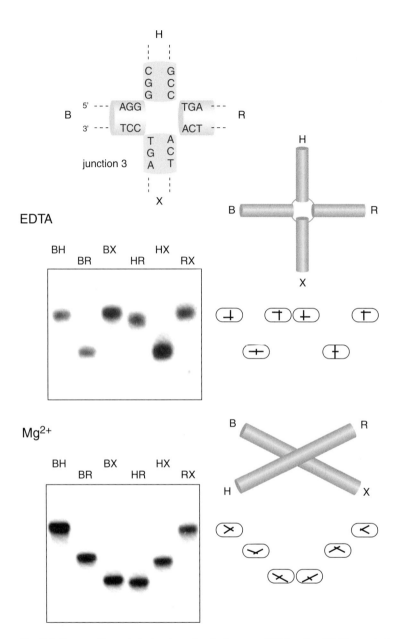

Fig. 5.4 Comparative gel electrophoresis of a four-way DNA junction [3]. The central sequence of junction 3 [3] is shown at the top, with the B, H, R, and X arms indicated. The six radioactively labeled, long-short arm species of junction 3 were electrophoresed in an polyacrylamide gel in 90 mM Tris-borate (pH 8.3) containing either 1 mM EDTA (upper gel) or 100 µM Mg^{2+} (lower gel). The slow-fast-slow-slow-fast-slow pattern of mobility in the presence of EDTA is interpreted in terms of the extended square structure shown (and the same as that shown in Fig. 5.3), giving the four species with 90° angles subtended between the long arms (slow species) and two with 180° angles (fast species). The interpretation is indicated by the schematics adjacent to the gel in this and

helical pairs was A-form in geometry [44]. The second was an all-DNA structure that contained a number of non-Watson-Crick base pairs [45]. Finally, the structure of a perfectly base-paired, all-DNA four-way structure was presented by Ho and coworkers [46]. The structure revealed all the expected features of the stacked X-structure, with perfect coaxial stacking of helices and an antiparallel conformation. Naturally, the atomic resolution structures provide much new detail that could not be obtained from lower resolution approaches, but the important overall structure was successfully solved by comparative gel electrophoresis, more than a decade sooner!

Comparative gel electrophoresis also revealed that when the central sequence was altered the junction could adopt equivalent alternative stacking conformers, which differed in the choice of stacking partner [3]; this was also later confirmed by FRET (see Chapter 8) for the same sequences [39]. This of course raised the interesting possibility that the two conformers might be in a dynamic exchange in free solution, biased one way or the other depending on central sequence. Enzymatic [47] and nuclear magnetic resonance (NMR, see Chapter 12) [48] experiments pointed toward a dynamic exchange between stacking conformers in solution. Comparative gel electrophoresis of one junction in particular was especially revealing. This junction, with a new central sequence, gave an electrophoretic pattern not previously observed, in which all six long-short arm species had virtually equal mobility [49]. This pattern would be consistent with a tetrahedral disposition of the four arms, yet that that seemed rather improbable from a structural perspective. A more plausible hypothesis is that the two stacking conformers were in rapid exchange compared to the rate of equilibration in the gel pores and equally populated. Confirmation of this had to await the advent of single-molecule FRET methods (see Chapter 8), but this did eventually provide a direct demonstration of conformer exchange in this junction [50]. But once again the simple polyacrylamide gel got there first by a number of years.

We also found that the conformer bias in DNA junctions was very sensitive to the charge on central phosphates, which could be manipulated by selective substitution by electrically neutral methyl phosphonates [51]. A novel form of the electrophoretic pattern was observed in the presence of Mg^{2+} ions in which each track contained two bands that could be interpreted in terms of the simultaneous presence of both alternative stacking conformers. Recovery of these individual species from the gel and detailed analysis revealed that the species differed in the chirality of the methyl phosphonate linkage, and these were assigned by ^{1}H-NMR experiments (see Chapter 12). We concluded that when the *pro*-S oxygen of a phosphate at the point of strand exchange is replaced by a methyl group, this destabilizes that conformer and thus

---

Fig. 5.4 (*Continued*)
subsequent figures. The pattern seen in the presence of Mg^{2+} ions is quite different; it is slow, intermediate, fast, fast, intermediate, slow. This is interpreted in terms of the stacked X-structure, giving two species with acute angles between the long arms (slow), two with obtuse angles (intermediate), and two with colinear long arms (fast).

alters the conformer bias such that the alternative form where the modified phosphate lies on a continuous strand is now stabilized. In the process of this analysis, we noted that the electrophoretic separation of four-way junctions with arms of unequal length could provide a powerful way of resolving the diastereomers of potentially very long DNA molecules containing single methyl phosphonate linkages. We subsequently exploited this to generate species used in mechanistic analysis of a junction-resolving enzyme [52].

We have also used comparative gel electrophoresis on the complexes of DNA junctions specifically bound by proteins. The protein is bound to the conventional set of species with two long arms in the different combinations and electrophoresed side by side as previously. The complexes are significantly retarded in the gel, giving good separation between free DNA and the complexes. The different long-short arm species of the complexes may then be compared and the pattern of mobilities analyzed just as for the free junctions. We have applied this method to a number of junction-resolving enzymes, nucleases that are selective for the structure of four-way junctions [53], required to resolve the junctions into duplex species as the penultimate step of recombination. We found that the enzymes Cce1 of yeast mitochondrial and the phage enzymes T4 endonuclease VII and T7 endonuclease I all generate new electrophoretic patterns, showing that the binding of the proteins distorts the structure of the junction. The pattern for Cce1 was similar to that of free junction in the absence of metal ions, suggesting an opening into a square, unstacked structure [54]. The opening was supported by the chemical reactivity of thymine bases located at the point of strand exchange and by the enhanced fluorescence of 2-aminopurine bases incorporated into the junction center [55]. T4 endonuclease VII [56] and T7 endonuclease I [57] each gave rather different electrophoretic patterns, but in both cases, we were able to analyze these and suggest a global conformation of the complexes. Subsequent crystal structures of complexes of each protein bound to DNA junctions [58,59] confirmed both structures. Thus the method works well on DNA-protein complexes. It has not yet been extended to RNA-protein complexes, but there is no reason to suppose this would not be successful.

5.6 Analysis of the structure of four-way RNA junctions

Although comparatively rare, fully base-paired 4H junctions exist in a number of significant RNA species, including the hairpin ribozyme and the spliceosomal U1 snRNA. Comparative gel electrophoresis showed that RNA 4H junctions are relatively polymorphic [60]. The junctions undergo pairwise coaxial stacking of helices under all conditions, but the angle subtended between the two axes varies with ionic conditions. At low salt concentrations, the conformation is parallel. Upon addition of moderate concentrations of divalent metal ions, the electrophoretic pattern changed to one that could be interpreted in terms of a stacked structure with perpendicular axes. Subsequent single-molecule FRET studies (see Chapter 8) showed that this results from a rapid equilibrium between parallel and antiparallel conformations [61]. Increasing the concentration of metal ions further resulted in electrophoretic patterns indicative of a bias toward the antiparallel structure. Steady-state FRET experiments indicated that the transition between parallel and antiparallel forms was driven by the noncooperative binding of ions [62].

5.7 The 4H junctions of the U1 snRNA and the hairpin ribozyme

Comparative gel electrophoresis analysis of the 4H junction of U1 snRNA showed this junction adopted a coaxially stacked structure with perpendicular axes [60] (Fig. 5.5A). This was recently confirmed by a crystallographic structure at low resolution [63]. The hairpin ribozyme is essentially built on a scaffold that comprises a 4H junction, and we have made a detailed study of it, both as an isolated junction [64] and in situ in the ribozyme [65,66]. Comparative gel electrophoresis showed that the junction adopted an antiparallel stacked X-structure in the presence of divalent metal ions (Fig. 5.5B). The natural stacking bias of the isolated junction together with the antiparallel conformation juxtaposed the two internal loops that are required to interact to generate the active form of the ribozyme. All these conclusions were verified when Ferré-D'Amaré and coworkers [67] solved the X-ray crystal structure of the ribozyme. The dynamic nature of this junction was later demonstrated using single-molecule FRET (see Chapter 8), with transitions between antiparallel and parallel forms, and between alternative stacking conformers [61]. These studies showed that the folding of the ribozyme into its active form is greatly accelerated by the presence of the four-way junction [68]. The chirality of the isolated junction of the hairpin ribozyme was deduced to be opposite to that of the complete ribozyme using an electrophoretic phasing experiments [36].

5.8 A more complex junction found in the HCV IRES

Perfectly paired 4H RNA junctions are actually in the minority, and most natural four-way RNA junctions contain one or more formally unpaired nucleotides at the point of strand exchange. The extra nucleotides would be expected to alter the structural and dynamic properties of the junctions. The $2HS_2 2HS_1$ four-way junction found in the hepatitis C virus (HCV) internal ribosome entry site (IRES) element has been studied

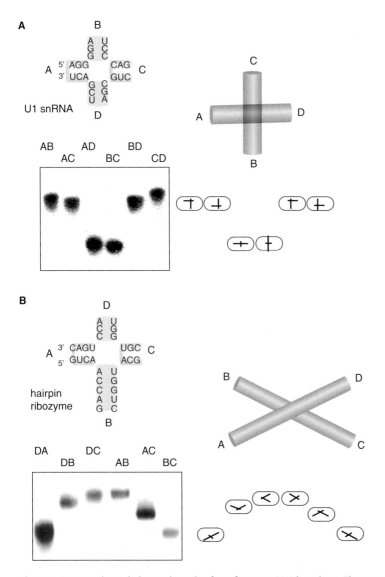

Fig. 5.5 Comparative gel electrophoresis of two four-way RNA junctions. The central sequences are shown, and in both cases, the arms designated sequentially A, B, C, and D. In both cases, electrophoresis is carried out in an 8% polyacrylamide gel in the presence of 90 mM Tris-borate (pH 8.3) and the indicated concentration of Mg^{2+} ions. (A) The U1A junction [60]. The pattern of electrophoresis of the six long-short arm species in the presence of 1 mM Mg^{2+} ions is shown. The mobility pattern of the six species is slow, slow, fast, fast, slow, slow. The simplest interpretation (shown on the right-hand side) is that of a stacked structure based on A on D and B on C coaxial stacking, with the axes nearly perpendicular. The pattern would also be consistent with a rapid exchange between nearly equal populations of parallel and antiparallel forms, but a recent crystal structure has found a perpendicular stacked structure for this RNA junction [63]. (B) The four-way junction of the hairpin ribozyme [64]. The pattern of electrophoresis of the six long-short arm species in the presence of

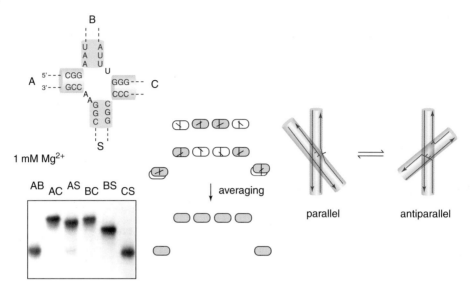

Fig. 5.6 Analysis of the 2HS₁HS₂ four-way junction of the HCV IRES by comparative gel electrophoresis [69]. The sequence of the junction around the point of strand exchange is shown. Comparative gel electrophoresis in a 10% polyacrylamide gel was performed in the presence of 90 mM Tris-borate (pH 8.3), 1 mM Mg^{2+}, using the six long-short arm species. The fast-slow-slow-slow-slow-fast pattern of mobilities is interpreted in terms of a rapid exchange between approximately equal populations of parallel and antiparallel conformations as shown, with strand polarities indicated for clarity.

using a combination of comparative electrophoresis and FRET (see Chapter 8) [69]. In marked contrast to the 4H RNA junctions, it was found that the IRES junction lost coaxial stacking in the absence of added metal ions, adopting an extended structure. On addition of divalent metal ions, the junction folded by pairwise coaxial stacking of arms, adopting the stacking conformer that places the extra nucleotides onto the exchanging strands. The pattern of electrophoretic mobilities of the six long-short arm species was consistent with a perpendicular axes (Fig. 5.6). However, an alternative interpretation of the pattern was suggested by time-resolved FRET (see Chapter 8) studies, involving a rapid exchange between approximately equal populations of parallel and antiparallel conformations. If the exchange is fast on the timescale of the gel electrophoresis (which is perhaps a little difficult to define precisely), this should generate an average of the mobility patterns of the two forms, consistent with the pattern obtained. Clearly we have to be alive to more complex interpretations of electrophoretic results in some cases, and this shows the advantage of combining

---

Fig. 5.5 (*Continued*)
2 mM Mg^{2+} ions is shown. The fast-intermediate-slow-slow-intermediate-fast pattern of mobilities is interpreted in terms of the antiparallel X-shaped structure shown, based on coaxial stacking of A on D and B on C arms. In the complete ribozyme, this conformation of the junction juxtaposes the loops of the A and B arms to facilitate their intimate association.

comparative gel electrophoresis with other biophysical methods such as FRET. An earlier crystal structure of the IRES junction had shown that it crystallized in a parallel conformation [70]; suggesting that the crystallization process has frozen out this component of the dynamic equilibrium.

5.9 Analysis of the structure of three-way RNA junctions

Three-way junctions are extremely common in natural RNA molecules [71]. Perfectly paired 3H junctions are extremely rare; almost all include one or more formally unpaired nucleotides in the regions connecting the helices. Applying the comparative gel electrophoretic method to a three-way junction requires the preparation of only three junctions, with two long and a single short arm.

5.9.1 A three-way junction of the HCV IRES element

The HCV IRES element contains a three-way junction that is required for its role in the internal initiation of translation. The junction is formally a $3HS_4$ junction, with the possibility of two alternative stacking conformers. However, in principle the junction can also undergo two steps of branch migration that would form $2HS_1HS_3$ and $2HS_2HS_2$ junctions. Comparative gel electrophoresis shows that the junction undergoes a structural transition induced by the addition of metal ions [72] (Fig. 5.7). The three helices of the IRES junctions are labeled C, D, and E, and following our normal convention, the long-short arm species are named by their long arms. Thus, species CD is shortened in arm D, for example. In the absence of added metal ions, the three species migrate with relative rates of DE > CE > CD. However, upon addition of metal ions, the pattern of electrophoretic mobility is significantly changed, with relative migration rates CE > DE > CD. The fast mobility of the CE species would be consistent a predominant species present in solution whose structure is based on coaxial stacking of the C and E helices, with the e strand (the strand with its 5' end located in the E helix) passing continuously between the two helices through the junction. This was confirmed by steady-state FRET experiments (see Chapter 8). The relative mobilities of the remaining species indicated that helix D is directed into the same quadrant as helix C. Repetition of the electrophoretic analysis in the present of different metal ions indicated that the folded structure is closely similar in 1 mM Mg^{2+}, 25 mM Na^+, and the trivalent ion $[Co(NH_3)_6]^{3+}$. It is therefore unlikely that specific ion binding is required to fold the junction, but rather just charge neutralization by diffuse binding of ions. Single-molecule FRET studies failed to find any fraction of a minor conformer based an alternative choice of helical stacking partners and provided no evidence for structural transitions [72].

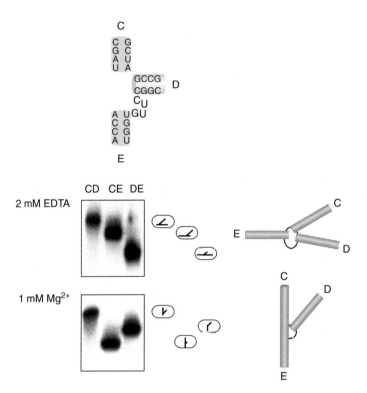

Fig. 5.7 Comparative gel electrophoresis of the conformation adopted by the three way junction of the HCV IRES in different ionic conditions [72]. The central sequence of the junction is shown; the component helices are labeled C, D, and E. Although the junction has been drawn here as a 3HS$_4$ junction, the sequence can be rearranged in a kind of branch migration process to give 2HS$_1$HS$_3$ and 2HS$_2$HS$_2$ junctions. The three long-short species were analyzed by comparative gel electrophoresis 10% polyacrylamide in 90 mM Tris-borate (pH 8.3) with either 2 mM EDTA to chelate metal ions or 1 mM Mg^{2+} ions. Our interpretation of the structures adopted is shown on the right, together with the anticipated structures and mobilities of the long-short arm junction species. In the presence of EDTA, the slow-intermediate-fast pattern is interpreted in terms of the extended structure, while the slow-fast-intermediate pattern observed in the presence of metal ions is interpreted in terms of a structure in which helices C and E are coaxially stacked, with an acute angle subtended between helices C and D.

5.9.2 Three-way junctions are the key architectural elements of the VS ribozyme

The core of the Varkud satellite (VS) ribozyme comprises five helices (numbered II through VI) that are organized by two three-way junctions to generate an H-shaped secondary structure (Fig. 5.8). Helices III, IV, and V create an HS$_1$HS$_5$HS$_2$ junction, while helices II, III, and VI form a 2HS$_5$HS$_2$ junction. Both junctions undergo two-state folding induced by the noncooperative binding of divalent metal ions, and their

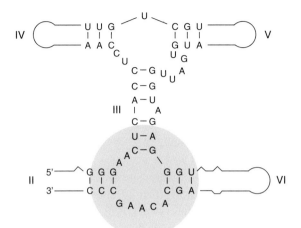

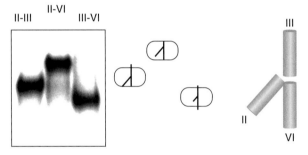

Fig. 5.8 Analysis of a three-way RNA junction of the VS ribozyme by comparative gel electrophoresis [73]. A scheme showing the secondary structure of the *trans*-acting ribozyme is shown, with the sequences of the two three-way junctions shown. The data presented here are for the II-III-VI junction (circled). The three long-short species were analyzed by comparative gel electrophoresis in a 10% polyacrylamide gel in the presence of 90 mM Tris-borate (pH 8.3), 5 mM Mg^{2+} ions. The intermediate-slow-fast pattern of the junction folded in the presence of divalent metal ions is shown on the right.

structures have been analyzed as free junctions by comparative gel electrophoresis. The II-III-VI junction folds by the coaxial stacking of arms III and VI, with an acute angle subtended between helices II and VI [73] (Fig. 5.8). The III-IV-V junction was found to fold by the coaxial stacking of arms IV and III, with an acute angle between helices V and III [74]. These structures are in good agreement with FRET measurements (see Chapter 8) on the isolated junctions and also with small angle X-ray scattering (SAXS, see Chapter 9) data from the isolated junctions and the complete ribozyme [75]. The two junctions are connected via the common helix III, creating a continuous coaxial alignment of helices IV, III, and VI. The dihedral angle subtended between helices II and V was determined by a variant of the helical phasing experiment, and from this a model of the complete *trans*-acting ribozyme was constructed [74]. This was in reasonable agreement with the SAXS-derived model [75].

5.9.3 The hammerhead ribozyme is a complex three-way helical junction

The hammerhead ribozyme provides an example of a still-more elaborated junction, where the junction itself is the catalytic center of the ribozyme. It is an $HS_1HS_7HS_3$ junction, with three helices named I, II, and III (Fig. 5.9). The global structure and folding was analyzed by electrophoresis, coincidentally at about the same time as its structure was solved by X-ray crystallography (see Chapter 13). Using comparative gel electrophoresis [76], we found that in the presence of divalent metal ions, the hammerhead junction folded by the coaxial stacking of helices III on II, with an acute angle subtended between helices I and II (Fig. 5.9), and this was in complete agreement with the structures observed in the crystal [77,78]. However, the folding of the hammerhead junction was rather more complex than had been found for simpler junctions and occurred in two stages with addition of increasing concentrations of divalent metal ions. Analysis of the electrophoretic mobility patterns as a function of Mg^{2+} ion concentration revealed the existence of a distinct species at intermediate Mg^{2+} concentration in which helices II and III were coaxially stacked, but helix I was directed into the same quadrant as helix III [76]. This scheme was confirmed by FRET studies [79] (see Chapter 8) and could be understood in terms of the crystal structures [77,78] (see Chapter 13), which showed the formation of two distinct domains within the structure. The electrophoretic and FRET results of the natural hammerhead sequence, together with those of sequence substitutions that blocked folding at different stages [80,81], were consistent with the sequential formation of these two domains as the metal ion concentration increased.

It was later shown that tertiary contacts between elements within helices I and II that were missing in all the original constructs are important in the function of the ribozyme in physiological conditions [82,83] and induce a different structure in the core [84]. We found that inclusion of these elements led to folding in a single step at significantly lower Mg^{2+} concentration [85].

5.10 Some final thoughts

Electrophoresis is a simple, very powerful but underused method for studying the global conformation of kinks and helical junctions in RNA. Yet it is fairly unquantitative and low resolution and gives a small fraction of the detail available from crystallography. So what is its value? Gel electrophoresis is much simpler and faster than crystallography, and of course, it does not require a crystal. As we have seen, comparative gel electrophoresis uncovered the stacked X-structure of the Holliday DNA junction a full decade ahead of the crystallographers and gave strong clues to the dynamic behavior before single-molecule methods. At the same time, gel electrophoresis provides a physical separation of different species that can then be recovered

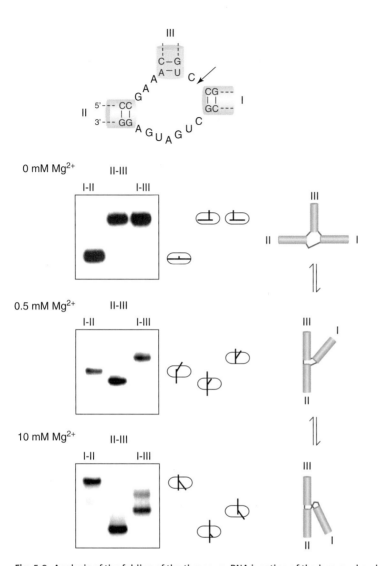

Fig. 5.9 Analysis of the folding of the three-way RNA junction of the hammerhead ribozyme as a function of Mg^{2+} ion concentration by comparative gel electrophoresis [76]. The sequence of the ribozyme core is shown, with the position of ribozyme cleavage arrowed. The arms are labeled I, II, and III. The electrophoretic analysis is shown for three different Mg^{2+} ion concentrations in 8% polyacrylamide gels in the presence of 90 mM Tris borate (pH 8.3). In the absence of Mg^{2+} ions, the junction is extended, with a large angle subtended between helices I and II. The mobility patterns in the presence of added metal ions are interpreted in terms of coaxial stacking of helices II and III (the species with shortened helix I is consistently the fastest), but the angle subtended by helix I depends on the Mg^{2+} ion concentration as shown. The final shape of the ribozyme in the presence of 10 mM Mg^{2+} is in good agreement with the structure found in the crystal by X-ray diffraction [77,78].

and studied by other means. The method has provided significant structural insights into a number of different systems in RNA. It has generally done so ahead of crystallography, but when crystal structures have become available, they have always confirmed the conclusions from gel electrophoresis. In fact, to date the gel electrophoresis has never got the wrong answer. So is there any serious disadvantage to the method? It is only really applicable to a small segment of larger species, essentially a single junction or similar, so that it means taking an inherently "divide and conquer" approach. Taking small elements out of their natural environment may result in the loss of larger-scale tertiary interactions that may in turn significantly influence the local structure of the element that is studied. Therefore not all structures may be suitable for analysis by comparative gel electrophoresis, and the method needs to be applied judiciously. However, when so used, gel electrophoresis pays great dividends.

5.11 Acknowledgments

Work in this laboratory is funded by Cancer Research UK, the Wellcome Trust, and HFSP. Many people in this laboratory have contributed to the applications of gel electrophoresis to nucleic acid structure, especially Gerald Gough, Derek Duckett, Alastair Murchie, Richard Grainger, the late Gurminder Bassi, Terry Goody, Jia Liu, Richard Pöhler, Daniel Lafontaine, and Sonya Melcher.

References

1. Gough GW, Lilley DMJ. DNA bending induced by cruciform formation. Nature. 1985;313: 154–156.
2. Cooper JP, Hagerman PJ. Gel electrophoretic analysis of the geometry of a DNA four-way junction. J. Mol. Biol. 1987;198:711–719.
3. Duckett DR, Murchie AIH, Diekmann S, von Kitzing E, Kemper B, Lilley DMJ. The structure of the Holliday junction and its resolution. Cell. 1988;55:79–89.
4. de Gennes PG. Reptation of a polymer chain in the presence of fixed obstacles. J. Chem. Phys. 1971;55:572–578.
5. Lumpkin OJ, Zimm BH. Mobility of DNA in gel electrophoresis. Biopolymers. 1982;21: 2315–2316.
6. Gohlke C, Murchie AIH, Lilley DMJ, Clegg RM. The kinking of DNA and RNA helices by bulged nucleotides observed by fluorescence resonance energy transfer. Proc. Natl. Acad. Sci. U. S. A. 1994;91:11660–11664.
7. Lerman LS Frisch, HL. Why does the electrophoretic mobility of DNA in gels vary with the length of the molecule. Biopolymers. 1982;21:995–997.
8. Lumpkin OJ, Dejardin P, Zimm BH. Theory of gel electrophoresis of DNA. Biopolymers. 1985; 24:1573–1593.
9. Levene SD, Zimm BH. Understanding the anomalous electrophoresis of bent DNA molecules: a reptation model. Science. 1989;245:396–399.

10. Lumpkin O, Levene SD, Zimm BH. Exactly solvable reptation model. Phys. Rev. A. 1989;39: 6557–6566.
11. Duke TA, Semenov AN, Viovy JL. Mobility of a reptating polymer. Phys. Rev. Lett. 1992;69: 3260–3263.
12. Barkema GT, Marko JF, Widom B. Electrophoresis of charged polymers: simulation and scaling in a lattice model of reptation. Phys. Rev. E Stat. Phys. Plasmas Fluids Relat. Interdiscip. Top. 1994;49:5303–5309.
13. Deutsch JM. Theoretical studies of DNA during gel electrophoresis. Science. 1988;240: 922–924.
14. Heuer DM, Yuan C, Saha S, Archer LA. Effect of topological asymmetry on the electrophoretic mobility of branched DNA structures with and without single-base mismatches. Electrophoresis. 2005;26:64–70.
15. Saha S, Heuer DM, Archer LA. Electrophoretic mobility of linear and star-branched DNA in semidilute polymer solutions. Electrophoresis. 2006;27:3181–3194.
16. Goody TA, Melcher SE, Norman DG, Lilley DMJ. The kink-turn motif in RNA is dimorphic, and metal ion dependent. RNA. 2004;10:254–264.
17. Klein DJ, Schmeing TM, Moore PB, Steitz TA. The kink-turn: a new RNA secondary structure motif. EMBO J. 2001;20:4214–4221.
18. Moore T, Zhang Y, Fenley MO, Li H. Molecular basis of box C/D RNA-protein interactions; cocrystal structure of archaeal L7Ae and a box C/D RNA. Structure. 2004;12:807–818.
19. Hamma T, Ferré-D'Amaré AR. Structure of protein L7Ae bound to a K-turn derived from an archaeal box H/ACA sRNA at 1.8 Å resolution. Structure. 2004;12:893–903.
20. Szewczak LB, Gabrielsen JS, Degregorio SJ, Strobel SA, Steitz JA. Molecular basis for RNA kink-turn recognition by the h15.5K small RNP protein. RNA. 2005;11:1407–1419.
21. Youssef OA, Terns RM, Terns MP. Dynamic interactions within sub-complexes of the H/ACA pseudouridylation guide RNP. Nucleic Acids Res. 2007;35:6196–6206.
22. Vidovic I, Nottrott S, Hartmuth K, Luhrmann R, Ficner R. Crystal structure of the spliceosomal 15.5 kD protein bound to a U4 snRNA fragment. Mol. Cell. 2000;6:1331–1342.
23. Wozniak AK, Nottrott S, Kuhn-Holsken E, et al. Detecting protein-induced folding of the U4 snRNA kink-turn by single-molecule multiparameter FRET measurements. RNA. 2005;11:1545–1554.
24. Montange RK, Batey RT. Structure of the S-adenosylmethionine riboswitch regulatory mRNA element. Nature. 2006;441:1172–1175.
25. Blouin S, Lafontaine DA. A loop loop interaction and a K-turn motif located in the lysine aptamer domain are important for the riboswitch gene regulation control. RNA. 2007;13:1256–12567.
26. Heppell B, Lafontaine DA. Folding of the SAM aptamer is determined by the formation of a K-turn-dependent pseudoknot. Biochemistry. 2008;47:1490–1499.
27. Smith KD, Lipchock SV, Ames TD, Wang J, Breaker RR, Strobel SA. Structural basis of ligand binding by a c-di-GMP riboswitch. Nat. Struct. Mol. Biol. 2009;16:1218–1223.
28. Liu J, Lilley DMJ. The role of specific 2'-hydroxyl groups in the stabilization of the folded conformation of kink-turn RNA. RNA. 2007;13:200–210.
29. Wang J, Fessl T, Schroeder KT, et al. Single-molecule observation of the induction of k-turn RNA structure on binding L7Ae protein Biophysical. J. 2012;103:2541–2548.
30. Drak J, Crothers DM. Helical repeat and chirality effects on DNA gel electrophoretic mobility. Proc. Natl. Acad. Sci. U. S. A. 1991;88:3074–3078.
31. Bhattacharyya A, Murchie AIH, Lilley DMJ. RNA bulges and the helical periodicity of double-stranded RNA. Nature. 1990;343:484–487.
32. Tang RS, Draper DE. Bulge loops used to measure the helical twist of RNA in solution. Biochemistry. 1990;29:5232–5237.

33. Tang RS, Draper DE. On the use of phasing experiments to measure helical repeat and bulge loop-associated twist in RNA. Nucleic Acids Res. 1994;22:835–841.
34. Riordan FA, Bhattacharyya A, McAteer S, Lilley DMJ. Kinking of RNA helices by bulged bases, and the structure of the human immunodeficiency virus transactivator response element. J. Mol. Biol. 1992;226:305–310.
35. Grainger RJ, Murchie AIH, Norman DG, Lilley DMJ. Severe axial bending of RNA induced by the U1A binding element present in the 3' untranslated region of the U1A mRNA. J. Mol. Biol. 1997;273:84–92.
36. Goody TA, Lilley DMJ, Norman DG. The chirality of a four-way helical junction in RNA. J. Am. Chem. Soc. 2004;126:4126–4127.
37. Lilley DMJ. Structures of helical junctions in nucleic acids. Q. Rev. Biophys. 2000;33:109–159.
38. Lilley DMJ, Clegg RM, Diekmann S, Seeman NC, von Kitzing E, Hagerman P. Nomenclature Committee of the International Union of Biochemistry: a nomenclature of junctions and branchpoints in nucleic acids. Recommendations 1994. Eur. J. Biochem. 1995;230:1–2.
39. Murchie AIH, Clegg RM, von Kitzing E, Duckett DR, Diekmann S, Lilley DMJ. Fluorescence energy transfer shows that the four-way DNA junction is a right-handed cross of antiparallel molecules. Nature. 1989;341:763–766.
40. Clegg RM, Murchie AIH, Zechel A, Carlberg C, Diekmann S, Lilley DMJ. Fluorescence resonance energy transfer analysis of the structure of the four-way DNA junction. Biochemistry. 1992;31:4846–4856.
41. Clegg RM, Murchie AIH, Zechel A, LIlley DMJ. The solution structure of the four-way DNA junction at low salt concentration; a fluorescence resonance energy transfer analysis. Biophys. J. 1994;66:99–109.
42. Cooper JP, Hagerman PJ. Geometry of a branched DNA structure in solution. Proc. Natl. Acad. Sci. U. S. A. 1989;86:7336–7340.
43. von Kitzing E, Lilley DMJ, Diekmann S. The stereochemistry of a four-way DNA junction: a theoretical study. Nucleic Acids Res. 1990;18:2671–2683.
44. Nowakowski J, Shim PJ, Prasad GS, Stout CD, Joyce GF. Crystal structure of an 82 nucleotide RNA-DNA complex formed by the 10–23 DNA enzyme. Nat. Struct. Biol. 1999;6:151–156.
45. Ortiz-Lombardía M, González A, Erijta R, Aymamí J, Azorín F, Coll M. Crystal structure of a DNA Holliday junction. Nat. Struct. Biol. 1999;6:913–917.
46. Eichman BF, Vargason JM, Mooers BHM, Ho PS. The Holliday junction in an inverted repeat DNA sequence: sequence effects on the structure of four-way junctions. Proc. Natl. Acad. Sci. U. S. A. 2000;97:3971–3976.
47. Murchie AIH, Portugal J, Lilley DMJ. Cleavage of a four-way DNA junction by a restriction enzyme spanning the point of strand exchange. EMBO J. 1991;10:713–718.
48. Miick SM, Fee RS, Millar DP, Chazin WJ. Crossover isomer bias is the primary sequence-dependent property of immobilized Holliday junctions. Proc. Natl. Acad. Sci. U. S. A. 1997;94:9080–9084.
49. Grainger RJ, Murchie AIH, Lilley DMJ. Exchange between stacking conformers in a four-way DNA junction. Biochemistry. 1998;37:23–32.
50. McKinney SA, Déclais A-C, Lilley DMJ, Ha T. Structural dynamics of individual Holliday junctions. Nat. Struct. Biol. 2003;10:93–97.
51. Liu J, Déclais A-C, McKinney SA, Ha T, Norman DG, Lilley DMJ. Stereospecific effects determine the structure of a four-way DNA junction. Chem. Biol. 2005;12:217–228.
52. Liu J, Déclais AC, Lilley, DMJ. Mechanistic aspects of the DNA junction-resolving enzyme T7 endonuclease I. Biochemistry. 2006;45:3934–3942.

53. Déclais AC, Lilley DMJ. New insight into the recognition of branched DNA structure by junction-resolving enzymes. Curr. Opin. Struct. Biol. 2008;18:86–95.
54. White MF, Lilley DMJ. The resolving enzyme CCE1 of yeast opens the structure of the four-way DNA junction. J. Mol. Biol. 1997;266:122–134.
55. Déclais A-C, Lilley DMJ. Extensive central disruption of a four-way junction on binding CCE1 resolving enzyme. J. Mol. Biol. 2000;296:421–433.
56. Pöhler JRG, Giraud-Panis M-JE, Lilley DMJ. T4 endonuclease VII selects and alters the structure of the four-way DNA junction; binding of a resolution-defective mutant enzyme. J. Mol. Biol. 1996;260:678–696.
57. Déclais A-C, Fogg JM, Freeman A, et al. The complex between a four-way DNA junction and T7 endonuclease I. EMBO J. 2003;22:1398–1409.
58. Biertümpfel C, Yang W, Suck D. Crystal structure of T4 endonuclease VII resolving a Holliday junction. Nature. 2007;449:616–620.
59. Hadden JM, Déclais A-C, Carr S, Lilley DMJ, Phillips SEV. The structural basis of Holliday junction resolution by T7 endonuclease I. Nature. 2007;449:621–624.
60. Duckett DR, Murchie AIH, Lilley DMJ. The global folding of four-way helical junctions in RNA, including that in U1 snRNA. Cell. 1995;83:1027–1036.
61. Hohng S, Wilson TJ, Tan E, Clegg RM, Lilley DMJ, Ha T. Conformational flexibility of four-way junctions in RNA. J. Mol. Biol. 2004;336:69–79.
62. Walter F, Murchie AIH, Duckett DR, Lilley DMJ. Global structure of four-way RNA junctions studied using fluorescence resonance energy transfer. RNA. 1998;4:719–728.
63. Pomeranz-Krummel DA, Oubridge C, Leung AKW, Li J, Nagai K. Crystal structure of human spliceosomal U1 snRNP at 5.5 Å resolution. Nature. 2009;458:475–480.
64. Walter F, Murchie AIH, Lilley DMJ. The folding of the four-way RNA junction of the hairpin ribozyme. Biochemistry. 1998;37:17629–17636.
65. Murchie AIH, Thomson JB, Walter F, Lilley DMJ. Folding of the hairpin ribozyme in its natural conformation achieves close physical proximity of the loops. Mol. Cell. 1998;1:873–881.
66. Walter F, Murchie AIH, Thomson JB, Lilley DMJ. Structure and activity of the hairpin ribozyme in its natural junction conformation; effect of metal ions. Biochemistry. 1998;37:14195–14203.
67. Rupert PB, Ferré-D'Amaré AR. Crystal structure of a hairpin ribozyme-inhibitor complex with implications for catalysis. Nature. 2001;410:780–786.
68. Tan E, Wilson TJ, Nahas MK, Clegg RM, Lilley DMJ, Ha T. A four-way junction accelerates hairpin ribozyme folding via a discrete intermediate. Proc. Natl. Acad. Sci. U. S. A. 2003;100:9308–9313.
69. Melcher SE, Wilson TJ, Lilley DMJ. The dynamic nature of the four-way junction of the hepatitis C virus IRES. RNA. 2003;9:809–820.
70. Kieft JS, Zhou K, Grech A, Jubin R, Doudna JA. Crystal structure of an RNA tertiary domain essential to HCV IRES-mediated translation initiation. Nat. Struct. Biol. 2002;9:370–374.
71. Lescoute A, Westhof E. Topology of three-way junctions in folded RNAs. RNA. 2006;12:83–93.
72. Ouellet J, Melcher SE, Iqbal A, Ding Y, Lilley DMJ. Structure of the three-way helical junction of the hepatitis C virus IRES element. RNA. 2010;16:1597–1609.
73. Lafontaine DA, Norman DG, Lilley DMJ. Structure, folding and activity of the VS ribozyme: importance of the 2-3-6 helical junction. EMBO J. 2001;20:1415–1424.
74. Lafontaine DA, Norman DG, Lilley DMJ. The global structure of the VS ribozyme. EMBO J. 2002;21:2461–2471.
75. Lipfert J, Ouellet J, Norman DG, Doniach S, Lilley DMJ. The complete VS ribozyme in solution studied by small-angle X-ray scattering. Structure. 2008;16:1357–1367.

76. Bassi G, Møllegaard NE, Murchie AIH, von Kitzing E, Lilley DMJ. Ionic interactions and the global conformations of the hammerhead ribozyme. Nat. Struct. Biol. 1995;2:45–55.
77. Pley HW, Flaherty KM, McKay DB. Three-dimensional structure of a hammerhead ribozyme. Nature. 1994;372:68–74.
78. Scott WG, Finch JT, Klug A. The crystal structure of an all-RNA hammerhead ribozyme: a proposed mechanism for RNA catalytic cleavage. Cell. 1995;81:991–1002.
79. Bassi GS, Murchie AIH, Walter F, Clegg RM, Lilley DMJ. Ion-induced folding of the hammerhead ribozyme: a fluorescence resonance energy transfer study. EMBO J. 1997;16:7481–7489.
80. Bassi GS, Murchie AIH, Lilley DMJ. The ion-induced folding of the hammerhead ribozyme: core sequence changes that perturb folding into the active conformation. RNA. 1996;2:756–768.
81. Bassi GS, Møllegaard NE, Murchie AIH, Lilley DMJ. RNA folding and misfolding of the hammerhead ribozyme. Biochemistry. 1999;38:3345–3354.
82. Khvorova A, Lescoute A, Westhof E, Jayasena SD. Sequence elements outside the hammerhead ribozyme catalytic core enable intracellular activity. Nat. Struct. Biol. 2003;10:1–5.
83. Canny MD, Jucker FM, Kellogg E, Khvorova A, Jayasena SD, Pardi A. Fast cleavage kinetics of a natural hammerhead ribozyme. J. Am. Chem. Soc. 2004;126:10848–10849.
84. Martick M, Scott WG. Tertiary contacts distant from the active site prime a ribozyme for catalysis. Cell. 2006;126:309–320.
85. Penedo JC, Wilson TJ, Jayasena SD, Khvorova A, Lilley DMJ. Folding of the natural hammerhead ribozyme is enhanced by interaction of auxiliary elements. RNA. 2004;10:880–888.

Alexander McPherson
6 Virus RNA structure deduced by combining X-ray diffraction and atomic force microscopy

6.1 Introduction

Current thinking on the conformations of encapsidated and free viral RNAs originated from two primary sources. The first was the early proposals on the likely structures assumed by free, single-stranded RNA (ssRNA) molecules [1–6]. These posited that long ssRNA molecules, through intramolecular base pairing, would initially form sequential arrays of local stem-loop structures, and that these might then be joined by longer range tertiary interactions that would promote condensation. The proposals appear to be in accord with the results of X-ray crystallographic analyses of small icosahedral viruses [7–9], the second principal source of current ideas. These X-ray studies focused especially on viruses of triangulation numbers T = 1 and 3 (reviewed by [10–15]) that correspond with particles having diameters ranging from 17 nm to about 30 nm and containing ssRNA molecules of between 850 and about 8,000 nucleotides.

A principal objective of this chapter is to illustrate how atomic force microscopy (AFM) contributed to our understanding of viral RNA structure; thus it is necessary to review some X-ray diffraction results and explore further the formative ideas that provided the bases according to which AFM images have been interpreted. Particularly important in that regard is the investigation of Satellite Tobacco Mosaic Virus (STMV) [16,17], a T = 1 virus of 17 nm diameter that contains as its genome an ssRNA molecule of 1,058 nucleotides [18], and whose X-ray analysis visualized nearly 45% of its mass [19–21]. In addition to STMV, other examples such as Turnip Yellow Mosaic Virus (TYMV) [22], Desmodium Yellow Mottle Virus (DYMV) [23], Satellite Panicum Mosaic Virus (SPMV), Satellite Tobacco Necrosis Virus (STNV) (STMV, SPMV, and STNV reviewed in [10]), and several other viruses such as Bean Pod Mottle Virus [24], Flock House Virus [25], and Pariacoto Virus [26] also become pertinent to the discussion. Because of its pronounced influence on our conception of viral RNA and how it is packaged, however, STMV serves as a touchstone throughout this chapter.

6.2 Why don't we learn more about RNA from X-ray crystallography?

There was initially little expectation that X-ray crystallography would reveal significant detail regarding the structure of encapsidated RNA. This, however, proved not to be the case, and as more virus structures were solved, the number of those in which segments of RNA could be clearly resolved accumulated [15,24,27–30]. Some explanation is required. Protein capsids of spherical viruses possess icosahedral symmetry (532 symmetry). As a consequence, particles have 60 orientations that allow them to superimpose upon themselves, and they may be incorporated into a crystal in any of those equivalent orientations and still maintain consistency with the space group symmetry of the unit cell of the crystal. The RNA inside the capsid, however, has no symmetry. It therefore enters the crystal lattice in any one of 60 nonequivalent orientations. As a consequence, the electron density of the RNA is averaged over the 60 orientations making the RNA density close to zero and basically unintelligible.

An exception is if some segments of the RNA are consistent with the icosahedral symmetry of the protein capsid. Then those segments will appear in electron density maps. While one might expect that this rarely occurs, this is not necessarily the case. Frequently, because of protein–nucleic acid associations, icosahedral symmetry is imposed or arises mutually for some segments of the RNA, and those segments emerge in Fourier syntheses.

6.3 X-ray studies revealing RNA

There are now a number of diffraction analyses in which portions of the genomic RNA, in spite of 60-fold randomization of orientation, appear in electron density maps. Parts of the RNA molecule exhibit local symmetry. The symmetry cannot be exact at the base sequence level since the primary structure can have no such symmetry; thus the symmetry is actually pseudosymmetry. Secondary structural elements of RNA, helical stems and loops, or pseudoknots can possess pseudosymmetry. RNA helixes have pseudo–twofold symmetry axes perpendicular to their axes. Thus, double helical stems are often found lying on icosahedral dyad axes. The pseudo-twofold axis passing through and in the plane of the central base pair of a helical segment may then coincide with an icosahedral twofold axis of the capsid. Other arrangements have been found as well, such as the trefoils in Bean Pod Mottle Virus [15,24] and the fivefold and pseudo-sixfold distributions of helical stems in TYMV [29] and SPMV [30] (Fig. 6.1).

Some interesting findings [31] emerged recently from the X-ray structure of an STNV-like particle produced by recombinant techniques in *Escherichia coli*. The particle was very similar in structure to the wild-type virus. The investigators concluded that the ssRNA consisted of multiple, short stem loops, with the loops, as in STMV, largely disordered. However, they found that the RNA interactions with the coat protein and the disposition of the RNA with respect to the capsid interior were not like that in either STMV or SPMV.

The appearance of stem loops in electron density maps suggests that this is likely to be a common structural motif for encapsidated ssRNA. It places stringent spatial

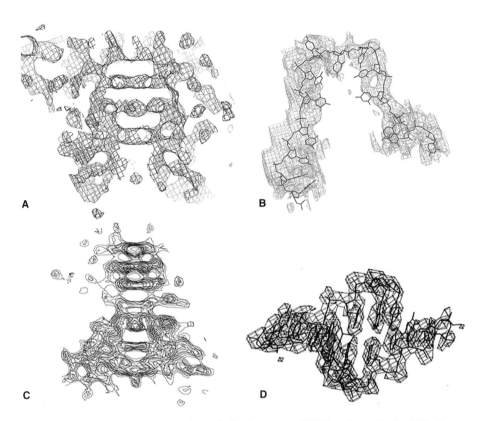

Fig. 6.1 (A) The electron density for a double helical segment of RNA found near the fivefold axis (view is perpendicular to the axis) of SPMV. At least 3 base pairs are clearly evident. (B) The electron density for an open ssRNA loop that appears within the pseudo-sixfold capsomeres of TYMV. (C) The electron density observed for a double helical stem of at least 7 base pairs within a pentameric capsomere of TYMV. (D) The electron density of the double helical portion, containing 7 to 9 base pairs, of a stem loop observed at an icosahedral dyad of STMV. By symmetry, the RNA segments appear at all symmetry equivalent positions.

constraints on any proposed folding pathway. Any scheme must accommodate the presence of 30 helical stems, and the RNA must fold so as to place them in their proper orientation at the 30 twofold axis positions of the icosahedral shell, inside the capsid, but also in contact with dimers of the coat protein. Other distributions of the stems, such as around fivefold axes as is seen in TYMV, impose equally rigorous constraints.

A core problem in deducing RNA three-dimensional structure is uncertainty regarding secondary structure. Base-pairing arrangements of no long (>200 nucleotides) ssRNA molecules are known, with the exception of ribosomal RNAs [32,33]. Thus, the problem of deducing the folding of an encapsidated viral ssRNA, at least in the case of STMV, reduces to finding a secondary structural pattern for the RNA that can then be folded into a highly compact spherical mass that exhibits 30 double helical segments on its exterior, all so oriented that their pseudodyad axes can be superimposed on icosahedral twofold axes.

6.4 Secondary structure prediction

The problem of deducing the secondary structures of long RNAs is particularly complex because of uncertainty in what base can pair with what other base, the way the bases might be paired (i.e. their hydrogen bonding arrangement), and the special accommodations that ssRNA molecules make to prolong opportunities for base pairing [34]. Nonstandard base pairs other than G–U have been inferred to be in helical elements of transfer RNA (tRNA) [35–39] and observed directly in helical RNA using X-ray crystallography [40–43]. Even earlier, Kaper [1–485] concluded that "the considerable variability allowed in interbase hydrogen bonding [44] makes the possibility of the occurrence of unusual base pairs in the double helix stretches of ssRNA seem almost unavoidable." It is not widely agreed what contribution these base pairs make to the stability of a short RNA helix [45,46], given the imposition of water molecules on the structure (i.e. it is unclear whether they marginally increase stability or whether they are neutral, allowing a helix to continue where it might otherwise be interrupted). In any case, acceptance of such pairs in helical regions of ssRNA would imply that helices would be longer than predicted using only canonical base pairing. Continuous Watson-Crick base pairing, plus G–U pairs, essentially establishes minimum length.

Another construction observed in ssRNA that also tends to lengthen helical regions is the occurrence of nonstandard base pairs at the ends of helices where geometry relaxes and may include the stacking or interleaving of successive bases. Although such features do not extend helices in a rigorous sense, they could help explain why average helical stem length observed in electron density maps appears longer than might be predicted by Watson-Crick base pairing alone. Original work of Fresco, Alberts, and Doty [1] also proposed that nucleotides, or even stretches of nucleotides, could bulge from otherwise base-paired regions, thereby allowing a longer, though more irregular helix to form.

Secondary structure prediction procedures [47–49] generally demonstrate that viral ssRNAs have the potential for extensive base pairing and formation of helical stems when folded back upon themselves in complicated ways. The fold, which those procedures find to maximize the pairing, and meet certain other criteria as well, such as high neighborhood correlation and small loop sizes, are usually assumed to be the minimum energy states of the molecule. The procedures are insufficient to approximate, let alone specify, the tertiary fold of the RNA and therefore its three-dimensional structure.

The RNA of STMV, for example, is predicted to have a minimum energy secondary structure where 67% of the total nucleotides are base paired. Folding an RNA molecule having this secondary structure into a spherical mass of 100 Å diameter and meeting the icosahedral symmetry constraints for helical stems may not be impossible, but no one has so far accomplished it. We believe this would not be useful to attempt because in the virion, the RNA likely has no such secondary structure. For minimum energy secondary structures for ssRNAs from other viruses that exhibit icosahedral distributions of helical segments, the task is no easier, and for T = 3 viruses with thousands of nucleotides, it is even more daunting.

There are persuasive reasons why encapsidated RNAs fail to assume the minimum energy structures predicted by most algorithms. Foremost is that algorithms generally assume that the RNA is fully synthesized when folding begins, so that any base on the chain could pair with any other, even if widely separated in the sequence. This is unlikely. Proteins begin to fold as soon as they emerge from the ribosomal tunnel and possibly while they are still in it. RNA molecules undoubtedly initiate folding as well, as they emerge from a replication complex.

If folding proceeds as synthesis is in progress, then stem loops and possibly other secondary structures such as pseudoknots [50] will form in the initial 5' portions of the polynucleotide while the latter 3' parts have yet to be made. Once a region of three or four consecutive base pairs has formed, it is energetically difficult to separate it once again into two single strands and reform the secondary structure in a different way, even if that second arrangement is energetically more stable. Thus, the folding pathway for a long RNA is littered with vast and deep kinetic traps, and progressive equilibration toward a true energy minimum for RNAs, unlike proteins, seems unlikely. In addition, as the RNA is being synthesized, coat proteins are abundantly present. These we might expect to initiate binding of available double helical elements as they are presented and, in so doing, further stabilize the kinetic intermediates.

Finally, a teleological point: Once the RNA is decapsidated inside the host cell, it must undergo or participate in a variety of biological processes. Among these are replication, translation, and transport between cells. These require that the RNA be unwound and in a nearly extended state. If the RNA were in its minimum energy state, it would be extraordinarily stable and virtually impossible to utilize biologically. Thus, we might expect that the folded, encapsidated RNA would not be in its minimum energy state when presented to the cell, but some metastable state that might readily be altered conformationally by viral or host proteins. Witz and Strazielle [6] and Kaper [13] concluded just this for viruses in general.

6.5 Generalized ssRNA secondary structural motifs

Helixes and loops in encapsidated RNA could in principle be compatible with many secondary patterns. These can be coarsely divided into three categories (Fig. 6.2). The distinguishing characteristic is the degree to which long-range pairing interactions

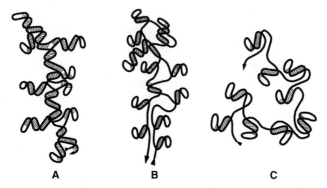

Fig. 6.2 Three possible motifs for a long RNA molecule of roughly 1,000 nucleotides, which are capable, by intermolecular base pairing, to fold back upon themselves and effect condensation. (A) Extensive base pairing between regions far from one another in the sequence produces a rigorously defined conformation that is highly constrained by pairing interactions. (B) Some distant intermolecular interactions impose more modest constraints on the overall folding, but most base pairing is localized in stem-loop elements. (C) The ssRNA exhibits the most unconstrained conformation, reflecting a sequence of strictly local stem-loop substructures. The stem-loop substructures in (C) are only metastable, and restructuring of a specific conformation to other similar arrangements could occur in a fluid manner. While (A) provides the greatest degree of base pairing and the lowest energy conformation, it also produces the most constrained three-dimensional structure. On the other hand, (C) may yield the least total base pairs and a higher energy conformation, but it also provides the greatest flexibility in terms of the three-dimensional structure it may assume.

are allowed involving bases far distant in the sequence. In an extreme case (Fig. 6.2A), the structure is highly self-involved and intricate with virtually every part of the structure interdependent upon every other. Nucleotide stretches hundreds apart, due to complicated folding back of the polynucleotide upon itself, pair to generate double-stranded regions. This is the kind of structure predicted by current algorithms on the basis of energy considerations. Implicit, however, is that the RNA molecule exists in its entirety when it folds into its three-dimensional structure. This, however, is probably not true for a viral RNA daughter strand produced from a replication complex [51]. An RNA's interactions with proteins as it folds further reduce the likelihood of such a final arrangement. The second category (Fig. 6.2B) allows only limited interaction of disparate parts of the RNA molecule, alleviating much interdependence within the molecule, as the overall structure arises from more localized substructures. This conformation is analogous to protein structure that forms local secondary elements, like helices and ribbons, which are then woven into supersecondary patterns by tertiary interactions.

The third category is exemplified by Fig. 6.2C. Here the RNA forms a series of stem-loop elements dependent upon pairing of stretches of bases relatively close to one another in the sequence. It is inherently simple, imposes no rigorous constraints on the overall conformation of the RNA, provides fluidity, and has one additional virtue. It does not require the RNA be entirely synthesized or transcribed before it begins assuming this folded state. The secondary structural pattern in Fig. 6.2C could plausibly arise as an RNA molecule was being synthesized from a replication complex and was reversibly stabilized by binding proteins.

From purely architectural principles, folding of a viral ssRNA molecule similar to that in Fig. 6.2A is unlikely. Not only is it difficult to reconcile the myriad structural elements and their organization with constraints imposed by X-ray crystallography, but such a structure would be intrinsically unfavorable from a mechanical standpoint. It would exemplify, as pointed out by Crane [52], what engineers term an "overdetermined structure," a brittle structure, one exhibiting too many internal interdependencies, like a framework with so many struts and welds, that it cannot tolerate stress. It is too rigid. An "overdetermined structure" in the case of encapsidated ssRNA seems particularly improbable. If the structure must assume other conformations when released from the virion or be transported across membranes and between cells in the plant, then it must possess some fluidity. Furthermore, it has been shown that in the wild there are many mutants (nontype strains) of most viruses. There are certainly examples for STMV, which contains the most highly constrained encapsidated ssRNA that we know of [53–55]. These include both single and multiple point mutations, and even some with substantial deletions [54,56]. Were the conformation of the encapsidated RNA closely dependent on precise long distant pairing arrangements, as in Fig. 6.2A, it could not likely tolerate these alterations.

6.6 The folding of RNA in STMV

In the case of STMV, reconciliation of secondary structure with crystallographic results would require placing contiguous stem-loop elements on the 30 twofold symmetry axes of the icosahedral net (Fig. 6.3). When that planar pattern is folded by connecting all edges, it produces a T = 1 icosahedron. It is difficult, from a topological perspective, to envision how a highly interdependent entity, a nonrepetitive arrangement of helical segments and single-stranded regions, could be efficiently folded in a manner consistent with the icosahedral grid. It is, however, possible, and Fig. 6.3A is an example. The RNA forms a hairpin with systematically alternating helices, stem loops, and single-stranded links. This quasi-linear structure can be threaded from icosahedral edge to edge to yield a pattern consistent with the X-ray structure (Fig. 6.3B). Structural constraints for even this relatively simple case are, nonetheless, considerable. The RNA molecule would, of necessity, have to be highly evolved and sophisticated, in terms of sequence, to reconcile genetic coding functions with structural responsibilities.

The structure exemplified by Fig. 6.2B shares some of the complexities of that in Fig. 6.2A but is sufficiently accommodating that it could withstand limited mutation and evolution and possess some conformational fluidity. It is difficult to rule out Fig. 6.2B. Importantly, however, it too would require the RNA molecule to be

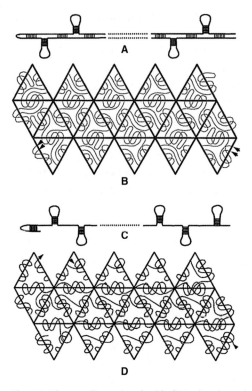

Fig. 6.3 The two-dimensional grid of 20 triangles, when folded so that adjacent edges meet, produces a three-dimensional, T = 1 icosahedron. The two-, three-, and fivefold symmetry elements of the icosahedron are indicated. Identifying efficient paths that connect all twofold symmetry operators may be considered in terms of their connectivity on this two-dimensional diagram. Thus, the most efficient way of placing linearly arrayed RNA helical segments at all dyads of a T = 1 icosahedron can be considered in terms of the most efficient way (shortest path) of connecting the twofold axes on this lattice. If an ssRNA were folded at its center and the two resulting antiparallel strands allowed to base pair in periodic intervals to form a long hairpin conformation, then it could, if it assumed the strictly defined secondary structural pattern in (A), be made to connect in a linear manner, shown in (B), the twofold axes in the icosahedral diagram. A series of linearly connected RNA stem-loop structures, like the RNA molecule in (C) can readily be passed by many similar, but equally efficient paths among the 30 twofold axis positions of the icosahedral diagram. That shown in (D) is among the simplest configurations and one that produces the shortest overall path. It begins at one fivefold apex at the lower right of the diagram and terminates at the most distant fivefold axis at the upper left.

substantially or even completely synthesized before folding proceeded, or it would require significant restructuring of any initial conformation. An additional mark against both Fig. 6.2A and Fig. 6.2B is that the periodic distribution of stem loops seen in the X-ray structure suggests, though doesn't absolutely require, a periodic distribution of base-paired elements in the secondary structure. Motifs such as in Fig. 6.2A,B, in general, lack that feature.

The RNA motif of Fig. 6.2C most readily conforms to the tertiary structure of the encapsidated RNA. It would do so in a straightforward and uncomplicated way, and it would place the least constraint on the overall secondary and tertiary structure of the RNA. The same conclusion may be reached from thermodynamic considerations as well [1], as discussed also by Kaper [6,13]. For STMV, the pattern illustrated in Figs. 6.2C and 6.3C meets the criterion of Occam's razor and, as shown in Fig. 6.3D, provides a plausible means of placing 30 helices at the twofold axes of the virus and generating both the loop elements and the connecting links. It could configure itself in an uninterrupted manner as a single daughter strand of RNA emerged from the replication complex. It is compatible with the suggestions and hypotheses of earlier investigators of RNA structure [1,3–6]. It would not lack conformational fluidity or the capacity to transform in a cooperative manner into alternate conformations. The RNA, further, has the property that, as it is synthesized, it presents a series of binding sites for the coat protein that could direct, in an ordered manner, assembly of the viral capsid.

6.7 Atomic force microscopy

Unlike most kinds of microscopy, AFM does not rely on the passage through or reflection from the sample by particles (electrons) or waves (light) but depends on aggregate atomic forces between a physical probe and the surface of the sample [57–60]. Hence it falls in the category of what is termed *probe microscopy*. The principle of AFM (Fig. 6.4) could hardly be simpler, being little more than that of a blind man with a cane or stick feeling his way through the physical world of hard objects. AFM, however, does this at the nanometer scale, and it applies the principle in a very systematic way and with unusual sensitivity [61–65].

A stylus with a very sharp tip (~20 nm radius) is drawn back and forth in a raster manner over a substrate (glass, plastic, mica) on which a sample has been deposited. The area on the substrate that is scanned establishes the magnification of the image. At close intervals along a line, the vertical deflection of a low spring constant cantilever, from which the stylus protrudes at one end, is recorded. When traces from all successive lines are assembled, then one has a topographical map of the entire plane. Deflections occur because of physical encounters between the tip and whatever object might occupy a specific point. The tip simply rides up and over and down the other side of structural features as they appear before it. The topographical map then is an image of the objects spread across the substrate plane.

The tip of the cantilever does not actually touch, in the common physical sense, the object beneath it but passes over and interacts with it by what are termed *aggregate atomic forces*, whose exact nature remains somewhat obscure, particularly for soft, deformable samples. A consequence of this "soft touch" interaction is that the sample is not physically perturbed by the motion of the tip (intentional physical

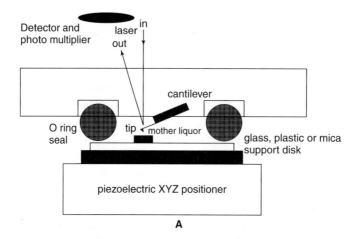

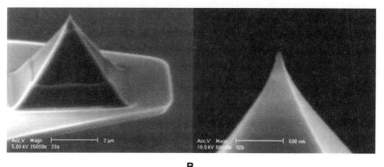

Fig. 6.4 (A) A schematic drawing illustrating the principles of an atomic force microscope. The vertical deflection that the cantilever tip experiences upon encountering some topographical feature on a specimen is amplified through a reflected laser beam, which is tracked and reported by a split diode photoelectric detector. Scanning takes place in a fluid-filled (or dry if preferred) cell of about 75 μL volume. The sample is translated in a raster manner by a piezoelectric positioner upon which the cell is mounted. Scan lines are accumulated in a serial manner and assembled into a three-dimensional image. (B) Scanning electron micrographs of AFM cantilever tips.

manipulation using the tip provides a useful investigative tool for some applications [61,65], but it is avoided when imaging). The scanning may be carried out in the dry state in air, or in fluids. This last feature is of great advantage for biological samples, including living cells and tissues.

There are two principal modes in which an AFM may be operated, and these are "contact mode" and "tapping mode." In contact mode, the stylus is simply "dragged" across the sample without losing contact with the surface. This, however, creates lateral friction, which is of little consequence if the sample being scanned is materially "hard," but damaging if the sample is "soft." Biological samples are characteristically "soft." In tapping mode, the AFM blind man taps his stick, the cantilever tip, as he moves ahead, lifting it slightly between each tap. This eliminates most lateral friction. The tapping mode [59,60] is used almost exclusively with biological samples such as tissues, cells, and their substituents, including macromolecules. The tip is not really tapped in this mode but constantly oscillates in the vertical plane. With soft biological samples where scanning must be relatively slow in order to capture detail, the acquisition time for a single image is usually from 1 to 4 minutes. More detailed procedures for scanning with AFM are provided in the references [61–65].

In defining the resolution of the method, it is necessary to discriminate between lateral resolution, points in the *xy* plane, and vertical resolution, the heights of features above the *xy* plane. Because the image is a convolution of the tip shape and the shape of the object beneath it (Fig. 6.5), the lateral resolution is modest, usually in the range of 1 to 2 nm at best. Lateral resolution is essentially limited by the acuity of the tip. Vertical resolution, however, is less dependent on tip shape or sharpness and is on the order of several angstroms. Thus, in making quantitative measures of dimensions, vertical values are used whenever possible. Spherically and cylindrically symmetrical objects are, therefore, straightforward to size accurately, as are center-to-center distances in arrays.

AFM achieves such extraordinary precision using such a crude principle as a tapping stick because of its ability to position an object, or determine its position, with extreme accuracy and its ability to amplify minute deflections experienced by the cantilever tip. The first of these is based on the application of piezoelectric positioning technology, which in practice provides "stepping motor" action at the

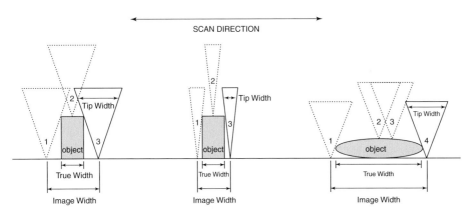

Fig. 6.5 Schematic illustration of the convolution of the shape of the AFM tip with the shape of the feature or particle being scanned. The side of the cantilever tip contacts the object and begins to produce a deflection of the cantilever before the tip apex actually reaches the object. Similarly, the opposite side of the tip is still in contact with the object even after the apex itself has passed. Thus, the total deflection implies a virtual lateral dimension for the object greater than its actual dimension. The difference between the virtual and actual dimension is a function of the width of the cantilever tip. The sharper the tip, the more accurate the observed dimensions and the greater the resolution attainable. As is evident in this illustration, the broadening effect becomes less significant as the width of the object becomes greater and its height less.

subnanometer scale. To detect small vertical fluctuations in tip position, a laser beam is reflected from the top surface of the cantilever as it moves over the substrate surface. Small deflections in the tip position are translated into large deflections, some distance away, of the reflected laser beam. These are tracked and recorded by a split diode, photomultiplier device

When dealing with any AFM sample, it is essential that it be made immobile on the substrate. If you touch an object and it moves or deforms, you cannot specify its position or shape. Thus, it is necessary that the object under study, including macromolecules like RNA, be firmly fixed to the substrate. With biological samples such as cells and viruses or large cellular organelles, if the substrate is first coated with poly-L-lysine, then the specimen will adhere tenaciously to the substrate. For specimens, such as nucleic acids, pretreatment of the substrate with the salts of divalent metal ions such as $MgCl_2$ suffices to hold RNA or DNA in place [66–69].

Scanning in fluids is technically more difficult for several reasons, as one might well imagine. For living specimens and for inanimate specimens as well, structure is certainly better preserved when it is fully hydrated. Some resolution, however, is sacrificed by scanning in fluids. Scanning in air is usually carried out with cantilever tips made of silicon, and these can be sharpened more efficiently than the tips made of silicon nitride that are generally used in fluids. Scanning of air-dried samples is common, considerably easier, faster, and yields higher resolution images in most cases. In the AFM images presented in this chapter, height above substrate is indicated by increasingly lighter color. Thus, points very close to the substrate are dark, and those well above the substrate are white.

6.8 Preparation of viral RNA samples for AFM

The preparation of viral RNA for AFM depends very much on the source and the intentions of the investigator. We utilized a wide variety of viruses ranging from $T = 1$ satellite viruses like STMV and SPMV to $T = 3$ viruses, including TYMV and poliovirus, to the helical, rod-shaped tobacco mosaic virus (TMV). The stability of the RNA from each virus was different, the structural forms it assumed varied, and the ease with which conformational transitions proceeded, or could be induced, differed dramatically. Transitions between structural forms ranged from facile in the case of poliovirus to very difficult in the case of the satellite viruses [66,67]. The ssRNA from poliovirus, for example, unraveled from a condensed mass into long chains of secondary structural elements over the course of an hour at room temperature. ssRNA from most $T = 3$ viruses transformed similarly over a few hours at moderate temperatures of around 50°C. The RNA from STMV, however, was inordinately stable and required extreme conditions to transform it into extended states [67].

Preparation of icosahedral virus RNA frequently began with standard extraction by phenol and chloroform [66]. Icosahedral virus RNA could be transformed from the conformation initially present from the phenol extraction to a still condensed but more linear form by exposure to temperatures between 25°C and 75°C. However, much secondary structure remained, and the RNA was clearly not in fully extended form, implying that the stability of local helical substructures was considerable. In the case of STMV, we were successful in further melting the RNA by heating it to 80°C in the presence of ethylenediaminetetraacetic acid (EDTA) or 3 M sodium chloride. The effectiveness of EDTA suggested that divalent metal ions continued to play a role in maintaining RNA structure even when it became linearized.

AFM of STMV RNA illustrated the major technical problems encountered in studies designed to visualize ssRNA. A feature of ssRNA that complicates analysis is the marked correlation between structure and stability. When ssRNA molecules in the range of 500 to 8,000 nucleotides assume structures that are highly base paired, contain mostly stem loops and other base-paired regions, and as a result are condensed, as they are when encapsidated, then they are remarkably stable to heat, pH, ribonucleases, and other physical and chemical agents. On the other hand, once unraveled, extended single-stranded segments appear, and either molecules begin aggregating severely through intermolecular base pairing using the newly available bases, or single-stranded segments quickly hydrolyze.

Disruption of viruses to reveal RNA was also dependent on the virus. The most extensive set of experiments was carried out on TYMV, a T = 3 icosahedral virus [70]. Among the physical agents used were temperature, dehydration, and freeze thawing. Chemical approaches included variation of pH, detergents, reducing agents, EDTA, a variety of proteases, and even low concentrations of phenol. Most approaches produced emission of RNA from TYMV, but only the most vigorous combinations of these agents induced the recalcitrant STMV to release nucleic acid. In the case of STMV, high pH, heating to 80°C and higher, and combination of the two had little effect on the intact virus, although STMV treated with proteinase K at 60°C did produce degradation as shown by light scattering [71]. In the end, we found that the best method for disruption of STMV that also permitted some degree of control was exposure of virus to buffer containing low concentrations of phenol in association with increases in temperature.

6.9 Atomic force microscopy of viral ssRNAs

Most technical details of the AFM analyses of viruses and their RNA molecules have been presented elsewhere [61, 66, 67, 70, 72] and will not be repeated here, but some complexities particular to ssRNA deserve further attention. Because bases are free to associate with complements in W–C base pairs, or with noncomplements in other kinds of base pairs (Hoogsteen, reverse W–C, etc.) [44], and because there are five almost unrestrained backbone angles to provide flexibility [73], an enormous range of secondary structures is possible for a long RNA molecule. Secondary structures will not utilize all pairing possibilities, and given the inherent conformational flexibility of single-stranded polynucleotides, a vast collection of possibilities exists for the formation of tertiary structures as well. In addition, RNA molecules may form complexes with a variety of cations, as well as proteins, and that further complicates prediction of their final folding pattern. These features create technical difficulties in visualizing ssRNA with electron microscopy or AFM. In the absence of divalent metal ions, ssRNAs rapidly aggregate through intermolecular hydrogen bonding; in their presence, the aggregation is even worse. Additionally, unlike DNA, an extended RNA chain is very susceptible to hydrolysis and easily fragments and degrades, particularly at elevated pH and temperature.

A question that occasionally arises in AFM analysis is whether an observed macromolecule is DNA or RNA. This is a problem when host cells or their contents, which contain all of these molecules, may be present or may inject contaminants into the sample. To definitively establish the identity of a macromolecule, pancreatic RNase A can be added to the sample [74]. If the molecules in question are RNA, then they are degraded into small fragments (Fig. 6.6C,D). If, on the other hand, the molecules are DNA, then they bind the RNase A along their length but remain uncleaved (Fig. 6.6A,B). As a consequence of the binding of protein, however, the diameter of the DNA fibers increases dramatically from 1.5 to 2 nm above the substrate to 3 or 4 nm heights.

An investigation of RNAs extracted from several different icosahedral viruses was carried out using AFM [66] that included poliovirus, Brome Mosaic Virus, TYMV, and

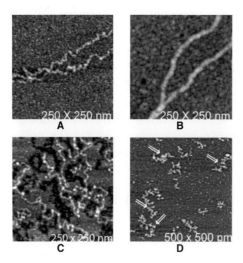

Fig. 6.6 (A) Two crossing strands of double-stranded calf thymus DNA as visualized by AFM. (B) A strand of calf thymus DNA after exposure to RNase A. The protein binds to the nucleic acid and results in a thick cord of the protein–nucleic acid complex. (C) The RNA extracted from poliovirus is visualized by AFM on the mica substrate. Both extended single-stranded segments and chains of dense secondary structural domains are apparent. (D) The same RNA has been treated with RNase A. The single-stranded regions are degraded, leaving only segments containing secondary structure. Double arrows indicate double-strand-containing regions that remain intact.

STMV, along with TMV. The viruses were extracted with conventional phenol/chloroform-based procedures and subsequently imaged over time as they passed through various conformational states or unfolded.

RNAs from poliovirus and STMV initially exhibited highly condensed, fairly uniform spherical shapes with diameters consistent with those expected from the interiors of their respective capsids. RNAs, when extracted with phenol from the capsids, appeared to maintain the same conformations as they had while encapsidated. At 25°C, poliovirus RNA gradually transformed into chains of globular domains having the appearance of thick, irregularly segmented fibers (Fig. 6.7). Poliovirus ultimately unwound further to reveal segmented portions of the fibers connected by single strands 0.5 nm to 1 nm in thickness. Virtually the same transformations were shown by the T = 3 virus RNAs included in the study. As tertiary structure was lost, it became evident that secondary structural elements were arranged in a sequential manner along the polynucleotide chains. STMV RNA (Fig. 6.8) was substantially more stable but exhibited an initial unfolding pattern that was similar to that of poliovirus RNA.

As a control, the RNA of TMV, which presumably has no secondary structure within the virion, was examined. TMV RNA exhibited no spherical mass upon extraction but, as expected, existed as a fully extended, naked thread of nucleic acid 0.5 nm to 1 nm height above the AFM substrate (Fig. 6.9). With time, at room temperature,

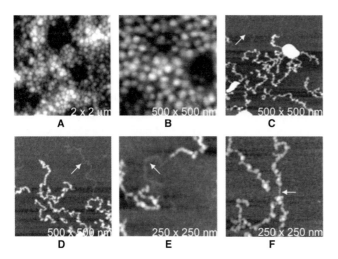

Fig. 6.7 In (A), and at higher magnification in (B), is genomic RNA extracted from poliovirus using phenol that was thawed after freezing and immediately imaged by AFM after spreading on magnesium treated mica. The RNA exists as uniform, compact spheres having the diameter of the inside of the virus capsid. With time, at room temperature, the spheres transform into the extended chains of structural domains seen in (C) and (D). (D) A segment of an unwound strand of RNA appears and is marked by an arrow. The very high, white objects in (C) are RNA spheres that have not yet fully transformed into strands. In some higher magnification images of the poliovirus RNA, like those in (E) and (F), segments of the chains of structural domains can be seen joined to others by brief intervals of completely unwound ssRNA. The connecting links are indicated by arrows.

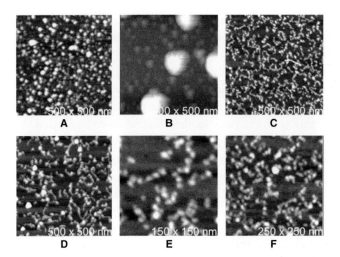

Fig. 6.8 In (A), and at higher magnification in (B), is RNA extracted using phenol from STMV. Like poliovirus RNA, it appears as uniform, compact spheres, but of diameter 10 nm, the diameter of the inside of the STMV capsid. These spherical particles appear to be stable almost indefinitely at room temperature. (C–F) Various samples of STMV RNA after heating to 65°C. The spheres transform into thick chains of structural domains. Because the STMV genomes are much shorter than the poliovirus genomes, the chains are significantly shorter. They are also in a higher state of condensation than the poliovirus RNA.

the TMV RNA strands began taking on the appearance of the unfolded poliovirus RNA as the TMV ssRNA acquired local secondary structures (Fig. 6.10), just as predicted [1–6]. As more time passed, the extent of condensed structure increased with the formation of multidomain chains. Presumably, tertiary structural interactions imposed themselves as well. Thus, by the time an hour or more had passed, the TMV RNA was significantly condensed. The folding of TMV RNA was essentially the reverse of the unfolding of the poliovirus RNA.

6.10 AFM results for extended STMV RNA

The open conformation of STMV RNA likely reflects the loss of some important tertiary interactions, but not the loss of significant secondary structural interactions [67]. As shown by AFM, the RNA at 65°C exists as short, thick chains of still highly condensed nucleic acid (Fig. 6.11), suggesting that secondary structure is maintained, but also significant tertiary interactions as well. The more extensive elongation of RNA molecules seen by AFM at higher temperatures, and in the presence of EDTA, suggests that residual tertiary interactions likely involve the participation of divalent cations. With elevated temperature and EDTA exposure, additional interactions were lost

AFM studies of Viral RNA Structure — 141

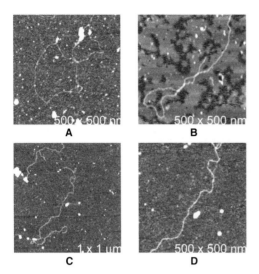

Fig. 6.9 (A–D) AFM images of RNA molecules extracted from the helical, rod-shaped virus TMV. The RNA shows no evidence of secondary structure, reflecting its conformation in the virion.

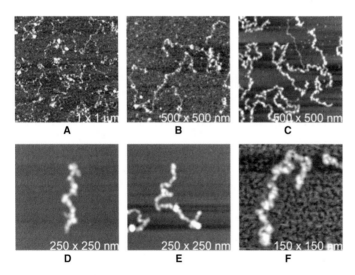

Fig. 6.10 With time in (A–C), local secondary folding domains begin to appear spontaneously along the previously extended strands of TMV RNA. The strands begin to assume the appearance of the RNA extracted from the icosahedral viruses. Ultimately, TMV RNA single strands in (D–F) coalesce into thick fibers of condensed folding domains very similar to those seen in the initial unfolding of icosahedral virus RNA, the transformation of spherical RNA masses into extended chains.

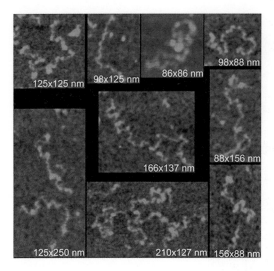

Fig. 6.11 Single molecules of STMV RNA that were phenol extracted from STMV and then heated to 75°C to 85°C in the presence of EDTA. The variation in light and dark along the chains indicates a height range of 0 to about 2.5 nm. The globular domains along the molecules presumably are stem-loop elements.

and the RNA extended to maximum lengths of about 280 nm. This, nonetheless, still represents a substantial compression. The molecules remain intact at those lengths. Single-stranded polynucleotides hydrolyze rapidly under the conditions required to attain this extended state; thus most of the linear chain must still be involved in secondary structure resistant to cleavage.

Cautious treatment of virus with phenol at elevated temperature promoted the emergence of linear chains of RNA from otherwise intact particles. Presumably, one or more pentameric capsomeres were lost or loosened enough to permit an end of the encapsidated RNA chain to escape the virion. The emergence of the RNA to make "pigtails" (Fig. 6.12A,B) was not a stochastic process, as all virus particles in the sample produced the RNA "pigtails" in a synchronous fashion, and all of the "pigtails," at least initially, were of about the same length. Some "pigtails" spontaneously elongated and produced extended chains as long as 200 nm (Fig. 6.12C,D). The fact that the capsids opened some portal to allow emission of an end of the RNA suggests that there may be some unique pentamer or other structural feature immediately adjacent to one end of the RNA molecule when it is encapsidated. Alternatively, it could be that the close proximity of the wandering end on the inside of the capsid weakens or unlocks a capsomere and causes it to dissociate from the virion (i.e. the RNA forces its way out).

We do not know if the free end is the 5' or the 3' end, but we favor 3'. The 3' end contains a tRNA-like structure, and in AFM images, we observe an enlargement at the distal terminus of the "pigtail." We also know from previous experiments that the tRNA-like structure is accessible on phenol-extracted RNA since it can be amino

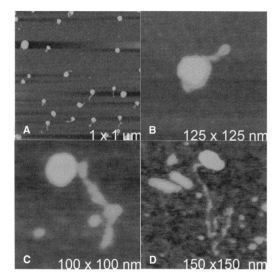

Fig. 6.12 When STMV particles were exposed to low concentrations of phenol at 75°C to 85°C, they developed "pigtails" of emerging RNA as seen in (A). The lengths of the pigtails within a population are roughly the same. This suggests that the initiation and rate of emission of the RNA is the same among particles. (B) A pigtail has just started to emerge from a disrupted virion. Note the bulge at the end of the strand. (C) A pigtail has elongated, but swollen areas indicative of structural domains are evident along its length. (D) The RNA has assumed fairly lengthy forms. The "pigtail" in (D) has a contour length of about 200 nm.

acylated with histidine [75]. Thus, it must be on the outside of the condensed, spherical mass of RNA in the encapsidated conformation.

Fig. 6.11 shows examples of STMV RNA molecules extracted from virions with phenol and then heated in the presence of EDTA to 80°C. The RNA has become much less compact and more extended, but it still exhibits a significant variation in height (or thickness) along its length. A fully extended ssRNA would have an average distance between consecutive phosphate groups of about 0.55 nm. Thus, a fully extended STMV RNA would have an expected length of approximately $1{,}058 \times 0.55$ nm = 582 nm. The maximum contour length of the STMV RNA in the AFM images is about half that. In addition, from previous work [74], we know how extended RNA appears in AFM images and the height to which it rises above the substrate plane. The RNA molecules illustrated in Fig. 6.11, even those of greatest extent, do not have the length, height, or appearance of fully extended RNA. The molecules we visualized still contain a substantial amount of secondary structure and are still condensed by about 50%.

The most plausible explanation is that the RNA exists as a chain of substructures (e.g. stem loops and pseudoknots that appear in a linear manner along the length of the RNA) (Fig. 6.3). If these small domains persisted, then they would explain the length, compression, and the heights above the substrate of the RNA molecules. The molecules seen in the AFM images (Fig. 6.11) also exhibit no prominent branches or accessory domains. Were branches of significant size present, they would likely be visible in AFM images. There are no branches suggestive of major domains, not only in the phenol-extracted RNAs (Fig. 6.8) but also in the long strands of RNA emerging from disrupted virions (Fig. 6.12).

6.11 ssRNA in T = 3 icosahedral viruses

As noted previously, ordered RNA has been observed in a number of T = 3, 30 nm diameter virus particles as well as T = 1 virions. Among those have been Bean Pod Mottle Virus [24], Flock House Virus [25], Pariacoto Virus [76], SPMV [30], and DYMV [23]. The most extensive arrangement of ssRNA, however, was that of TYMV [29]. A remarkable display of density appeared in TYMV difference electron density maps (Fig. 6.13A). Ultimately, based on details of the density masses (Fig. 6.13B) and application of icosahedral symmetry, nearly 70% of the RNA in the virus could be accounted for. Again, it was not possible to align nucleotide sequence with observed density, but the arrangement was consistent with a linear sequence of local secondary structures.

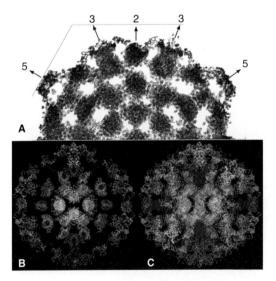

Fig. 6.13 (A) The difference electron density present within the capsid of TYMV and representing icosahedrally ordered ssRNA. (B,C) The distribution of secondary structural elements within the TYMV capsid as deduced from the electron density. Associated with each of the 60 icosahedral operators are two helical segments perpendicular to and intersecting dyad axes, and two lying parallel with and perpendicular to the fivefold axes. Two clusters of single-stranded loops occupy the pentameric and hexameric capsomeres. It is not apparent from the difference Fourier maps how these structural elements are linked. (A) All the secondary structural elements lying in a plane through the coplanar two-, three-, and fivefold axes. (B) All elements in the virus.

In a series of experiments in which TYMV capsids were systematically disrupted by a number of different physical and chemical factors, three kinds of results were obtained. In some experiments, the capsids simply burst by expansion, and the RNA was forced quickly from the capsid. In these cases, the RNA appeared as a spider-like formation of strands connected to a central mass still associated with the disrupted capsid (Fig. 6.14A). In some other experiments, the protein capsid was split into two or a few large fragments composed of many capsomeres. The fragments were joined by thick cables of significantly condensed RNA (Fig. 6.14B,C).

The third result (Fig. 6.14D–F) displayed the ssRNA in its most extended form, which closely resembled that of released poliovirus RNA and extended STMV RNA. Again, the molecules were linear arrays of thick and thin regions of arbitrary heights in the AFM images, and again the entire linear arrangement was substantially shorter than an extended RNA molecule would be. As with poliovirus RNA and STMV RNA, and clearly unlike TMV RNA, the simplest interpretation was that the molecule consisted of a linear sequence of secondary structural elements. TYMV is not believed to exhibit strong capsid protein–RNA interactions as do most ssRNA viruses, including STMV, possibly due to its high content of polyamines (77). In addition, TYMV appears to have the unusual capacity to form complete, but empty capsids in vivo [13,78]. Thus, it is unlikely that its assembly is like that of STMV. Most other T = 3 viruses, however, do exhibit strong electrostatic interactions between the interior of the capsid and the nucleic acid, and these may utilize similar assembly mechanisms.

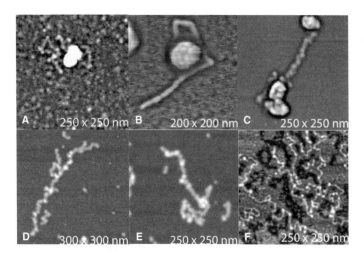

Fig. 6.14 AFM images of TYMV RNA from viruses disrupted by a variety of combinations of physical and chemical agents. (A) Virions that burst suddenly release spider-like splashes of nucleic acid. (B) A single thick cable of RNA emerges from a virion that has apparently lost one or two capsomeres. (C) A virion has split into two parts, but the parts remain connected by a cable of RNA. (D–F) The RNA molecules are seen in more extended form and appear very similar in AFM images to poliovirus and STMV RNA.

Intriguing findings emerged from work on Flock House Virus [26]. Using X-ray diffraction, stem loops of ssRNA were visualized as lying on icosahedral twofold axes in native particles. The helical segments were about 10 base pairs in length. Virus particles were then constructed with the same protein capsid, but containing not Flock House Virus RNA, but heterologous RNA of another unrelated virus, Pariacoto Virus. When X-ray diffraction analysis was repeated on crystals of the hybrid virus, stem loops of RNA were again found at the same locations and still forming a dodecahedral cage within the capsid. The heterologous RNA had a different nucleotide sequence, and almost certainly a different secondary structure, yet it still provided the stem-loop elements necessary to assure assembly of the virus and arrival at an ordered native structure.

An attempt was made in the author's laboratory (unpublished results) to use a "sliding window" [50,79] to predict the stem loops that would occur along the STMV RNA polynucleotide chain. Along the 1,058 nucleotide sequence, about 30 stem loops of appropriate sizes did indeed appear. If some conservative relaxation of restrictions on base pairing was applied, then stems were even more frequent and greater in length. It was subsequently discovered by randomization of nucleotide sequence, however, that almost any ssRNA of 1,000 nucleotides, having any arbitrary sequence, would similarly exhibit the capacity to form about 30 stable stem loops. That is, virtually any ssRNA will form a stem loop about every 30–35 nucleotides regardless of sequence. This explains the results obtained in the hybrid experiments involving Flock House Virus and Pariacoto Virus. Stem loops associated with capsid protein at the dyad sites are not specific stem loops, nor need they be of a particular size or connected together in the same way. So long as they exist as a linear chain of stem loops (Fig. 6.15) that can satisfy the protein-binding sites, then they will suffice.

Other recent work [80] on the assembly of STNV, which has a genome length of 1,239 nucleotides, suggests that the assembly mechanism of STNV may be similar to that proposed here for STMV. Assembly of STNV required ssRNA as an obligatory

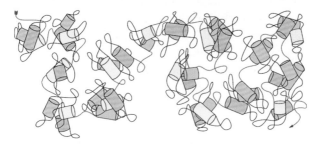

Fig. 6.15 X-ray diffraction evidence and AFM observations are most consistent with viral ssRNA molecules that exist inside virions as linear arrays of secondary structural domains like those illustrated schematically here. Each domain is a small ensemble of double helical stems and loops. The domains are connected by single-stranded linking regions.

component. In this case, it was the messenger (mRNA) transcript for the coat protein. Through identification of RNA aptamers bound by the coat protein, Bunka et al. [80] deduced that aptamers bound by coat protein existed in stem-loop substructures. Finally, they concluded that their results overall "were most consistent with an assembly mechanism based on kinetically driven folding of the RNA" [80].

There are additional reports of the encapsidation of heterologous ssRNAs by different viruses, chiefly T = 3 icosahedral viruses, and there are other interesting cases involving quite different viruses as well. The yeast retrotransposon Ty3, which is in a sense a proto-retrovirus, has an RNA genome. If the mRNA for its capsid protein is introduced into *E. coli*, then spherical virus-like particles are produced, and these contain the mRNA, not the natural genomic RNA [81]. Again, this points up the relative structural nonspecificity in many cases of the protein-nucleic acid interactions involved in the assembly of virus particles.

6.12 A model for assembly of STMV inspired by crystallography and AFM

The conformations of a viral RNA when it is encapsidated and when it unfolds probably mirrors the transient conformational states that it traverses as it folds into the encapsidated conformation. Indeed, there is experimental evidence for this [67]. The missing component is the coat protein that assists in the folding process. A sequential array of stem loops and/or pseudoknot substructures that interact directly with coat protein dimers and are ultimately marshaled into the icosahedral virion places rigorous and meaningful constraints on the virus assembly mechanism. The assembly of the virus, however, must be placed in the context of the cellular events that transpire during replication, translation, and packaging. These processes, though spatially confined, have a temporal sequence, and this must also enter into consideration.

As illustrated schematically (Fig. 6.16), upon infection and decapsidation the RNA assumes physiological conformations that support creation of replicative double-stranded RNA complexes, and, subsequently, translation into capsid protein (for a more detailed discussion, see Primrose and Dimmock [51] and Symons [82]). Eventually, the processes become competitive. The RNA molecules in the absence of capsid protein, however, maintain conformations appropriate to their roles in replication. If the RNA remains inclined toward replicative conformations, which cannot be encapsidated, then some agent must intervene and divert RNA folding from physiologically useful conformations to the sequence of stem-loop structures that can be encapsidated. The only plausible candidate is the capsid protein. STMV makes no other protein, and there is no evidence or likelihood that either the host cell or the helper virus provides assistance. The model for replication coupled with assembly (Fig. 6.16) is virtually the same as described for single-stranded DNA (ssDNA) phages [51].

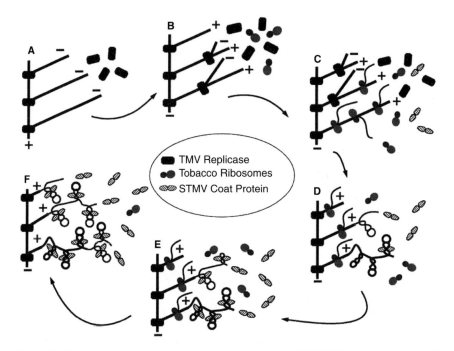

Fig. 6.16 Diagram illustrating the replication and packaging of STMV RNA as best understood from current data. (A) After infection, the positive-stranded viral RNA molecule is complexed by the TMV replicase, and multiple negative strands are synthesized. (B) Replicase molecules engage negative RNA strands transcribed from the positive strands and by an alternate process produce additional positive strands. Ribosomes begin to appear as a consequence of the accumulation of translatable RNA and compete with the replicase for positive-strand RNA. (C) Bound ribosomes synthesize coat protein even while the replicase continues its activity. Coat protein begins to accumulate.
(D) Ribosomes and coat protein provide the primary competition for positive-strand RNA. Ribosomes bind to translation initiation sites, while coat protein binds to metastable stem-loop elements along the RNA. (E) Coat protein binding to stem-loop elements causes positive-strand RNA to be removed through encapsidation. (F) In the late stages of replication and virus production, coat protein is dominant and proceeds to bind and encapsidate all remaining positive-stranded RNA and that newly produced by transcription from the remaining negative RNA strands, which are not packaged.

As the pool of capsid protein increases in the cell (Fig. 6.16), monomers associate as dimers, which are the binding species for RNA stem loops [21,28,83]. RNA strands emerging from the replicative complex form local, possibly transient stem-loop substructures, and these are bound and stabilized by protein dimers. While metastable stem-loop substructures might unravel in the absence of protein, in its presence stem loops are secured, and the production of new double-stranded replicative form RNA is terminated. Protein dimers bound to stem loops along an RNA molecule, maintained in close proximity to one another, experience protein-protein and additional protein–nucleic acid interactions and cooperatively coalesce into an icosahedral capsid (Fig. 6.17). This "cocondensation process" is in accordance with previous proposals [84].

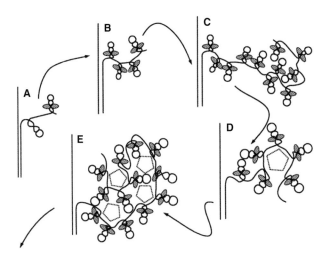

Fig. 6.17 Schematic diagram illustrating a path of assembly consistent with the structural model. Coat protein dimers begin binding the first stem-loop element to appear on the positive-strand RNA (A) and proceed binding contiguously along the RNA molecule, as in (B). When a sufficient number of stem loops, stabilized by the binding of coat protein, have been formed, then the complexes coalesce into a fluid, but disordered aggregate (C). (D) The complexes of RNA and coat protein sort themselves out through cooperative intramolecular interactions and the pentameric nucleus of a virion forms. (E) Additional RNA-protein subcomplexes add to the nucleus until the entire RNA-protein mass condenses into the icosahedral virion (protein outside, RNA inside).

Based on X-ray crystallographic and AFM evidence, a model was constructed of the encapsidated RNA of STMV [21] (Fig. 6.18). The model uses only A–U base pairs; thus it contains no nucleotide-specific secondary structure information. More recently, however, a secondary structure for STMV RNA was proposed that incorporated significant chemical probing data into a sliding window prediction algorithm [79,85]. The model (Fig. 6.18) was subsequently modified to incorporate that secondary structure, and it was further refined using energy minimization [86]. Currently, this is the only complete, though admittedly still speculative, model for an entire virus that includes both protein and RNA.

According to the assembly mechanism and model proposed above, the RNA and protein are coconspirators in a cooperative process with neither partner dominant. The RNA specifies, by formation of metastable substructures, the order of binding of protein dimers and their disposition along the nucleic acid. The protein subunits, through associations with one another, direct the overall organization of the RNA, guiding the formation of its tertiary structure, and the shaping of an icosahedral capsid around it. While the RNA may not direct capsid structure by specifying icosahedral interactions, it is certainly complicit in the assembly process. Protein subunits, even dimers so far as we know, do not coalesce into capsids in the absence of RNA. As for many ssRNA phages [84], it appears that only complex formation between a protein dimer and an

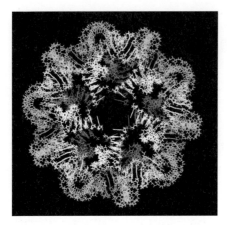

Fig. 6.18 A model using only A–U base pairs, shown here, was constructed that was consistent with the X-ray diffraction results having 30 stem loops with stems of 7 to 9 base pairs and disposed with their central pseudodyad axes coincident with icosahedral twofold axes. The stem loops are distributed in a linear fashion as deduced from the AFM images. The arrangement uses all 1,058 nucleotides and has a final, roughly spherical shape of diameter about 100 Å, the inside diameter of the STMV virion.

RNA stem loop can structurally complete a capsid unit and activate it, through creation of appropriate new interfaces, for aggregation and assembly into the icosahedral virion.

The proposed assembly pathway is consistent with the current RNA model [86], with the known biological events that occur in the replication of the virus [82,87–91], and with the properties of the interacting protein and RNA molecules [28,83]. It is also analogous to the sequence of events in the life cycles of ssRNA phages (see [92]) and some other viruses, for example, the ssDNA-containing fd bacteriophage.

6.13 AFM of large ssRNA viruses

Although most of the AFM work to date has been on small icosahedral viruses, some investigation has been carried out of large ssRNA viruses. These ssRNAs substantially magnify the problems of folding and compaction because of their extraordinary lengths. In the case of coronavirus, the RNA molecule is nearly a megabase in length and comprises the greater part of the mass of the entire virus particle. The ssRNA, complexed in some presently unknown way with proteins, is essentially encapsidated within a lipid bilayer extensively decorated with peplomers. The difficulty is in conceptualizing how a long RNA molecule with an extraordinary propensity to self-associate can be condensed into a relatively small particle. It must result in the same structure every time, and it must be brought about without the formation of snarls, tangles, knots, or intermolecular aggregates along the way.

It is unlikely that condensation is driven in an ordered manner exclusively through the formation of base-paired secondary structures that subsequently interact among themselves and with cations to fold in a directed manner to the final arrangement. This is expecting too much of RNA and the process of self-assembly. ssRNA, as we noted previously, is the molecular equivalent of double-sided "sticky tape." How it can be folded into a small package without falling into a plethora of intermediate kinetic traps is extraordinary indeed. How is the RNA protected from itself?

The explanation undoubtedly lies in the fact that the ssRNA is thoroughly complexed with proteins. This is exemplified by the rod-shaped TMV, in which coat proteins bind bases of the RNA and prevent intramolecular base pairing. A similar strategy must prevail in very long ssRNAs as well, but in a much more complex and sophisticated fashion. In TMV, the RNA is essentially linear; in coronavirus, it must be folded in some intricate manner so as to both prevent base pairing and yet produce profound condensation. The question then is what does this protein–nucleic acid complex look like and how does it come about.

Although we are far from answering that question, it is being addressed with AFM. Fig. 6.19 presents some early images of coronavirus RNA expelled by disrupted virions. They show that the encapsidated RNA, initially appearing as a spherical entity, in fact more closely resembles a "lock washer" with a distinct gap that separates the ends of a crescent shaped mass (Fig. 6.20). Coronavirus "pinwheels" that remain structurally

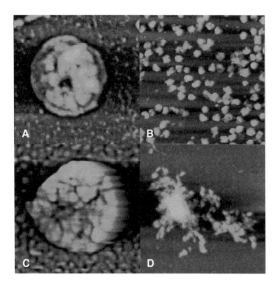

Fig. 6.19 (A) An intact coronavirus with the central mass of ssRNA encapsidated within a membrane envelope visualized by AFM. If the envelope is disrupted or removed by detergent, then the more or less spherical masses of ssRNA complexed with protein are seen in AFM images of (B) and (C). In some cases, vigorously disrupted virions release a splatter of the nucleic acid–protein complex as seen in (D). Coronavirus for these AFM studies was grown and purified in the laboratory of Professor Michael Buchmeier by Megan Angelini.

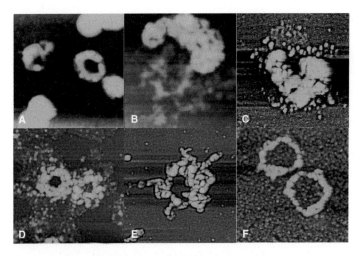

Fig. 6.20 Two forms were commonly observed in AFM images of coronavirus disrupted by various combinations of physical and chemical agents. The crescent shapes in (A), (B), and (C) indicate that the complex is originally a thick but linear mass that has folded upon itself around a center point to form a "lock washer" or toroidal form. (D,E) The central lumen of the "lock washer" expands to create "pinwheel" forms. (F) Large circular figures of the protein-RNA complex are observed.

consistent with the crescent forms. In all of these images, it is clear that the RNA is heavily decorated with proteins, again emphasizing the profound roles that proteins play in shaping the ultimate conformations assumed by encapsidated RNA molecules.

References

1. Fresco JR, Alberts BM, Doty P. Some molecular details of the secondary structure of ribonucleic acid. Nature. 1960;188:98–101.
2. Onoa B, Tinoco I Jr. RNA folding and unfolding. Curr Opin Struct Biol. 2004;14:374–9.
3. Spirin AS. Some problems concerning the macromolecular structure of ribonucleic acids. Prog Nucleic Acid Res. 1963;1:301–45.
4. Studier FW. Conformational changes of single stranded DNA. J Mol Biol. 1969;41:189–97.
5. Tinoco I Jr, Uhlenbeck OC, Levine MD. Estimation of secondary structure in ribonucleic acids. Nature. 1971;230:362–7.
6. Witz J, Strazielle C. Viral RNAs. In: Fasman GD, Timasheff SN, eds. Subunits in Biological Systems Part B. New York: Marcel Dekker; 1973:100–85
7. Caspar DLD. Viral and Rickettsial Infections of Man. New York: Lippencott; 1965.
8. Caspar DLD, Klug A. Physical principles in the construction of regular viruses. Cold Spring Harb Symp Quant Biol. 1962;1962:1–24.
9. Horne RW, Wildy P. Symmetry in virus architecture. Virology. 1961;15:348–73.
10. Ban N, Larson SB, McPherson A. Structural comparison of the plant satellite viruses. Virology. 1995;214:571–83.

11. Carrillo-Tripp M, Shepard CM, Borelli IA, et al. VIPERdb2: an enhanced and web API enabled relational database for structural virology. Nucleic Acids Res. 2009;37:D436–42.
12. Chiu W, Burnett RM, Garcea R. Structural Biology of Viruses. Oxford: Oxford University Press; 1997.
13. Kaper JM. The chemical basis of virus structure, dissociation and reassembly. In: Neuberger A, Tatum EL, eds. The Frontiers of Biology Series. Amsterdam: North-Holland; 1975:163–205.
14. Rossmann MG, Johnson JE. Icosahedral RNA virus structure. Annu Rev Biochem. 1989; 58:533–73.
15. Schneemann A. The structural and functional role of RNA in icosahedral virus assembly. Annu Rev Microbiol. 2006;60:51–67.
16. Valverde RA, Dodds JA. Some properties of isometric virus particles which contain the satellite RNA of TMV. J Gen Virol. 1987;68:965–72.
17. Valverde RA, Dodds JA. Evidence for a satellite RNA associated naturally with the U5 strain and experimentally with the U1 strain of TMV. J Gen Virol. 1987;67:1875–84.
18. Mirkov TE, Mathews DM, Du Plessis DH, Dodds JA. Nucleotide sequence and translation of satellite tobacco mosaic virus RNA. Virology. 1989;170:139–46.
19. Larson SB, Day JS, Nguyen C, Cudney B, McPherson A. Progress in the development of an alternative approach to macromolecular crystallization. Cryst Growth Des. 2008;8:3038–52.
20. Larson SB, Koszelak S, Day J, Greenwood A, Dodds JA, McPherson A. Three-dimensional structure of satellite tobacco mosaic virus at 2.9 A resolution. J Mol Biol. 1993;231:375–91.
21. Larson SB, McPherson A. Satellite tobacco mosaic virus RNA: structure and implications for assembly. Curr Opin Struct Biol. 2001;11:59–65.
22. Canady MA, Larson SB, Day J, McPherson A. Crystal structure of turnip yellow mosaic virus. Nat Struct Biol. 1996;3:771–81.
23. Larson SB, Day J, Canady MA, Greenwood A, McPherson A. Refined structure of desmodium yellow mottle tymovirus at 2.7 A resolution. J Mol Biol. 2000;301:625–42.
24. Chen ZG, Stauffacher C, Li Y, et al. Protein-RNA interactions in an icosahedral virus at 3.0 A resolution. Science. 1989;245:154–9.
25. Fisher AJ, Johnson JE. Ordered duplex RNA controls capsid architecture in an icosahedral animal virus. Nature. 1993;361:176–9.
26. Tihova M, Dryden KA, Le TV, et al. Nodavirus coat protein imposes dodecahedral RNA structure independent of nucleotide sequence and length. J Virol. 2004;78:2897–905.
27. Johnson JE, Rueckert RR. Packaging and release of the viral genome. In: Chiu W, Burnett RM, Garcea R, eds. Structural Biology of Viruses. Oxford: Oxford University Press; 1997:269–87.
28. Larson SB, Koszelak S, Day J, Greenwood A, Dodds JA, McPherson A. Double-helical RNA in satellite tobacco mosaic virus. Nature. 1993;361:179–82.
29. Larson SB, Lucas RW, Greenwood A, McPherson A. The RNA of turnip yellow mosaic virus exhibits icosahedral order. Virology. 2005;334:245–54.
30. Makino DL, Day J, Larson SB, McPherson A. Investigation of RNA structure in satellite panicum mosaic virus. Virology. 2006;351:420–31.
31. Lane SW, Dennis CA, Lane CL, et al. Construction and crystal structure of recombinant STNV capsids. J Mol Biol. 2011;413:41–50.
32. Noller HF. Structure of ribosomal RNA. Annu Rev Biochem. 1984;53:119–62.
33. Noller HF. RNA structure: reading the ribosome. Science. 2005;309:1508–14.
34. Xin Y, Olson WK. BPS: a database of RNA base-pair structures. Nucleic Acids Res. 2009; 37:D83–8.
35. Dirheimer G, Baranowski W, Keith G. Variations in tRNA modifications, particularly of their queuine content in higher eukaryotes. Its relation to malignancy grading. Biochimie. 1995;77:99–103.

36. Michel F, Westhof E. Slippery substrates. Nat Struct Biol. 1994;1:5-7.
37. Simons RW, Grunberg-Manago M, eds. RNA Structure and Function. Cold Spring Harbor, NY: Cold Spring Harbor Press; 1998.
38. Leontis NB, Santa Lucia JJ. Modeling and structure determination of nucleic acids. ACS Symposium Series. 1998;682:285-303.
39. Varani G, Pardi A. Structure of RNA. In: I Mattaj and K Nagai, eds. RNA–Protein Interactions. Oxford: IRL Press; 1994:120-50.
40. Cate JH. Crystal structure of a group I ribozyme domain: principles of RNA packing. Science. 1996;20:1678-85.
41. Holbrook EL. Structure of an RNA internal loop consisting of tandem C-A+ base pairs. Biochemistry (Mosc). 1998;37:11726-31.
42. Holbrook SR, Cheong C, Tinoco I Jr, Kim SH. Crystal structure of an RNA double helix incorporating a track of non-Watson-Crick base pairs. Nature. 1991;353:579-81.
43. Hunter DD, Castranova V, Stanley C, Dey RD. Effects of silica exposure on substance P immunoreactivity and preprotachykinin mRNA expression in trigeminal sensory neurons in Fischer 344 rats. J Toxicol Environ Health A. 1998;53:593-605.
44. Voet D, Rich A. The crystal structures of purines, pyrimidines and their intermolecular complexes. Prog Nucleic Acid Res Mol Biol. 1970;10:183-265.
45. Porschke D, Eigen M. Co-operative non-enzymic base recognition. 3. Kinetics of the helix-coil transition of the oligoribouridylic – oligoriboadenylic acid system and of oligoriboadenylic acid alone at acidic pH. J Mol Biol. 1971;62:361-81.
46. Craig ME, Crowther DM, Doty P. Relaxation kinetics of dimer formation by self complementary oligonucleotides. J Mol Biol. 1971;62:383-401.
47. Matthews D. Revolutions in RNA secondary structure prediction. J Mol Biol. 2006;359:526-32.
48. Olson WK, Zhurkin VB. Working the kinks out of nucleosomal DNA. Curr Opin Struct Biol. 2011;21:348-57.
49. Zuker M, Stiegler P. Optimal computer folding of large RNA sequences using thermodynamics and auxiliary information. Nucleic Acids Res. 1981;9:133-48.
50. Gultyaev AP, Pleij CW, Westhof E, eds. RNA Structure: Pseudoknots. Chichester, UK: John Wiley & Sons; 2005.
51. Primrose SB, Dimmock NJ. Introduction to Modern Virology. Oxford: Blackwell Scientific; 1980.
52. Crane HR. Principles and problems of biological growth. Sci Monthly. 1950:376-89.
53. Domingo E, Holland JJ. High error rates, population equilibrium and evolution of RNA replication systems. In: Holland J, Domingo E, Alquist P, eds. RNA Genetics. Boca Raton, FL: CRC Press; 1988:3-35.
54. Kurath G, Heick JA, Dodds JA. RNase protection analyses show high genetic diversity among field isolates of satellite tobacco mosaic virus. Virology. 1993;194:414-8.
55. Kurath G, Rey ME, Dodds JA. Analysis of genetic heterogeneity within the type strain of satellite tobacco mosaic virus reveals variants and a strong bias for G to A substitution mutations. Virology. 1992;189:233-44.
56. Kurath G, Dodds JA. Mutation analyses of molecularly cloned satellite tobacco mosaic virus during serial passage in plants: evidence for hotspots of genetic change. RNA. 1995;1:491-500.
57. Binning G, Quate CF. Atomic force microscope. Phys Rev Lett. 1986;56:930-3.
58. Bustamante C, Keller D. Scanning force microscopy in biology. Phys Today. 1995;48:32-8.
59. Hansma HG, Hoh JH. Biomolecular imaging with the atomic force microscope. Annu Rev Biophys Biomol Struct. 1994;23:115-39.
60. Hansma HG, Pietrasanta L. Atomic force microscopy and other scanning probe microscopies. Curr Opin Chem Biol. 1998;2:579-84.

61. Baclayon M, Roos WH, Wuite GJ. Sampling protein form and function with the atomic force microscope. Mol Cell Proteomics. 2010;9:1678–88.
62. Bragga PC, Ricci D. Atomic Force Microscopy in Biomedial Research. New York: Humana Press; 2011.
63. Goldsbury CS, Scheuring S, Kreplak L. Introduction to atomic force microscopy. In: AFM in Biology; Curr. Protoc Protein Sci, chapter 17, 2009;1–19.
64. Jena BP, Heinrich Horber JK, eds. Atomic Force Microscopy in Cell Biology. New York: Academic Press; 2002.
65. Morris VJ, Kirby AR, Gunning AP. Atomic Force Microscopy for Biologists. 2nd ed. London: Imperial College Press; 2010.
66. Kuznetsov YG, Daijogo S, Zhou J, Semler BL, McPherson A. Atomic force microscopy analysis of icosahedral virus RNA. J Mol Biol. 2005;347:41–52.
67. Kuznetsov YG, Dowell JJ, Gavira JA, Ng JD, McPherson A. Biophysical and atomic force microscopy characterization of the RNA from satellite tobacco mosaic virus. Nucleic Acids Res. 2010;38:8284–94.
68. McPherson A, Kuznetsov YG. Imaging of cells, viruses, and virus-infected cells by atomic force microscopy. In: Mendez-Vilas A, ed. Current Microscopy Contributions to Advances in Science and Technology. Badajoz, Spain: Formatex Research Center; 2012:540–8.
69. Kuznetsov YG, Malkin AJ, McPherson A. AFM studies on the mechanisms of nucleation and growth of macromolecular crystals. J Cryst Growth. 1999;196:489–502.
70. Kuznetsov YG, McPherson A. Atomic force microscopy investigation of turnip yellow mosaic virus capsid disruption and RNA extrusion. Virology. 2006;352:329–37.
71. Day J, Kuznetsov YG, Larson SB, Greenwood A, McPherson A. Biophysical studies on the RNA cores of satellite tobacco mosaic virus. Biophys J. 2001;80:2364–71.
72. Kuznetsov YG, McPherson A, eds. Atomic force microscopy (AFM) in the imaging of viruses and virus infected cells. Microbiol Mol Biol Rev. 2011;75:268–85.
73. Berman HM, Olson W, Beveridge DL, Westbrook J, Gelbin A, Deneny T, Hsieh S-H, Srinivasan AR, Schneider B. The RNA data base. Biophys. J. 1992;63:751–9.
74. Kuznetsov YG, McPherson A. Identification of DNA and RNA from retroviruses using ribonuclease A. Scanning. 2006;28:278–81.
75. Felden B, Florentz C, McPherson A, Giege R. A histidine accepting tRNA-like fold at the 3'-end of satellite tobacco mosaic virus RNA. Nucleic Acids Res. 1994;22:2882–6.
76. Tihova M, Dryden KA, Thuc-vy LL, et al. Nodavirus coat protein imposes dodecahedral RNA structure independent of nucleotide sequence and length. J Virol. 2004;78:2897–905.
77. Beer SV, Kosuge T. Spermidine and spermine – polyamine components of turnip yellow mosaic virus. Virology. 1970;40:930–8.
78. Michels B, Leimkuhler M, Lechner MD, Adrian M, Lorber B, Witz J. Polymorphism of turnip yellow mosaic virus empty shells and evidence for conformational changes occurring after release of the viral RNA. A differential scanning calorimetric study. Eur J Biochem. 1999;264:965–72.
79. Bleckley S, Schroeder SJ. Incorporating global features of RNA motifs in predictions for an ensemble of secondary structures for encapsidated MS2 bacteriophage RNA. RNA. 2012;18:1309–18.
80. Bunka DHJ, Lane SW, Lane CL, et al. Degenerate RNA packaging signals in the genome of satellite tobacco necrosis virus: implications for the assembly of a T = 1 capsid. J Mol Biol. 2011;413:51–65.
81. Larsen LS, Kuznetsov Y, McPherson A, Hatfield GW, Sandmeyer S. TY3 GAG3 protein forms ordered particles in Escherichia coli. Virology. 2008;370:223–7.

82. Symons RH, Haseloff J, Visvader JE, et al. On the Mechanism of Replication of Viroids, Virusoids, and Satellite RNAs. Orlando, FL: Academic Press; 1985.
83. Larson SB, Day J, Greenwood A, McPherson A. Refined structure of satellite tobacco mosaic virus at 1.8 A resolution. J Mol Biol. 1998;277:37–59.
84. Knolle P, Hohn T. Morphogenesis of RNA Phages. Huntington, NY: Cold Spring Harbor Press; 1975.
85. Reeder J, Steffen P, Giegerich R. pknotsRG: RNA pseudoknot folding including near-optimal structures and sliding windows. Nucleic Acids Res. 2007;35:W320–4.
86. Zeng Y, Larson SB, Heitsch CE, McPherson A, Harvey SC. A model for the structure of satellite tobacco mosaic virus. J Struct Biol. 2012;180:110–16
87. Burke DC, Russell WC. Control Processes in Virus Multiplication. Cambridge: Cambridge University Press; 1975.
88. Dodds JA. Structure and function of the genome of satellite tobacco mosaic virus. Can J Plant Pathol. 1991;13:192–5.
89. Maramorosch K. Viroids and Satellites. Molecular Parasites at the Frontier of Life. Boca Raton, FL: CRC Press; 1991.
90. Martin SJ. The Biochemistry of Viruses. Cambridge: Cambridge University Press; 1978.
91. Pritsche C, Mayo X. Satellites of plant viruses. In: Mandahar CL, ed. Plant Viruses. Boca Raton, FL: CRC Press; 1989:289–321.
92. Casjens S. Nucleic acid packaging by viruses. In: Casjens, S. Virus Structure and Assembly. Boston: Jones and Bartlett; 1985:76–147.

Mathilde Bercy, Pierre Mangeol, Thierry Bizebard, N. Kyle Tanner,
Josette Banroques, and Ulrich Bockelmann

7 Investigating RNA structure and folding with optical tweezers

7.1 Introduction

A new class of techniques to investigate the structure and dynamics of biomolecules and their assemblies has developed rapidly over the past 20 years. In contrast to more classical approaches of molecular biology, these techniques probe single molecules, do not require synchronization, and avoid averaging over molecular ensembles, which are likely to be heterogeneous in behavior or in composition. Temporal fluctuations and molecule-to-molecule variability thus become directly observable and thereby give access to details of the physical mechanisms underlying a biological process. Typical techniques for probing single molecules include force measurements, fluorescence measurements, and electrical measurements of translocation through nanopores. Among these techniques, force measurement has the specificity to allow direct manipulation of the molecule of interest while measuring its response. By using atomic force microscopy (see Chapter 6), magnetic tweezers, or optical tweezers, one can pull on individual molecules and measure forces in a typical range of 0.1–100 pN. Initially, force measurements have mostly been used to investigate DNA mechanical properties [1] and stepping mechanisms of molecular motors like myosins and kinesins [2, 3]. More recently, single-molecule studies of RNA-based systems have become increasingly popular, with prominent examples including the mechanical unfolding of RNA structures [4–8], the observation of viral RNA helicases activity [9, 10], and the first studies on the assembly and function of the ribosome [11–14]. A number of reviews have been published on single-molecule force measurements of DNA and RNA [15–19].

RNA is found to be included in a large and increasing number of cellular events. While its role in protein translation has been known for a few decades, its catalytic activities, through the spliceosome or ribozymes, or regulatory activities, through microRNAs, small interfering RNAs, messenger RNAs, or other noncoding RNAs, have been discovered recently, and they have been identified as important roles of RNA in the cell. Mechanistic details of these functions are mostly due to the specific structures that RNA adopts, which makes the study of its folding of prime interest. However, although the study of RNA folding is not new, traditional techniques give only an average picture of the RNA behavior, and they are not likely to give access to the multiplicity of conformations in which RNA can fold, or to transient states and their kinetics; that is, the folding process is largely unknown, and marginal structures are unlikely to be observed, which gives only a partial view of the status of the RNA. Moreover, since proteins are invariably associated with RNA folding and structure in vivo, understanding how RNA folds in the presence of proteins and how RNA interacts with proteins appears to be equally important for unraveling the complexities of the RNA world. The effect of protein binding on RNA stability or the mechanistic details of RNA-protein assembly are examples that underlie RNA folding in a cellular environment and are very difficult to examine with traditional biochemistry techniques.

This chapter is devoted to showing how the application of optical tweezers can be used to tackle some of the difficult challenges faced when investigating RNA structure and folding. In Section 7.2, we provide background on single-molecule force measurements and present a typical experimental implementation of force measurements on RNA with double-optical tweezers. In Section 7.3, we illustrate the possibilities and limitations of the technique with two selected examples. We show how optical tweezers are used to probe the structure and the folding dynamics of RNA hairpins (Section 7.3.1), as well as to study a more complex system, consisting of a fragment of the 23S ribosomal RNA (rRNA) of *Escherichia coli* in the presence of a protein specifically binding this structure (Section 7.3.2). Section 7.4 contains the concluding part of the chapter.

7.2 Single-RNA force measurements with optical tweezers

In order to probe a single RNA molecule, one needs a machine able to access its scale. With a base-pair distance smaller than a nanometer, its folding process occurring in the millisecond range at energies comparable to the thermal energy, a single RNA can only be studied with extremely precise tools.

Over the past 20 years, many groups have been developing and improving such tools, and now single RNA molecules are mainly probed mechanically with optical-tweezers-based setups. Optical tweezers originate from the seminal work of Arthur Ashkin [20] where he found that micron-sized dielectric particles could be trapped at the focus of a converging laser beam. The beam trapping the particle could then easily be steered and thereby act as microscopic tweezers. It was then shown that forces in the piconewton range could be measured acting upon a particle trapped with optical tweezers [21], which opened the way to the mechanical study of single molecules.

Currently, the most precise and most stable configuration of optical tweezers is called double-optical tweezers [22]. This configuration consists of focusing two laser beams (originating usually from a single laser source) by a single microscope objective. The resulting two focal points can then be used to trap micron-size beads. Force measurement on these beads is often done by an interferometric method, known as back-focal-plane interferometry [23]; this enables sub-piconewton force measurements with millisecond time resolutions. Sub-nanometer displacement between the two optical tweezers is achieved with very stable steering devices, such as piezo-electric mirror mounts operating in a closed regulation loop with integrated position sensors. Additional stability in the measurement is achieved by shifting the frequency of one of the two laser beams [24], which eliminates the effect on force measurement of the interference naturally occurring when using a single laser source to create the two optical tweezers.

To probe a single RNA molecule with a double-optical tweezers, one has to isolate it and pull on its extremities by attaching the latter to two beads trapped in the double-optical tweezers (Fig. 7.1). Usually the RNA molecule of interest is too small to be easily manipulated. Therefore, the fragment of RNA is incorporated into a much larger construction (i.e. a few kilobases-long nucleic acid molecule). The attachment of the extremities of the molecule on two beads has to be done through chemical modification; direct modifications on RNA are technically difficult, and it is therefore quite natural that groups have used DNA, which is much more versatile.

Such a molecular construction can be built in the following way:

- A few kilobases-long RNA sequence, including the fragment of interest, is transcribed with an RNA polymerase from an appropriate DNA template.
- Two double-stranded DNA molecules, having sequences matching the sequence of the RNA molecule outside the fragment of interest, are synthesized by PCR. For each of these DNAs, one ensures that a chemical modification to be used later on is added (typical modifications are biotin and digoxigenin that are recognized, respectively, by functionalized beads covered with streptavidin and anti-digoxigenin). For one of these DNAs, the modification can be incorporated directly by using a modified primer in the PCR. For the other one, the modification can be incorporated after the PCR by creating a 5' overhang on the DNA molecule with an appropriate restriction enzyme and then enzymatically filling this overhang with chemically modified nucleotides.
- One hybridizes the single-stranded RNA with its complementary DNA strand from each PCR. This creates a molecular construct, where the RNA sequence of interest is flanked on both sides by double-stranded nucleic acid "handles," each one a few kilobase-pairs long.

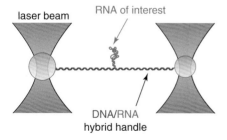

Fig. 7.1 Principle of a double-optical tweezers assay. A molecular construction consisting of the RNA of interest flanked by two DNA/RNA hybrids is attached at its extremities to two beads held by double-optical tweezers. The two beads can be of different size in order to facilitate the experiment.

A schematic view of the resulting construction attached between two beads is presented in Fig. 7.1. In practice, the molecular construction is attached to the functionalized beads by simple mixing. When the two different beads are close enough to a molecule in solution, then the specific attachment between the beads can form. Once all the described steps have been performed successfully, one can start to probe a single RNA molecule. Protocols of the steps have been published elsewhere [7, 25].

7.3 Probing RNA and RNA-protein interactions: selected examples

7.3.1 Probing the structure and the folding dynamics of RNA hairpins

An RNA hairpin is the simplest secondary structure formed by an RNA molecule. It is a nucleic acid chain that contains two complementary sequences, which can thus form a double helix on itself. It contains two parts: a double-stranded stem and a single-stranded loop (Fig. 7.2). The size of each part is fixed by the nucleotide sequence of the initial chain. A hairpin represents a model system for studying the self-assembly of

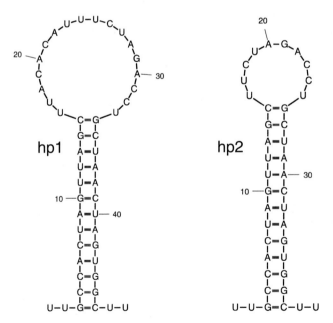

Fig. 7.2 The two hairpins used have the same stem (double-stranded helix), but their loops are different. The loops contain 18 bases in hp1 and only 10 in hp2. The energies of the two hairpins have been calculated (with mFold), and they are close: $\Delta G_{hp1} = 90$ kJ/mol and $\Delta G_{hp2} = 92$ kJ/mol ($\pm 5\%$).

nucleic acid chains. In addition, hairpins are very useful in single-molecule studies since they can be unfolded and refolded repeatedly.

As explained in the previous section, an RNA hairpin can be inserted between two DNA/RNA-hybrid nucleic acid handles and captured with a double-optical-tweezers device. The construct can thus be stretched and released several times, and the unfolding and refolding of the hairpin recorded. Two different RNA hairpins, hp1 and hp2, were used here. They are presented in Fig. 7.2.

In this subsection, we take the two hairpins as a model system to illustrate the information that can be extracted from single-molecule force measurements, the corresponding theoretical background, and the procedures for data analysis. We will consider the relation between the energy landscape determined by the RNA sequence and the measured force signals, the difficulties that arise from the out-of-equilibrium character of certain force-induced structural transitions, and the fitting procedures used to determine thermodynamic as well as kinetic parameters. We first consider force measurements performed at various time-independent distances between the two optical traps and then afterward the measurements performed where the traps are moved with respect to each other with constant velocity. Parameters of the force-induced unfolding/folding transitions that will be extracted are as follows: (1) the free energy of the hairpin structure, (2) the change in length of the RNA, (3) the folding and unfolding rates, and (4) the position of the transition barriers to folding and unfolding. To the best of our knowledge, no other single technique exists that is able to provide this combined information.

In addition, these single-molecule experiments give access to statistical distributions, (i.e. information that goes beyond the average values that are typically obtained by classical ensemble measurements). Of course, the precision of statistical values, like averages and mean-square deviations, derived from an ensemble of individual measurements increases with the size of the statistical ensemble. It would certainly be possible to improve the precision of some of the values derived in this subsection by increasing the number of individual measurements, but this does not affect the description of the principles and of the data analysis methods that represents our main goal.

The energy landscapes of the hairpin structures under external load are schematically represented in Fig. 7.3. As the sequence of the stem is identical for hp1 and hp2, the first part of their landscapes is the same. The variation in this part reflects the different binding energies of the different base pairs composing the stem. The wider loop of hp1 leads to the longer decrease of the hp1 energy landscape as compared to the one of hp2. The landscapes tilt with increasing force F, since the external mechanical load subtracts a work of $F \cdot l$ from the zero-force energy landscape, with the change in length l corresponding to the horizontal axis of the schema.

If the energy barrier separating the folded and unfolded states is not too high compared to the thermal energy $k_B T$, then the hairpin is able to spontaneously flip between the two states. Such flips thus correspond to the "thermal breathing" of

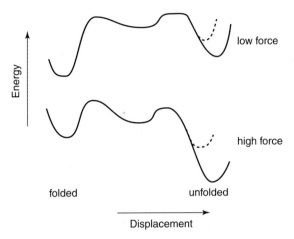

Fig. 7.3 Schematic representation of the energy landscapes of hairpin structures hp1 (solid) and hp2 (dashed). At low force the local minima of the folded states (left) lie below the local minima of the unfolded states, while the opposite is true at high force.

the hairpin at thermal equilibrium. They are observed on hairpin hp2, as shown in Fig. 7.4. The experimental trace corresponds to a succession of unfolding and folding events occurring at constant extension. In this case, thermal fluctuations are sufficient to trigger both the unfolding and refolding of the hairpin. There is a value of the extension where the probabilities of the two states are close. If they are equal, then the energies of the folded and unfolded states coincide. For a somewhat greater (respectively lower) extension, the folded state is energetically above (below) the unfolded state. As shown in Fig. 7.4, at smaller extensions the folded state is indeed more occupied than the unfolded one, while the opposite is true at higher extensions.

At any force, there is a nonzero probability that thermal fluctuations trigger the unfolding or folding of the hairpin, but the spontaneous event is not observed if the corresponding dwell time significantly exceeds the timescale of the measurement. For hp1, no force value exists where spontaneous unfolding and folding are both observed. As shown in Fig. 7.5, spontaneous unfolding of this hairpin can occur with a dwell time as long as 5 s when the construct is maintained at a fixed extension corresponding to a force of about 12 pN. To allow refolding of hp1 with typical dwell times of a few seconds, the tension has to be reduced below 11 pN. In practice, the duration of an individual measurement is limited to a few minutes because of temporal drift and the finite stability of both the construct and the attachments to the beads.

We now come to the case where the mobile optical trap is moved away from the fixed trap with a constant velocity. In the experiments presented in this chapter, we always used a displacement velocity of 50 nm/s. Although this value implies a relatively long duration of several seconds for a typical stretch/release cycle, and thus belongs to what is considered a slow displacement in the field of single-molecule

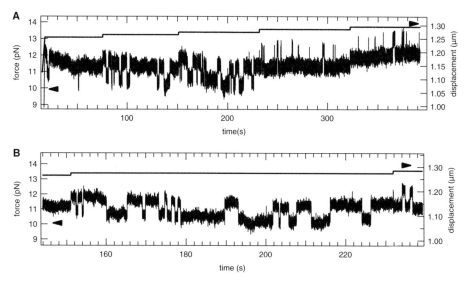

Fig. 7.4 Force flipping between the folded and unfolded states at constant extension as a function of time (hairpin hp2). (A) The extension is increased by steps (upper curve). At the beginning of the experiment, the folded state is favored. For the intermediate extensions, the hairpin flips rapidly between the two states. At the greater extension, the unfolded state is favored. (B) An enlargement of the zone of rapid flips. The dwell times on each state are exponentially distributed.

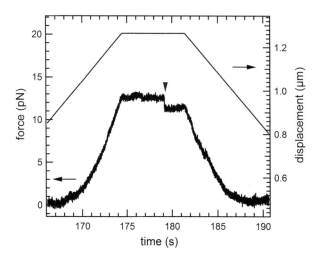

Fig. 7.5 Force response of hairpin hp1 versus time during a particular extension sequence (thin, upper curve). The construct is first stretched (at 50 nm/s) to a length where it remains folded and maintained to this position. After about 5 s, the hairpin opens (force drop, arrow). The traps are then approached again. During the release of the construct, the refolding of the hairpin is observed.

force measurements, it can already lead to nonequilibrium effects as will be shown subsequently.

Typical force versus extension curves are presented in Fig. 7.6. The first part of the curves corresponds to the elastic response when stretching the handles. Force increases with an increasing slope until, at around 13–16 pN, a sudden drop in force (by about 1 pN) is observed. The latter corresponds to unfolding of the hairpin. After the sudden event, the force increases again. For the hairpins used here, no intermediate states are observed during the unfolding event. This is attributed to the stem sequence, which begins with three G-C pairs. When the force becomes sufficiently high to unfold this stable initial sequence at the bottom of the stem, then the whole stem unfolds in a cooperative manner. When the stress is released by bringing the traps closer to each other, the hairpin can refold, and the construct returns to its initial structure. The force versus extension curves that are measured during unloading are similar in shape to the ones measured upon stretching, but a hysteresis is observed between the two types of curves. This hysteresis shows that the unfolding and folding do not occur at thermal equilibrium (i.e. for typical dwell times of the transitions, the change in force induced by the displacement is not negligible). As explained previously, hp1 exhibits longer dwell times due to its longer loop, and hence a more pronounced hysteresis is observed for hp1 than for hp2. This result qualitatively agrees with unfolding measurements on DNA hairpins, where it has been shown that the magnitude of the hysteresis increases with increasing stability of the stem for constant

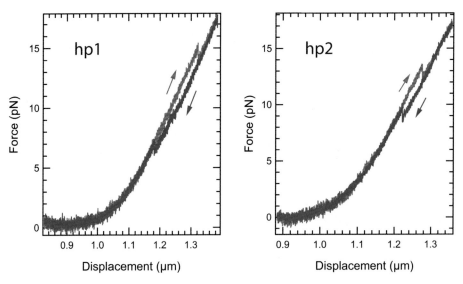

Fig. 7.6 Force versus extension curves at 50 nm/s of the hairpins hp1 (left) and hp2 (right). The orange curves represent the stretching of the construct when the traps are moved apart from each other. The force drop corresponds to the unfolding of the hairpin. The green curves represent the reverse process: the stress is released, and the hairpin refolds. The hysteresis is notably greater for hp1 than for hp2 due to its longer loop.

loop size and with increasing loop size for fixed stability of the stem [26]. In our case, the hysteresis is significantly smaller for the RNA hairpin hp2 (4.5 pN) than for hp1 (7.5 pN). Although not explicitly shown here, it follows from the previously discussed consideration that the hysteresis increases with increasing displacement velocity.

Variability of the unfolding and folding forces is seen after performing repeated stretching and release cycles on the same hairpin molecule. For hp1, for example, the opening force varies by about 1 pN around its average value. When an ensemble of different molecules of the same hairpin is measured, the observed dispersion is slightly broader; but there is a minimum force below which the unfolding probability is so small that we never observed an unfolding event. Representing a histogram as a function of the hysteresis rather than the absolute values of the unfolding and refolding forces enables us to decouple the intrinsic variations caused by the stochastic nature of the process from force offsets that can be caused by small differences between the experimental conditions of different individual measurements (e.g. by a dispersion in size and optical index of the beads). As shown in Fig. 7.7, the measured histograms of the hysteresis can be fitted to Gaussian curves of sizable width, reflecting the stochastic nature of the thermally activated process.

We now describe two different methods to estimate the free-energy change associated with unfolding and folding of a secondary structure. The first method relies directly on the force-measurement curve of the unfolding transition (Fig. 7.8) and is applicable when the experiment is done close to equilibrium. In the initial state i, when the hairpin is still folded, the total free energy G_i of the system can be written

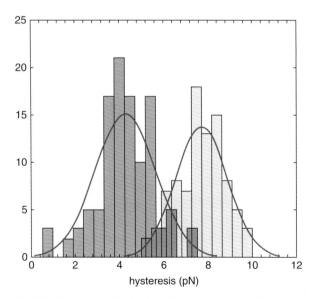

Fig. 7.7 Histogram of the hysteresis measured on hp1 (gray) and hp2 (purple). The histograms are fitted with a Gaussian curve. Due to the loop length difference of the two hairpins, the hysteresis is notably greater for hp1 (~7.5 pN) than for hp2 (~4.5 pN).

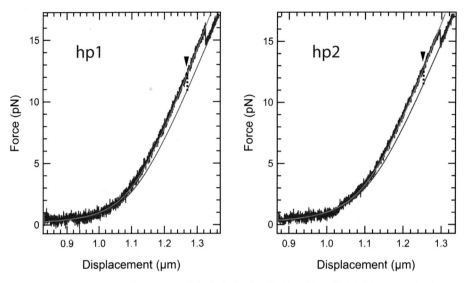

Fig. 7.8 Force versus extension curves of the hairpins hp1 (left) and hp2 (right) (blue curves). The solid lines are fits of the data. Orange curve: Odijk solution of the worm-like chain model [27]; purple: approximate solution taking the unfolded single-stranded RNA hairpin into account [14, 28]. The arrow indicates the average of the unfolding and folding forces.

as the sum of two terms, $G_i = El_i + G_{RNA,i}$, where El_i is the total elastic energy of the system including the trapped beads and the molecular construct, and $G_{RNA,i}$ is the free energy retained in the hairpin. In the final state f, just after unfolding the hairpin, the free energy reads $G_f = El_f + G_{RNA,f}$, with the total elastic energy El_f and the free energy $G_{RNA,f}$ retained in the unfolded RNA. El_f includes the elastic energy of the newly formed and stretched single-stranded RNA. We have $El_i > El_f$ because of the force drop at the opening of the hairpin, and we have $G_{RNA,f} > G_{RNA,i}$ because of the opening of the base pairs of the hairpin stem. The relative displacement of the traps is negligible during the sudden folding and unfolding events. When the opening of the hairpin occurs close to equilibrium, the free energies of the initial and final states can be considered as equal $G_i = G_f$. Consequently, the free-energy change of the hairpin $\Delta G_{hairpin} = G_{RNA,f} - G_{RNA,i}$ is given by $\Delta G_{hairpin} = El_i - El_f$. Therefore, $\Delta G_{hairpin}$ can be estimated as the area between the force/displacement curves observed before and after unfolding.

The precision of this free-energy estimation relies on the quality of the fits of the two parts of the curve. The first part, corresponding to the elastic response of the DNA/RNA hybrid handles, is well described by the worm-like chain (WLC) model, commonly used to describe the elasticity of double-stranded DNA. This model gives the mean extension $<x>_{WLC}$ of the construct as a function of force [27]:

$$<x>_{WLC} = L_0 \left(1 - \frac{1}{2}\sqrt{\frac{k_B T}{F l_p}} + \frac{F}{K} \right) \tag{7.1}$$

where L_0 is the crystallographic length of the construct, l_p is its persistence length, K is its elastic modulus, and k_BT is the thermal energy. In our measurements, the displacement of the beads in the traps has to be taken into account. This is easily achieved by considering an additional term in the mean extension $<x> = <x>_{WLC} + F/k_{trap}$, where k_{trap} is the equivalent stiffness of the traps. To theoretically describe the second part of the curve, the elastic response of the unfolded RNA structure should be taken into account. Although, to the best of our knowledge, no full theoretical description exists for the elasticity of complex, single-stranded RNA structures, approximate solutions exist that can be used to fit the data [14, 28].

This first method can be applied to any secondary structure of interest, provided that the experiment can be performed sufficiently close to equilibrium. If the structure unfolds in different substeps, it can be applied on each step.

The force versus extension curves can also be used to derive the length Δx released when opening the hairpin. It is given by the length difference of the construct after and before the hairpin unfolds, at a given force (more details are given in the next section). This leads to $\Delta x_{hp1} = 23\pm4$ nm and $\Delta x_{hp2} = 16\pm4$ nm. Taking into account the width of the double helix stem in the folded state (about 2 nm), these values correspond to a length of the stretched RNA of 0.48 and 0.44 nm/nucleotide, respectively.

The second method to derive free-energy changes relies on the fact that a hairpin behaves like a two-state system in an external field at finite temperature, the states being folded and unfolded. After a series of stretching and releasing is applied to a hairpin construct, it is possible to trace the probability of opening versus force (Fig. 7.9). At equilibrium, the probability of one of the states of the two-state system is given by [4, 28]

$$p(F) = \frac{1}{1+\exp\left(\frac{E(F)}{k_BT}\right)} \tag{7.2}$$

where $E(F) = (F - F_{1/2})\Delta x$. The force $F_{1/2}$ corresponds to the midpoint of the transition, $p(F_{1/2}) = 1/2$. The additional contour length released during unfolding Δx can either be measured as described previously or obtained by fitting the experimental data to the probability function $p(F)$. The resulting change in free energy $\Delta G(F_{1/2}) = F_{1/2}\Delta x$ includes the free-energy change corresponding to stretching the generated single strand from $F = 0$ to $F_{1/2}$.

In Table 7.1, we present the free-energy changes estimated by the two methods and theoretical predictions obtained with the mFold program (http://mfold.rna.albany.edu). For the calculations, we used [K$^+$] = 50 mM; [Mg^{2+}] = 5 mM (corresponding to the salt conditions used in the measurements) and a temperature of 29°C (corresponding to our sample temperature [14]). The salt correction is calculated for a single-stranded DNA molecule of equivalent base sequence and then applied to the values obtained with RNA. The ΔG values calculated for hp1 and hp2 are very close because

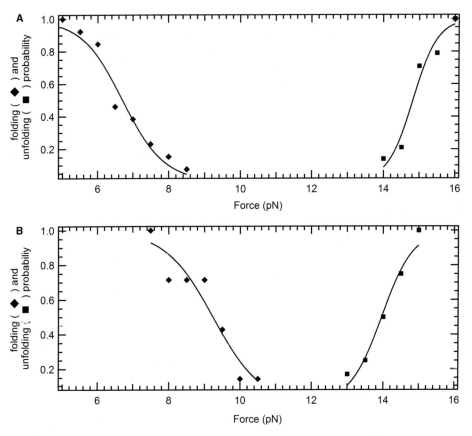

Fig. 7.9 Folding (♦) and unfolding (■) probability versus force for the hairpins hp1 (A) and hp2 (B). The solid lines are the fits with a two-state system probability. (A) The data are from 14 consecutive stretch/release cycles applied to the same molecule. The fit of the unfolding probability leads to $\Delta G_{hp1} = 106 \pm 24$ kJ/mol and $\Delta x_{hp1} = 12 \pm 3$ nm. (B) The data are from 12 stretch/release cycles on three different molecules in the same experiment. The fit of the unfolding probability leads to $\Delta G_{hp2} = 79 \pm 11$ kJ/mol and $\Delta x_{hp1} = 9 \pm 1$ nm.

the two hairpins have the same stem. Indeed, the length and GC content of the stem are the parameters controlling the free-energy changes of the hairpin, whereas the loop length has little influence at these sizes.

When the first method is directly applied to the curves presented in Fig. 7.8, we obtain $\Delta G_{hp1} = 134$ kJ/mol and $\Delta G_{hp2} = 115$ kJ/mol for hp1 and hp2, respectively. These values are higher than the theoretical predictions. This apparent discrepancy sheds light on the importance of the close to equilibrium conditions. The measured curves are recorded at 50 nm/s upon unfolding the hairpin. However, as shown in Fig. 7.7, a significant hysteresis is observed between the folding and unfolding curves. By using in the analysis the average of the folding and unfolding forces rather than the

Table 7.1 Free-energy changes predicted by mFold and estimated by the two methods presented. First method: area measurement on the force versus extension curve at the average of the folding and unfolding forces. Second method: fit of the unfolding probability.

	ΔG_{hp1} (kJ/mol)	ΔG_{hp2} (kJ/mol)
Prediction	90 ± 6	92 ± 5
First method	86 ± 10	84 ± 10
Second method	106 ± 24	79 ± 11

unfolding force (points indicated by arrows in Fig. 7.8), we can make an estimation of the equilibrium case. This modified analysis leads to the ΔG values presented in Table 7.1 that are closer to the predictions.

Fitting both ΔG and Δx with the second method to the unfolding measurements, we obtain ΔG_{hp1} = 106±24 kJ/mol and Δx = 12±3 nm for hp1, and ΔG_{hp2} = 79±11 kJ/mol and Δx = 9±1 nm for hp2. The free energies obtained with this method are close to the predicted ones. However, the fitted lengths, Δx, are significantly smaller than the lengths directly measured (cf. values discussed previously). As shown in Fig. 7.9, the width of the transition that determines the fitting parameter Δx is not the same for the folding and unfolding events. This difference again indicates nonequilibrium effects in the measurements. Fitting the width of the transition thus does not appear to be a reliable manner to determine Δx. An increased variability in the force of the transition leads to an underestimation in Δx. The measured variability is stronger for folding than for unfolding and stronger for hp1 than for hp2. Both trends are indeed expected when the measurement is performed out of equilibrium.

Sometimes the experiment cannot be carried out close to equilibrium. Unfolding may involve large structures, and hopping between folded and unfolded states is not observed. Using out-of-equilibrium theory to retrieve equilibrium parameters is then necessary. Several successful attempts to obtain equilibrium parameters from out-of-equilibrium data have been published recently [29, 30]. It has to be noted that these methods require averaging over a large data set, with very low experimental noise at least for the use of Jarzynski equality [31], since the evaluation includes exponential averaging.

Observation of force flipping, on the other hand, is a clear sign of a close-to-equilibrium situation [32]. In this case, it is possible to determine folding and unfolding rates as well as the position of the transition barriers to folding and unfolding, as we will present now.

Both the folding rate $k_{u \to f}$ and the unfolding rate $k_{f \to u}$ depend on the extension imposed on the construct. The dwell times τ_u and τ_f in the unfolded and folded states can be obtained from the force versus time measurement at constant extension (Fig. 7.4). The probability distribution of the lifetimes in a given state at a given

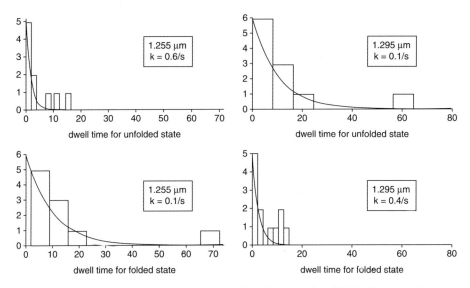

Fig. 7.10 Histograms of the dwell times in the unfolded (upper part) and folded (lower part) states, at two different extensions (1.255 μm and 1.295 μm), obtained from the flip measurements at constant extension with the hp2 construct. Solid lines are exponential fits with the following function: $N_{f \to u}(t) = N_0 k_{f \to u} \exp(-k_{f \to u} \cdot t)$ where $N_{f \to u}(t)$ is the number of counts of the folded → unfolded transition, and N_0 is the total number of counts. A similar expression holds for the unfolded → folded transition. The obtained transition rates are the following:
At 1.255 μm, $k_{u \to f} = 0.6\ s^{-1}$ and $k_{f \to u} = 0.1\ s^{-1}$
At 1.295 μm, $k_{u \to f} = 0.1\ s^{-1}$ and $k_{f \to u} = 0.4\ s^{-1}$

extension for a two-state system is exponentially distributed [4]. The probability distribution $P_{f \to u}(t)$ for an initially folded hairpin to unfold after a dwell time t is given by $P_{f \to u}(t) = k_{f \to u} \exp(-k_{f \to u} \cdot t)$. A similar expression holds for the probability distribution $P_{u \to f}(t)$ of the reverse reaction, with the kinetic constant $k_{u \to f}$.

A fit of these distributions gives the transition rates at the considered extension. Fig. 7.10 illustrates this analysis for two different extensions (1.255 and 1.295 μm), corresponding to forces of $F_{folded} = 11.6$ pN, $F_{unfolded} = 10.2$ pN and $F_{folded} = 13.6$ pN, $F_{unfolded} = 12.8$ pN, respectively. At the larger extension (right: 1.295 μm), the unfolded state is favored: the dwell times of this state are notably longer than the ones of the folded state, and the kinetics appears faster for opening than closing, as reflected by the transition rates ($k_{u \to f} = 0.1\ s^{-1} < k_{f \to u} = 0.4\ s^{-1}$). The smaller extension (left: 1.255 μm), on the other hand, favors the folded state, and the characteristics of the histograms are thus reversed. The transition rates are $k_{u \to f} = 0.6\ s^{-1} > k_{f \to u} = 0.1\ s^{-1}$.

At an intermediate extension that corresponds to $F_{1/2}$, the transition rates are equal, $k_{u \to f} = k_{f \to u}$, and the occupation probability of the folded and unfolded states is the same. This point of equal probability can also be obtained by plotting the average dwell time of the folded and unfolded states, as shown in Fig. 7.11. The crossing of the two dwell times defines the extension at which the two states are equally probable.

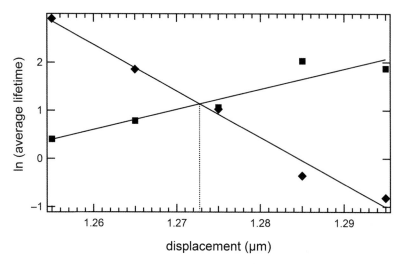

Fig. 7.11 Average lifetimes of the folded (♦) and unfolded (■) states for the RNA hairpin hp2. The average lifetimes vary exponentially with the extension of the molecular construct. Solid lines are linear fits of the data.

In this particular case, we obtain $x = 1.273$ μm, corresponding to the force values $F_{folded} = 14.0$ pN and $F_{unfolded} = 12.6$ pN, respectively.

Fig. 7.11 also allows us to estimate the spatial separation Δx_{fTu} between the folded state and the transition state to unfolding as well as the separation Δx_{uTf} between the unfolded state and the transition state to folding. The following expression gives the effect of force on the unfolding rate at thermal equilibrium:

$$k_{f \to u}(F) = k_{f \to u}(F_{ref}) \exp\left(\frac{(F - F_{ref})\Delta x_{fTu}}{k_B T}\right) \quad (7.3)$$

where $k_{f \to u}(F_{ref})$ is the unfolding rate at a reference force F_{ref} that should be chosen sufficiently close to the range of interest where the transition occurs. In the present case of a displacement-controlled experiment, the experimentally imposed change in displacement is converted into the change in force of the theoretical description by using the local stiffness k of the system, a parameter that is influenced by both the stiffness of the measuring device and the molecular stiffness. This local stiffness, $k = 0.07$ pN/nm in the present case, is obtained by a linear interpolation of the force versus displacement curve in the vicinity of the transition. The spatial separation Δx_{fTu} is then obtained from the slope of the decreasing line of Fig. 7.11. We obtain $\Delta x_{fTu} = 5.5$ nm. From the slope of the increasing line, we obtain the thermally averaged distance between the unfolded state and the transition state to folding, $\Delta x_{uTf} = 2.5$ nm.

We thus find that the energy landscape of hairpin hp2 (Fig. 7.3) is asymmetric in the sense $\Delta x_{fTu} \neq \Delta x_{uTf}$. Interestingly, both distances are significantly shorter than the length of the stem. For the unfolding transition, this result can be understood from the fact that the first five base pairs of the hairpin stem are particularly stable (four GC among these five base pairs). Once the energy barrier of this initial part is surmounted, the rest of the stem offers less resistance to unfolding. For the folding transition, the closing of the base pairs close to the loop is critical. The small value of Δx_{uTf} derived from the measurements suggests that the entire RNA hairpin hp2 folds as soon as the first pairs at the loop side are formed and that the reduction in force required for this to happen is small.

7.3.2 Exploring the folding dynamics of complex RNA structures in presence of proteins

The study of complex RNA structures comes with new challenges. While single hairpins have a rather well-known structure, natural structures can include multiple hairpins, kissing loops, pseudoknots, and numerous complex tertiary interactions. Therefore, natural RNAs often display very complex three-dimensional structures that are difficult, if not impossible, to predict nowadays. Moreover, the folding hierarchy of RNA and how RNA-binding proteins modify this hierarchy are not well known, essentially because classical biochemistry methods do not easily provide access to transient folding states.

Single-molecule force measurements with double-optical tweezers are a very powerful tool to directly probe these complex structures and give access to their thermodynamics and kinetics. We show here how such a complex RNA structure can be studied with the example of an irregular stem in domain II of *E. coli* 23S rRNA (Fig. 7.12).

By pulling on such a structure, one can observe that contrary to the previous examples, it unfolds in several steps (Fig. 7.13), reflecting the stability of different RNA substructures. After the structure is fully unfolded, the construction is relaxed by moving the beads back together. After a rather short time, usually shorter than a second, one pulls again on the refolded structure; interestingly, the RNA seems to unfold by visiting systematically the same transient states (Fig. 7.13), since pauses in the unfolding process follow only a few curves of elasticity corresponding to a small number of molecular configurations.

One can then map these different transient states on the RNA structure by measuring their molecular lengths obtained during the experiment and comparing them with the total length of the RNA. Practically, one measures the length of the construction at a given force from its fully folded states to the transient states; let this length be l_i. Since the total length of the construction can be known from the fully unfolded state, let this length be L_t; the number of bases unfolded is simply derived from the ratio l_i/L_t. By taking into account the fact that the initial state and all intermediate

7 Investigating RNA structure and folding with optical tweezers — 173

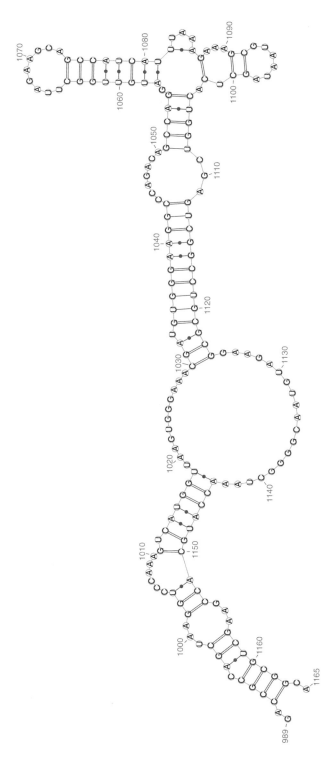

Fig. 7.12 Secondary structure of the 989–1,165 23S rRNA fragment from *E. coli*.

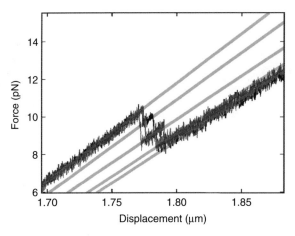

Fig. 7.13 Force versus displacement (at 50 nm/s) curve of the rRNA fragment. Three successive measurements (in red, black, and gray) on the same RNA molecule without L20C show reproducible intermediate states (from Mangeol et al. [14]).

states exhibit, at least, one RNA helix, while the fully unfolded state is a linear RNA single strand, L_t is the sum of the molecular length measured directly on the curves of elasticity from the fully folded to fully unfolded states with the RNA diameter (2.2 nm). Altogether, the number b_i of bases unfolded when the ith intermediate state is reached is given by the ratio $b_i = N_0 l_i / L_t$, where N_0 is the total number of bases of the RNA to be unfolded. The output of this calculation is presented on the RNA structure shown in Fig. 7.14A (the bar indicated on the structure is the mean position of the intermediate state, and the colored area neighboring the bar corresponds to the standard deviation of this position). Interestingly, the standard deviation on the position of intermediate states (approximately three bases) is smaller than the standard deviation of the results of many experiments, suggesting that the intermediate states are probably rather similar from one experiment to another but nevertheless differ by a few bases.

As a ribosomal RNA, this RNA fragment interacts with several ribosomal proteins in the ribosome structure. L20 in particular interacts with the RNA fragment at the base of the stem. We choose to study the interaction of the RNA fragment with this protein. Since the full-length L20 protein easily forms large aggregates, we used a truncated version of L20 consisting of the C-terminal half of the protein, L20C, interacting similarly with the stem of the RNA than the full-length protein (Fig. 7.14).

We repeated the pulling experiment presented before but in the presence of the protein. The concentration of L20C is chosen such that at equilibrium the large majority of RNA molecules are bound by L20C. When unfolding the structure, one can clearly observe that the RNA response to force is very different compared to the case without protein (Fig. 7.15).

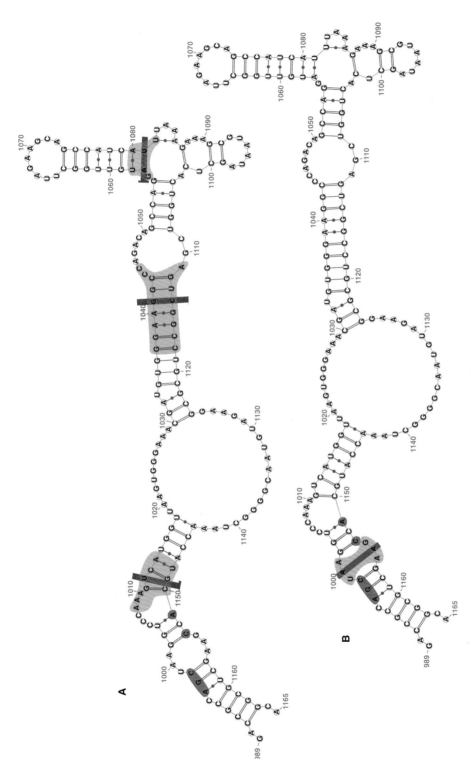

Fig. 7.14 Positions of the intermediate states on the rRNA structure without (A) and with (B) L20C. Bars indicate the mean positions of unfolding intermediates, and the colored regions represent the standard deviations around this mean (from Mangeol et al. [14]). Nucleotides that contact L20 in the ribosome are in gray [33, 34].

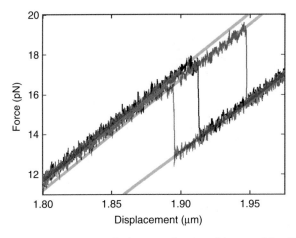

Fig. 7.15 Force versus displacement (at 50 nm/s) curve of the rRNA fragment in the presence of L20C. Three successive measurements (in blue, black, and gray) on the same RNA molecule with L20C show reproducible intermediate states (from Mangeol et al. [14]).

Although the RNA still unfolds in multiple steps, these steps are very different. The RNA mainly unfolds in one small step followed by a much larger step, and both occur at higher forces than observed without protein. When repeating the previous analysis, which led us to understand how the RNA was hierarchically unfolded, we could assign each substep in the force measurement to a substructure in the RNA fragment; by doing so, we found that the base of the RNA stem is strongly stabilized by L20C binding (Fig. 7.14B). Interestingly, this part of the RNA corresponds precisely to the part preceding the binding site of the protein. We can localize the binding site of the protein with a precision of three bases. Most importantly, thanks to this method, we are able to report the direct effects of the protein on the RNA structure: L20C acts as a clamp that stabilizes an essential subdomain of the 23S RNA. To the best of our knowledge, it is not possible to observe this direct effect by another technique than the single-molecule force measurement.

Since these measurements are done mostly out of equilibrium, it is rather difficult to obtain all the associated thermodynamic parameters. However, when the transition to an intermediate state displays flips, such as in the hairpin example of Section 7.3.1, the transition to this state is achieved close to equilibrium and the methods used previously to obtain the free energy of the structure are applicable. The first substep of the RNA unfolding with and without protein displays such a behavior (Fig. 7.16), and the corresponding free energies were therefore calculated. We find that the first substructure unfolded in the experiment without protein has a free energy of 49.4±10.5 kJ/mol, which agreed well with the predicted energy (52.7±5.4 kJ/mol; from mFold). Similarly, we find that the first substructure unfolded in the experiment with protein has

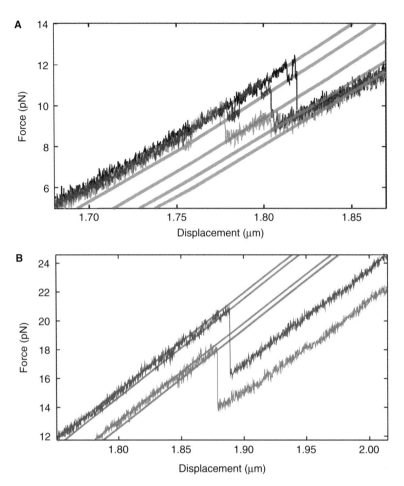

Fig. 7.16 Force flipping in stretching the rRNA fragment in the absence (A) and in the presence (B) of L20C. In both cases, these transitions occur most often during the first substep. (A) Three experiments are shown. (B) Two experiments are shown; for clarity, the lower trace has been shifted downward by 2 pN (from Mangeol et al. [14])

a free energy of 47.8±10.0 kJ/mol, (predicted energy 50.6±5.0 kJ/mol). To quantify the effect of the protein on the RNA structure, we applied the same method, but keeping in mind that this process is out of equilibrium. Therefore, this time, instead of obtaining the free energy of the structure, we estimate the work necessary to extend unfolding from 18 to 36 bases (i.e. behind the L20C-binding site) by extrapolating to higher forces the elasticity curve of the first intermediate in the absence of L20C (36 bases unwound). Finally, the work necessary to unfold the first 36 bases in the presence of L20C was 99.2±18.4 kJ/mol instead of 49.4±10.5 kJ/mol in its absence. The difference, ca. 50 kJ/mol, is the energy necessary to remove L20C from its site.

7.4 Conclusion

Single-molecule force measurements, and in particular the optical-tweezers technique, can give access to information that is otherwise very difficult if not impossible to obtain. As illustrated in this chapter, thermodynamic and kinetic parameters of an RNA structure can be determined. This even holds when an RNA domain is embedded in a complex three-dimensional structure that includes tertiary interactions. The fact that the measured structural transitions often occur out of equilibrium makes the quantitative analysis of the tweezers data challenging. This has stimulated fundamental research in statistical physics. This chapter also illustrates the possibility of directly observing and quantifying the effect of proteins on RNA structures. While classical methods represent the most important and powerful tools to study RNA, single-molecule experiments, and in particular force measurements with double-optical tweezers, can provide a complementary point of view on the RNA molecule, its enzymatic activities, and its interactions with proteins.

7.5 Acknowledgments

This work was funded by the Centre National de la Recherche Scientifique and Agence Nationale pour la Recherche (1503 01 and 1503 02 [HelicaRN]). Laboratoire de Nanobiophysique is associated with CNRS (UMR 7083 Gulliver).

References

1. Smith SB, Finzi L, Bustamante C. Direct mechanical measurements of the elasticity of single DNA molecules by using magnetic beads. Science. 1992;258:1122.
2. Block SM, Goldstein LSB, Schnapp BJ. Bead movement by single kinesin molecules studied with optical tweezers. Nature. 1990;348:348–52.
3. Svoboda K, Schmidt CF, Schnapp BJ, Block SM. Direct observation of kinesin stepping by optical trapping interferometry. Nature. 1993;365:721–7.
4. Liphardt J, Onoa B, Smith SB, Tinoco I Jr, Bustamante C. Reversible unfolding of single RNA molecules by mechanical force. Science. 2001;292:733.
5. Onoa B, Dumont S, Liphardt J, Smith SB, Tinoco I Jr, Bustamante C. Identifying the kinetic barriers to mechanical unfolding of the *T. thermophila* ribozyme. Science. 2003;299:1892–5.
6. Harlepp S, Marchal T, Robert J, et al. Probing complex RNA structure by mechanical force. Eur Phys J E. 2003;12:605.
7. Mangeol P, Côte D, Bizebard T, Legrand O, Bockelmann U. Probing DNA and RNA single molecules with a double optical tweezers. Eur Phys J E. 2006;19:311–7.
8. Greenleaf WJ, Frieda KL, Foster DAN, Woodside MT, Block SM. Direct observation of hierarchical folding in single riboswitch aptamers. Science. 2008;319:630–3.
9. Dumont S, Cheng W, Serebrov V, et al. RNA translocation and unwinding mechanism of HCV NS3 helicase and its coordination by ATP. Nature. 2006;439:105–8.

10. Cheng W, Dumont S, Tinoco I Jr, Bustamante C. NS3 helicase actively separates RNA strands and senses sequence barriers ahead of the opening fork. Proc Natl Acad Sci U S A. 2007;104:13954–9.
11. Wen JD, Lancaster L, Hodges C, et al. Following translation by single ribosomes one codon at a time. Nature. 2008;452:598–603.
12. Qu X, Wen JD, Lancaster L, Noller HF, Bustamante C, Tinoco I Jr. The ribosome uses two active mechanisms to unwind messenger RNA during translation. Nature. 2011;475:118–21.
13. Kaiser CM, Goldman D, Chodera JD, Tinoco I Jr, Bustamante C. The ribosome modulates nascent protein folding. Science. 2011;334:1723–7.
14. Mangeol P, Bizebard T, Chiaruttini C, Dreyfus M, Springer M, Bockelmann U. Probing ribosomal protein-RNA interactions with an external force. Proc Natl Acad Sci U S A. 2011;45:18272.
15. Bustamante C, Bryant Z, Smith S. Ten years of tension: single molecule DNA mechanics. Nature. 2003;421:423–7.
16. Allemand JF, Bensimon D, Croquette V. Stretching DNA and RNA to probe their interactions with proteins. Curr Opin Struct Biol. 2003;13:266–74.
17. Bockelmann U. Single molecule manipulation of nucleic acids. Curr Opin Struct Biol. 2004;14:368–73.
18. Neuman KC, Block SM. Optical trapping. Rev Sci Instrum. 2004;75:2787.
19. Li PT, Vieregg J, Tinoco I Jr. How RNA unfolds and refolds. Annu Rev Biochem. 2008;77:77–100.
20. Ashkin A, Dziedzic J M, Bjorkholm JE, Chu S. Observation of a single-beam gradient force optical trap for dielectric particles. Opt Lett. 1986;11:288–90.
21. Block SM, Blair DF, Berg HC. Compliance of bacterial flagella measured with optical tweezers. Nature. 1989;338:514–8.
22. Lang MJ, Asbury CL, Shaevitz JW, Block SM. An automated two-dimensional optical force clamp for single molecule studies. Biophys J. 2002;83:491–501.
23. Gittes F, Schmidt CF. Interference model for back-focal-plane displacement detection in optical tweezers. Opt Lett. 1998;23:7–9.
24. Mangeol P, Bockelmann U. Interference and crosstalk in double optical tweezers using a single laser source. Rev Sci Instrum. 2008;79:83–103.
25. Cissé I, Mangeol P, Bockelmann U. DNA unzipping and force measurements with a dual optical trap. In: Peterman EJG, Wuite GJL, eds. Single Molecule Analysis: Methods and Protocols. New York, Humana Press; 2011:783–45.
26. Woodside MT, Anthony PC, Behnke-Parks WM, Larizadeh K, Herschlag D, Block SM. Direct measurement of the full, sequence-dependent folding landscape of a nucleic acid. Science. 2006;314:1001–4.
27. Odijk T. Stiff chains and filaments under tension. Macromolecules 1995;28:7016–8.
28. Bustamante C, Marko JF, Siggia ED, Smith S. Entropic elasticity of lambda-phage DNA. Science. 1994;265:1599–600.
29. Liphardt J, Dumont S, Smith SB, Tinoco I Jr, Bustamante C. Equilibrium information from nonequilibrium measurements in an experimental test of Jarzynski's equality. Science. 2002;296:1832–5.
30. Collin D, Ritort F, Jarzynski C, Smith SB, Tinoco I Jr, Bustamante C. Verification of the Crooks fluctuation theorem and recovery of RNA folding free energies. Nature. 2005;437:231–4.
31. Jarzynski C. Nonequilibrium equality for free energy differences. Phys Rev Lett. 1997;78:2690–3.
32. Bockelmann U, Thomen P, Essevaz-Roulet B, Viasnoff V, Heslot F. Unzipping DNA with optical tweezers: high sequence sensitivity and force flips. Biophys J. 2002;82:1537–53.
33. Schuwirth BS, Borovinskaya MA, Hau CW, et al. Structures of the bacterial ribosome at 3.5 A resolution. Science. 2005;310:827–34.
34. Selmer M, Dunham CM, Murphy FV IV, et al. Structure of the 70S ribosome complexed with mRNA and tRNA. Science. 2006;313:1935–42.

Airat Gubaev and Dagmar Klostermeier

8 Fluorescence resonance energy transfer as a tool to investigate RNA structure and folding

8.1 An introduction to fluorescence resonance energy transfer

Fluorescence resonance energy transfer (FRET) is a nonradiative transfer of energy from a donor fluorophore to an acceptor fluorophore via a through-space interaction of their transition dipoles (Fig. 8.1A). The efficiency of this process depends on the distance separating the fluorescent dyes, and the experimental determination of the FRET efficiency therefore allows for the determination of this distance. The theoretical framework for the quantitative description of FRET and its distance dependence has been formulated by Theodor Förster [1]. Stryer and Haugland [2] demonstrated the applicability to biological molecules by measuring FRET efficiencies between donor and acceptor dyes attached to polyproline helices at different distances and have coined the term *molecular ruler*. Since then, innumerous studies have validated FRET as a powerful fluorescence technique to measure intramolecular distances in macromolecules and to monitor distance changes caused by conformational changes. FRET is particularly valuable for studying large, flexible, multidomain or multicomponent entities that are not readily amenable to nuclear magnetic resonance (NMR; see Chapter 12) or X-ray crystallography (see Chapter 13). FRET is therefore well suited for the structural characterization of RNAs and RNA/protein complexes.

As a consequence of the interaction of the donor and acceptor transition dipoles during FRET, a donor fluorophore in the excited state returns to the ground state without emitting a photon and transfers its excitation energy to the acceptor. The acceptor changes into the excited state and emits fluorescence upon return to the ground state (Fig. 8.1B). The rate constant k_t for energy transfer depends on the inverse sixth power of the distance r separating the donor and the acceptor dyes (equation 8.1).

$$k_t = \frac{1}{\tau_D} \cdot \left(\frac{R_0}{r}\right)^6 \tag{8.1}$$

R_0 is the so-called Förster distance (see equation 8.21), and τ_D is the fluorescence lifetime of the donor. FRET competes with other radiative and nonradiative processes deactivating the excited state of the donor fluorophore (Fig. 8.1B). The FRET efficiency E is determined by the rate constant k_t relative to the rate constants of all deactivation processes (equation 8.2):

$$E = k_t / (k_{D,r} + k_{D,nr} + k_t) \tag{8.2}$$

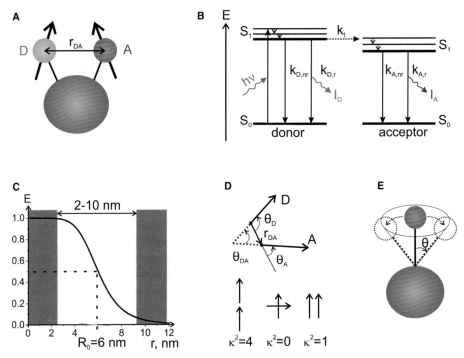

Fig. 8.1 Fluorescence resonance energy transfer (FRET). (A) FRET is a distance-dependent, nonradiative, through-space interaction between the transition dipoles (arrows) of a donor (D, green) and an acceptor (A, red) fluorophore. From the FRET efficiency, the inter-dye distance r_{DA} can be deduced. The blue sphere (not to scale) represents the biomolecule the dyes are attached to. (B) Jablonski diagram for transitions of donor and acceptor fluorophores. Excitation of the donor fluorophore (blue arrow) leads to a transition from the electronic ground state (S_0) to the first excited electronic state (S_1). Deactivation of the excited state of the donor can occur via nonradiative ($k_{D,nr}$) or radiative processes ($k_{D,r}$; I_D, green arrow), or by energy transfer to the acceptor (k_t). Energy transfer leads to the transition of the acceptor to the excited state that can subsequently be deactivated by fluorescence emission ($k_{A,r}$; I_A, red arrow) or nonradiative processes ($k_{A,nr}$). Fine lines and arrows indicate higher vibronic levels and the relaxation to the vibronic ground state of S_1. (C) Distance dependence of the energy transfer efficiency E. If the donor and acceptor fluorophores are separated by the Förster distance (R_0, here 6 nm), the FRET efficiency is 0.5. With a donor/acceptor pair with an R_0 of 6 nm, distances between 2 and 10 nm can be determined from experimental FRET efficiencies (highlighted region). (D) The orientation factor κ^2 is a measure for the relative orientations between donor and acceptor transition dipoles (D, A). It can be calculated from the angles θ_D and θ_A that the transition dipoles of donor D and acceptor A form with the joining vector r_{DA} and from the angle θ_{DA} they form with respect to each other. The resulting values for κ^2 for three different relative orientations of the transition dipoles (arrows) are illustrated below, including the extreme possible values of $\kappa^2 = 0$ and $\kappa^2 = 4$. (E) Segmental flexibility of fluorophores attached to biomolecules. The half-angle θ of the cone that is accessible to the linker between biomolecule (blue sphere) and fluorophores (red sphere) is a measure for the local flexibility.

where $k_{D,r}$ is the rate constant for radiative deactivation processes and $k_{D,nr}$ is the rate constant for nonradiative deactivation processes of the donor. Substitution of the rate constants $k_{D,r}$ and $k_{D,nr}$ for the donor fluorescence lifetime τ_D according to equation 8.3

$$\frac{1}{\tau_D} = k_{D,r} + k_{D,nr} \tag{8.3}$$

and combination of equations 8.1 and 8.2 yields the dependence of E on the donor-acceptor distance r (equation 8.4).

$$E = \frac{R_0^6}{R_0^6 + r^6} \tag{8.4}$$

With common, commercially available fluorophores, donor-acceptor distances in the range of ~2–10 nm can be measured (Fig. 8.1C). The transfer efficiency E reflects the fraction of all photons absorbed by the donor that are transferred to the acceptor fluorophore. Experimentally, this ratio can be determined by exciting the donor and measuring the donor and acceptor fluorescence intensities I_D and I_A (equation 8.5):

$$E = \frac{I_A}{I_D + I_A} \tag{8.5}$$

However, because of different quantum yields of the donor and acceptor fluorophores and different detection sensitivities for donor and acceptor emission, the FRET efficiency calculated from measured fluorescence intensities is an apparent value. It does not reflect true intramolecular distances but reports on relative distances and is therefore often called a proximity ratio. The calculation of corrected FRET efficiencies is described in Section 8.4.2.

8.2 Introduction of donor and acceptor fluorophores into RNAs and RNA/protein complexes

For FRET experiments, donor and acceptor fluorophores have to be attached site-specifically to the molecule of interest. Fluorophores are frequently introduced into proteins by the specific modification of cysteines with maleimide derivatives of the donor and acceptor dyes. The N-terminus can be specifically labeled with amino-reactive reagents. Alternatively, N-terminal cysteines can be labeled in reactions with thioester dye derivatives [3]. Similarly, labeling at the C-terminus can be performed by derivatizing it to a thioester and reacting it with a thiol-modified dye [4]. The variety of functional groups that can be introduced into proteins has been expanded with the incorporation of non-natural amino acids [5–7]. Expressed protein ligation [8–9] of a recombinantly produced protein and a chemically synthesized peptide, or of two recombinantly produced proteins, further expands the available repertoire of fluorescent labeling. RNA, in contrast, lacks suitable functional groups for fluorescent modification. Fluorophores can be incorporated directly into RNA by use of fluorescently labeled building blocks during chemical synthesis, either at the 5'- or 3'-end or internally. Alternatively, amino- or other functional groups can be introduced during RNA synthesis and are modified after synthesis with succinimidyl esters or other derivatives of the dyes. For larger RNA beyond the scope of chemical synthesis, fluorophores can be introduced at the 5'-end by using modified nucleotides for initiation of *in vitro* transcription [10]. Large RNAs site-specifically labeled with donor and acceptor fluorophores can be generated by ligation of shorter, synthesized fragments or *in vitro* transcribed RNA. Both direct RNA ligation and ligation using a DNA splint [11] have been described (reviewed by Moore and Query [12]). Another approach introduces loop regions into large RNAs that are generated by *in vitro* transcription, and fluorophores are introduced by hybridization of oligonucleotides complementary to the loop sequences that carry the fluorescently moiety [13]. Recent strategies of modifying RNA via chemical biology and click chemistry approaches are described in Chapter 4.

8.3 Ensemble FRET

FRET experiments can be performed on an ensemble of a large number of molecules in solution or on the level of single molecules. Although single-molecule FRET experiments require more elaborate instrumentation, recent advances have afforded a routine use of either method with commercially available instruments. In ensemble FRET experiments, FRET efficiencies can be calculated from donor and acceptor fluorescence intensities or from donor fluorescence lifetimes measured for a large ensemble of non-identical molecules. These FRET efficiencies therefore reflect average distances over all molecules, weighted by their relative populations. In time-resolved FRET experiments, distributions of inter-dye distances for different conformational subensembles are obtained from donor fluorescence decays in the absence and presence of the acceptor.

8.3.1 Steady-state FRET

In the simplest FRET experiment, fluorescence emission spectra of donor/acceptor-labeled molecules are measured in a conventional steady-state fluorescence spectrometer with continuous illumination of the sample. The excitation source is usually a xenon lamp, and fluorescence is detected by a photomultiplier (Fig. 8.2A). To determine the FRET efficiency for a donor/acceptor-labeled RNA or RNA/protein complex, reference spectra are measured for the donor-labeled and for the acceptor-labeled

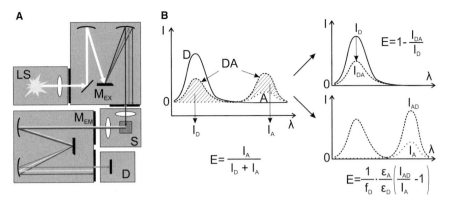

Fig. 8.2 Steady-state FRET. (A) Scheme of a steady-state fluorescence spectrometer. Light from the continuous light source (LS) is split into different wavelengths by a monochromator (M_{EX}), and the sample (S) is illuminated with light of the required wavelength. The fluorescence emission is filtered by a second monochromator (M_{EM}), and light of the wavelength of interest is detected by a photomultiplier (D). (B) Spectra for molecules carrying only a donor (D, line) or an acceptor fluorophore (A, dotted line), and for molecules carrying donor and acceptor fluorophores (DA, broken line). The calculation of the FRET efficiency from the integrated fluorescence intensities for donor and acceptor, I_D and I_A (shaded, left), or from donor quenching (upper right) or sensitized acceptor emission (bottom right) is illustrated. Left: I_D, integrated donor fluorescence; I_A, integrated acceptor fluorescence, corrected for direct excitation of the acceptor (highlighted areas). Upper right: I_D, I_{DA}, donor fluorescence in the absence and presence of the acceptor fluorophore. Bottom right: I_A, I_{AD}, fluorescence intensity of the acceptor in the absence and presence of the donor fluorophore; ε_A, ε_D, extinction coefficient of the donor and acceptor at the excitation wavelength of the donor f_D, fraction of molecules carrying only a donor fluorophores.

molecules upon donor excitation, yielding the fluorescence intensity of the donor in the absence of FRET and the acceptor fluorescence intensity due to direct excitation. From the fluorescence spectrum of the donor/acceptor-labeled molecules upon donor excitation, the donor and acceptor fluorescence intensities in the presence of FRET can then be extracted (Fig. 8.2B). The spectrum is corrected for direct excitation of the acceptor by subtracting the reference spectrum, and donor and acceptor contributions to the emission are determined by describing the spectrum as a linear combination of the reference spectra, which yields the relative contributions (equation 8.6).

$$F_{DA}(\lambda) = f_D \cdot F_D(\lambda) + f_A \cdot F_A(\lambda) \tag{8.6}$$

The apparent FRET efficiency is calculated from the relative contributions of donor and acceptor fluorescence, f_D and f_A, according to equation 8.7:

$$E = \frac{f_A}{f_D + f_A} \tag{8.7}$$

As outlined for equation 8.5, this FRET efficiency is a proximity ratio due to different quantum yields and detection sensitivities for donor and acceptor fluorescence. Errors can arise from the presence of incompletely labeled species and from different concentrations of donor and acceptor or concentration differences between reference and donor/acceptor-labeled samples. These errors can be circumvented when FRET efficiencies are calculated from quenching of the donor fluorescence or from the appearance of acceptor fluorescence (sensitized emission, Fig. 8.2B). To determine the FRET efficiency from donor quenching, the fluorescence intensity of the donor in the absence (I_D, donor-only-labeled molecules) and in the presence of the acceptor (I_{DA}, donor/acceptor-labeled molecules) is compared, and E is determined according to equation 8.8 (Fig. 8.2B):

$$E = 1 - \frac{I_{DA}}{I_D} \tag{8.8}$$

This procedure is also possible if a non-fluorescent acceptor is used. For fluorescent acceptors, the calculation of E from the sensitized acceptor emission can be performed from the spectra of the acceptor-labeled and the donor/acceptor-labeled molecules according to equation 8.9 (Fig. 8.2B):

$$E = \frac{1}{f_D} \cdot \frac{\varepsilon_A}{\varepsilon_D} \left(\frac{I_{AD}}{I_A} - 1 \right) \tag{8.9}$$

where f_D is the fraction of molecules carrying only a donor moiety, ε_A and ε_D are the extinction coefficients of the acceptor and the donor fluorophores at the excitation wavelength (of the donor), and I_{AD} and I_A are the fluorescence intensities of the acceptor in the presence and absence of the donor fluorophore. The acceptor emission has to be corrected for contributions from the fluorescence of the donor, using the reference spectrum from molecules carrying a donor only. A more detailed discussion of the advantages and disadvantages of these approaches has been reported by Clegg [14].

Steady-state FRET experiments measure donor and acceptor fluorescence intensities over a large ensemble of molecules. For a heterogeneous sample, the calculated FRET efficiencies thus reflect average distances over all fluorescent molecules, and they represent a population-weighted average over all species. The measured fluorescence intensities are sensitive to errors in fluorophores concentrations and to fluorescent impurities. Data analysis is complicated when labeling efficiencies are low, and contributions of molecules carrying only a donor or only an acceptor have to be taken into account.

8.3.2 Time-resolved FRET

In time-resolved FRET experiments, the fluorescence lifetime of the donor in the absence and presence of the acceptor is measured in a lifetime spectrometer, either in the time domain (Fig. 8.3A-C) or in the frequency domain (Fig. 8.3D-F). Time-domain lifetime spectrometers have a pulsed excitation source, such as a pulsed laser or laser diode, or a flash lamp. The detector is a microchannel plate photomultiplier or a streak camera that can accurately determine the arrival time of photons (Fig. 8.3A). In this configuration, fluorescence decays (Fig. 8.3B) are constructed from the arrival time of emitted photons by time-correlated single-photon counting [15]. Donor fluorophores

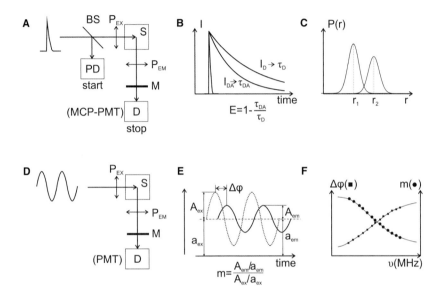

Fig. 8.3 Time-resolved FRET. (A) Instrumentation for lifetime measurements in the time domain. The sample S is excited by pulsed (laser) light that is linearly polarized, and the fluorescence decay is reconstructed from the arrival times of emitted photons by time-correlated single-photon counting. A beamsplitter (BS) diverts a fraction of the excitation light to a photodiode as a reference for the time of excitation (start). The first photon emitted from the sample that is detected by the MCP-PMT detector (MCP-PMT, D) provides the stop signal, and the difference yields the arrival time. M, monochromator; P_{EX}, polarizer to establish excitation with vertically polarized light; P_{EM}, polarizer set to 54.7° to detect isotropic emission. (B) Fluorescence decays for molecules carrying only a donor fluorophores (I_D, τ_D) and for molecules carrying donor and acceptor fluorophores (I_{DA}, τ_{DA}), and calculation of the FRET efficiency E from the donor fluorescence lifetimes in the absence and presence of the acceptor, I_D and I_{DA}. (C) Distance distributions $P(r)$ determined from analyzing the fluorescence decay of the donor-only- and donor/acceptor-labeled molecules according to equations 8.11, 8.14, and 8.15. The mean distances r_1 and r_2 for two different conformers are indicated. The area under the curve reflects the relative population of each conformer. (D) Instrumentation for lifetime measurements in the frequency domain. The sample S is excited by intensity-modulated light. The intensity-modulated emitted light is detected by a photomultiplier (PMT, D). M, monochromator; P_{EX}, polarizer to establish excitation with vertically polarized light; P_{EM}, polarizer set to 54.7° to detect isotropic emission. (E) Intensity-modulated excitation light (dotted line) and emitted light (line). The phase shift $\Delta\phi$ of the emitted light and the demodulation m due to the fluorescence lifetime are indicated. (F) Determination of lifetimes from frequency-domain data. From the phase shift $\Delta\phi$ and the demodulation m as a function of the modulation frequency, the fluorescence lifetime(s) can be obtained.

in the sample are excited by a short light pulse, and the arrival time of the first photon emitted from the sample is detected. The excitation/detection cycle is repeated (typically 10^5–10^6 times), and the fluorescence decay is obtained as the envelope of the histogram of photons by their arrival times (Fig. 8.3B). Alternatively, lifetimes can be determined in the frequency domain. Here, the donor fluorophore is excited by intensity-modulated light (Fig. 8.3D), and the lifetime can be calculated from the phase shift and demodulation of the intensity-modulated emission (Fig. 8.3E,F). Both spectrometer types are commercially available. While frequency-domain experiments avoid fluorophore damage due to short, high-intensity laser pulses, the advantage of time-domain spectrometers lies with the larger number of data points that allow for a more accurate determination of lifetimes, particularly when decays are bi- or multiphasic because of more than one lifetime.

The fluorescence lifetime is the average time a fluorophore spends in the excited state before it returns to the ground state. From donor fluorescence decays after pulsed excitation in the absence and presence of the acceptor, the FRET efficiency E can be calculated as

$$E = 1 - \frac{\tau_{DA}}{\tau_D} \tag{8.10}$$

where τ_D is the donor lifetime in the absence and τ_{DA} in the presence of the acceptor (Fig. 8.3B). However, many fluorophores, especially when attached to biomolecules, show multiphasic fluorescence decays and multiple fluorescence lifetimes. In these cases, the FRET efficiency can be calculated from amplitude-averaged lifetimes, $\langle\tau_D\rangle$ and $\langle\tau_{DA}\rangle$, but its significance may be difficult to judge if the origin of the individual lifetimes is unknown.

In a more sophisticated implementation of time-resolved FRET, donor fluorescence decays in the presence of the acceptor can be analyzed in terms of multiple inter-dye distances. Here, a model is needed that describes the donor-acceptor distances in the system studied, either as different, but discrete distances or as one, two, or more Gaussian or other distributions. The fluorescence decay of the donor, $I_D(t)$, can be described as a sum of exponentials according to equation 8.11:

$$I_D(t) = \sum_i \alpha_i \exp\left(-\frac{t}{\tau_i}\right) \tag{8.11}$$

with the donor lifetimes τ_i and the corresponding amplitude factors α_i. In the presence of the acceptor fluorophores, the donor decay $I_{DA}(t)$ is described as

$$I_{DA}(t) = \sum_i \alpha_i \exp\left(-\frac{t}{\tau_i}\cdot\left(1+\frac{R_0^6}{r^6}\right)\right) \tag{8.12}$$

with the Förster distance R_0 and the discrete inter-dye distance r, or as

$$I_{DA}(t) = \int_0^\infty P(r) \sum_i \alpha_i \exp\left(-\frac{t}{\tau_i}\cdot\left(1+\frac{R_0^6}{r^6}\right)\right) dr \tag{8.13}$$

for distance distributions $P(r)$ (Fig. 8.3C). In many cases, distance distributions can be modeled as a (weighted) Gaussian distribution [16] according to equation 8.14:

$$P(r) = 4\pi r^2 c \cdot e^{-a(r-b)^2} \tag{8.14}$$

where c is a normalization factor for the integral of the distribution, and a and b describe the width and mean distance of the distribution. Altogether, the donor decay in the presence of the acceptor for multiple lifetimes and multiple species n with relative populations f_n and donor-acceptor distance distributions $P_n(r)$ is described by

$$I_{DA}(t) = \sum_n f_n \int P_n(r) \sum_i \alpha_i \exp\left(-\frac{t}{\tau_i}\cdot\left(1+\frac{R_0^6}{r^6}\right)\right) dr \tag{8.15}$$

Time-resolved FRET experiments require measuring the donor fluorescence decay in the absence of the acceptor to determine the intrinsic donor lifetime(s). The intrinsic donor lifetimes are fixed to the predetermined values when the donor decay in the presence of the acceptor is analyzed according to equation 8.15. Such an analysis retrieves the mean distance and the width of the distribution determined for each conformational state, as well as the relative population of these species. Data analysis requires the determination of the Förster distance for the dye pair used (see Sections 8.4.2.3 and 8.4.2.4).

Time-resolved FRET experiments are largely independent of fluorophore concentrations and less background sensitive compared to intensity-based steady-state FRET experiments. Fractions of molecules carrying a donor fluorophore only can readily be taken into account in data analysis. From time-resolved FRET experiments, two or more species in the sample can be identified, and their populations can be quantified, allowing for the calculation of equilibrium constants and thermodynamic data (ΔG^0) from the relative populations (see Section 8.5.2). It has to be noted, however, that the correct model for the distance distribution is often not clear. Data analysis becomes difficult for more heterogeneous samples with more than two species due to the large number of free parameters in equation 8.15.

8.4 Single-molecule FRET

Single-molecule FRET determines the FRET efficiency of a large number of single molecules, one at a time, and directly reveals FRET distributions. In single-molecule FRET experiments, microscopic techniques (Fig. 8.4) are used to detect fluorescence intensities or fluorescence lifetimes for single molecules that are either diffusing

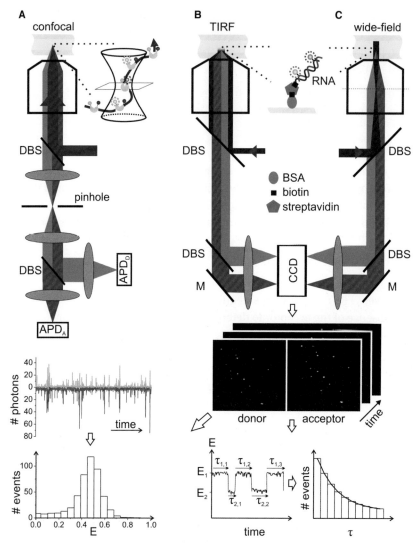

Fig. 8.4 Single-molecule FRET. (A) Confocal microscopy. Excitation light (blue) is focused into the sample by an objective. Fluorescence of molecules traversing the confocal volume (inset) is excited, and the emitted fluorescence (green, red) is collected by the objective, filtered by a pinhole for selection of the detection plane, split into donor (green) and acceptor (red) contributions by a dichroic beam splitter (DBS), and focused onto avalanche photodiode detectors (APD_D, APD_A). The raw data are time traces of donor and acceptor fluorescence (center panel: green, donor signal; red, the acceptor signal, inverted for clarity). Each burst of fluorescence corresponds to a single-molecule event. FRET efficiencies for each detected single molecule can be calculated from donor and acceptor intensities of individual bursts. The FRET histogram (bottom panel) directly reports on the distribution of species with different FRET efficiencies. (B) Total internal reflection microscopy. The excitation light (blue) is coupled into the objective off axis, such that the beam impinges on the slide/sample interface at an angle above the critical angle for total internal reflection. On the distant side of the slide, an evanescent field (blue gradient) is generated that allows for selective excitation of fluorescent molecules immobilized on the slide surface (inset).

freely in solution (Fig. 8.4A) or immobilized on surfaces (Fig. 8.4B,C). To determine different species and their distance distributions, a large number of single molecules is observed, either sequentially (in solution) or simultaneously (immobilized on a surface). The FRET efficiency for each single molecule is calculated, and the distribution of FRET efficiencies becomes visible in FRET histograms (Fig. 8.4A). The observation of (immobilized) single molecules over extended periods of time allows the calculation of dwell times in different FRET states and the determination of rate constants for the inter-conversion between these states (Fig. 8.4B,C).

8.4.1 Instrumentation and experimental procedure

Single-molecule FRET experiments require a confocal, total internal reflection (TIRF), or wide-field microscope (Fig. 8.4). Experiments on donor/acceptor-labeled molecules diffusing freely in solution are performed in a confocal microscope (Fig. 8.4A). Depending on the experiment type, the excitation source is either a pulsed or a continuous wave laser or laser diode. The excitation light is focused into a femtoliter excitation volume, the confocal volume, through a high-numerical aperture objective. The emitted fluorescence is collected by the same objective, split into the spectral contributions of donor and acceptor fluorescence, and detected by avalanche photodiodes (APDs) with single-photon sensitivity. The raw data obtained are fluorescent time traces of donor and acceptor fluorescence (Fig. 8.4A). Bursts of photons originate from one fluorescently labeled molecule traversing the confocal volume due to Brownian motion. Their duration is determined by the diffusion time of the labeled molecule and is usually on the order of a few milliseconds. From the intensities of the donor and acceptor fluorescence, integrated over the duration of each fluorescence

---

Fig. 8.4 (*Continued*)
Fluorescence (green, red) is collected by the objective, split into donor (green) and acceptor (red) contributions by a dichroic beam splitter (DBS), and focused onto adjacent areas of a charge-coupled device (CCD) camera chip. The raw data are images (or a series of images) of the surface in terms of donor and acceptor fluorescence (center panel). Each spot in the images represents a single molecule. From a pair of images, FRET efficiencies can be calculated from the donor and acceptor intensities for individual molecules, and FRET histograms can be created (bottom left). From the FRET efficiency of a single molecule in a series of images taken at different time points, FRET time traces can be constructed (bottom center), and dwell times τ of molecules in each state can be determined. $\tau_{1,n}$ denotes dwell times in state 1 (high FRET efficiency, E_1); $\tau_{2,n}$ denotes dwell times in state 2 (low FRET efficiency, E_2). Dwell-time histograms (bottom right) reveal rate constants for transitions between the different conformers in states 1 and 2. (C) Wide-field microscopy. The excitation light (blue) is focused into the back-focal plane (broken line) of the objective, leading to parallel light illuminating the sample. Fluorescence (green, red) from molecules immobilized on the surface (inset) and above is collected by the objective, split into donor (green) and acceptor (red) contributions by a dichroic beam splitter (DBS), and focused onto adjacent areas of a CCD chip. The raw data are images (or a series of images) of the surface in terms of donor and acceptor fluorescence (center panel). FRET histograms, time traces, and dwell-time histograms can be constructed as described in B).

burst, the FRET efficiency for each molecule can be calculated according to equation 8.5. Histograms of these single-molecule FRET efficiencies directly reveal FRET distributions and thus populations of different species. If pulsed excitation is used, fluorescence decays can be constructed from the measured photon arrival times for each photon detected within a burst, and fluorescence lifetimes are obtained on a single-molecule basis. These FRET efficiencies are independent of different detection efficiencies and can also be used to generate FRET histograms. Two-dimensional plots of FRET efficiencies (calculated from intensities) and donor lifetimes allow for distinction of changes in FRET caused by distance differences from changes because of altered donor or acceptor spectral properties [17]. Due to the small number of photons within a fluorescent burst, however, the accuracy of the lifetime determination is limited.

From the time-dependent changes of fluorescence intensities and thus FRET efficiencies during a single burst, rapid conformational fluctuations on the millisecond timescale can be observed. However, large-scale conformational changes of RNAs or RNA/protein complexes usually occur in seconds to minutes and will thus not be detected by confocal microscopy. The observation time can be extended to the relevant timescale when surface-immobilized single molecules are imaged. Biotinylated RNAs and RNA complexes can be immobilized on quartz slides functionalized with streptavidin (see the introduction into surface immobilization strategies for single-molecule FRET experiments by Lamichhane et al. [18]). Surface-immobilized single molecules are imaged by confocal scanning, wide-field, or TIRF microscopy (Fig. 8.4B,C). In laser scanning confocal microscopy, the confocal volume is scanned across the surface. The donor and acceptor fluorescence intensities from each spot are detected by a donor and an acceptor APD and used to construct images of the surface. In wide-field microscopy (Fig. 8.4C), the complete surface and the area above is illuminated by parallel light, generated by focusing the laser beam into the back-focal plane of the objective. The fluorescence is either detected from above or collected by the objective and imaged onto the chip of a charge-coupled device (CCD) camera. Contributions from donor and acceptor fluorescence are separated and imaged in adjacent regions of the CCD chip. In TIRF microscopy (Fig. 8.4B), the laser beam is coupled into the objective off axis (objective-type TIRF), such that it impinges onto the glass/solution interface above a critical angle, and is totally reflected. On the distant side of the interface, an evanescent field is generated that decays exponentially with increasing distance from the surface, leading to the selective excitation of molecules at the surface. Alternatively, the total internal reflection conditions can be realized using a prism (prism-type TIRF). As with wide-field microscopy, the emitted fluorescence is detected either from the same or the distant side of the illumination. The spectral contributions of donor and acceptor emission are separated and imaged onto adjacent regions of the chip of a CCD camera. Single-molecule FRET experiments on surface-immobilized molecules thus generate a pair of images, one in terms of donor fluorescence and a second in terms of acceptor fluorescence (Fig. 8.4B,C). Each spot reflects the (donor or acceptor) fluorescence of a single molecule. FRET efficiencies for each molecule can be calculated by aligning the two images and integrating the donor and acceptor fluorescence intensities for each spot reflecting the same molecule. For time-dependent experiments, a pair of images is measured at different time points. The time resolution is determined by the response time of the CCD camera and can be as short as 30 ms. Time traces of FRET efficiencies for each molecule are constructed by calculating the FRET efficiency for each pair of images. The observation time in this case is mainly determined by the photostability of the donor and acceptor fluorophores used (reviewed by Kapanidis and Weiss [19]). Bleaching and blinking due to transitions from the excited singlet state to the triplet state can be suppressed by using oxygen scavengers and triplet-state quenchers [20], affording observation times of up to several minutes.

8.4.2 Data analysis

From donor and acceptor intensities during fluorescence bursts detected in single-molecule FRET experiments, apparent FRET efficiencies can be calculated from equation 8.5. This proximity ratio does not take into account instrument non-idealities and will not correctly reflect inter-dye distances but may be suitable to interpret distance changes. To obtain corrected FRET efficiencies that can be converted into inter-dye distances, thorough correction procedures are required.

8.4.2.1 Identifying single-molecule events

The first step of data analysis requires the identification of individual single-molecule events. For measurements on freely diffusing molecules, raw data are time traces of donor and acceptor fluorescence (Fig. 8.4A). Burst identification is achieved by applying a relative threshold criterion to the count rate (> n-fold above background), in combination with an absolute threshold (> n photons, typically $50 < n < 100$). The background in donor and acceptor channels has to be determined from the times when no molecule is in the confocal volume and is subtracted from the measured intensities for subsequent calculations. Similarly, intensity spots in images obtained from surface-immobilized molecules (Fig. 8B,C) are identified by threshold criteria, and pairs of spots are assigned to one single molecule. Donor and acceptor intensities for each spot are then corrected for background, usually determined from the same image outside single-molecule spots.

8.4.2.2 Correction for instrument non-ideality

Once single-molecule events have been identified, background-corrected donor and acceptor fluorescence intensities are further corrected for non-idealities of the experimental setup: (1) cross-talk of donor photons into the acceptor channel and of acceptor photons into the donor channel (α, β), (2) the different quantum yields and detection efficiencies for donor and acceptor fluorescence (γ), and (3) the direct excitation of the acceptor (δ). The values for α, β, γ, and δ can be determined experimentally from the spectral properties of the dyes. As the spectral properties of the dyes in turn are influenced by the local environment at the attachment site, correction parameters have to be determined individually for each construct used. These parameters may differ for the donor/acceptor and acceptor/donor configurations and may change upon ligand binding. Therefore, they have to be determined for both dye configurations and in the absence and presence of ligands.

For single-molecule experiments on freely diffusing molecules, correction parameters for donor/acceptor-labeled constructs are determined using the corresponding constructs labeled only with donor ($D/-$, $-/D$ species) and only with acceptor ($A/-$, and $-/A$ species) [21]. The absorbance of these solutions at the donor excitation wavelength is measured (A_D, A_A) in a standard absorbance spectrometer and corrected for buffer contributions (A_D^{buffer}, A_A^{buffer}). The correction parameter δ as a measure for the fraction of direct excitation of the acceptor is calculated from these values according to equation 8.16:

$$\delta = \frac{A_A - A_A^{buffer}}{A_D - A_D^{buffer}} \cdot \frac{c_D}{c_A} \tag{8.16}$$

where c_D and c_A are the concentrations of the donor- and acceptor-labeled molecules. Concentration determination of labeled biomolecules is complicated by possible changes in dye absorbance upon attachment to the protein or RNA. To minimize these effects, labeled proteins and RNA molecules can be degraded with proteinase K or RNase, respectively. The dye concentrations c_D, c_A can then be calculated from the absorbance of this solution with extinction coefficients for the free dye as provided by the manufacturer.

The remaining correction parameters α, β, and γ can be determined from counting rates directly measured in the confocal microscope (equations 8.17 to 8.19). α is determined from the counting rates of the donor-labeled protein in donor and acceptor channels (I_D^{donor}, I_A^{donor}), corrected for buffer contributions (I_D^{buffer}, I_A^{buffer}):

$$\alpha = \frac{I_A^{donor} - I_A^{buffer}}{I_D^{donor} - I_D^{buffer}} \tag{8.17}$$

β is determined analogously with the acceptor-labeled protein:

$$\beta = \frac{I_D^{acceptor} - I_D^{buffer}}{I_A^{acceptor} - I_A^{buffer}} \tag{8.18}$$

For identical concentrations of donor- and acceptor-labeled proteins, γ is calculated as

$$\gamma = \frac{1}{\delta} \cdot \frac{I_A^{acceptor} - I_A^{buffer}}{I_D^{donor} - I_D^{buffer}} \tag{8.19}$$

With these correction parameters, the FRET efficiency can then be calculated after equation 8.20 [22]:

$$E_{FRET} = \frac{(1+\beta\gamma\delta)\cdot\left(I_A - \frac{\alpha+\gamma\delta}{1+\beta\gamma\delta}\cdot I_D\right)}{(1+\beta\gamma\delta)\cdot\left(I_A - \frac{\alpha+\gamma\delta}{1+\beta\gamma\delta}\cdot I_D\right) + (\gamma+\gamma\delta)\cdot(I_D - \beta I_A)} \qquad (8.20)$$

This corrected FRET efficiency is free from effects caused by instrumental characteristics or the environment of the dyes in a specific construct and thus reports on true inter-dye distances. The experimental error of these FRET efficiencies (including the uncertainty in the determination of the correction parameters) is <10% [23]. The correction parameters α, β, γ, and δ can be determined directly from each experiment if alternating laser excitation (ALEX) of donor and acceptor fluorophores is used (see Section 8.5.3 [24, 25]).

For FRET data obtained from a series of images of surface-immobilized molecules, the correction parameters can also be determined directly from the single-molecule experiments. α can be calculated from the intensities in donor and acceptor images from molecules carrying only a donor, and β from the intensities of molecules carrying only an acceptor. γ is obtained from time traces of donor and acceptor fluorescence intensities for single molecules where the acceptor bleaches before the donor, by comparing the loss of acceptor fluorescence with the increase in donor fluorescence [26].

8.4.2.3 The Förster distance R_0

The Förster theory relates the rate constant of energy transfer, k_t, to the spectral properties of the donor and acceptor dyes between which energy transfer occurs. The relatively complex equation obtained from quantum-mechanical considerations can be simplified by summarizing all parameters except the donor fluorescence lifetime τ_D and the inter-dye distance r into the constant R_0, the so-called Förster distance. The Förster distance corresponds to the inter-dye distance at which the probability for energy transfer is 50% and depends on the spectral properties of the donor and acceptor dyes (equation 8.21). R_0 (in nm) can be calculated as

$$R_0^6 = 8.785\cdot 10^{-11}\frac{\kappa^2 \phi_D J}{n^4} \qquad (8.21)$$

with the orientation factor κ^2, the quantum yield of the donor ϕ_D, the overlap integral J (in $M^{-1}\,cm^{-1}\,nm^4$; see equation 8.22), and the refractive index n. The refractive index n is usually set to 1.33 for water. This approximation is valid even though the biomolecule is present between the dyes because of the three-dimensional through-space interaction of the dipoles. The overlap integral J can be determined from the fluorescence emission spectrum of the donor, $F_D(\lambda)$, normalized to an integral of 1, and the absorbance spectrum of the acceptor fluorophores, converted to the extinction coefficient $\varepsilon_A(\lambda)$. These spectra can be measured in conventional absorbance and fluorescence spectrometers, according to equation 8.22:

$$J = \int_0^\infty F_D(\lambda) \cdot \varepsilon_A(\lambda) \cdot \lambda^4 d\lambda \approx \sum_i F_D(\lambda_i) \cdot \varepsilon_A(\lambda_i) \cdot \lambda_i^4 \tag{8.22}$$

with ε in $M^{-1}\,cm^{-1}$ and λ in nm. κ^2 is often approximated by 2/3, assuming that the donor and acceptor dyes are flexibly attached such that they sample all possible relative orientations on the timescale of the experiment (see Section 8.4.2.4). The quantum yield of the donor is determined relative to a standard fluorophore [27, 28].

The spectral properties of the dyes are affected by the local environment of the dyes and hence by the attachment site. Similar to the correction parameters α, β, γ, and δ, the Förster distance therefore needs to be determined for both possible dye configurations (D/A vs. A/D), for each construct, and in the absence and presence of ligands.

8.4.2.4 The orientation factor κ^2

The efficiency of energy transfer depends not only on the distance separating the donor/acceptor pair but also on their relative orientation that is described by the orientation factor κ^2 (equation 8.23).

$$\kappa^2 = (\cos\theta_{DA} - 3\cos\theta_D \cdot \cos\theta_A)^2 \tag{8.23}$$

θ_{DA} is the angle between donor and acceptor transition dipoles, and θ_D and θ_A the direction of the donor and the acceptor transition dipole relative to a vector joining their origins (Fig. 8.1D). κ^2 can assume any value from zero to four, has a value of 2/3 if both dyes are flexibly attached, and can sample all possible relative orientations during the timescale of the fluorescence decay through unrestricted motion. The flexible attachment of dyes, on the other hand, may lead to problems in the extraction of distance information from FRET efficiencies. The effect of the uncertainty of κ^2 on distances determined from FRET efficiencies has been treated extensively [29–32].

The mobility of donor and acceptor fluorophores attached to biomolecules can be estimated from their fluorescence anisotropy. As a rule of thumb, a steady-state anisotropy of <0.2 is an indication of the validity of the $\kappa^2 = 2/3$ assumption [33]. A more rigorous approach requires the measurement of fluorescence anisotropy decays. The fluorescence of the labeled molecule is excited with a pulse of linearly polarized light, and the anisotropy decay $r(t)$ is constructed from the fluorescence decays of the components of the emitted light polarized parallel (p) or perpendicular (s) to the polarization of the excitation light (equation 8.24).

$$r(t) = \frac{I_p(t) - I_s(t)}{I_p(t) + 2I_s(t)} \quad (8.24)$$

Anisotropy decays can be described by a sum of exponential functions (equation 8.25):

$$r(t) = \sum_i \beta_i e^{-t/\tau_{c,i}} \quad (8.25)$$

with the rotational correlation times $\tau_{c,i}$, and the corresponding amplitudes β_i. For a dye covalently attached to a biomolecule via a flexible linker, the anisotropy decay often contains two components: the fast, segmental mobility of the dye, characterized by a small correlation time $\tau_{c,1}$, and the global tumbling of the dye with the macromolecule, with a larger correlation time $\tau_{c,2}$. From the corresponding amplitudes β_1 and β_2 of these two decay components, an order parameter S can be calculated according to equation 8.26:

$$S = \sqrt{\frac{\beta_2}{\beta_1 + \beta_2}} \quad (8.26)$$

The order parameter S is related to the half opening angle θ of a cone within which the dye can move (equation 8.27, Fig. 8.1E):

$$\theta = \arccos\left(\frac{1}{2}\left(\sqrt{1 + 8S} - 1\right)\right) \quad (8.27)$$

Values for θ of >60° indicate a nearly unrestricted motion of the dyes, and the distribution of κ^2 around 2/3 will be narrow, justifying the 2/3 approximation [34]. Smaller values for θ indicate that the mobility of the donor and acceptor fluorophores is restricted. In these cases, the possibility range of values for κ^2 and their probability $P(\kappa^2)$ can be determined from time-resolved fluorescence anisotropy measurements and have to be taken into account when fitting the donor decay in the presence of the acceptor [33, 35–37].

The error introduced in distances by an incorrect assumption of $\kappa^2 = 2/3$ is generally <26% and will not exceed 10% as long as one of the dyes rotates freely [33].

8.4.2.5 Analysis of FRET histograms

A general treatment of photon statistics in single-molecule FRET experiments can be found in [38]. FRET histograms can often be described by one or more Gaussian distributions. Using probability distribution analysis [39] or proximity ratio histogram analysis [40], the shape of FRET histograms can be described quantitatively with high precision, either model-based or in combination with maximum entropy methods [41] without *a priori* assumptions. For a detailed discussion of these approaches, the reader is referred to the original references.

8.4.3 FRET data and RNA folding

Steady-state FRET experiments with donor/acceptor-labeled RNA molecules allow to characterize folded and unfolded states and can be used to assess the effect of metal ions, small molecules, or proteins on RNA folding and stability. With time-resolved FRET, folded and unfolded states, and possibly folding intermediates, can be identified, and their relative populations can be determined under different experimental conditions (e.g. as a function of temperature or metal ion concentration), yielding equilibrium constants and free-energy differences. Finally, FRET histograms from single-molecule FRET experiments reveal the distribution of different conformers directly. From the analysis of FRET efficiency time traces, the rate constants of inter-conversion of these conformers can be determined. The order of appearance of different conformers allows for distinction of on- and off-pathway intermediates. When all the information obtained is combined, folding energy landscapes can be constructed that describe the overall energetics and kinetics of folding, and thus the pathways from the unfolded RNA to the native structure.

8.4.4 From FRET data to structural models of RNA and RNA/protein complexes

Experimentally determined FRET efficiencies can be converted into distances separating the dyes attached to the biomolecule or biomolecular complex of interest. These distance restraints define global conformations and relative domain rearrangements. Together with internal reference points, a coordinate system can be constructed that allows for the modeling of the overall architecture of large complexes. Examples ranging from manual rigid-body docking of two entities to sophisticated modeling approaches can be found in the literature (see Section 8.5 for illustrative examples). By manual docking, the two partners can be arranged such that the distances in the resulting model are within the range determined from FRET efficiencies. More systematic approaches (reviewed by Brunger et al. [42]) yield a three-dimensional probability distribution for the dye position and take into account experimental errors and linker dynamics [43, 44]. While these approaches recover accurate molecular models [45], they require detailed knowledge of dye attachment geometry, reliable structural information, and a large number of distance restraints. As FRET is not restricted by the size of the molecule studied, these approaches are complementary to existing structural biology techniques and can provide valuable insight into the molecular architecture of large complexes.

8.5 Selected examples

The examples in the following sections have been selected to illustrate questions that can be answered with FRET approaches and the insight gained from different types of FRET experiments.

8.5.1 Steady-state FRET: ribozymes, rRNA, and RNA polymerase transcription complexes

RNA helical junctions are important building blocks of large RNA molecules, and their folding has been studied extensively to understand the folding mechanism of large RNAs (reviewed by Lilley [46]). Hammerhead ribozymes are a class of small self-cleaving RNAs that are involved in rolling circle replication in viroids and satellite RNAs [47, 48]. They consist of three base-paired stems (helices 1–3, Fig. 8.5A) that are connected by conserved unpaired nucleotides. The conserved central bases are essential for ribozyme activity, which requires one or more catalytic divalent metal ions [47, 48]. A structural model for the hammerhead ribozyme has been obtained from a FRET study that placed donor (carboxyfluorescein) and acceptor (tetramethylrhodamine) fluorophores on the end of helix 3 and different positions on helix 1, on the end of helix 2 and different positions on helix 3, or on the ends of helices 1 and 2 [49] (Fig. 8.5A). The steady-state FRET efficiencies, calculated from measurements of acceptor sensitization, in conjunction with geometric considerations of regular A-RNA, were consistent with a Y-shaped structure in which helices 2 and 3 are coaxially stacked, and helix 1 forms an acute angle with helix 2. This structural model was largely confirmed by the crystal structure [50] (Fig. 8.5A) that also revealed the local arrangement of the nucleotides forming the catalytic site. Ion-dependent folding of the hammerhead ribozyme has also been studied by FRET with ribozymes carrying donor (fluorescein) and acceptor (Cy3) fluorophores on the ends of the helical arms [51]. FRET efficiencies were calculated from sensitized acceptor emission. Measuring the end-to-end distances for all constructs as a function of the Mg^{2+} concentration allowed delineation of the folding pathway [51]. While the hammerhead ribozyme is extended in the absence of Mg^{2+}, it forms an intermediate at low Mg^{2+} concentrations of ~500 µM, in which the helical arms 2 and 3 are coaxially stacked, and helix 1 forms an acute angle with helix 3. At high Mg^{2+} concentrations (10 mM), a second Mg^{2+}-binding site becomes occupied, and helix 1 moves toward helix 2 (Fig. 8.5A), leading to the global conformation observed in the crystal structure [50].

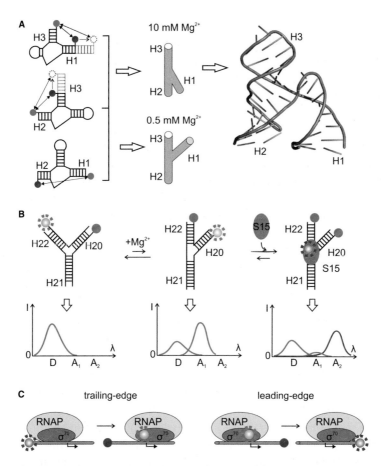

Fig. 8.5 Steady-state FRET experiments to determine global conformations and conformational changes and to monitor movement. (A) Determination of the global conformation of the hammerhead ribozyme from steady-state FRET efficiencies. Donor (green) and acceptor (red) fluorophores were introduced in different positions of the helical arms H1, H2, and H3 [49]. Distances determined from FRET efficiencies (double-headed arrows) were consistent with a model for the global conformation of the hammerhead ribozyme (center top) in which H2 and H3 are coaxially stacked, and H1 forms an acute angle (right, PDB-ID 1mme [50]). At low Mg^{2+} concentrations, the ribozyme adopts a different conformation with H1 closer to H3 (center bottom) [51]. (B) A three-fluorophore FRET assay to simultaneously monitor a conformational change in a ribosomal RNA three-way helical junction and binding of a ribosomal protein. The three-way junction in the 16S rRNA, formed by helical arms H20, H21, and H22, adopts a global conformation with H21 and H22 forming a pseudocontinuous helix, and with H22 at an acute angle from H22 when Mg^{2+} or the ribosomal protein S15 (gray sphere) bind. By placing a donor (D, green) on H22, one acceptor (A_1, orange) on H20, and a second acceptor (A_2, red) on S15, the three different states of the RNA, unfolded (left, green fluorescence), folded (middle, orange fluorescence), and protein-bound (right, red fluorescence), can be distinguished [63]. (C) Monitoring movement of RNA polymerase on DNA by FRET. By placing a donor fluorophore (green) on RNA polymerase (RNAP)–bound sigma factor 70 (σ^{70}), and an acceptor fluorophore (red) upstream (left) or downstream (right) of RNAP, movements of the trailing edge (left) and the leading edge (right) can be monitored as a decrease or increase in FRET (increase/decrease in donor fluorescence), respectively [67].

These studies illustrate how steady-state FRET allows for the determination of global RNA conformations in folded and unfolded states, and of folding intermediates. Together with the observation of structural changes as a function of the Mg^{2+} concentration, folding pathways can be delineated.

The hairpin ribozyme is a self-cleaving element that occurs in the satellite RNA of tobacco ringspot and other plant viruses [52, 53] (reviewed by Fedor [54]). It is required for RNA processing during rolling circle replication. Two helical regions (helices A and B) carry the loops A and B that form the catalytic site. In the satellite RNA, these two helical arms occur in the context of a four-way helical junction (helices A–D, Fig. 8.6A). Similar to the studies with the hammerhead ribozyme, the global conformation of the hairpin ribozyme has been deduced from steady-state FRET experiments that monitored the end-to-end distance of the helical arms on ribozymes labeled with donor (fluorescein) and acceptor (Cy3)-fluorophores [55]. The FRET efficiencies were consistent with a flat X-like structure of the four-way junction in the absence of Mg^{2+}, with equal distances between the ends of the four arms. In the presence of Mg^{2+} ions, the end-to-end distances between helices A and B, and C and D were reduced, whereas the distances between arms A and C, and B and D were increased, in agreement with the formation of a more compact, antiparallel form of the ribozyme, in which helices A and D, and B and C are coaxially stacked, and the two loops carrying the conserved nucleotides of the catalytic site are close [55, 56]. The docked structure is a prerequisite for self-cleavage [56, 57]. The high-resolution structure of the hairpin ribozyme [58] (Fig. 8.6A) confirmed the global conformation deduced from FRET and revealed the architecture of the catalytic site. The formation of the antiparallel, docked structure is promoted by other divalent ions, but not by monovalent ions [55, 57]. Interestingly, manganese ions favor the formation of a docked structure but do not support cleavage, pointing toward minor structural differences between the Mg^{2+}- and Mn^{2+}-stabilized conformations that do not lead to different FRET efficiencies. A loopless junction in which the loops are replaced by base-paired regions also folds into a similar structure [55, 59], demonstrating that the four-way junction is an important structural scaffold for ribozyme activity. The influence of the loops and the junction, and contributions of specific nucleotides, on hairpin ribozyme folding has been dissected in further steady-state [59] and time-resolved FRET studies (see Section 8.5.2).

These studies highlight how global conformations of folded and unfolded states defined in steady-state FRET experiments help correlate structure with activity and allow us to dissect contributions of ions and individual nucleotides to tertiary structure stability.

Ribosomes are large RNA/protein complexes that are assembled in a hierarchical process. In the assembly pathway of prokaryotic ribosomes, one of the first steps is binding of the protein S15 to a three-way helical junction in the central domain of the 16S rRNA, formed by helices 20–22 [60] (Fig. 8.5B). Binding of S15 to the junction stabilizes a conformation in which helical arms 20 and 22 are coaxially stacked, and

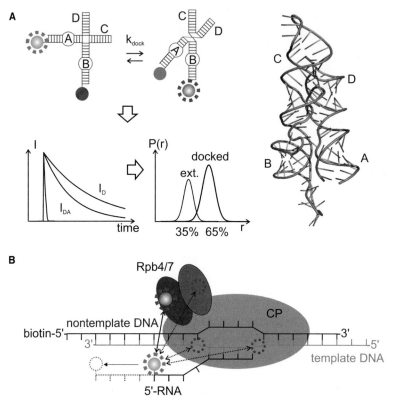

Fig. 8.6 Application of time-resolved and single-molecule FRET to determined conformational distributions of RNA and the architecture of RNA/protein complexes. (A) Determination of the conformations of the hairpin ribozyme. The four helical arms A, B, C, and D of the hairpin ribozyme form a four-way helical junction. By placing a donor (green) and acceptor fluorophore (red) on the end of helical arms A and B, the docking of the extended hairpin ribozyme (low FRET) into a compact state (high FRET) with interactions between loops A and B (see crystal structure, right, PDB-ID 1hp6 [58]) can be followed. From measurements of the donor fluorescence lifetime in the absence (I_D) and presence (I_{DA}) of the acceptor, distance distributions for the docked and extended (ext.) form of the ribozyme can be obtained [73]. The relative populations can be extracted, and the equilibrium constant for docking can be calculated. By comparing the equilibrium constant K_{dock} for the natural ribozyme and variants lacking individual interactions, ΔG^0 and $\Delta\Delta G^0$ values can be obtained that quantify contributions of these interactions to the stabilization of the docked state [36, 75]. (B) Determination of the pathway of the 5'-end of transcribed RNA on the RNA polymerase. A donor fluorophore (green) was placed on the transcribed RNA, and acceptor fluorophores (red) were introduced on the DNA template or RNA polymerase subunits (Rpb4/7) bound to the core polymerase (CP). From (single-molecule) measurements of the FRET efficiencies between the donor and the acceptor in the different positions, inter-dye distances were extracted (double-headed arrows). The position of the donor fluorophore, and thus the 5'-region of the transcribed mRNA, was then determined by triangulation [89].

helix 21 forms an ~60° angle with helix 22 [61]. The S15-induced conformation can also be induced by Mg^{2+} binding [61–63] (Fig. 8.5B). The sequence of the 16S rRNA in this region differs from the corresponding region of the human ribosome, rendering the S15-binding site a possible target for antibiotics that interfere with prokaryotic ribosome assembly. Using three fluorophores (fluorescein, tetramethylrhodamine, and Texas Red), the first two attached to the ends of helical arms 21 and 22, and the third on the S15 protein, the conformational change of the RNA and binding of the S15 protein can be monitored simultaneously [63] (Fig. 8.5B). FRET between fluorescein and tetramethylrhodamine, attached to the three-way junction, reports on the conformation of the RNA (unfolded = low FRET; folded = high FRET). S15 binding introduces the Texas Red fluorophore into the complex that acts as an acceptor for energy transferred from fluorescein and tetramethylrhodamine, and the resulting Texas Red emission is a measure for complex formation. Using this three-color-FRET assay, small molecule compounds were tested for their potential to interfere with RNA folding or S15 protein binding. Due to the high sensitivity of fluorescence techniques, such an assay can be used for high-throughput screening of large libraries of small molecules. Ion-induced folding of this helical junction has later been studied at the single-molecule level, demonstrating that this RNA can be used as an Mg^{2+} sensor [64] and unraveling its folding pathway and the underlying kinetics [62].

In bacterial transcription initiation (reviewed by Young et al. [65]), RNA polymerase, in complex with initiation factors, binds to the promoter region and forms a so-called RNA polymerase/promoter closed complex. Subsequently, a transcription bubble with ~14 disrupted base pairs is generated that covers the region –11 to +3 relative to the start of transcription, forming the open complex (reviewed by Murakami and Darst [66]). After a phase of abortive transcription with the repeated synthesis of short transcripts, promoter escape allows for processive elongation of transcription. By monitoring FRET between a donor fluorophore (tetramethylrhodamine) in different positions on the transcription factor σ^{70}, bound to the RNA polymerase, and an acceptor (Cy5) on the upstream end ("trailing-edge FRET") or the downstream end ("leading-edge FRET") of the DNA template, the translocation of the σ^{70}/RNA polymerase complex along the DNA has been followed [67] (Fig. 8.5C). DNA templates were constructed that lacked guanine until position 12, 13, 14, 15, or 16 from the transcription start site, such that transcription stopped at this position when CTP was omitted from the reaction. Changes in FRET efficiency upon incorporation of the limited number of nucleotides were measured. Complementary FRET experiments were performed with a donor fluorophore (fluorescein) on the RNA polymerase core and an acceptor (tetramethylrhodamine) on σ^{70}. From these experiments, it was shown that σ^{70} is not released during the transition from transcription initiation to elongation but remains bound to and translocates with RNA polymerase during transcription [67]. FRET experiments in solution and in a polyacrylamide gel matrix yielded similar results. Distances derived from FRET experiments with different positions of the label on σ^{70} were consistent with all regions of σ^{70} moving in a concerted fashion with the RNA polymerase. These findings were later confirmed in single-molecule FRET experiments on freely diffusing [68] and surface-immobilized molecules [25].

A subsequent FRET study, combined with distance-constrained docking, has delineated the structural organization of the bacterial RNA polymerase and of the polymerase/promoter open complex [69]. Here, the RNA polymerase was labeled with fluorescein at four different sites. Cy5 was incorporated into several positions of the DNA template. σ^{70} was labeled with tetramethylrhodamine in different positions. FRET experiments were performed on RNA polymerase and on RNA polymerase/promoter open complexes, such that the complexes contained a fluorescein/tetramethylrhodamine or a tetramethylrhodamine/Cy5 donor/acceptor pair. Distances were derived from the measured FRET efficiencies. The large number of distances obtained (105 for RNA polymerase, 66 for the polymerase/promoter open complex) afforded the generation of structural models by distance-constrained docking of the RNA polymerase core, the DNA, and the σ^{70} segment. Starting models were generated by combining crystallographic data on RNA polymerase [70] and a σ^{70} fragment contacting the –10 promoter region [71] with cross-linking data that had defined the position of the DNA segment within the open complex [72], and with a homology model of a σ^{70} fragment that contacts the –35 promoter region [69]. These initial models were then modified in a grid search toward agreement of experimental FRET efficiencies, and FRET efficiencies were calculated for the resulting models. Linker lengths were taken into account by adding a corresponding value to the calculated FRET efficiencies. The resulting structural model with the lowest deviation between experimental and calculated FRET values was further refined in Markov-chain Monte Carlo simulations [69]. Comparison of the final models for the RNA polymerase and the polymerase/promoter open complex showed that the largest part of σ^{70} is in the same position and conformation in both complexes. Its N-terminus is inserted into the cleft bearing the catalytic center of RNA polymerase but has left the active site in the open complex, pointing toward a role in coordinating individual steps in transcription initiation [69].

In conjunction, these studies illustrate how the careful selection of reference points enables monitoring relative movements of a protein along a nucleic acid track, how the fate of individual components of a complex can be followed, and how multiple distance restraints from steady-state FRET experiments can be employed to map the topology of a large multisubunit complex.

8.5.2 Time-resolved FRET: the hairpin ribozyme

Tertiary structure folding and stability of the hairpin ribozyme has been studied in detail, probing global conformations (see 8.5.1) and the contribution of structural elements and individual bases to tertiary structure. As such, the hairpin ribozyme has served as a model system to understand the role of helical junctions in folding of large RNA molecules. Time-resolved FRET allowed the characterization of the equilibrium between the open, inactive form and the docked, active conformation under conditions where both forms coexist and afforded the quantification of the two populations. Using hairpin ribozymes based on two-way, three-way, and four-way helix junctions, labeled with donor (fluorescein) and acceptor (tetramethylrhodamine) on the end of the helical arms containing the loops and the bases forming the active site, it was demonstrated that the junction determines the fraction of the hairpin ribozyme that adopts the active conformation [73] (Fig. 8.6A). While the hairpin ribozyme is completely docked already at low Mg^{2+} (100 μM) concentrations in the context of the natural four-way junction, 100-fold higher Mg^{2+} concentrations are required to dock the two-way junction minimal ribozyme that had been investigated toward gene therapeutic applications. In context of a three-way junction, only ~50% of the ribozymes are docked even at high (millimolar) Mg^{2+} concentrations [73]. A subsequent study dissected the thermodynamic basis for tertiary structure stabilization of the three different hairpin ribozymes by determining the docked fraction from FRET experiments as a function of temperature, combined with UV melting curves that yielded information on the fraction of ribozymes that had already dissociated into single strands at these temperatures [74]. The parameters describing the distance distribution for the released donor/acceptor single strand were determined in control FRET experiments. The corresponding parameters were then kept constant during the fit of the donor decay in the presence of the acceptor with three distance distributions (docked, extended, and single strand). From the fractions of docked and extended species, the equilibrium constant of docking was calculated for each temperature, and the enthalpic and entropic contributions were extracted from van't Hoff plots [74]. The results showed that the higher tertiary structure stability of the four-way junction ribozyme is mainly due to a lower entropic cost for docking, consistent with the junction as a scaffold for ribozyme folding. By quantifying the docked and extended form of modified ribozymes, energetic contributions of individual bases and functional groups have been dissected in time-resolved FRET experiments [36, 75]. The energetic contributions of individual hydrogen bonds in a ribose zipper connecting loops A and B have been probed by comparing ΔG^0 values for docking in the wild-type ribozyme (calculated from the equilibrium constant K as the ratio of extended and docked forms, Fig. 8.6A) and in ribozymes lacking individual 2'-OH groups involved in the zipper motif. While none of the individual hydrogen bonds provided a significant stabilization of the tertiary structure in the four-way junction ribozyme, removal of each hydrogen bond led to a large destabilization of the minimal two-way junction ribozyme. The results demonstrated that the two-way junction construct is not an appropriate model system to probe tertiary interactions in the natural hairpin ribozyme, and that the energetic contribution of hydrogen bonds to RNA tertiary structure stability is context dependent. In similar probing experiments of individual hydrogen bonds in a hydrogen bond network at the docking interface [58] using modified ribozymes that lack one or more of the individual interactions, elimination of a single hydrogen bond resulted in a destabilization by ~2 kJ/mol [75]. In contrast, the energetic contribution per hydrogen bond was smaller when two or more hydrogen bonds were removed, pointing to a cooperative interaction network. Hairpin ribozyme folding has later also been studied at the single-molecule level [76, 77]. Altogether, these studies illustrate how the identification and quantification of different conformers allows for the delineation of tertiary and secondary structure transitions, and the dissection of entropic and enthalpic contributions and contributions from individual interactions to RNA stability.

8.5.3 Single-molecule FRET: folding of large ribozymes and transcription by RNA polymerases

While the folding of small RNA molecules often occurs in a two-state process, folding of large RNAs can occur in more complex pathways, involving one or more folding intermediates. Group I and group II introns (reviewed by Cech [78] and Lambowitz and Zimmerly [79]) are large, multidomain RNA elements that are removed from RNA in a self-splicing reaction. Their folding has been studied extensively to extract the principles governing folding of large RNAs in general. The group II intron folds in two steps. In the first step, an obligatory intermediate is generated by a rate-limiting compaction of domain 1 [80]. The second step of domain assembly has been analyzed in single-molecule FRET experiments [81]. To this end, Cy3- and Cy5-labeled DNA oligonucleotides were hybridized to two loops introduced into domains 1 and 4, and the ribozyme was surface-immobilized for TIRF experiments via a biotinylated DNA oligonucleotide hybridized to the 3'-end. Time traces of donor/acceptor fluorescence and FRET efficiencies revealed transitions between a low, an intermediate, and a high FRET state. The observed FRET states were consistent with a folding pathway that involved the conversion of the already known, "extended" intermediate into a more compact, "folded" intermediate and a subsequent conversion into the native ribozyme [81]. The rate constants for each step were derived from dwell-time analyses. The native state is rarely populated in the absence of substrate and has therefore not been detected in previous bulk experiments. *In vivo*, the DEAD-box helicase Mss116 promotes intron splicing [82]. DEAD-box proteins catalyze the adenosine triphosphate (ATP)–dependent structural rearrangement of RNAs and RNA/protein complexes by local duplex stabilization (reviewed by Hilbert et al. [83]). To understand the role of Mss116 in folding of the group II intron, single-molecule FRET experiments were performed in the presence of Mss116 [84]. While the intron exists predominantly as the extended intermediate state in the absence of Mss116, transitions between all three folding states (extended intermediate, compact intermediate, and native state) were observed when Mss116 was present. Mss116-facilitated folding was ATP-dependent and specific: The compact intermediate was stabilized by Mss116 in the absence of ATP, or by other RNA-binding proteins, but the formation of the native state [84] required Mss116 and ATP. Thus, Mss116 appears to promote group II intron folding by ATP-dependent unwinding of kinetic traps.

These studies demonstrate that single-molecule FRET experiments afford the detection of rare conformers. FRET time traces reveal the inter-conversion kinetics, delineate RNA folding pathways, and allow the distinction between on- and off-pathway intermediates.

Single-molecule FRET has also been used to study the phase of abortive transcription by RNA polymerase in more detail, using the "leading edge-FRET" approach (see Section 8.5.1) to monitor movement of the polymerase toward the downstream end of the transcribed DNA [25]. Here, TIRF microscopy with ALEX (reviewed by Kapanidis et al. [85]) was used. In this type of experiment, donor and acceptor are excited in alternation by two individual lasers, and emission from both fluorophores is detected. The stoichiometry of donor and acceptor dyes can be calculated for each molecule in addition to the FRET efficiency. As a result, molecules can be sorted, and data analysis can be restricted to those species that carry a donor/acceptor pair, facilitating data interpretation in heterogeneous systems. The sorting also allows for the extraction of correction parameters to calculate FRET efficiencies from measured fluorescence intensities (see Section 8.4.2.2) directly from the experimental data [25]. These experiments showed that the leading edge of RNA polymerase moves forward during abortive transcription by a distance that correlates with the length of the RNA product. Subsequent studies provided evidence for RNA polymerase movement in a "DNA scrunching" mechanism, in which the polymerase remains located at the promoter and compacts the downstream DNA by pulling it toward its active site [86]. The accumulated stress has been suggested to aid in overcoming interactions between the polymerase and the promoter and to contribute to promoter escape. In combination, these studies illustrate how ALEX allows for species sorting and species-specific data analysis and for a detailed analysis of complex molecular machines.

8.5.4 Single-molecule FRET and modeling of complex structures

Due to the distance dependence of the FRET efficiency, FRET serves as a "molecular ruler" and enables the determination of inter- and intramolecular distances within RNAs and RNA/protein complexes. Although steady-state FRET has been used successfully to generate a structural model for the RNA polymerase/promoter open complex (see Section 8.5.1), averaging over all conformations usually prevents the extraction of meaningful inter-dye distances, due to either heterogeneity because of incomplete labeling or conformational heterogeneity. Single-molecule approaches afford the determination of FRET efficiencies for individual species, corrected for instrument nonideality, that reflect intra- or intermolecular distances between dyes (see Section 8.4.2). In principle, a number of distances from FRET experiments with species carrying the donor and acceptor dyes in different positions can be used for modeling global conformations and to delineate the topology of large complexes. Distance restraints from single-molecule FRET experiments have been used to generate a structural model for the DEAD-box helicase YxiN in solution [87]. YxiN consists of a conserved helicase core and a C-terminal RNA-binding domain. Distances were determined from single-molecule FRET experiments with helicases carrying donor (Alexa488) and acceptor (Alexa546) fluorophores within the helicase core and on the RNA-binding domain. The RNA-binding domain was then manually positioned relative to the helicase core, such that the experimental distances are fulfilled [87], yielding the first structural model of a full-length DEAD-box helicase. The model is consistent with biochemical data and with the structure of the RNA-binding domain in complex with its RNA ligand [88].

Multiple distances from single-molecule FRET have also been employed to map the pathway of messenger RNA (mRNA) in RNA polymerase elongation complexes. The donor fluorophore tetramethylrhodamine was placed at the 5'-end of the transcribed mRNA, and the acceptor Alexa647 was attached to different positions either on the surface-attached DNA template strand or on a protein subunit of the polymerase [89] (Fig. 8.6B). In single-molecule TIRF experiments with mRNAs of different lengths (17 to 35 nucleotides), inter-dye distances were determined, and the 5'-end of the mRNA exiting from the polymerase was determined by triangulation [89]. It was shown that RNA detaches from the enzyme surface after leaving the polymerase through the exit tunnel. Once it has reached a certain length, it reconnects to the dock domain and a linker domain. The triangulation method has been further developed into a "nanopositioning system" [43], which uses one "antenna dye" on the moiety whose position is to be determined, and a number of "satellite dyes" on the structurally known part. The approach delivers a probability distribution for the position of the antenna dye, taking into account experimental errors in FRET efficiencies and Förster distances, and the length and segmental flexibility of the linker used for fluorophore attachment [43]. In an application to the RNA polymerase elongation complex, antenna dyes were introduced at various positions onto the non-template DNA strand. Satellite dyes were introduced into the template strand, RNA polymerase subunits, or the mRNA. The most probable positions of the antenna dyes were determined from FRET experiments, and the non-template DNA strand was traced within the elongation complex [90]. Subsequent experiments have mapped a minimal open promoter complex and revealed the dynamics of DNA entering and exiting the active site, and conformational changes during the transition from initiation to elongation [91].

8.6 Perspectives

FRET is a powerful tool to study the folding and stability of RNA, and the architecture of RNA/protein complexes. More complex systems can be addressed by extending the classical dual-fluorophore-FRET approach to energy transfer between three [92, 93] and more [94] fluorophores or by using switchable dyes [95]. These approaches afford the simultaneous measurement of multiple distances, such that the dynamics of large entities can be followed. In the last decade, novel FRET-based techniques were applied to unravel the architecture and dynamics of heterogeneous, large, and increasingly complex systems. These approaches include specific excitation schemes, such as ALEX, to sort differently labeled species [25, 86, 96]), and the use of zero-mode wave guides [97, 98]. In combination with improved modeling approaches, these developments will help unravel the architecture of large complexes, the pathways of their assembly, and their dynamics. With a combination of FRET with complementary techniques, such as AFM (see Chapter 6) [99] or optical [100–102] (see Chapter 7) and magnetic tweezers [103], conformational changes and mechanical properties can be determined at the same time. The combination of these techniques holds great promise toward a comprehensive understanding of the structure, dynamics, and mechanism of RNA/protein complexes.

8.7 Acknowledgments

Work in the authors' laboratory was funded by the VolkswagenStiftung, the Swiss National Science Foundation, and the Deutsche Forschungsgemeinschaft.

References

1. Förster T. Zwischenmolekulare Energiewanderung und Fluoreszenz. Ann Phys. 1948; 437:55–75.
2. Stryer L, Haugland RP. Energy transfer: a spectroscopic ruler. Proc Natl Acad Sci U S A. 1967;58:719–26.
3. Schuler B, Pannell LK. Specific labeling of polypeptides at amino-terminal cysteine residues using Cy5-benzyl thioester. Bioconjug Chem. 2002;13:1039–43.
4. Tolbert TJ, Wong, CH. Intein-mediated synthesis of proteins containing carbohydrates and other molecular probes. J Am Chem Soc. 2000;122:5421–8.
5. Wang L, Xie J, Schultz PG. Expanding the genetic code. Annu Rev Biophys Biomol Struct. 2006;35:225–49.
6. Ryu Y, Schultz PG. Efficient incorporation of unnatural amino acids into proteins in *Escherichia coli*. Nat Methods. 2006;3:263–5.
7. Wang L, Zhang Z, Brock A, Schultz PG. Addition of the keto functional group to the genetic code of *Escherichia coli*. Proc Natl Acad Sci U S A. 2003;100:56–61.
8. Scheibner KA, Zhang Z, Cole PA. Merging fluorescence resonance energy transfer and expressed protein ligation to analyze protein-protein interactions. Anal Biochem. 2003;317:226–32.
9. Severinov K, Muir TW. Expressed protein ligation, a novel method for studying protein-protein interactions in transcription. J Biol Chem. 1998;273:16205–9.
10. Pitulle C, Kleineidam RG, Sproat B, Krupp G. Initiator oligonucleotides for the combination of chemical and enzymatic RNA synthesis. Gene. 1992;112:101–5.
11. Moore MJ, Sharp PA. Site-specific modification of pre-mRNA: the 2'-hydroxyl groups at the splice sites. Science. 1992;256:992–7.

12. Moore MJ, Query CC. Joining of RNAs by splinted ligation. Methods Enzymol. 2000;317:109–23.
13. Smith GJ, et al. Efficient fluorescence labeling of a large RNA through oligonucleotide hybridization. RNA. 2005;11:234–9.
14. Clegg RM. Fluorescence resonance energy transfer and nucleic acids. Methods Enzymol. 1992;211:353–88.
15. O'Connor D, Phillips D. Time-Correlated Single Photon Counting. London: Academic Press; 1984.
16. Haas E, et al. Distribution of end-to-end distances of oligopeptides in solution as estimated by energy transfer. Proc Natl Acad Sci U S A. 1975;72:1807–11.
17. Sisamakis E, et al. Accurate single-molecule FRET studies using multiparameter fluorescence detection. Methods Enzymol. 2010;475:455–514.
18. Lamichhane R, et al. Single-molecule FRET of protein-nucleic acid and protein-protein complexes: surface passivation and immobilization. Methods. 2010;52:192–200.
19. Kapanidis AN, Weiss S. Fluorescent probes and bioconjugation chemistries for single-molecule fluorescence analysis of biomolecules. J Chem Phys. 2002;117:10953–64.
20. Rasnik I, McKinney SA, Ha T. Nonblinking and long-lasting single-molecule fluorescence imaging. Nat Methods. 2006;3:891–3.
21. Andreou AZ, Klostermeier D. Conformational changes of DEAD-box helicases monitored by single molecule fluorescence resonance energy transfer. Methods Enzymol. 2012;511:75–109.
22. Theissen B, et al. Cooperative binding of ATP and RNA induces a closed conformation in a DEAD box RNA helicase. Proc Natl Acad Sci U S A. 2008;105:548–53.
23. Gubaev A, Hilbert M, Klostermeier D. The DNA gate of *Bacillus subtilis* gyrase is predominantly in the closed conformation during the DNA supercoiling reaction. Proc Natl Acad Sci U S A. 2009;106:13278–83.
24. Lee NK, et al. Accurate FRET measurements within single diffusing biomolecules using alternating-laser excitation. Biophys J. 2005;88:2939–53.
25. Margeat E, et al. Direct observation of abortive initiation and promoter escape within single immobilized transcription complexes. Biophys J. 2006;90:1419–31.
26. Ha T, et al. Single-molecule fluorescence spectroscopy of enzyme conformational dynamics and cleavage mechanism. Proc Natl Acad Sci U S A. 1999;96:893–8.
27. Parker CA, Rees WT. Correction of fluorescence spectra and measurement of fluorescence quantum efficiency. Analyst. 1960;85:587–600.
28. Magde D, Wong R, Seybold PG. Fluorescence quantum yields and their relation to lifetimes of rhodamine 6G and fluorescein in nine solvents: improved absolute standards for quantum yields. Photochem Photobiol. 2002;75:327–34.
29. Dale RE, Eisinger J, Blumberg WE. The orientational freedom of molecular probes. The orientation factor in intramolecular energy transfer. Biophys J. 1979;26:161–93.
30. Wu P, Brand L. Orientation factor in steady-state and time-resolved resonance energy transfer measurements. Biochemistry. 1992;31:7939–47.
31. van der Meer BW. Kappa-squared: from nuisance to new sense. J Biotechnol. 2002;82:181–96.
32. VanBeek DB, et al. Fretting about FRET: correlation between kappa and R. Biophys J. 2007;92:4168–78.
33. Haas E, Katchalski-Katzir E, Steinberg IZ. Effect of the orientation of donor and acceptor on the probability of energy transfer involving electronic transitions of mixed polarization. Biochemistry. 1978;17:5064–70.
34. Parkhurst LJ, et al. Time-resolved fluorescence resonance energy transfer studies of DNA bending in double-stranded oligonucleotides and in DNA-protein complexes. Biopolymers. 2001;61:180–200.
35. Albaugh S, Steiner RF. Determination of distance distribution from time domain fluorometry. J Phys Chem. 1989;93:8013–16.

36. Klostermeier D, Millar DP. Tertiary structure stability of the hairpin ribozyme in its natural and minimal forms: different energetic contributions from a ribose zipper motif. Biochemistry. 2001;40:11211–18.
37. Klostermeier D, Millar DP. Time-resolved fluorescence resonance energy transfer: a versatile tool for the analysis of nucleic acids. Biopolymers. 2001;61:159–79.
38. Gopich I, Szabo A. Theory of photon statistics in single-molecule Forster resonance energy transfer. J Chem Phys. 2005;122:14707.
39. Antonik M, et al. Separating structural heterogeneities from stochastic variations in fluorescence resonance energy transfer distributions via photon distribution analysis. J Phys Chem B. 2006;110:6970–8.
40. Nir E, et al. Shot-noise limited single-molecule FRET histograms: comparison between theory and experiments. J Phys Chem B. 2006;110:22103–24.
41. Brochon JC. Maximum entropy method of data analysis in time-resolved spectroscopy. Methods Enzymol. 1994;240:262–311.
42. Brunger AT, et al. Three-dimensional molecular modeling with single molecule FRET. J Struct Biol. 2010;173:497–505.
43. Muschielok A, et al. A nano-positioning system for macromolecular structural analysis. Nat Methods. 2008;5:965–71.
44. Wozniak AK, et al. Single-molecule FRET measures bends and kinks in DNA. Proc Natl Acad Sci U S A. 2008;105:18337–42.
45. Sindbert S, et al. Accurate distance determination of nucleic acids via Forster resonance energy transfer: implications of dye linker length and rigidity. J Am Chem Soc., 2011;133:2463–80.
46. Lilley DM. Folding of branched RNA species. Biopolymers. 1998;48:101–12.
47. Symons RH, et al. Self-cleavage of RNA in the replication of viroids and virusoids. J Cell Sci Suppl. 1987;7:303–18.
48. Forster AC, Symons RH. Self-cleavage of plus and minus RNAs of a virusoid and a structural model for the active sites. Cell. 1987;49:211–20.
49. Tuschl T, et al. A three-dimensional model for the hammerhead ribozyme based on fluorescence measurements [erratum appears in Science. 1995;267:1581]. Science. 1994;266:785–9.
50. Scott WG, Finch JT, Klug A. The crystal structure of an all-RNA hammerhead ribozyme: a proposed mechanism for RNA catalytic cleavage. Cell. 1995;81:991–1002.
51. Bassi GS, et al. Ion-induced folding of the hammerhead ribozyme: a fluorescence resonance energy transfer study. EMBO J. 1997;16:7481–9.
52. Buzayan JM, Hampel A, Bruening G. Nucleotide sequence and newly formed phosphodiester bond of spontaneously ligated satellite tobacco ringspot virus RNA. Nucleic Acids Res. 1986;14:9729–43.
53. DeYoung M, et al. Catalytic properties of hairpin ribozymes derived from chicory yellow mottle virus and arabis mosaic virus satellite RNAs. Biochemistry. 1995;34:15785–91.
54. Fedor MJ. Structure and function of the hairpin ribozyme. J Mol Biol. 2000;297:269–91.
55. Walter F, Murchie AIH, Lilley DMJ. Folding of the four-way RNA junction of the hairpin ribozyme. Biochemistry. 1998;37:17629–36.
56. Murchie AI, et al. Folding of the hairpin ribozyme in its natural conformation achieves close physical proximity of the loops. Mol Cell. 1998;1:873–81.
57. Walter F, et al. Structure and activity of the hairpin ribozyme in its natural junction conformation: effect of metal ions. Biochemistry. 1998;37:14195–203.
58. Rupert PB, Ferre-D'Amare AR. Crystal structure of a hairpin ribozyme-inhibitor complex with implications for catalysis [see comments]. Nature. 2001;410:780–6.
59. Zhao ZY, et al. The folding of the hairpin ribozyme: dependence on the loops and the junction. RNA. 2000;6:1833–46.

60. Mizushima S, Nomura M. Assembly mapping of 30S ribosomal proteins from *E. coli*. Nature. 1970;226:1214–18.
61. Orr JW, Hagerman PJ, Williamson JR. Protein and Mg^{2+}-induced conformational changes in the S15 binding site of 16 S ribosomal RNA. J Mol Biol. 1998;275:453–64.
62. Kim HD, et al. Mg2+-dependent conformational change of RNA studied by fluorescence correlation and FRET on immobilized single molecules. Proc Natl Acad Sci U S A. 2002;99:4284–9.
63. Klostermeier D, et al. A three-fluorophore FRET assay for high-throughput screening of small-molecule inhibitors of ribosome assembly. Nucleic Acids Res. 2004;32:2707–15.
64. Ha T, et al. Ligand-induced conformational changes observed in single RNA molecules. Proc Natl Acad Sci U S A. 1999;96:9077–82.
65. Young BA, Gruber TM, Gross CA. Views of transcription initiation. Cell. 2002;109:417–20.
66. Murakami KS, Darst SA. Bacterial RNA polymerases: the wholo story. Curr Opin Struct Biol. 2003;13:31–9.
67. Mukhopadhyay J, et al. Translocation of sigma(70) with RNA polymerase during transcription: fluorescence resonance energy transfer assay for movement relative to DNA. Cell. 2001;106:453–63.
68. Kapanidis AN, et al. Retention of transcription initiation factor sigma70 in transcription elongation: single-molecule analysis. Mol Cell. 2005;20:347–56.
69. Mekler V, et al. Structural organization of bacterial RNA polymerase holoenzyme and the RNA polymerase-promoter open complex. Cell. 2002;108:599–614.
70. Zhang G, et al. Crystal structure of *Thermus aquaticus* core RNA polymerase at 3.3 A resolution. Cell. 1999;98:811–24.
71. Malhotra A, Severinova E, Darst SA. Crystal structure of a sigma 70 subunit fragment from *E. coli* RNA polymerase. Cell. 1996;87:127–36.
72. Naryshkin N, et al. Site-specific protein-DNA photocrosslinking. Analysis of bacterial transcription initiation complexes. Methods Mol Biol. 2001;148:337–61.
73. Walter NG, Burke JM, Millar DP. Stability of hairpin ribozyme tertiary structure is governed by the interdomain junction. Nat Struct Biol. 1999;6:544–9.
74. Klostermeier D, Millar DP. Helical junctions as determinants for RNA folding: origin of tertiary structure stability of the hairpin ribozyme. Biochemistry. 2000;39:12970–8.
75. Klostermeier D, Millar DP. Energetics of hydrogen bond networks in RNA: hydrogen bonds surrounding G+1 and U42 are the major determinants for tertiary structure stability of the hairpin ribozyme. Biochemistry. 2002;41:14095–102.
76. Bokinsky G, Zhuang X. Single-molecule RNA folding. Acc Chem Res. 2005;38:566–73.
77. Bokinsky G, et al. Single-molecule transition-state analysis of RNA folding. Proc Natl Acad Sci U S A. 2003;100:9302–7.
78. Cech TR. Self-splicing of group I introns. Annu Rev Biochem. 1990;59:543–68.
79. Lambowitz AM, Zimmerly S. Group II introns: mobile ribozymes that invade DNA. Cold Spring Harb Perspect Biol. 2011;3:a003616.
80. Su LJ, Waldsich C, Pyle AM. An obligate intermediate along the slow folding pathway of a group II intron ribozyme. Nucleic Acids Res. 2005;33:6674–87.
81. Steiner M, et al. Single-molecule studies of group II intron ribozymes. Proc Natl Acad Sci U S A. 2008;105:13853–88.
82. Halls C, et al. Involvement of DEAD-box proteins in group I and group II intron splicing. Biochemical characterization of Mss116p, ATP hydrolysis-dependent and -independent mechanisms, and general RNA chaperone activity. J Mol Biol. 2007;365:835–55.
83. Hilbert M, Karow AR, Klostermeier D. The mechanism of ATP-dependent RNA unwinding by DEAD box proteins. Biol Chem. 2009;390:1237–50.

84. Karunatilaka KS, et al. Single-molecule analysis of Mss116-mediated group II intron folding. Nature. 2010;467:935–9.
85. Kapanidis AN, et al. Alternating-laser excitation of single molecules. Acc Chem Res. 2005;38:523–33.
86. Kapanidis AN, et al. Initial transcription by RNA polymerase proceeds through a DNA-scrunching mechanism. Science. 2006;314:1144–7.
87. Karow AR, Klostermeier D. A structural model for the DEAD box helicase YxiN in solution: localization of the RNA-binding domain. J Mol Biol. 2010;402:629–37.
88. Hardin JW, Hu YX, McKay DB. Structure of the RNA binding domain of a DEAD-box helicase bound to its ribosomal RNA target reveals a novel mode of recognition by an RNA recognition motif. J Mol Biol. 2010;402:412–27.
89. Andrecka J, et al. Single-molecule tracking of mRNA exiting from RNA polymerase II. Proc Natl Acad Sci U S A. 2007;105:135–40.
90. Andrecka J, et al. Nano positioning system reveals the course of upstream and nontemplate DNA within the RNA polymerase II elongation complex. Nucleic Acids Res. 2009;37:5803–9.
91. Treutlein B, et al. Dynamic architecture of a minimal RNA polymerase II open promoter complex. Mol Cell. 2012;46:136–46.
92. Hohng S, Joo C, Ha T. Single-molecule three-color FRET. Biophys J. 2004;87:1328–37.
93. Haustein E, Jahnz M, Schwille P. Triple FRET: a tool for studying long-range molecular interactions. ChemPhysChem. 2003;4:745–8.
94. Lee J, et al. Single-molecule four-color FRET. Angew Chem Int Ed Engl. 2010;49:9922–5.
95. Uphoff S, et al. Monitoring multiple distances within a single molecule using switchable FRET. Nat Methods. 2010;7:831–6.
96. Kapanidis AN, et al. Fluorescence-aided molecule sorting: analysis of structure and interactions by alternating-laser excitation of single molecules. Proc Natl Acad Sci U S A. 2004;101:8936–41.
97. Levene MJ, et al. Zero-mode waveguides for single-molecule analysis at high concentrations. Science. 2003;299:682–6.
98. Petrov A, et al. Dynamics of the translational machinery. Curr Opin Struct Biol. 2011;21:137–45.
99. He Y, et al. Manipulating protein conformations by single-molecule AFM-FRET nanoscopy. ACS Nano. 2012;6:1221–9.
100. Comstock MJ, Ha T, Chemla YR. Ultrahigh-resolution optical trap with single-fluorophore sensitivity. Nat Methods. 2011;8:335–40.
101. Zhou R, Schlierf M, Ha T. Force-fluorescence spectroscopy at the single-molecule level. Methods Enzymol. 2010;475:405–26.
102. Hohng S, et al. Fluorescence-force spectroscopy maps two-dimensional reaction landscape of the Holliday junction. Science. 2007;318:279–83.
103. Shroff H, et al. Biocompatible force sensor with optical readout and dimensions of 6 nm^3. Nano Lett. 2005;5:1509–14.

Alexey G. Kikhney, Sebastian Doniach, and Dmitri I. Svergun
9 RNA studies by small angle X-ray scattering in solution

9.1 Introduction to SAXS

Small angle scattering (SAS) of X-rays (SAXS) is a powerful method for the analysis of structure and interactions of biological macromolecules in solution. SAXS provides unique insights about proteins, nucleic acids, and their complexes, as well as their macromolecular folding, unfolding, aggregation, shape, conformation, and assembly processes in solution. Over the years, major progress has been made in instrumentation and computational methods leading to exciting developments in the application of SAXS to structural biology [1]. An increasing number of laboratories are now working to advance ab initio and rigid body modeling methods, the calculation of theoretical scattering curves from atomic models and the characterization of quaternary structure and intrinsic flexibility. In addition, advances in the automation of data collection and analysis make high-throughput applications of SAXS experiments possible [2–4].

For a successful structural study by SAXS, a good-quality sample is a prerequisite. Similar to the high-resolution methods, macromolecular X-ray crystallography (MX; see Chapter 13) and nuclear magnetic resonance (NMR; see Chapter 12), SAXS requires milligram amounts of highly pure, monodisperse material that remains soluble at high concentration. With the recent advances in SAXS instrumentation, the requirements on the sample amounts and concentrations are relaxed such that modern SAXS needs less material than MX and NMR (sub-milligrams at sub-mM concentrations). However, the most distinct advantage of SAXS as compared to the other two methods is the speed of both data collection and sample characterization. No additional crystallization step is necessary. On high-brilliance third-generation synchrotron radiation sources, scattering data can be collected in seconds, allowing an almost immediate characterization of sample and sample quality through the evaluation of the radially averaged scattering pattern. In addition, in contrast to NMR, SAXS enables structural studies in an extremely broad range of molecular sizes from a few kDa to GDa and in experimental environments from extreme (e.g. high pressure or cryofrozen) to nearly native. SAXS is also able to determine structures in a time-resolved manner yielding unique information about the kinetics and dynamics of interactions. SAXS can thus be used as a method for the rapid screening of samples in various conditions, including, for example, identification and optimization of crystallization conditions or complex formation [5,6].

SAXS can probe structure on an extremely broad range of macromolecular sizes [1]. Small proteins and polypeptides in the range of 1–10 kDa, macromolecular complexes and large viral particles up to several hundred MDa can all be measured with modern instrumentation under near native conditions. It is often attractive to laboratory-based researchers as the amount of material required for a complete study is relatively low (typically below 1–2 mg), and almost any biologically relevant sample conditions can be used. The effect of changes to the sample environment (pH, temperature, salt concentration, and ligand/cofactor titration) can be easily measured, and, moreover, at high-brilliance synchrotron beamlines, time-resolved experiments can be conducted [7,8].

9.2 SAXS experiment

The following paragraphs describe the typical experimental setup for a SAXS study, sample requirements, and the progress of sample evaluation. In a typical SAXS experiment, solution scattering occurs when a monochromatic and collimated X-ray beam irradiates a macromolecular sample in solution and is scattered (Fig. 9.1). For dilute homogeneous and monodisperse solutions without strong interactions between the particles, the scattering pattern from the sample is proportional to the scattering of a single particle averaged over all orientations, since the individual particles can be found in all orientations in the solution. This results in a two-dimensional isotropic scattering pattern that can be radially averaged, thus representing the scattering intensity as a function of the angle (i.e. function of the momentum transfer $s = 4\pi\sin\theta/\lambda$, where λ is the beam wavelength and 2θ is the scattering angle). Though the observed scattering intensity is often measured on a relative scale (depending on the detector collecting the scattered photons), it depends on the size of the particles (larger particles scatter stronger at small angles than smaller ones) and on the solute concentration.

9.2.1 Sample preparation

On modern third-generation synchrotron sources, precise SAXS patterns can be collected in a fraction of a second (undulator beamline) or within a few minutes (bending

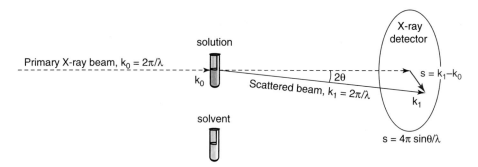

Fig. 9.1 Schematic representation of a SAXS experiment.

magnet beamline); on a home source, the exposure time can be up to 1 hour. Depending on the experimental setup, a single SAXS measurement requires between a few and a few dozen microliters of sample. Each measurement of the macromolecular solution requires two measurements of the corresponding buffer (before and after the sample). The buffer should not contain excessive amounts of additives (e.g. not more than 0.5 M NaCl, not more than 5% glycerol, not more than 5 mM adenosine triphosphate, etc.).

Sample monodispersity should be higher than 90% and needs to be verified by native gel filtration, ultracentrifugation, dynamic light scattering, and so forth. If the sample is aggregated, the scattering data will be difficult or even impossible to interpret. Typically, for each sample a concentration series (e.g. 1, 2, 4, 8 mg/mL) has to be measured. The sample concentrations should be determined as precisely as possible (accuracy better than 10% is required), since the concentrations are necessary to normalize the scattering data.

Since X-rays interact mainly with the electrons, they may affect the integrity of the measured sample by ionization of the solute or solvent and subsequent radiation damage due to free radicals. The radiation damage problem is often treated by adding scavengers such as about 2 mM dithiothreitol or protectants like glycerol to the sample before the measurement.

The scattering experiment is based on the difference (contrast) in the average electron density between the solvent and the dissolved particles (Tab. 9.1). Typically, first the pure solvent (usually dialysis buffer) is measured, followed by the sample solution (solvent and macromolecules). Subsequently, the scattering pattern of the pure solvent is subtracted from the scattering pattern of the sample solution to obtain the scattering pattern of the macromolecules. This pattern is proportional to the scattering from a single particle averaged over all orientations and to the sample concentration.

9.2.2 Form and structure factor: particle interactions

In the general case, the experimental scattering pattern results from a correlation of the shape of the individual molecules (form factor) and the arrangement of the

Tab. 9.1 Average electron density of selected molecules.

	Average electron density, e/nm^3
tRNA	568
5 S rRNA	567
Bovine serum albumin	429
Lysozyme	438
Water	334

molecules in the solution (structure factor). An experimental scattering pattern can thus be expressed as

$$I_{exp}(c, s) = I(s)SF(c, s) \tag{9.1}$$

where $I_{exp}(c, s)$ is the experimental (measured) scattering, $I(s)$ is the form factor, and $SF(c, s)$ is the structure factor. To construct a model for a biological macromolecule, it is important to limit the effects of the structure factor to get a "pure" form factor (essentially to make $SF[c, s] = 1$). At low concentrations (typically below a few mg/mL), these effects are usually small and interactions between individual molecules may be neglected. Further, scattering patterns from different concentrations can be compared and, if differences are observed, extrapolated to infinite dilution. As an example, a processed synchrotron X-ray scattering pattern from aminoacyl–transfer RNA (tRNA) in solution [9] is presented in Fig. 9.2.

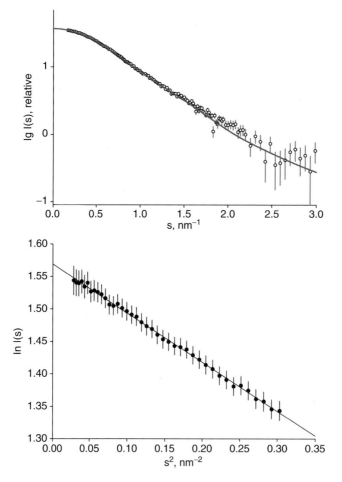

Fig. 9.2 Processed X-ray scattering data from aminoacyl-tRNA (open circles with error bars) and the theoretical scattering profile calculated from the atomic model (solid line), PDB ID: 4TRA [92]. The Guinier plot displayed in the insert yields the radius of gyration R_g of 2.3 nm (note that the linear fit is preformed in the range $s < 1.3/R_g$).

9.3 Methods

9.3.1 Distance distribution function

The shape of a particle of uniform density can be described by the correlation function $\gamma(r)$ (first introduced by Debye and Bueche in 1949), which illustrates the probability of two points within the particle being situated at a distance r. Thus, for $r < D_{max}$, $\gamma(r)$ is positive, and for $r \geq D_{max}$, $\gamma(r) \equiv 0$, where D_{max} is the maximum intraparticle distance. The scattering intensity $I(s)$ is actually the Fourier transform of the distance distribution function $p(r) = r^2 \gamma(r)$:

$$I(s) = 4\pi \int_0^{D_{max}} p(r) \frac{\sin(sr)}{sr} dr \qquad (9.2)$$

or, inversely,

$$p(r) = \frac{r^2}{2\pi^2} \int_0^{\infty} \frac{s^2 I(s) \sin(sr)}{sr} ds \qquad (9.3)$$

Despite this simple relation, obtaining the $p(r)$ function from the scattering pattern directly is not trivial. While calculating the $p(r)$ directly from $I(s)$ requires integration from zero to infinity and is strongly influenced by the noise in the data, an opposite "indirect" procedure is much more stable, and the relation of $I(s)$ to $p(r)$ is utilized in indirect transform methods [10,11]. Here, one parameterizes the $p(r)$ and computes the parameters yielding the transformed intensity that agrees with the experimental $I(s)$. The $p(r)$ function of aminoacyl-tRNA computed from the experimental X-ray data in Fig. 9.2 is presented in Fig. 9.3.

9.3.2 Overall parameters: radius of gyration, molecular mass, and volume

From the radially averaged one-dimensional curves, several important parameters can be directly obtained providing information about the size, oligomeric state, and overall shape of the molecule. At very low angles, the scattering pattern can be described with two parameters that correspond to the molecular mass (MM) and the size of the particle. The first parameter, the intensity at zero angle or forward scattering $I(0)$, is proportional to the molecular mass corresponding to in-phase scattering from all electrons in the particle. The second parameter is the radius of gyration (R_g), which corresponds to the averaged squared distance from the center of mass of the particle weighted by the scattering length density. These two parameters are related by the so-called Guinier law [12] as $I(s) = I(0)\exp(-s^2 R_g^{2/3})$ for scattering

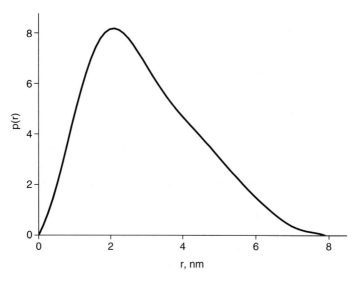

Fig. 9.3 Distance distribution function $p(r)$ of tRNA computed from the experimental data in Fig. 9.2 using GNOM [11].

angles that satisfy the relation $sR_g < 1.3$. Note that for highly anisotropic particles, the range over which this holds is quite limited. Forward scattering $I(0)$ cannot be measured directly because it follows the direction of the primary beam, which is masked before reaching the detector by a beamstop, such that available data range starts from a value s_{min}. Thus, the Guinier law can be used as an approximation to extrapolate the intensity to zero scattering angle. Note that the Guinier law is valid for smaller particles for a longer s range than for larger ones. Depending on the sample-detector distance, this may lead to problems when calculating the $I(0)$ and R_g for larger (or very elongated) particles as the number of experimental points may not be sufficient for an accurate estimation (i.e. the range $s_{min} < s < 1.3/R_g$ does not contain sufficient numbers of experimental points). Linearity of the Guinier plot is often considered as an indicator of absence of significant interparticle interactions (aggregation or strong repulsion); see an example for aminoacyl-tRNA in the insert of Fig. 9.2. In some cases, when the Guinier approximation is difficult to apply, the R_g may still be calculated from the $p(r)$, essentially using the information from the full scattering curve. The $p(r)$ function can be calculated by an indirect transformation, which has a somewhat relaxed requirement on the minimum momentum transfer, namely, $s_{min} < \pi/D_{max}$.

Another parameter that can be computed from the small angle portion of the data ($s < 2.5$ nm^{-1}) is the excluded volume of the hydrated particle. It is obtained using the Porod equation [13]:

$$V_p = \frac{2\pi^2 I(0)}{\int_0^\infty [I(s) - K_4] s^2 ds} \qquad (9.4)$$

where K_4 is a constant determined to ensure the asymptotical intensity decay proportional to s^{-4} at higher angles following the so-called Porod's law for globular particles of uniform density. For real macromolecules, the electron density is of course not uniform; however, at sufficiently high MM (> 30 kDa), the subtraction of an appropriate constant from the scattering data generates a reasonable approximation to the scattering of the corresponding homogenous body. The particle volume V_p allows one to make an alternative estimate of the MM with the added advantage that this estimate is independent of errors in the sample concentration. Typically, for proteins V_p (in nm³) is about 1.7 the MM (in kDa), and for RNA V_p is approximately equal to MM (this is explained by significant difference in partial specific volumes between proteins and RNA). While this estimate is a crude approximation only, with prior knowledge of the expected size/state of the sample, a rough indication of the homogeneity of the sample can be made directly after the measurement.

9.4 Modeling

In addition, to the overall parameters, one-dimensional solution scattering profiles $I(s)$ allow one to meaningfully analyze three-dimensional (3-D) structures. As a direct structure reconstruction is impossible, the main approach when interpreting SAXS data is creating structural models such that the computed scattering pattern matches the observed experimental pattern.

9.4.1 Ab initio modeling

As was demonstrated in the 1990s [14–16], low-resolution particle shapes can be reconstructed ab initio (i.e. without any prior information) from the scattering patterns. Subsequently, we shall briefly consider some of the shape reconstruction techniques, which are applicable to different types of macromolecular solutions.

9.4.1.1 Bead models

The low-resolution ab initio shape reconstruction programs DAMMIN [16] and DAMMIF [17] represent the particle as a collection of several thousands of densely packed beads and employ simulated annealing to search for a compact interconnected model that fits the low-resolution portion of the data (usually to about 2 nm resolution) minimizing discrepancy:

$$\chi^2 = \frac{1}{N-1}\sum_j \left[\frac{I(s_j) - cI_{calc}(s_j)}{\sigma(s_j)}\right]^2 \quad (9.5)$$

where N is the number of experimental points, c is a scaling factor, and $I_{calc}(s_j)$ and $\sigma(s_j)$ are the calculated intensity and the experimental error at the momentum transfer s_j, respectively. The volume of the resulting model is an important parameter characterizing the particle under investigation. For oligomeric particles consisting of identical subunits, symmetry can be taken into account as a rigid constraint so that the symmetrically related beads are added/removed to the model simultaneously.

9.4.1.2 Dummy residue models

An alternative higher-resolution ab initio model can be constructed using the higher angles of scattering data by the program GASBOR [18] representing the protein as an ensemble of dummy residues forming a "chain-compatible" model. The spatial positions of these residues aim at approximately corresponding to those of the Cα atoms in the protein structure, and the number of residues should be equal to that in the protein. Although this technique was specifically developed for protein modeling, it may with some caveats be applied ab initio to modeling of RNA assuming that one nucleotide can be represented by about 2.5 residues. For large particles, GASBOR is significantly slower than DAMMIN or DAMMIF; therefore, it should be started only if sufficient resources are available.

9.4.1.3 Multiphase models

For multicomponent particles (e.g. RNA-protein complexes), there is a possibility to extend the approach to a multiphase bead modeling where the phase index varies up to the number of components describing not only the overall shape but also the internal organization of the composite molecule. The prerequisite for this is availability of the scattering profiles from the individual components (or their combinations) so that they are fitted simultaneously with the data of the intact particle. A hypothetical example is a nucleoprotein complex, where the scattering patterns from the protein, from the RNA, and from the entire complex are available. This approach is implemented in a multiphase modeling program MONSA [19], which, for this case, would build a bead model depicting the protein and RNA parts inside the complex in such a way that all three scattering patterns are fitted simultaneously. The concept of the multiphase modeling is especially powerful for neutron scattering (SANS), where multiple scattering patterns can be collected in solutions containing different amounts of heavy water (contrast variation) [20] and are interpreted in terms of the shapes of the components.

9.4.1.4 Comparison of multiple models

Due to the intrinsic ambiguity of the SAXS data interpretation, multiple ab initio modeling runs always produce somewhat different models yielding nearly identical scattering patterns. These models can be superimposed and compared to obtain the most probable model and an averaged model, which is done automatically in the program package DAMAVER [21]. The latter repeatedly runs the program SUPCOMB [22] to align and compare pairs of models represented by beads. Such analysis allows one to assess the reliability of the ab initio models: multiple reconstructions from same input data should not demonstrate significant deviations from each other. This is expressed in terms of the so-called normalized spatial discrepancy (NSD) (values of NSD about unity indicate that the two shapes are similar). Another approach to compare multiple models is to group them into clusters using the program DAMCLUST [23]. When considering the ab initio models, it should always be remembered that the enantiomorphous structures (i.e. mirror images) would produce exactly the same scattering intensity. This ambiguity is automatically taken into account by the comparison algorithms discussed previously, but the handedness of the averaged models is still arbitrary.

9.4.2 SAXS and complementary methods

SAXS becomes much more powerful when combined with other methods. High-resolution models determined by MX (see Chapter 13) and NMR (see Chapter 12) can be used to build models of complexes by fitting the SAXS data (granted that no major conformation change occurs). If the interface between the components of the complex is known (e.g. from mutagenesis studies, fluorescence resonance energy transfer [FRET] (see Chapter 8), or NMR), this information can be used to reduce the ambiguity of the models created from SAXS. Residual dipolar couplings (RDCs) from NMR can determine the relative orientation of one complex component in relation to the others [24]. Electron microscopy models, especially on large and highly symmetric particles, may help to define the search volume for the ab initio methods.

9.4.2.1 High-resolution models

When a high-resolution model of the macromolecule is available, it is possible to analytically compute the scattering pattern and to compare it to the experimental SAXS data. In particular, possible RNA conformations can be rapidly screened to identify the biologically relevant structure.

The theoretical scattering curve of an atomic model can be calculated and compared to the experimental data. The computation of the theoretical scattering includes the atomic scattering in vacuo, the excluded volume (the volume occupied by the macromolecule that is inaccessible to solvent), and the scattering from the solvation shell. If the theoretical scattering of a model fits the experimental data well, then such a model is considered a valid description of the macromolecule measured in solution.

There are several approaches to computation of theoretical scattering curves from atomic models, which differ in the way that the atomic scattering intensity is calculated and the solvent excluded volume is subtracted, and in how the hydrated surface with a higher solvent density is approximated. Usually the Debye formula [25] is employed to compute the atomic scattering; however, the time required for computation increases proportionally to the square of the number of atoms and makes this approach ineffective for large RNAs and complexes. The spherical harmonics approximation on the other hand is probably the most efficient approach to the calculation of the scattering intensity. In the program CRYSOL [26], using this approach the computation time scales linearly with the size of the model, and it is highly accurate up to $s < 5.0$ nm^{-1} but has also been used up to much higher resolutions (up to $s = 10–15$ nm^{-1}). CRYSOL employs a Gaussian sphere approximation for the calculation of the excluded volume [27] and spherical harmonics for the calculation of an envelope of variable scattering length density at the surface of the atomic model in order to approximate the solvation shell. Based on a combined SAXS/SANS study [28], one can estimate the scattering density of the solvation shell to be about 10% higher than that of the bulk solvent. When fitting the experimental data, the excess scattering density of the solvation shell is adjusted as a fitting parameter to obtain the best fit to the scattering data up to a resolution of about 0.5 nm. Fig. 9.2 displays that the scattering pattern computed from the high-resolution structure of yeast tRNA agrees well with the experimental data aminoacyl-tRNA in solution.

Alternative methods for the approximation of the solvation shell include the addition of an explicit layer of water molecules in a solvent density matching that of the bulk solution [29] or the calculation of nonuniformly hydrated atomic models [30]. Some recent programs are now available as web servers (e.g. FoXS [31] or AXES [32]). All the present methods are utilizing the concept of the solvation shell around the macromolecule, and most of them are applicable not only for proteins but also for RNA.

9.4.2.2 Rigid body modeling

Macromolecular complexes can be assembled by fitting the individual components into ab initio shapes [33]. However, since SAXS provides only low-resolution models and because of ambiguities in shape reconstruction, such complexes are usually modeled through direct refinement against the scattering data. A wide variety of interactive or automated programs exist to model the positions and orientations of complex subunits based on SAXS data [34–37]. If the structures of the separate subunits are known, the theoretical scattering of a complex can be easily calculated using an application of the spherical harmonics formalism. This formalism, utilized in the previously described program CRYSOL [26], is implemented in the program SASREF for automated rigid body modeling [35]. SASREF allows for the use of symmetry, orientation constraints (e.g. from RDCs measured by NMR; see Chapter 12), interresidue contacts (e.g. from mutagenesis or cross-linking experiments), and intersubunit distances (e.g. from Fourier transform infrared spectroscopy (FTIR) and FRET; see Chapter 8). The program starts by arbitrarily positioning a subunit that is then randomly moved and rotated as rigid body while taking any user-defined constraints into account. Through simulated annealing, SASREF searches for the best fit of the computed complex scattering to the experimental data by minimizing the discrepancy and the penalty term. The latter may include restraints from other methods such as those mentioned previously but always observes constraints so that the models generated are interconnected and feature no steric clashes.

9.4.3 Flexible systems

Recently, SAXS also became a powerful technique for the analysis of flexible biological systems. For these systems, interpretation of the scattering data in terms of a single model is not feasible because of significant conformational polydispersity. Quantitative analysis of such systems can be performed using an ensemble optimization method (EOM) [38], which allows for a coexistence of multiple conformers in solution. In EOM, the analysis of a potentially flexible system using scattering data is carried out in two steps. First a large pool of random configurations is generated, utilizing high-resolution models for regions of known structure when available. Then ensembles of candidate conformations are selected from this pool by a genetic algorithm such that the average computed scattering over the ensemble fits the experimental scattering data. If the R_g distribution of the models in the selected ensembles is as broad as that in the initial random pool, the macromolecule of interest is likely to be flexible; obtaining a narrow R_g distribution peak suggests that the system is rigid [39]. Note, however, that for the resulting ensemble to be thermodynamically admissible, constraints must be applied in the selection of candidate conformations so that a reasonable approximation to a Boltzmann ensemble is generated [40]. Other methods to analyze systems of flexible macromolecules in terms of ensembles have recently also become available [41–44].

9.4.4 Mixtures

Another important application of SAXS to rapidly characterize solutions of biological macromolecules is the quantitative description of mixtures. For homogeneous polydisperse solutions, the scattering pattern is a sum of the contributions from each component $I_k(s)$ weighted by the component's volume fraction v_k:

$$I(s) = \sum_{k=1}^{K} v_k I_k(s) \tag{9.6}$$

There are a number of methods developed for the analysis of equilibrium mixtures using SAS [1,45–47]. If the number of components in the mixture is unknown but a series of measurements of the sample containing different amounts of the components is available (e.g. sample at different stages of an assembly process), a model-independent estimate of the number of components can be obtained using singular value decomposition [48,49]. If the scattering patterns of the individual components are known (e.g. calculated by CRYSOL from known models) the volume fractions can be computed by the program OLIGOMER [47]. Recently, a multivariate curve resolution method [50] was proposed to determine scattering patterns from components in oligomeric mixtures.

A powerful technique for the analysis of mixtures is time-resolved SAXS, which can be used for the study of RNA folding under near native conditions of temperature and pH [7,8] and of complex formation and dissociation. Third-generation synchrotrons provide the necessary flux for recording SAXS data with a good signal-to-noise ratio using millisecond exposures; modern fast readout detectors and rapid mixing devices are installed at many SAXS beamlines making time-resolved scattering a perfect tool for investigating kinetic processes in complex systems of interacting molecules, allowing one to link structure to biological function.

9.5 Resolution and ambiguity of SAXS data interpretation

The maximum s value in the scattering pattern determines the nominal resolution of the experiment as $d = 2\pi/s_{max}$. Because of the information loss due to the spherical average, the nominal resolution range of a scattering experiment is not directly related to the resolution of the models obtained. The spherical average leads moreover to the intrinsic ambiguity of the SAXS data, whereby multiple models can fit the data equally well. For some simple objects (e.g. a perfect sphere), such ambiguity does not exist, but real molecules, especially biological macromolecules, have complicated shapes, and this ambiguity poses one of the most important problems when interpreting the SAXS curves. The minimum s value is defined by the beamline configuration. For a reliable R_g estimation using the Guinier approximation, it is required that the range between s_{min} and $1/R_g$ contains a sufficient number of experimental points. A somewhat softer restriction is required for a stable reconstruction of the $p(r)$ function using an indirect transformation technique, namely, $s_{min} < \pi/D_{max}$.

9.6 Practical applications

In this section, we shall present some applications of SAXS to study RNA molecules and their complexes with proteins. In principle, the experiments and also analysis approaches for the nucleic acids are similar to those for solutions of proteins, but there are also important RNA- specific aspects. On a negative side, the RNA-containing samples are often more susceptible to radiation damage than the protein samples, and care must be taken to avoid these effects, especially with high-brilliance synchrotron radiation sources (addition of radioprotectants, sample flow during exposure, etc.). On a positive side, for X-rays, RNA has about twice the contrast of proteins in aqueous solutions (i.e. four times higher scattering intensity for the same MM). The RNA samples therefore usually provide a better signal/noise ratio compared to protein samples, and less material is required for the SAXS studies. RNAs are much more charged macromolecules compared to proteins, and their properties are critically dependent on electrostatic interactions. Therefore, the ionic strength of the solution usually plays a very significant role in the studies of RNA solutions. Subsequently, we shall give typical examples of the solution studies of RNA-containing samples highlighting both similarities with the analysis of protein systems and specific approaches utilized for nucleic acids.

9.6.1 Ab initio shape determination

The ab initio bead modeling methods (like the programs DAMMIN and DAMMIF described previously) represent the particle shape using finite volume elements, and in this sense there is no difference between the shape determination for proteins and of nucleic acids. Therefore the low-resolution structure of RNA molecules, similar to that of proteins, can be determined ab initio from the scattering data. Fig. 9.4 presents the shape analysis of a free *Thermus flavus* 5S ribosomal RNA in solution [51] (Fig. 9.4). This study was among the first applications of ab initio shape determination procedures to RNA molecules, and simultaneously, the publication contained one of the first systematic analyses of the ambiguity of the shape reconstruction. A dozen independent shape determination runs by DAMMIN on the experimental data in Fig. 9.4A provided models with similar overall appearance and differences in finer detail (Fig. 9.4B). These were superimposed and averaged to obtain the final consensus shape in Fig. 9.4C. Based on the results of this study, an automated analysis and averaging procedure was subsequently developed [21], which is now widely employed in practice of shape determination. The average shape of the 5S rRNA depicts a bent elongated molecule with a compact central region and two projecting arms, similar to those of tRNA. Interestingly, the reconstructed model displayed a good agreement with the overall shape of the theoretically predicted structure of the 5S rRNA from *Xenopus laevis* [52]. The theoretical model was somewhat more extended, but both models were displaying two elongated arms (helices). A tentative model of 5S rRNA

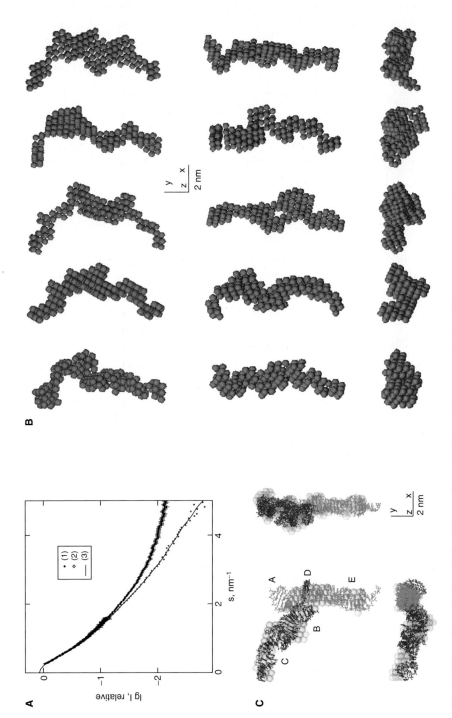

Fig. 9.4 Shape determination of a free *Thermus flavus* 5S ribosomal RNA in solution [51]. (A) X-ray scattering data: (1) experimental curve; (2) shape scattering curve after subtraction of a constant to force the s^{-4} decay; (3) scattering from an ab initio bead model. (B) Low-resolution models

was constructed from the crystal structure of the A-D-E domain [53] and of the B-C domains as predicted [52], which can be well positioned within the shape (Fig. 9.4C). Not unexpectedly, an a posteriori comparison with the 5S rRNA inside the ribosome [54,55] indicated that the ribosomal RNA becomes essentially more compact in the context of the assembled 50S ribosomal subunit.

The ab initio shape determination by SAS became a regular tool to characterize RNA molecules in solution, and numerous recent applications are presented in [56]. Coarse-grained approaches specifically tailored to model the tertiary structure of RNAs based on SAXS data are being developed [57]. Very often, SAS is used for the structure prediction in combination with the other low-resolution techniques including, together with standard biophysical techniques, methods like hydroxyl radical footprinting and multiplexed hydroxyl radical cleavage [58].

9.6.2 Analysis of RNA flexibility

Many approaches in SAXS data analysis were initially developed for proteins and became useful for RNA molecules. The ensemble fitting approach described previously [38] is also applicable to characterize flexibility of RNA, as demonstrated by the study of *Escherichia coli* RNA chaperone Hfq. After providing insights into the flexibility and function of the C terminus of the Hfq [59], its complex with a 34-nucleotide-long natural substrate small RNA, DsrA$_{34}$, was analyzed in a subsequent study [60]. As the NMR (see Chapter 12), SAXS, and enzymatic studies suggested that DsrA$_{34}$ is unfolded, extended, and lacks defined secondary structure when bound to Hfq, the ensemble fitting approach was used to assess the flexibility and the accessible conformational space of the complexes. The major insight from this approach was that DsrA$_{34}$ in the complex is unfolded, as the experimental data could not be fitted by compact RNA structures. In the complex, the protein maintained its doughnut-like structure, whereas the extended DsrA$_{34}$ was found to explore one hemisphere in

---

Fig. 9.4 (*Continued*)
of the 5S rRNA obtained in five independent shape determination runs (from left to right). The middle and bottom rows are rotated counterclockwise by 90° around the *y*- and *x*-axes, respectively. (C) The final overall shape of the 5S rRNA (semitransparent beads) superimposed with the atomic models of the domains A-D-E (dark dots) and of a fragment containing domains B-C (gray dots). The upper left panel is displayed in the orientation as in (B) rotated counterclockwise by 55° in the figure plane. The right and bottom panels are rotated counterclockwise by 90° around the *y*- and *x*-axes with respect to that in the upper left panel. Originally published in J. Biol. Chem. by Funari, S. S., Rapp, G., Perbandt, M. et al. (2000) Structure of free *Thermus flavus* 5 S rRNA at 1.3 nm resolution from synchrotron X-ray solution scattering, 275, 31283–8. © American Society for Biochemistry and Molecular Biology.

conformational space contacting a broad area of the protein structure in agreement with the NMR chemical shifts data. It is speculated that the structural flexibility of RNA ligands bound to Hfq stochastically may facilitate base pairing providing the foundation for the RNA chaperone function inherent to Hfq.

Another example of accounting for flexibility in the RNA modeling is the study of solution structures of three phylogenetically distinct bacterial ribonuclease P (RNase P) RNAs from *E. coli*, *Agrobacterium tumefaciens* and *Bacillus stearothermophilus* [61]. The RNA component of the ribonucleoprotein enzyme RNase P processes tRNAs by cleavage of precursor tRNAs. The crystal structures of the bacterial RNase P, including full-length RNAs and a ternary complex with substrate are available, but the solution structures of free RNAs are unknown. The authors employed SAXS using a combination of homology modeling, normal mode analysis, and molecular dynamics

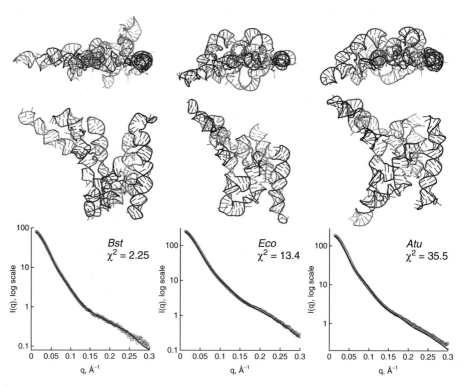

Fig. 9.5 Solution structures of three phylogenetically distinct bacterial RNase P RNAs from *E. coli*, *A. tumefaciens*, and *B. Stearothermophilus* [61]. *(Top)* All-atom models of the RNase P RNA variants are shown. RNA models were generated in silico by taking into account available crystallographic and phylogenetic information. (Bottom) Agreement between theoretical scattering profiles calculated from the models (solid line) and experimental SAXS data (open circles). Bst, *B. stearothermophilus*; Eco, *E. coli*; Atu, *A. tumefaciens*. Originally published in RNA by Kazantsev, A.V. et al. (2011) Solution structure of RNase P RNA, 17, 1159–71. © RNA Society.

with selective 2′-hydroxyl acylation analyzed by primer extension. The approach was used to refine the available RNA structures in solution under the high-ionic-strength conditions required for catalytic activity (Fig. 9.5).

9.6.3 Nonstochiometric RNA-protein mixtures and complex formation

One of the problems arising in the studies of weakly bound protein-RNA complexes is that excess of one of the components is required for the full binding, and the solution can no longer be considered monodisperse. Yadavalli et al. [62] investigated a complex of tRNA with human mitochondrial phenylalanyl-tRNA synthetase (mtPheRS) by SAXS. To achieve the maximum amount of the 1:1 complex, a 1.5:1 tRNA:mtPheRS was analyzed, and in the subsequent analysis, the excess tRNA was taken into account as an additional component. The SAXS data suggested that the enzyme exists in two conformations: a "closed" state when not bound to the tRNA and an "open" state upon tRNA binding [63]. This indicated that for aminoacylation a rearrangement of the tRNA-binding and catalytic domains of mtPheRS is required. To test this hypothesis, the "closed" form of mtPheRS was generated by cross-linking the catalytic and tRNA-binding domains that still catalyzed Phe activation but was no longer able to transfer Phe to tRNA. SAXS experiments confirmed the presence of both the closed and open forms of mtPheRS in solution. Thus, SAXS studies underscored the importance of conformational flexibility of the two domains in mtPheRS that is essential for enzyme activity and provided a model for tRNA^{Phe} recognition. Interestingly, a similar modular setup with independently active domains is found in other aminoacyl-tRNA synthetases highlighting the evolutionary development of these enzymes [64,65]. This structural and evolutionary adaptability of aminoacyl-tRNA synthetases makes them fit for functional expansion as housekeeping or regulatory proteins [66,67].

The direct modeling of the protein-tRNA complex was impossible because of the excess tRNA. Given that the high-resolution structure of the tRNA was available and a tentative model of the protein was built, a set of possible models of the complex was generated such that the contact conditions between the tRNA and the protein were observed. The scattering patterns of these models and of the tRNA structure alone were calculated by CRYSOL [26] and taken in a linear combination to fit the experimental pattern using OLIGOMER [47] (see Section 9.4.4). Here, the volume fractions of the two components were determined by a least-squares procedure to minimize the discrepancy between the experimental and calculated scattering patterns. The overall best model provided the optimal fit to the experimental data for the mixture of 53% 1:1 tRNA-protein complex and 47% of the free tRNA.

9.6.4 Structural studies of spliceosome function assisted by SAXS measurements

Most eukaryotic genes are expressed as precursor messenger RNAs (pre-mRNAs) that are converted to mRNA by splicing, an essential step of gene expression in which noncoding sequences (introns) are removed and coding sequences (exons) are ligated together. Alternative splicing is prevalent in higher eukaryotes, and it enhances their complexity by increasing the number of unique proteins expressed from a single gene.

Nuclear pre-mRNA splicing is catalyzed by the spliceosome, a multi-megadalton ribonucleoprotein complex. Both the conformation and composition of the spliceosome are highly dynamic, affording the splicing machinery its accuracy and at the same time flexibility. During spliceosome assembly, an intricate RNA-RNA interaction network is formed that is extensively rearranged during catalytic activation of the spliceosome and the catalytic steps of splicing (reviewed by Will and Luhrmann [68]). Whereas RNA-RNA secondary interactions in the spliceosome are, for the most part, well characterized, information about the nature and dynamics of RNA tertiary interactions is scarce.

In a recent study [69], a 111-nt RNA construct (hereafter referred to as U2/U6) was used to investigate the structure of the U2/U6 complex (Fig. 9.6). This construct contains the entire base-paired region of the *Saccharomyces cerevisiae* U2/U6 small nuclear RNA (snRNA) complex linked by a UUCG tetraloop on helix II to ensure proper stoichiometry of the U2 and U6 strands (Fig. 9.6). A novel method to generate structural models was used in which a large number of starting models generated with MC-Sym [70] were sorted by agreement with the experimental NMR (see Chapter 12) and SAXS data. The structural models in best agreement with both data sets were jointly refined against SAXS and NMR data by restrained molecular dynamics and energy minimization. The resulting U2/U6 complex has a well-defined fold that provides new insight into its role as an essential component of the spliceosome active site.

Based on SAXS data collected in 150 mM NaCl and 2 mM $MgCl_2$, an ab initio model of U2/U6 was calculated using DAMMIF [17]. The low-resolution model reveals that U2/U6 forms a "Y" shape in solution (Fig. 9.7) composed of three arms roughly the size of A-form helices.

To elucidate the locations of helices within the ab initio envelope, SAXS data were taken on RNA constructs containing 11- to 12-bp helical extensions of helix II or III with nonnative secondary structure. The helical extensions manifest as additional envelope density adjacent to the corresponding helix. Thus, the extended constructs allow identification of helix II and helix III, while the unperturbed helical feature likely belongs to the U6 intramolecular stem-loop (ISL).

The investigation of this relatively large molecule in solution using sparse NMR data and SAXS provides a determination of the free U2/U6 structure, which indicates the relative configuration of essential RNA components in the absence of proteins.

This work provides insight into the structural and functional role of other essential splicing factors necessary for formation of the spliceosomal active site.

9.6.5 How SAXS helps elucidate riboswitch structure-function relationships

Riboswitches provide straightforward examples of how small functional RNA molecules, which are coded in the noncoding RNA (ncRNA) parts of genome sequences, provide substrate-driven control of gene expression. All riboswitches found to date lie in bacterial genomes with the exception of the thiamine pyrophosphate (TPP) riboswitch for which homologs are found both in bacteria and in plants and fungi.

Riboswitches are structured functional RNA elements that are typically located at the 5' noncoding regions of mRNAs and comprise an aptamer domain and an expression platform domain. The aptamer domain functions as a sensor to recognize and bind small molecule targets with extremely high specificity. The role of the expression platform is to modulate gene expression in response to the binding events in the aptamer domain.

Elucidation of the structural changes, which follow on substrate binding, and how they influence gene expression is a subject of considerable interest, and a large

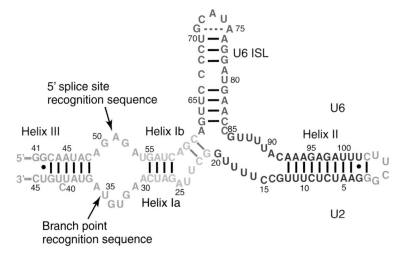

Fig. 9.6 Proposed secondary structure of a 111-nt RNA based on the *S. cerevisiae* U2/U6 RNA complex [69]. Structural features are helix I (green), helix II (purple), helix III, and an internal loop that binds to 5′ splice site and branch point (orange), U6 ISL (blue), U-rich loop (dark gray), and nonnative sequences (light gray). (Black lines or circles) NMR determined base pairing. Originally published in RNA by Burke, J.E. et al. (2012) Structure of the yeast U2/U6 snRNA complex, 18, 673–83. © RNA Society.

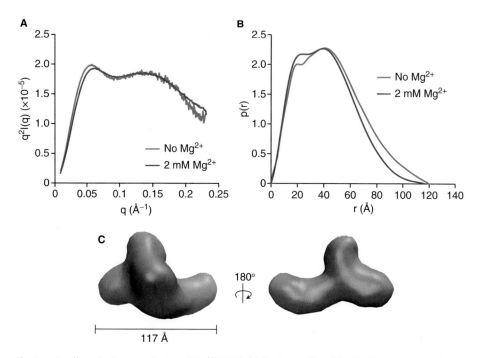

Fig. 9.7 Small-angle X-ray scattering of U2/U6 [69]. (A) Kratky profile of the U2/U6 complex in the absence and presence of Mg^{2+}. All experiments were conducted in 50 mM Tris (pH 7.0), 150 mM NaCl, and 0 or 2 mM $MgCl_2$. (B) Pair distance distribution function plot of U2/U6 in the absence and presence of Mg^{2+}. (C) Ab initio structure of U2/U6 in 2 mM $MgCl_2$. Twenty structures were generated using the program DAMMIF and then averaged with DAMAVER, yielding a normalized spatial discrepancy (NSD) of 0.85. Originally published in RNA by Burke, J.E. et al. (2012) Structure of the yeast U2/U6 snRNA complex, 18, 673–83. © RNA Society.

variability in mechanisms is found among the hundreds of riboswitches identified so far.

In general, the binding of a substrate to the aptamer can lead either to inhibition of transcription elongation via formation of a terminator structure or blockage of translation through sequestering of the ribosome-binding site. In some cases, the riboswitch can influence splicing activity. Riboswitches affect gene regulation by sequestering sequences within their ligand-bound tertiary structures that would otherwise be involved in secondary structure formation or by directly shielding regulatory sequences involved in transcription or translation.

In a recent report, Chen et al. [71] use a combination of SAXS and NMR (see Chapter 12) to study the function of the *S*-adenosyl-methionine-binding, SAM-II riboswitch, This is an interesting example since SAM-II is one of the only riboswitches combining the sequences of both the aptamer and the expression platform in a single compact structure in its substrate-bound state [72]. The SAXS data suggest that the RNA undergoes marked conformational changes upon Mg^{2+} compaction and SAM metabolite binding and that Mg^{2+} binding is essential to induce a state that is competent to bind metabolite. Only after metabolite binding does the formation of various secondary and tertiary structural elements including a pseudoknot occur to sequester the Shine-Dalgarno sequence (SD sequence) of the RNA.

Chemical probing data indicate that SAM-dependent stabilization of the L1-P2b tertiary interaction is likely to be responsible for translational repression via an occlusion mechanism rather than by secondary structural switching.

The combination of SAXS and NMR (Chapter 12) used by Chen et al. [71] demonstrates unambiguously that, in addition to the metabolite, the Mg^{2+} ion is required for the RNA to adopt its well-organized native folded structure. In the crystal structure of the SAM-II riboswitch, a magnesium ion is positioned in a pocket formed between the 50 termini from two RNA molecules. This magnesium ion–binding site is far away from the metabolite-binding pocket, suggestive of a delocalized ion-binding site. However, based on the NMR data, it is likely that there is a specific magnesium-binding site near the triple helix region where the ligand-binding pocket is formed.

These studies indicate that Mg^{2+} and metabolite binding may act together to populate a conformation that is competent ultimately for turning off translation. Thus, studies using both SAXS and NMR provide a unified view of the nature of the free and bound states of SAM-II that are consistent with previous X-ray crystallography (Chapter 13), chemical probing (Chapters 2, 3), and single-molecule FRET (Chapter 8) data. In summary, the SAM-II riboswitch appears to adopt multiple conformations along its folding pathway, and under physiological conditions, the RNA is found to be unstructured in the absence of the ligand. Addition of the metabolite induces the formation of a complicated binding pocket with the putative SD sequence of the RNA sequestered as well as extensive tertiary interactions between the RNA and the metabolite. Mg^{2+} ion is an essential cofactor for proper folding of the RNA and can facilitate the formation of a well-organized tertiary structure. Occlusion of the SD sequence is postulated to inhibit ribosomal binding and to prevent translation initiation.

This work complements and extends recent studies on SAM-II in two aspects. First, SAXS provides unambiguous evidence for at least four conformational states of varying compactions, with compaction induced by "Mg^{2+} only" comparable with compaction induced by "metabolite only" and "metabolite and Mg^{2+}," although the underlying secondary and tertiary structural arrangements are not completely identical (see also analogous findings for the glycine riboswitch [73]). This suggests that the Mg^{2+}-compacted state samples the metabolite- and Mg^{2+}-induced states in agreement with prevalent ideas about conformational selection as a basis for RNA molecular recognition [74–77]. This is in contrast to the TPP riboswitch in which a large-scale conformational change leads to occlusion of the ribosome-binding site [78].

NMR analysis provides greater insights into the nature of secondary and tertiary structural rearrangement of the RNA in the presence of Mg^{2+} and metabolite, suggesting that the metabolite binding drives the folded conformer to populate the final state competent for shutting down protein synthesis. Somewhat similar behavior has been found for a construct involving the apatamer of the adenine riboswitch, accompanied by its expression platform (M. Ali et al., unpublished).

9.6.6 Use of SAXS and ASAXS to study the influences of counterions on RNA folding

Many small functional RNAs fold into specific 3-D structures that are the key to their function. Unlike proteins where a hydrophobic core guides the formation of folding intermediates, RNA 3-D structures are stabilized both by counterion shielding of the coulomb forces induced by the phosphate-sugar backbone and by tertiary H-bond interactions. Anomalous SAXS (ASAXS) has proved to be a useful tool for studying the spatial distribution of this counterion cloud [79–81].

SAXS has been a workhorse for characterizing folding intermediates of small RNAs as a function both of salt concentration and of mutations introduced to study the free-energy perturbations induced by specific mutations. As an excellent example, the thermodynamics of tertiary interactions has been probed by studying the effects of mutations of the folding free-energy landscape of 195-nt *Azoarcus sp.* bacterial group I ribozyme using SAXS, ribozyme activity, hydroxyl radical footprinting, and native polyacrylamide gel electrophoresis (PAGE) [82].

9.6.7 Quantitation of free-energy changes estimated from SAXS 3-D reconstructions

In addition to characterizing the folding intermediates, SAXS measurements have also provided a way to generate semiquantitative estimates of the free-energy changes induced by counterion shielding.

In the particular case of the glycine riboswitch, Lipfert et al. [73] used low-resolution structural models derived from SAXS together with Poisson-Boltzmann (PB) theory to show that simple electrostatics can account for the observed data on salt dependence of folding and to obtain quantitative estimates of energetic factors in folding from the U to M states.

Low-resolution structures of the U, M, and B (unfolded, intermediate, and fully folded) states of the *Vibrio cholerae* VCI-II tandem aptamer were generated by ab initio structure reconstructions from SAXS data using DAMMIN [16]. The U and M states likely comprise ensembles of structures with flexibility in junction regions of the molecule. As SAXS is an ensemble technique, the scattering profiles reflect ensemble averages, and the low-resolution reconstructions represent the average electron density of the corresponding states.

The U state reconstruction shows an elongated shape, consistent with helices kept apart by electrostatic repulsion under the low-salt conditions. The M state adopts a more compact conformation, consistent with increased helical packing compared with the U state. The B state, however, is marked by a considerable rearrangement of electron density from the periphery toward the center of the structure, compared with the U and M states.

The reconstructed low-resolution models consist of dummy residues or beads. As there is no direct correspondence between beads and nucleic acid residues, total charge of the molecule ($Q_{tot} = 226e$, where e is the elementary charge) was divided equally among all N_{bead} beads, such that $Q_{bead} = Q_{tot}/N_{bead}$. PB energies were computed for these models as a function of salt concentrations using the software APBS [83].

PB predictions were compared with the experimentally observed fractional occupancies of the M state for Na^+ and Mg^{2+} titrations in the absence of glycine. The PB prediction of the salt dependence for the U-to-M transitions agrees well with the experimental observations for both the Na^+ and Mg^{2+} titrations parameterized using an empirical fit to the Hill equation. (Note that the Hill form for the dependence of the occupancy on the salt concentration is used for mathematical convenience, as an empirical equation used to describe the data. Unlike in its usual interpretation, the power law index does not correspond to the number of specifically bound ligands or ions but is rather a fitting parameter loosely describing the observed cooperativity of the transition.) The reasonable agreement between PB theory and the experimental data suggests that simple electrostatic screening is sufficient to account for all or nearly all of the energetics of the U-to-M transition as a function of increasing cation concentration.

The M-to-B transition involves specific ion binding. Glycine binding and the population of the glycine-bound B state are supported by Mg^{2+}, Mn^{2+}, and Ca^{2+}, but not Sr^{2+}, Ba^{2+}, and Zn^{2+}. The discrimination of at least 100-fold between different species of divalent ions could not be accounted for by simple electrostatic theories [84,85] and strongly suggested that formation of the B state involves specific ion site binding. This was subsequently borne out by crystallographic measurement [86].

9.7 Conclusions and outlook

From the point of view of structural biology, proteins dominated the molecular landscape in the second half of the twentieth century. However, the enormous advances in genome sequencing have led to large changes in molecular biology as it has become clear that only 2% of the human genome codes for proteins, while a larger fraction is probably transcribed as ncRNA [87]. At present this is a controversial subject [88]. Nevertheless we can say that one of the major tasks for structural biology in the first part of the twenty-first century will be to study the structure and function of RNA molecules. This is providing enormous challenges to structural biology as MX (Chapter 13), the technique of choice for protein structure determination, is much less effective for RNA molecules, which tend to be a lot more flexible and difficult to crystallize than proteins. It is clear that MX will keep playing a crucial role, but complementary techniques will become more and more important. Fortunately, the approaches developed to analyze protein structures in solution by SAXS appeared to be directly applicable or easily adaptable to the RNA studies. In this chapter, we provided several examples of the SAXS studies of RNA; more applications can be found in recent reviews describing the use of SAXS to study RNA shapes and flexibility [89], studies of interactions with ions [90], and time-resolved analysis of RNA folding [91]. Modern SAXS turns out to be an essential tool to help characterize the RNA molecules in solution, and the power of SAXS is further enhanced by the recent developments in instrumentation (most importantly, high-brilliance synchrotron sources) and the wider use of the advanced analysis computational methods. Undoubtedly, SAXS will be playing a significant role in helping to determine 3-D RNA structure, along with chemical footprinting (Chapter 2), NMR (Chapter 12), and a number of other techniques.

9.8 Acknowledgments

AK and DS would like to acknowledge the WeNMR project (European FP7 e-Infrastructure grant, contract No. 261572; www.wenmr.eu).

References

1. Feigin LA, Svergun DI. Structure Analysis by Small-Angle X-ray and Neutron Scattering. New York: Plenum Press; 1987.
2. Franke D, Kikhney AG, Svergun DI. Automated acquisition and analysis of small angle X-ray scattering data. Nuclear Instrum Methods Phys Res A. 2012;689:52–9.
3. Hura GL, Menon AL, Hammel M, et al. Robust, high-throughput solution structural analyses by small angle X-ray scattering (SAXS). Nat Methods. 2009;6:606–12.
4. Round AR, Franke D, Moritz S, et al. Automated sample-changing robot for solution scattering experiments at the EMBL Hamburg SAXS station X33. J Appl Crystallogr. 2008;41:913–7.
5. Bonnete F, Finet S, Tardieu A. Second virial coefficient: variations with lysozyme crystallization conditions. J Cryst Growth. 1999;196:403–14.
6. Hamiaux C, Perez J, Prange T, Veesler S, Ries-Kautt M, Vachette P. The BPTI decamer observed in acidic pH crystal forms pre-exists as a stable species in solution. J Mol Biol. 2000;297:697–712.
7. Lamb J, Kwok L, Qiu X, Andresen K, Park HY, Pollack L. Reconstructing three-dimensional shape envelopes from time-resolved small-angle X-ray scattering data. J Appl Crystallogr. 2008;41:1046–52.
8. Pollack L, Doniach S. Time-resolved X-ray scattering and RNA folding. In: Daniel H, ed. Methods in Enzymology. San Diego: Academic Press; 2009:253–68.

9. Bilgin N, Ehrenberg M, Ebel C, et al. Solution structure of the ternary complex between aminoacyl-tRNA, elongation factor Tu, and guanosine triphosphate. Biochemistry. 1998;37:8163–72.
10. Glatter O. A new method for the evaluation of small-angle scattering data. J Appl Crystallogr. 1977;10:415–21.
11. Svergun DI. Determination of the regularization parameter in indirect-transform methods using perceptual criteria. J Appl Crystallogr. 1992;25:495–503.
12. Guinier A. La diffraction des rayons X aux tres petits angles; application a l'etude de phenomenes ultramicroscopiques. Ann Phys (Paris). 1939;12:161–237.
13. Porod G. General theory. In: Glatter O, Kratky O, eds. Small-Angle X-ray Scattering. London: Academic Press; 1982:17–51.
14. Svergun DI, Stuhrmann HB. New developments in direct shape determination from small-angle scattering 1. Theory and model calculations. Acta Crystallogr. 1991;A47:736–44.
15. Chacon P, Moran F, Diaz JF, Pantos E, Andreu JM. Low-resolution structures of proteins in solution retrieved from X-ray scattering with a genetic algorithm. Biophys J. 1998; 74:2760–75.
16. Svergun DI. Restoring low resolution structure of biological macromolecules from solution scattering using simulated annealing. Biophys J. 1999;76:2879–86.
17. Franke D, Svergun DI. DAMMIF, a program for rapid ab-initio shape determination in small-angle scattering. J Appl Crystallogr. 2009;42:342–6.
18. Svergun DI, Petoukhov MV, Koch MHJ. Determination of domain structure of proteins from X-ray solution scattering. Biophys J. 2001;80:2946–53.
19. Svergun DI, Nierhaus KH. A map of protein-rRNA distribution in the 70 S *Escherichia coli* ribosome. J Biol Chem. 2000;275:14432–9.
20. Zaccai G, Jacrot B. Small angle neutron scattering. Annu Rev Biophys Bioeng. 1983;12:139–57.
21. Volkov VV, Svergun DI. Uniqueness of ab initio shape determination in small angle scattering. J Appl Crystallogr. 2003;36:860–4.
22. Kozin MB, Svergun DI. Automated matching of high- and low-resolution structural models. J Appl Crystallogr. 2001;34:33–41.
23. Petoukhov MV, Franke D, Shkumatov AV, et al. New developments in the ATSAS program package for small-angle scattering data analysis. J Appl Crystallogr. 2012;45:342–50.
24. Gabel F, Simon B, Nilges M, Petoukhov M, Svergun D, Sattler M. A structure refinement protocol combining NMR residual dipolar couplings and small angle scattering restraints. J Biomol NMR. 2008;41:199–208.
25. Debye P. Zerstreuung von Roentgenstrahlen. Ann Phys. 1915;46:809–23.
26. Svergun DI, Barberato C, Koch MHJ. CRYSOL – a program to evaluate X-ray solution scattering of biological macromolecules from atomic coordinates. J Appl Crystallogr. 1995;28:768–73.
27. Fraser RDB, MacRae TP, Suzuki E. An improved method for calculating the contribution of solvent to the X-ray diffraction pattern of biological molecules. J Appl Crystallogr. 1978;11:693–4.
28. Svergun DI, Richard S, Koch MHJ, Sayers Z, Kuprin S, Zaccai G. Protein hydration in solution: experimental observation by x-ray and neutron scattering. Proc Natl Acad Sci U S A. 1998;95:2267–72.
29. Yang S, Park S, Makowski L, Roux B. A rapid coarse residue-based computational method for X-ray solution scattering characterization of protein folds and multiple conformational states of large protein complexes. Biophys J. 2009;96:4449–63.
30. Poitevin F, Orland H, Doniach S, Koehl P, Delarue M. AquaSAXS: a web server for computation and fitting of SAXS profiles with non-uniformly hydrated atomic models. Nucleic Acids Res. 2011;39:W184–9.

31. Schneidman-Duhovny D, Hammel M, Sali A. FoXS: a web server for rapid computation and fitting of SAXS profiles. Nucleic Acids Res. 2010;38:W540–4.
32. Grishaev A, Guo LA, Irving T, Bax A. Improved fitting of solution X-ray scattering data to macromolecular structures and structural ensembles by explicit water modeling. J Am Chem Soc. 2010;132:15484–6.
33. Wriggers W, Chacon P. Using Situs for the registration of protein structures with low-resolution bead models from X-ray solution scattering. J Appl Crystallogr. 2001;34:773–6.
34. Konarev PV, Petoukhov MV, Svergun DI. MASSHA – a graphic system for rigid body modelling of macromolecular complexes against solution scattering data. J Appl Crystallogr. 2001;34:527–32.
35. Petoukhov MV, Svergun DI. Global rigid body modeling of macromolecular complexes against small-angle scattering data. Biophys J. 2005;89:1237–50.
36. Sun Z, Reid KB, Perkins SJ. The dimeric and trimeric solution structures of the multidomain complement protein properdin by X-ray scattering, analytical ultracentrifugation and constrained modelling. J Mol Biol. 2004;343:1327–43.
37. Boehm MK, Woof JM, Kerr MA, Perkins SJ. The Fab and Fc fragments of IgA1 exhibit a different arrangement from that in IgG: a study by X-ray and neutron solution scattering and homology modelling. J Mol Biol. 1999;286:1421–47.
38. Bernado P, Mylonas E, Petoukhov MV, Blackledge M, Svergun DI. Structural characterization of flexible proteins using small-angle X-ray scattering. J Am Chem Soc. 2007;129:5656–64.
39. Bernado P. Effect of interdomain dynamics on the structure determination of modular proteins by small-angle scattering. Eur Biophys J. 2010;39:769–80.
40. Doniach S, Lipfert J. Small and wide angle X-ray scattering from biological macromolecules and their complexes in solution. In: Egelman EH, ed. Comprehensive Biophysics. Oxford: Academic Press; 2012:376–97.
41. Pelikan M, Hura GL, Hammel M. Structure and flexibility within proteins as identified through small angle X-ray scattering. Gen Physiol Biophys. 2009;28:174–89.
42. Yang S, Blachowicz L, Makowski L, Roux B. Multidomain assembled states of Hck tyrosine kinase in solution. Proc Natl Acad Sci U S A. 2010;107:15757–62.
43. Rozycki B, Kim YC, Hummer G. SAXS ensemble refinement of ESCRT-III CHMP3 conformational transitions. Structure. 2011;19:109–16.
44. Marsh JA, Forman-Kay JD. Ensemble modeling of protein disordered states: experimental restraint contributions and validation. Proteins. 2012;80:556–72.
45. Fowler AG, Foote AM, Moody MF, et al. Stopped-flow solution scattering using synchrotron radiation: apparatus, data collection and data analysis. J Biochem Biophys Methods. 1983;7:317–29.
46. Koenig S, Svergun D, Koch MHJ, Hubner G, Schellenberger A. synchrotron radiation solution X-ray scattering study of the pH dependence of the quaternary structure of yeast pyruvate decarboxylase. Biochemistry. 1992;31:8726–31.
47. Konarev PV, Volkov VV, Sokolova AV, Koch MHJ, Svergun DI. PRIMUS – a Windows-PC based system for small-angle scattering data analysis. J Appl Crystallogr. 2003;36:1277–82.
48. Golub GH, Reinsh C. Singular value decomposition and least squares solution. Numer Math. 1970;14:403–20.
49. Konarev PV, Petoukhov MV, Volkov VV, Svergun DI. ATSAS 2.1, a program package for small-angle scattering data analysis. J Appl Crystallogr. 2006;39:277–86.
50. Blobel J, Bernado P, Svergun DI, Tauler R, Pons M. Low-resolution structures of transient protein-protein complexes using small-angle X-ray scattering. J Am Chem Soc. 2009;131:4378–86.
51. Funari SS, Rapp G, Perbandt M, et al. Structure of free *Thermus flavus* 5 S rRNA at 1.3 nm resolution from synchrotron X-ray solution scattering. J Biol Chem. 2000;275:31283–8.

52. Westhof E, Romby P, Romaniuk PJ, Ebel JP, Ehresmann C, Ehresmann B. Computer modeling from solution data of spinach chloroplast and of *Xenopus laevis* somatic and oocyte 5 S rRNAs. J Mol Biol. 1989;207:417–31.
53. Correll CC, Freeborn B, Moore PB, Steitz TA. Metals, motifs, and recognition in the crystal structure of a 5S rRNA domain. Cell. 1997;91:705–12.
54. Nissen P, Hansen J, Ban N, Moore PB, Steitz TA. The structural basis of ribosome activity in peptide bond synthesis [see comments]. Science. 2000;289:920–30.
55. Yusupov MM, Yusupova GZ, Baucom A, et al. Crystal structure of the ribosome at 5.5 A resolution. Science. 2001;292:883–96.
56. Rambo RP, Tainer JA. Bridging the solution divide: comprehensive structural analyses of dynamic RNA, DNA, and protein assemblies by small-angle X-ray scattering. Curr Opin Struc Biol. 2010;20:128–37.
57. Yang S, Parisien M, Major F, Roux B. RNA structure determination using SAXS data. J Phys Chem B. 2010;114:10039–48.
58. Parisien M, Major F. Determining RNA three-dimensional structures using low-resolution data. J Struct Biol. 2012;179:252–60.
59. Beich-Frandsen M, Vecerek B, Konarev PV, et al. Structural insights into the dynamics and function of the C-terminus of the *E. coli* RNA chaperone Hfq. Nucleic Acids Res. 2011;39:4900–15.
60. de Almeida Ribeiro E Jr, Beich-Frandsen M, Konarev PV, et al. Structural flexibility of RNA as molecular basis for Hfq chaperone function. Nucleic Acids Res. 2012;40:8072–84.
61. Kazantsev AV, Rambo RP, Karimpour S, Santalucia J Jr, Tainer JA, Pace NR. Solution structure of RNase P RNA. RNA. 2011;17:1159–71.
62. Yadavalli SS, Klipcan L, Zozulya A, et al. Large-scale movement of functional domains facilitates aminoacylation by human mitochondrial phenylalanyl-tRNA synthetase. FEBS Lett. 2009;583:3204–8.
63. Klipcan L, Levin I, Kessler N, Moor N, Finarov I, Safro M. The tRNA-induced conformational activation of human mitochondrial phenylalanyl-tRNA synthetase. Structure. 2008;16:1095–104.
64. Ibba M, Curnow AW, Soll D. Aminoacyl-tRNA synthesis: divergent routes to a common goal. Trends Biochem Sci. 1997;22:39–42.
65. Alexander RW, Schimmel P. Domain-domain communication in aminoacyl-tRNA synthetases. Prog Nucleic Acid Res Mol Biol. 2001;69:317–49.
66. Park SG, Kim HJ, Min YH, et al. Human lysyl-tRNA synthetase is secreted to trigger proinflammatory response. Proc Natl Acad Sci U S A. 2005;102:6356–61.
67. Ray PS, Arif A, Fox PL. Macromolecular complexes as depots for releasable regulatory proteins. Trends Biochem Sci. 2007;32:158–64.
68. Will CL, Luhrmann R. Spliceosome structure and function. Cold Spring Harb Perspect Biol. 2011;3:a003707.
69. Burke JE, Sashital DG, Zuo X, Wang YX, Butcher SE. Structure of the yeast U2/U6 snRNA complex. RNA. 2012;18:673–83.
70. Parisien M, Major F. The MC-Fold and MC-Sym pipeline infers RNA structure from sequence data. Nature. 2008;452:51–5.
71. Chen B, Zuo X, Wang YX, Dayie TK. Multiple conformations of SAM-II riboswitch detected with SAXS and NMR spectroscopy. Nucleic Acids Res. 2012;40:3117–30.
72. Gilbert SD, Rambo RP, Van Tyne D, Batey RT. Structure of the SAM-II riboswitch bound to S-adenosylmethionine. Nat Struct Mol Biol. 2008;15:177–82.
73. Lipfert J, Das R, Chu VB, et al. Structural transitions and thermodynamics of a glycine-dependent riboswitch from *Vibrio cholerae*. J Mol Biol. 2007;365:1393–406.

74. Hammes GG, Chang YC, Oas TG. Conformational selection or induced fit: a flux description of reaction mechanism. Proc Natl Acad Sci U S A. 2009;106:13737–41.
75. Zhang Q, Al-Hashimi HM. Domain-elongation NMR spectroscopy yields new insights into RNA dynamics and adaptive recognition. RNA. 2009;15:1941–8.
76. Wilson RC, Smith AM, Fuchs RT, Kleckner IR, Henkin TM, Foster MP. Tuning riboswitch regulation through conformational selection. J Mol Biol. 2011;405:926–38.
77. Haller A, Souliere MF, Micura R. The dynamic nature of RNA as key to understanding riboswitch mechanisms. Acc Chem Res. 2011;44:1339–48.
78. Ali M, Lipfert J, Seifert S, Herschlag D, Doniach S. The ligand-free state of the TPP riboswitch: a partially folded RNA structure. J Mol Biol. 2010;396:153–65.
79. Das R, Mills TT, Kwok LW, et al. Counterion distribution around DNA probed by solution X-ray scattering. Phys Rev Lett. 2003;90:188103.
80. Pabit SA, Finkelstein KD, Pollack L. Using anomalous small angle X-ray scattering to probe the ion atmosphere around nucleic acids. Methods Enzymol. 2009;469:391–410.
81. Kirmizialtin S, Pabit SA, Meisburger SP, Pollack L, Elber R. RNA and its ionic cloud: solution scattering experiments and atomically detailed simulations. Biophys J. 2012;102:819–28.
82. Behrouzi R, Roh JH, Kilburn D, Briber RM, Woodson SA. Cooperative tertiary interaction network guides RNA folding. Cell. 2012;149:348–57.
83. Baker NA, Sept D, Joseph S, Holst MJ, McCammon JA. Electrostatics of nanosystems: application to microtubules and the ribosome. Proc Natl Acad Sci U S A. 2001;98:10037–41.
84. Bai Y, Greenfeld M, Travers KJ, et al. Quantitative and comprehensive decomposition of the ion atmosphere around nucleic acids. J Am Chem Soc. 2007;129:14981–8.
85. Chu VB, Bai Y, Lipfert J, Herschlag D, Doniach S. Evaluation of ion binding to DNA duplexes using a size-modified Poisson-Boltzmann theory. Biophys J. 2007;93:3202–9.
86. Lipfert J, Sim AY, Herschlag D, Doniach S. Dissecting electrostatic screening, specific ion binding, and ligand binding in an energetic model for glycine riboswitch folding. RNA. 2010;16:708–19.
87. Pennisi E. Genomics. ENCODE project writes eulogy for junk DNA. Science. 2012;337:1159–61.
88. Graur D, Zheng Y, Price N, Azevedo RB, Zufall RA, Elhaik E. On the immortality of television sets: "function" in the human genome according to the evolution-free gospel of ENCODE. Genome Biol Evol. 2013;5(3):578–90.
89. Rambo RP, Tainer JA. Improving small-angle X-ray scattering data for structural analyses of the RNA world. RNA. 2010;16:638–46.
90. Pollack L. SAXS studies of ion-nucleic acid interactions. Annu Rev Biophys. 2011;40:225–42.
91. Pollack L. Time resolved SAXS and RNA folding. Biopolymers. 2011;95:543–9.
92. Westhof E, Dumas P, Moras D. Restrained refinement of two crystalline forms of yeast aspartic acid and phenylalanine transfer RNA crystals. Acta Crystallogr A. 1988;44(pt 2):112–23.

Jean-François Ménétret, Heena Khatter, Angelita Simonetti, Igor Orlov, Alexander G. Myasnikov, Srividhya Krishnagiri Venkatasubramanian, Sankar Manicka, Morgan Torchy, Kareem Mohideen, Anne-Sophie Humm, Isabelle Hazemann, Alexandre Urzhumtsev, and Bruno P. Klaholz

10 Integrative structure-function analysis of large nucleoprotein complexes

10.1 Summary

The structural analysis of large macromolecular complexes composed of nucleic acids (RNA or DNA) and proteins is a challenging task and requires the careful integration of a series of techniques. These comprise in particular biochemical preparation and biophysical characterization of samples, sequence analysis, X-ray crystallography, single particle cryo electron microscopy (cryo-EM), cryo electron tomography (cryo-ET), and fluorescence imaging. These techniques in principle allow gaining insights into molecular architectures from the atomic level to the molecular and cellular levels. Here we give an overview of some practical aspects of how we combine these in order to address the basis of the molecular mechanism of action of nucleic acid complexes such as ribosome complexes and DNA-bound nuclear receptors.

The structural analysis of large macromolecular nucleoprotein complexes requires the combination of a variety of complementary techniques that, depending on the application and the quality of the sample, provide information from the atomic level to the molecular and cellular levels. Spanning such a large range of resolution and scales as done in our laboratory on nucleoprotein complexes from ~50 kDa up to more than 100 MDa requires the integration of biochemistry, bioinformatics, X-ray crystallography (see Chapter 13), single particle cryo-EM, cryo-ET, and eventually molecular and cellular imaging. Cryo-EM/ET is situated in the middle of the resolution range and plays a key role in integrating the various structural data because it can resolve cellular and molecular features with increasingly improved resolution (in some cases up to the 3 Å resolution range where amino acid side chains can be resolved). We therefore give a brief introduction into some technical aspects of cryo-EM and cryo-ET, which are interesting to know, particularly to take into account possibilities and limitations when combining cryo-EM maps or cryo-ET tomograms (molecular scale) with crystal structures (atomic scale) or confocal series (cellular scale). In this context, it should be mentioned that the preparation of homogeneous samples is crucial to the same degree for crystallization and for single particle cryo-EM in order to obtain well-diffracting crystals or high-resolution cryo-EM maps. In the event that obtaining homogeneous samples is not possible, sorting structural states by cryo-EM and image processing can help with addressing sample heterogeneity to some extent and may, in this case, provide additional information on the dynamics of the macromolecular complex. Another technical limitation to take into account is the size and molecular weight of the complex: complexes in the range of 80–200 kDa may be difficult to crystallize, but they are also tricky to analyze by cryo-EM because of their relatively small size for this method. We provide here examples on nucleoprotein complexes such as ribosome complexes, translation factors, and DNA-bound nuclear receptors, which together address these various aspects.

Cryo-EM allows the observation of individual molecules of a macromolecular complex in a native, frozen-hydrated state [1]. To achieve this, a small amount of sample (5–30 μl at ~0.5 mg/ml) is required that has been purified to homogeneity and characterized beforehand by biophysical methods; these include, for example, gel filtration, native gel analysis, sucrose or glycerol gradients, analytical ultracentrifugation (AUC), MALDI and ESI mass spectrometry, dynamic light scattering (see Chapter 9), multiangle laser-light scattering (MALLS), and small angle X-ray scattering (SAXS); see Fig. 10.1 for examples of some nucleoprotein complexes.

We have recently published a new method to prepare a homogeneous sample of a complex between a ribosome and a messenger RNA (mRNA) [2] characterized by cryo-EM. The cryo-EM experiment comprises the following steps: the sample is deposited on a holey-carbon grid, the liquid excess is blotted with a filter paper, and then the sample is flash-frozen to obtain a frozen-hydrated sample surrounded by amorphous (noncrystalline) buffer, resulting in the sample being in a meniscus spanning the holes of the holey carbon (Fig. 10.2). The observation of the sample under cryo-conditions is thus done without fixation or staining, which could otherwise lead to observation artifacts. Alternatively, a thin carbon layer deposited over the holey carbon can be used, but with the possibility of adsorption effects on the surface. In any case, the buffer composition should be optimized for best sample stability (characterized biophysically), and it should contain low amounts of salt (preferably below 200 mM; this is where complexes are more stable anyway) and some pH buffer but few or no agents such as glycerol, sucrose, or detergents in order to obtain best image contrast (as opposed to the freezing of three-dimensional [3-D] crystals, which normally requires cryo-protectants). The flash-freezing has three advantages: (1) it maintains the sample in a hydrated, functional state; (2) it allows introducing the hydrated sample into the microscope, which operates under vacuum; and (3) it reduces radiation damage of the sample caused by the electron beam. Image data acquisition in the holes is performed nowadays semiautomatically using high-resolution cryo electron microscopes (equipped with a stable sample support device and with a field-emission gun) and recorded on film, charged couple device (CCD), or complementary metal oxide semiconductor (CMOS) cameras (the latter are based on the notably more

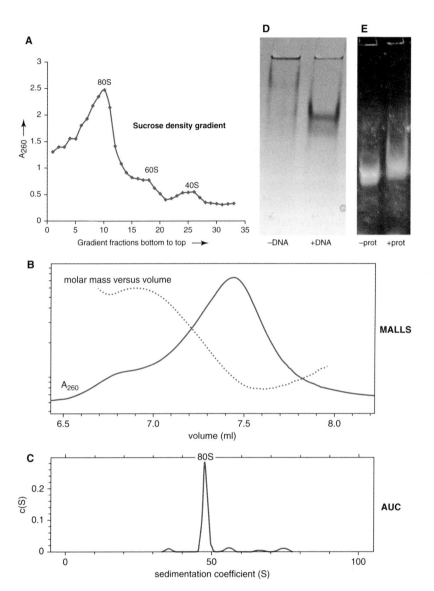

Fig. 10.1 Examples for the biophysical characterization of nucleoprotein complexes. (A) Sucrose density gradient centrifugation allows separating large nucleoprotein complexes based on their sedimentation coefficient (in Svedberg units S). This can be used at analytical or preparative scale. Here a continuous gradient profile is shown separating 80S ribosomes from 60S and 40S ribosomal subunits. (B) Multiangle laser light scattering (MALLS) is used for the biophysical characterization of a sample. It includes analytical gel filtration and a light scattering measurement of the eluent from which the molar mass of the sample can be calculated. Here, some sample heterogeneity is visible from the rising dotted line around the peak position of the profile, which ideally should have been flat. (C) Analytical ultracentrifugation (AUC) can be run under sedimentation velocity or equilibrium conditions, which allows measuring the precise molecular mass of macromolecular complexes

sensitive direct electron detection, which will certainly be the cameras used in the future and gives excellent results in our hands).

When images are acquired in the cryo electron microscope (at low temperature, i.e. <–140°C, and low electron dose condition, i.e. <25 e$^-$/Å^2), the electrons go through the sample (therefore this is called transmission electron microscopy) and generate a projection image of the object, which contains all internal features of the object and not just the contours/surface (Fig. 10.2). Particles embedded in the vitreous ice layer are in an immobilized state and therefore provide individual and distinct views of the specimen. Projection images representing different angular views of the objects can then be combined into a 3-D reconstruction (this is the principle of back projection, which works best if the particles have a near uniform angular distribution rather than some preferential views). Because cryo-EM images of hydrated biological specimens tend to be rather noisy (and the more, the smaller the object is), particles with similar angular views are usually averaged to form class averages (using software for multi-variate statistical analysis and classification), which are used as input images for the 3-D reconstruction. The angle assignment of the individual views is done based on common lines or by projection matching with a reference [3,4]; alternatively, random conical tilt [5] or orthogonal tilt [6] can be used to get an initial structure. In order to obtain a high-resolution cryo-EM map, forward projections from the 3-D reconstruction are calculated and then used as references to align the individual particles, and after angle assignment, a new 3-D reconstruction is obtained. Iteration of this process allows refining the cryo-EM structure to a resolution of 8–12 Å routinely, and to a resolution of 5 Å or higher for large, homogeneous data sets or highly symmetric objects. At a resolution of 8–12 Å, protein domains for which the crystal structure is known can be positioned, and secondary structure elements such as protein α-helices and RNA or DNA helices are visible (see Figs. 10.3 and 10.4). At 5 Å resolution, ß-strands may be distinguished, but a resolution of 3.5A or better is required to see amino acid side chains. The achievable resolution depends on various parameters including sample homogeneity, quality and stability of the electron microscope, image data set size, and image processing procedures, which may include structural sorting of particles. Heterogeneity can arise from variable composition or structural conformations, or from mixed functional states, which occurs more often for larger complexes because these tend to have many components such as DNA, RNA, transfer RNA (tRNA), mRNA,

---

Fig. 10.1 (*Continued*)
under equilibrium conditions, or the dissociation constant of a complex. Here it is used for checking the homogeneity of an 80S ribosome sample based on its sedimentation coefficient and molecular mass (measured value of 4.3 MDa is virtually identical to the one calculated from the molecular mass of the individual ribosomal RNAs [rRNAs] and ribosomal proteins), with some aggregation and subunit dissociation characterized by small peaks on either side of the main peak. (D,E) Native gel analysis of two DNA/protein complexes. In (D), adding DNA to the protein leads to a defined complex, which migrates properly under nondenaturing conditions, while the DNA-free protein is smearing. In (E), the addition of a protein to a preformed DNA/protein complex leads to a band shift revealing that the added protein indeed interacts with the DNA/protein complex.

10 Integrative structure-function analysis of large nucleoprotein complexes — 247

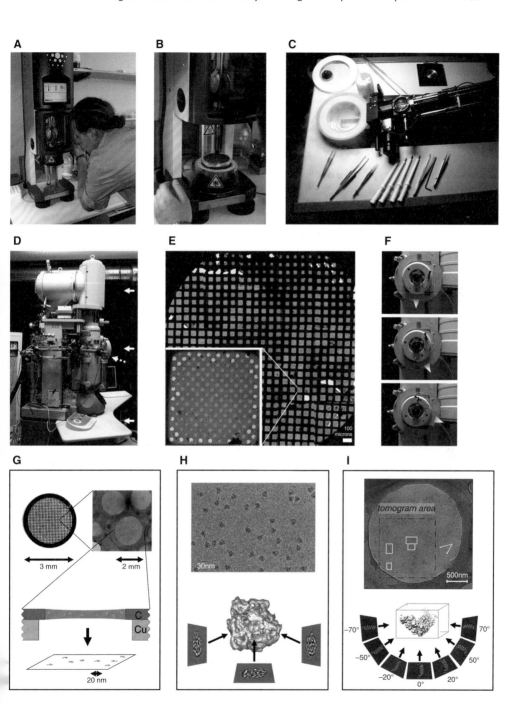

Fig. 10.2 Cryo-EM sample preparation and basic concepts of single particle and electron tomography. (A) Plunge freezing of the sample, shown is the deposition of a droplet (~3 µl) of the sample on the grid, which is then followed by the blotting of the liquid excess and plunge freezing (cooling rates are around 10^4 K/s). The protective box (Vitrobot, FEI) allows controlling temperature and humidity during the process. Cryo-EM provides specimen preservation in a functional state (no adsorption, no drying, no staining). (B) Transfer of the frozen EM grid, shown is the time point right after the plunge freezing into liquid ethane (cooled by liquid nitrogen around the central pot). The grid is then manually transferred to a storage box and kept in liquid nitrogen. (C) Transfer of the EM grid onto a sample holder (a cartridge onto which the grid is put and fixed with a clip ring). The sample is then loaded from the mounting table to the selection rod (right part) and put under vacuum, before the whole rod is transferred and mounted onto the cryo electron microscope. (D) Transmission electron microscope (here the on-site Polara F30, FEI), the horizontal arrows indicate at the top the electron source (a field emission gun, FEG), at the center the goniometer onto which the sample is fixed (from there it can be moved and rotated), and at the bottom the acquisition system (below the table). The dotted arrow indicates the position of the specimen rod transferred from the mounting table (C), which now serves to select and insert the sample into the microscope. (E) View of the cryo-EM grid at low magnification (stitched together from smaller images to form an overview of the entire grid), which serves to evaluate best areas of the grid in terms of ice thickness, particle distribution, low contamination with ice crystals, and so forth. The zoomed image at the bottom left shows preselected holes in the perforated carbon grid from which images can be acquired automatically; the holes marked in blue have already been recorded; those marked in green will be recorded later. (F) Different rotational positions of the goniometer (highlighted by a yellow mark), at zero tilt (i.e. the EM grid is in the horizontal position) from which most data are collected for single particle cryo-EM, at intermediate rotation angle (during goniometer rotation, illustrated by the red light on), and at 70° tilt angle (the maximum tilt angle used for electron tomography; for tomogram acquisition, incremental tilt angles of usually 2 to 3° or more are used, resulting in a series of more than 100 images). (G) Principle of cryo-EM: an EM-grid (copper, Cu, 300 mesh, i.e. 300 squares per inch [25.4 mm]), covered by a thin carbon layer, which has holes within which the sample forms a meniscus (about 100nm thick) when deposited on the grid and flash frozen (see panels A and B). The particles are immobilized (ideally with random orientations) within the layer of amorphous buffer and produce projection images when the electrons pass through the sample. Scales are indicated, going from the grid dimensions down to the particle size. (H) Principle of single particle cryo-EM image acquisition and processing. Top, a field of view with well-separated ribosomes; image acquired under low-dose conditions at –3 µm defocus, 59.000 magnification, and 150 kV acceleration voltage. Below is shown the principle of 3-D reconstruction based on back-projection of the individual particle images under their individual viewing angle. Each particle image provides a two-dimensional projection of the object from which the 3-D reconstruction can be obtained and which contains all internal features of the object. Precise image alignment and angle determination allows refining the 3-D reconstruction until a high-quality cryo-EM map is obtained (see Figs. 10.3 and 10.4 for examples). Obtaining high-resolution cryo-EM maps requires homogeneous samples, processing of several ten thousands to hundred thousands of particles, and high-quality data acquisition (sampling at 1 to 3 µm/pixel, drift free, etc.). (I) Principle of cryo-ET image acquisition and processing. Top, the region of the hole in the holey carbon from which a tilt series has been acquired (tomogram area); the white boxes indicate polyribosomes, and the arrows show the gold beads that serve as markers for the initial image alignment. Below, principle of tomogram reconstruction: because the incremental tilt angles are known from the data acquisition on the microscope (see panel F), the 3-D reconstruction can be obtained directly from the tilt series images (provided the goniometer is well calibrated and the images are properly aligned). The magnification used is usually in the range of 25,000 to 45,000 and defocus –3 to –10 µm depending on the sample thickness, angular increments around 2 to 3 degrees or higher, and sampling around 3 to 10 Å/pixel (these parameters depend on the sample type, e.g. single particle/molecular assembly or cell sections, and the resolution that may consequently be achieved in the 3-D reconstruction).

protein factors, ligands, and so forth, which can bind in substoichiometric amounts. In these cases, cryo-EM and particle sorting may allow simultaneously obtaining individual structures at improved resolution for several functional complexes that are in equilibrium with each other (see Fig. 10.3 for an example).

As the example of sample heterogeneity shows, averaging of too highly variable structures into a single 3-D reconstruction will limit the resolution or may just not be possible when the overall conformation of the complex is excessively variable. This is even more problematic in the case of molecules with unique structures, for example, large, flexible molecular assemblies or cell sections with numerous components. In these cases, cryo-ET can be applied, keeping in mind that the resolution will be usually limited to the 20–40 Å [7] range (see the example of polyribosome structures analyzed by cryo-ET, Fig. 10.3G). Cryo-ET consists of an acquisition of a series of images of the same object under different angular views by progressively rotating the object at each picture taken (see Fig. 10.2F). Practically, the specimen is rotated around a fixed axis by the goniometer of the electron microscope in angular increments of 2°–5°, covering a tilt range of ±70° or ±60° which is limited for practical reasons such as increasing sample thickness at high tilt angles. Since the angular increments are known, the 3-D object can be reconstructed directly from the tilt image series, thus producing a tomogram, which can be analyzed and interpreted in the same way as a cryo-EM map (with the limitation of a lower resolution and of distortion artifacts due to missing-wedge effects from the tilt series acquisition; the latter can be reduced by double-tilt cryo-ET [7]). Individual tilt images are much noisier than traditional cryo-EM images because the electron dose is at least ten times lower due to the fact that the dose has to be distributed over a series of images recorded from the same object rather than from many (e.g. for a tilt series of 120 images and a cumulative dose of 60 electrons/$Å^2$ [$e^-/Å^2$] this would be ~0.5 $e^-/Å^2$, compared to ~10–20 $e^-/Å^2$ for single particle cryo-EM). In the case when similar substructures can be found in the tomogram, these can be eventually averaged (this is called subtomogram averaging). Cryo-ET can be used for the 3-D analysis of various large objects such as cells, viruses, and bacteria (these are usually cryofixed, cryosubstituted, or plastic embedded and sectioned with an ultramicrotom or milled with a focused-ion beam), and also for cellular or extracted and purified assemblies such as ribosomes and RNA polymerases.

We will now describe several examples of nucleoprotein complexes that we are studying in our laboratory. These are all involved in the regulation of gene expression and thus are relevant for pharmaceutical applications (here these include in particular antibiotics, and potential drugs against cancers, psoriasis, diabetes, and other diseases). The complexes analyzed comprise (1) transcription complexes such as nuclear receptors and chromatin remodeling complexes, and (2) translation complexes such as functional complexes of the ribosome or of translation factors. The examples we selected here illustrate the combination and integration of the various tools discussed above.

10.2 Integrative structure-function analysis of nucleoprotein complexes, example 1: translation complexes

Regulation of protein synthesis represents one of the key steps in gene expression. The translation of the mRNA into a peptide chain is catalyzed by the ribosome, a molecular machinery that goes through phases of initiation, elongation, and termination during which regulatory factors are involved [8–14]. We are particularly interested in the initiation step of translation because it is the rate-limiting step in the process of peptide synthesis, and the strong regulation of this step by ribosomal factors makes it a potential target for antibiotics directed against the bacterial ribosome [11,12]. We undertook the characterization of intermediate states of translation initiation in bacteria, ranging from the preinitiation complex of the 30S small ribosomal subunit to the 30S initiation complex with mRNA, initiator fMet-tRNA^{fMet}, and the three initiation factors (IF1, IF2, and IF3), resulting in the initiation complex of the fully assembled 70S ribosome [15–17]. In eubacteria, initiation of translation requires three initiation factors (IF1, IF2, and IF3) for the correct selection and proper positioning of the initiator fMet-tRNA^{fMet} onto the mRNA bound to the ribosome. IF1 stabilizes IF2 on the 30S subunit, whereas IF3 has an editing activity with respect to the type and the positioning of the tRNA in the peptidyl (P) site. IF2 has a key role because it is present in the initiation complexes of the small 30S subunit and also in the 70S ribosome initiation complexes: IF2 promotes the recruitment of initiator tRNA and the joining of the subunits of the ribosome, and it has a ribosome-dependent guanosine triphosphatase (GTPase) activity, which is activated upon subunit joining.

The translation initiation complex comprising the 70S ribosome and initiation factor IF2 in the presence of either a nonhydrolyzable guanosine triphosphate (GTP) analogue or guanosine diphosphate (GDP) was reconstituted in vitro, and its structure was determined by cryo-EM [16]. This study showed the binding site of initiation factor IF2 on the ribosome, revealing significant conformational changes of the factor and of the ribosome related with GTP hydrolysis. Building upon this result, we extended our studies and directed our attention toward the translation preinitiation complex (where folded mRNAs have a key structural role in initiation [17,18]) and next to the 30S translation initiation complex, which contains the small ribosomal subunit and the initiator tRNA and the initiation factors. This complex was supposed to provide insights into the mechanism of initiator tRNA recruitment on the small subunit. Until we started this work, the study of such a complex did not seem feasible for technical reasons including the tendency for aggregation and heterogeneity of the 30S complex. However, we have been able to overcome these difficulties by developing a novel 3-D classification procedure [18], which has been used by other groups since [19,20]. By overcoming the problem of sample heterogeneity through advanced image processing (based on resampling and multivariate statistical analysis and classification in 3-D, 3D-SC), it became possible to determine several structures simultaneously from a single sample (Fig. 10.3; [21]).

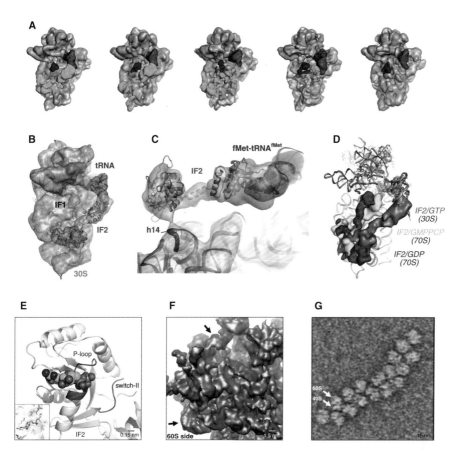

Fig. 10.3 Integrative structure-function analysis of nucleoprotein complexes, example 1: translation complexes. (A) Single particle cryo-EM and 3-D classification allowing the separation of multiple structural states in a sample of the *T. thermophilus* 30S translation initiation complex. Translation initiation factor 2 (IF2) is colored in green, initiator fMet-tRNA[fMet] in red, and initiation factor 1 (IF1) in blue [16,18,23]. When IF2 is absent, the cytosine-cytosine-adenine (CCA) region of the tRNA adopts a different conformation because there is no interaction with IF2 to stabilize it, in contrast to when IF2 is present (see panels B and C). (B) Detailed interpretation of the 30S initiation complex with the fMet-tRNA[fMet] and initiation factors IF1 and IF2 (using the cryo-EM map of the best-defined state 1, panel A) by fitting crystal structures of the individual components [18]. (C) Contact points of IF2 with the 30S subunit (helix h14) and with the initiator fMet-tRNA[fMet] (the CCA end is colored in blue), revealing the mechanism by which IF2 and the initiator tRNA cooperatively bind to the 30S subunit in order to initiate translation initiation in bacteria. (D) Comparison of the conformation of IF2 in 30S and 70S initiation complexes (combined analysis with our previous study [16]) showing IF2 in the GTP state (30S) and in the guanosine-5′-[(β,γ)-methyleno]triphosphate (GMPPCP) and GDP states (70S) when the 50S subunit has joined the 30S initiation complex. The tRNA changes conformation from a transient P/I state (green) on the 30S to the classical P state in the 70S when interactions with IF2 are released and IF2 is on the way to leaving the ribosome. (E) Crystal structure of IF2 [33]. A part of the structure is shown in order to highlight the binding site of the nucleotide next to the P-loop and the switch II region; the insert shows the electron density map at 2.3 Å resolution (X-ray diffraction data were collected at the Swiss Light Source). (F) Preliminary single particle cryo-EM reconstruction of the human 80S ribosome (Khatter et al., in preparation; see raw image in Fig. 10.2H), which already

The structural analysis of the 30S initiation complex in the presence of initiation factors IF1 and IF2 and initiator tRNA has allowed the localization of the factors on the ribosome [19,21–23] and provided a detailed analysis of the interactions and conformational changes associated with the ribosome and the associated proteins. The high-resolution 3-D reconstructions (9 Å) allowed the generation of molecular models of this key translation intermediate (the first gene-decoding complex during peptide synthesis) by fitting the crystal structures of the individual components [23–28]. The structure of the 30S initiation complex bound with mRNA, fMet-tRNA^{fMet}, and factors IF1 and IF2 revealed the molecular basis of cooperativity between the initiator tRNA and IF2 (Fig. 10.3; [18]): the acceptor stem of the fMet-tRNA is held and thus stabilized by IF2, whereas the tRNA anticodon stem is buried in the decoding site of the 30S subunit. The special position of the tRNA (which we called P/I because it is in the peptidyl site but in a conformation of initiation, similar to a 70S P/I state described by the group of J. Frank [19]) allows the recruitment of the large ribosomal subunit. By increasing the interface area between the ribosomal subunits, IF2 promotes subunit joining and formation of the fully assembled machinery ready to proceed to protein synthesis during the elongation phase. It should be noted that the position of the G domain of IF2 in the 30S complex is such that it is perfectly preoriented for activation by the GTPase-associated center of the large subunit when it docks on the 30S initiation complex [14, 16].

In order to address the role of IF2 during subunit joining and analyze the mechanism of GTP hydrolysis in more detail, we have also performed crystallization experiments on IF2. Crystals were obtained, and we determined the structure at 2 Å resolution using multiwavelength anomalous dispersion phasing of the seleno-methionine substituted protein and advanced refinement procedures ([30–32]; Fig. 10.3E). In combination with a series of functional data (point mutations in the catalytic site, kinetics analysis in the

---

Fig. 10.3 (*Continued*)
shows prominent secondary structure elements of the RNA (helices with some phosphate bumps are indicated by the arrows; view on the 60S subunit side). Data were collected on the in-house Polara Tecnai F30 electron microscope (acceleration voltage 100 kV, defocus range from −2 μm to −3 μm, magnification 59,000×, pixel size 1.82 Å, total electron dose 15 e$^-$/Å^2, 4k/4k CCD camera).
(G) Cryo-ET of large assemblies, as exemplified by our study of eukaryotic polyribosomes (wheat germ in vitro cell-free system, see polysomes in Fig. 10.2I). Shown is the central section of the reconstructed tomogram, with the ribosomal subunits indicated for one of the ribosomes. Tilt series were acquired from −60° to +60° with increments of 3° on the in-house Polara Tecnai F30 electron microscope (acceleration voltage 150 kV, defocus −5 μm, magnification 59,000×, pixel size 1.82 Å, total dose 60 e$^-$/Å^2, 4k/4k CCD camera, reconstruction performed in the Imod software package). To reduce distortion effects in the reconstruction, data may be also collected by dual-axis tomography involving an in-plane rotation of the sample and a second tilt series acquisition, which improves the polysome tomogram quality [7]. Approximate scale bars are included for panels E, F, and G to illustrate the integration across various scales (here more than two orders of magnitude). Panels A-D adapted from Simonetti et al., 2008 with permission from Nature Publishing.

context of 30S and 70S initiation complexes) and by determining the crystal structures of IF2 with GTP and GDP, we have recently been able to describe the molecular determinants for GTP hydrolysis by IF2 [33]. Taken together, this study illustrates the integration of biochemical preparation and biophysical characterization of IF2, X-ray crystallography (Chapter 13) of the isolated factor, fast kinetics and cryo-EM analysis of functional ribosome complexes, separation of multiple structural states, and a sequence-structure analysis, which allowed designing and performing new functional assays.

10.3 Integrative structure-function analysis of nucleoprotein complexes, example 2: transcription complexes

The other important example of a nucleoprotein complex, which we would like to detail more here, is a complex involved in transcription. Transcription regulation through steroid hormones and other natural ligands such as vitamin D is mediated by nuclear receptors (NRs). These are transcription regulators that bind to the promoter region of the target genes, thus forming DNA/protein complexes in which the NRs can bind to different types of DNA response elements. These are organized as direct repeats (DRs) or inverted repeats of hexa-nucleotide half-sites. NRs form either homodimers (in the case of steroids) or heterodimers (with retinoid X receptor RXR as a common partner). The superfamily of steroid/thyroid hormone/retinoid NRs comprises two well-conserved core domains, the DNA-binding domain (DBD) and the ligand-binding domain (LBD). DBDs and LBDs are connected through a hinge region, which is thought to be flexible in order to allow NRs to bind to various DNA response element types. For DR response elements, NRs bind as heterodimers with RXR located on the upstream half-site, with the exception of RAR/RXR/DR1 (RAR: retinoic acid receptor) and PPAR/RXR/DR1 (PPAR: peroxisome proliferator-activated receptor) [34] complexes whose polarity is reversed. In order to analyze the molecular mechanism of hormone-dependent transcriptional activation, we are studying the architecture of transcription complexes and regulatory complexes including NRs and coregulatory proteins involved in chromatin remodeling. These complexes are important therapeutic targets directly related with diseases such as cancer, osteoporosis, obesity, and type 2 diabetes. We recently focused our attention on the human RXR/VDR (vitamin D receptor) complex bound to a DR spaced by three nucleotides (DR3).

Because of the high pharmaceutical potential of VDR ligands, several crystal structures of the VDR LBD with ligands have been determined over the past years. The VDR LBD contains a dimerization interface with the RXR LBD, a 50-residue insertion of unknown function, and a ligand-dependent transcriptional activation region, AF-2. Agonist ligand binding induces a conformational change of the AF-2 helix (H12) that allows the recruitment of coactivators from the p160 family steroid coactivator that remodels chromatin, or of the DRIP/TRAP (DRIP: Vitamin D Receptor Interacting

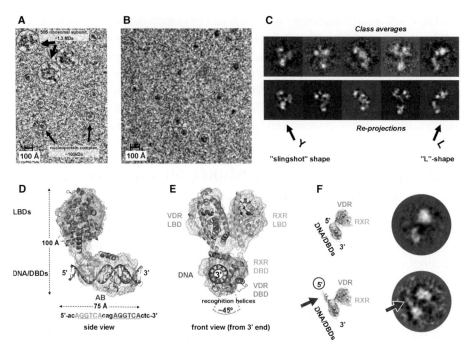

Fig. 10.4 Integrative structure-function analysis of nucleoprotein complexes, example 2: transcription complexes. (A) Cryo-EM image of single particles of a rather small DNA-bound nuclear receptor complex (molecular mass ~100 kDa) in comparison with the large ribosomal subunit (50S, molecular mass ~1.3 MDa). Single particles are indicated with red circles. Data were collected on the in-house Polara Tecnai F20 electron microscope (acceleration voltage 200 kV, defocus ~ −5 μm [for processing, data were collected at a defocus from −2.0 μm to −4.0 μm], magnification 50,000×, pixel size 2 Å, total electron dose 20 e$^-$/Å^2, on SO163 film [32]). (B) Improved image contrast of the single particles when reducing the acceleration voltage to 100 kV (Polara Tecnai F30 electron microscope [now possible to focus closer] defocus −2.5 μm, magnification 59,000×, pixel size 1.82 Å, total electron dose 20 e$^-$/Å^2, 4k/4k CCD camera). (C) Comparison of class averages with the corresponding reprojections of the 3-D reconstruction, revealing the characteristic shapes of the nucleoprotein complex. (D, E) Structure of the RXR/VDR/DNA nuclear receptor complex as determined by cryo-EM and interpreted by fitting the available crystal structures of the individual RXR and VDR ligand-binding and DNA-binding domains (LBD and DBD, respectively). The global architecture of the complex is opened, with the LBDs positioned off center on the 5' end of the DNA response element [32]. (F) Validation of the correctness of the structure and assignment of the DNA polarity within the complex using a DNA extended by 15 nucleotides on the 5' end. The class averages of the RXR/VDR complex with the longer DNA show an additional density on the 5' side of the complex.

Protein and TRAP: Thyroid Hormone Receptor-associated Proteins) mediator complexes that interact with the basal transcriptional machinery and help in recruiting the RNA polymerase to the transcription start site. While the crystal structures of the isolated LBDs and also the DNA-bound DBDs of VDR and/or RXR were known, there was no structure available for the full RXR/VDR complex. Therefore, we set out to address its structural organization, taking into account that this complex is a representative of the function of heterodimeric NRs binding to DR2, 3, 4, or 5 response elements.

The cryo-EM structure of the RXR/VDR/DR3 complex now allows identifying the relative positions of the DNA and DBDs with respect to the LBDs where the natural ligands and pharmaceutical drugs bind (Fig. 10.4; [36]). The LBDs are positioned in an asymmetric manner on the 5' end of the response element, shifted away from the center of the two half-sites. As a result, the LBDs display an unforeseen polarity with respect to the DNA. A key concept is provided by clarifying the binding mode of the full complex to the DNA, which explains the general role of RXR as a common partner within heterodimeric NR complexes (adaptability to different response elements), in contrast to the role of its variable partner (stabilization of the heterodimeric complex on a precise half-site direct repeat, here DR3 for VDR). The structure also reveals the important role of the conformationally well-defined hinge domains thought until now to be disordered: they stabilize an overall open conformation without requiring DBD/LBD interdomain contacts. Taken together, this study revealed an open conformation with respect to the DNA (as also observed for other complexes studied by solution SAXS [35] and in contrast to the compact crystal structure of the PPAR/RXR/DR1 complex [36]), in which the receptor is positioned asymmetrically on the response element without interdomain contacts but with a well-ordered hinge region, which stabilizes the complex in a precise conformation. This is functionally relevant for the recruitment mechanism of coregulators with chromatin-modifying activity, which thus can bind to the LBD and also to the hinge region.

This study illustrates the strong potential of single particle cryo-EM to address the structure-function relationship of full NR/DNA complexes, among the smallest macromolecular complexes studied to date by cryo-EM. Studying such relatively small complexes (~100 kDa) has been particularly challenging because of the low contrast of the particles that were imaged under pure cryo-conditions (unsupported holey carbon, no staining, no adsorption, no cross-linking; see Fig. 10.4A,B). Indeed, the generally accepted lower limit has so far been of the order of 200–300 kDa for the feasibility of a cryo-EM study. By developing new approaches in sample preparation (EM grid optimization for amorphous ice quality and particle distribution) and image processing, we have been able to determine the structure of a key nucleoprotein complex of 100 kDa at 10–12 Å resolution, which gives insights into the 3-D architecture of a complete nuclear receptor. This included a full validation of the structure determination procedures at the level of image processing and map interpretation through the fitting of the individual crystal structures of the isolated RXR and VDR LBDs and DBDs, and through the localization of the 15-nucleotide extension added at the 5' end of the response element (Fig. 10.4F). This work opens the way to the analysis of relatively small DNA and RNA protein complexes that are difficult to crystallize and have been judged too small for single particle cryo-EM until now.

10.4 Outlook

The structure-function relationship of nucleoprotein complexes – prepared either as subunits or as full complexes – increasingly benefits from an integrative approach including functional studies, high-quality sample preparation, and characterization by biochemical and biophysical methods (FRET [see Chapter 8], fluorescence anisotropy, AUC, MALLS, MALDI and ESI mass spectrometry, etc.), bioinformatics and modeling, conventional and low-resolution [37,38] crystallography, cryo-EM/cryo-ET, and cell imaging with fluorescent probes on endogenous complexes for correlative microscopy at the light and electron microscopy interfaces. The latter is a major challenge for the future in order to go towards integrated cellular structural biology, meaning that new tools will need to be designed for integrative bioinformatics and 3-D modeling. It is important to note that such intensively pursued integrative structural biology approaches require extensive expertise in each method, with potentials and limits of each in mind and validation tools in hand [39], in order to best combine their usage and integrate the results and relate them with functional data (kinetics, interactomics, mutations related with diseases, etc.). Moreover, the cost-intense state-of-the-art technologies such as high-resolution cryo electron microscopes, high-brilliance and low-divergence X-ray beam lines at the synchrotrons, and large and multimodal cell imaging facilities are set up preferably as large transnational infrastructures (e.g. Instruct, http://www.structuralbiology.eu/; FRISBI, http://www.frisbi.eu; http://www.eurobio imaging.eu/EuroBioImaging; etc.) that can provide project-based access and allow developing new cutting-edge approaches at the interface of these technologies. This greatly facilitates multiresolution data integration on gene-regulating complexes such as those discussed here. The synergy between these various approaches thus allows – in isolated form or in the cellular context – (1) the determination of the architecture of different functional macromolecular assemblies and the detailed interaction analysis within these; (2) the visualization of conformational changes of these macromolecules and of the associated partners and protein ligands; and (3) the localization and tracking in the cellular context by life cell imaging, superresolution imaging (PALM (photo-activated localization microscopy), STORM (stochastic optical reconstruction microscopy), STED (stimulated emission depletion) and GSD (ground-state depletion), etc.), fluorescence cross-correlation spectroscopy (FCCS), and other related techniques. The structural biology component of these approaches involves the combination of different, yet complementary techniques, in particular cryo-EM and 3-D reconstruction of single molecules; cryo-ET of single particle molecular assemblies or cell sections; SAXS of purified complexes in solution; crystallography (See Chapter 13); and NMR (See Chapter 12) to gain atomic insights into protein-protein and protein-nucleic acid (RNA/DNA) interactions and dynamics and their implications for understanding molecular mechanisms, which are key not only for human health and medical applications but also for exploring the beauty of nature at the molecular level and its evolution over time.

10.5 Acknowledgments

We would like to thank our colleagues from the Department of Integrated Structural Biology for fruitful interactions and collaborations over the years in this challenging field, notably Jean-Claude Thierry and Dino Moras who supported the idea of integrated structural biology from the early phase on; Isabelle Billas, Natacha Rochel, Jean Cavarelli, Marc Ruff, Marat Yusupov, Patrick Schultz, Olivier Poch, and many other colleagues; our enthusiastic collaborators and their team members Ali Hamiche, Stefano Marzi, Pascale Romby, Claudio Gualerzi, Vladimir Shirokov; the late Robert Schuster for his constant interest, Alexander Spirin, Gilbert Eriani, Philippe Dumas, and Jean-Luc Vonesch; and the members of our platforms and services (structural biology and genomics, mass spectrometry, cell culture, bioinformatics, imaging, sequencing, and workshop facilities and support at the SLS, ESRF and Soleil synchrotrons). This work was supported by the European Research Council (ERC Starting Grant), the European Molecular Biology Organization (EMBO) Young Investigator Programme (YIP), the Fondation pour la Recherche Médicale (FRM), the Association pour la Recherche sur le Cancer (ARC), the Alsace Region, the Agence National pour la Recherche (ANR, grants NRcoactiv, STIR, IRES-Histone, MITIC, Nucleoplat), Alsace-Russie/Ukraine (ARCUS), the Federation of European Biochemical Societies (FEBS), the European Commission as SPINE2-complexes (contract no LSHG-CT-2006–031220), the Centre National pour la Recherche Scientifique (CNRS), the Institut National pour la Recherche Médicale (Inserm), the Infrastructures en Biologie Santé et Agronomie (IBiSA), the Integrated Structural Biology Infrastructure for Europe (Instruct), and the French Infrastructure for Integrated Structural Biology (FRISBI) ANR-10-INSB-05-01.

References

1. Dubochet J, Adrian M, Chang JJ, et al. Cryo-electron microscopy of vitrified specimens. Q Rev Biophys. 1988;21:129–228.
2. Prongidi-Fix L, Schaeffer L, Simonetti A, et al. Rapid purification of ribosomal particles assembled on histone H4 mRNA: a new method based on mRNA-DNA chimeras. Biochem J. 2013;449(3):719–28.
3. Böttcher B, Wynne SA, Crowther RA. Determination of the fold of the core protein of hepatitis B virus by electron cryomicroscopy. Nature. 1997;386:88–91.
4. van Heel M. Angular reconstitution: a posteriori assignment of projection directions for 3D reconstruction. Ultramicroscopy. 1987;21:111–23.
5. Radermacher M, Wagenknecht T, Verschoor A, Frank J. Three-dimensional reconstruction from a single-exposure, random conical tilt series applied to the 50S ribosomal subunit of Escherichia coli. J Microsc. 1987;146(pt 2):113–36.
6. Chandramouli P, Hernandez-Lopez R, Wang HW, Leschziner AE. Validation of the orthogonal tilt reconstruction method with a biological test sample J Struct Biol. 2011;175:85–96.
7. Myasnikov AG, Afonina Z, Klaholz BP. Single particle and molecular assembly analysis of polyribosomes by single- and double-tilt cryo electron tomography. Ultramicroscopy. 2013;126:33–9.
8. Frank J, Gonzalez RL Jr. Structure and dynamics of a processive Brownian motor: the translating ribosome. Annu Rev Biochem. 2010;79:381–412.
9. Noller HF. RNA structure: reading the ribosome. Science. 2005;309:1508–14.
10. Klaholz BP. Molecular recognition and catalysis in translation termination complexes. Trends Biochem Sci. 2011;36(5):282–92.

11. Gualerzi CO, Pon CL. Initiation of mRNA translation in prokaryotes. Biochemistry. 1990;29: 5881–9.
12. Rodnina MV, Wintermeyer W. Recent mechanistic insights into eukaryotic ribosomes. Curr Opin Cell Biol. 2009;21:435–43.
13. Sonenberg N, Hinnebusch AG. Regulation of translation initiation in eukaryotes: mechanisms and biological targets. Cell. 2009;136:731–45.
14. Dunkle JA, Cate JH. Ribosome structure and dynamics during translocation and termination. Annu Rev Biophys. 2010;39:227–44.
15. Simonetti A, Marzi S, Jenner L, et al. A structural view of translation initiation in bacteria. Cell Mol Life Sci. 2009;66:423–36.
16. Myasnikov AG, Simonetti A, Marzi S, et al. Structure-function insights into prokaryotic and eukaryotic translation initiation. Curr Opin Struct Biol. 2009;19:300–9.
17. Myasnikov AG, Marzi S, Simonetti A, et al. Conformational transition of initiation factor 2 from the GTP- to GDP-bound state visualized on the ribosome. Nat Struct Mol Biol. 2005;12:1145–9.
18. Marzi S, Myasnikov AG, Serganov A, et al. Structured mRNAs regulate translation initiation by binding to the platform of the ribosome. Cell. 2007;130:1019–31.
19. Simonetti A, Marzi S, Myasnikov AG, et al. Structure of the 30S translation initiation complex. Nature. 2008;455:416–20.
20. Fischer N, Konevega AL, Wintermeyer W, Rodnina MV, Stark H. Ribosome dynamics and tRNA movement by time-resolved electron cryomicroscopy. Nature. 2010;466:329–33.
21. Papai G, Tripathi MK, Ruhlmann C, et al. TFIIA and the transactivator Rap1 cooperate to commit TFIID for transcription initiation. Nature. 2010;465:956–60.
22. Allen GS, Zavialov A, Gursky R, Ehrenberg M, Frank J. The cryo-EM structure of a translation initiation complex from Escherichia coli. Cell. 2005;121:703–12.
23. Julián P, Milon P, Agirrezabala X, et al. The cryo-EM structure of a complete 30S translation initiation complex from Escherichia coli. PLoS Biol. 2011;9:e1001095.
24. Steitz TA. A structural understanding of the dynamic ribosome machine. Nat Rev Mol Cell Biol. 2008;9:242–53.
25. Carter AP, Clemons WM Jr, Brodersen DE, et al. Crystal structure of an initiation factor bound to the 30S ribosomal subunit. Science. 2001;291:498–501.
26. Schluenzen F, Tocilj A, Zarivach R, et al. Structure of functionally activated small ribosomal subunit at 3.3 angstroems resolution. Cell. 2000;102:615–23.
27. Yusupov MM, Yusupova GZ, Baucom A, et al. Crystal structure of the ribosome at 5.5 Å resolution. Science. 2001;292:883–96.
28. Petry S, Brodersen DE, Murphy FV IV, et al. Crystal structures of the ribosome in complex with release factors RF1 and RF2 bound to a cognate stop codon. Cell. 2005;123:1255–66.
29. Roll-Mecak A, Cao C, Dever TE, Burley SK. X-ray structures of the universal translation initiation factor IF2/eIF5B: conformational changes on GDP and GTP binding. Cell. 2000;103:781–92.
30. Afonine PV, Grosse-Kunstleve RW, Urzhumtsev A, Adams PD. Automatic multiple-zone rigid-body refinement with a large convergence radius. J Appl Cryst. 2009;42:607–15.
31. Afonine PV, Echols N, Grosse-Kunstleve RW, et al. Toward automated crystallographic structure refinement with phenix.refine. Acta Crystallogr D. 2012;68:352–67.
32. Hendrickson WA, Horton JR, LeMaster DM. Selenomethionyl proteins produced for analysis by multiwavelength anomalous diffraction (MAD): a vehicle for direct determination of three-dimensional structure. EMBO J. 1990;9:1665–72.
33. Simonetti A, Marzi S, Fabbretti A, et al. Structure of the protein core of translation initiation factor 2 in apo, GTP-bound and GDP-bound forms. Acta Cryst D. 2013;69: 925–33
34. Chandra V, Huang P, Hamuro Y, et al. Structure of the intact PPAR-gamma-RXR- nuclear receptor complex on DNA. Nature. 2008;456:350–6.

35. Rochel N, Ciesielski F, Godet J, et al. Common architecture of nuclear receptor heterodimers on DNA direct repeat elements with different spacings. Nat Struct Mol Biol. 2011;18:564–70
36. Orlov I, Rochel N, Moras D, Klaholz BP. Structure of the full human RXR/VDR nuclear receptor heterodimer complex with its DR3 target DNA. EMBO J. 2012;31:291–300.
37. Lunin VY, Urzhumtsev A, Podjarny AD. An initio phasing of low-resolution Fourier syntheses. In: Arnold E, Himmel DM, Rossmann MG, eds. International Tables for Crysallography. Vol. F. Chichester, UK: John Wiley & Sons; 2011:437–42.
38. Urzhumtsev A, Afonine PA, Lunin VY. Crystallographic maps and models at low and at subatomic resolutions. In: Lunin V, Read R, Urzhumtsev A, eds. Advancing Methods for Biomolecular Crystallography. Dordrecht, The Netherlands: Springer; 2013:215–24.
39. Urzhumtseva L, Afonine PV, Adams PD, Urzhumtsev A. Crystallographic model quality at a glance. Acta Crystallogr D. 2009;65:297–300.

Ivan Krstic, Andriy Marko, Claudia M. Grytz, Burkhard Endeward, and Thomas F. Prisner

11 Structure and conformational dynamics of RNA determined by pulsed EPR

11.1 Introduction

Nucleic acids are very flexible molecules, which unlike proteins often adopt several tertiary structures for a given sequence. This conformational flexibility plays an important role in the many functions that RNAs can have in cells, ranging from gene regulation to transcription and catalysis [1]. Therefore, a detailed understanding of the dynamics goes hand in hand with the determination of tertiary structures of RNA molecules to elucidate their functional role as well as to develop suitable drugs for inhibition. Spectroscopic methods, such as nuclear magnetic resonance (NMR; Chapter 12), infrared (IR), fluorescence resonance energy transfer (FRET; Chapter 8), and electron paramagnetic resonance (EPR), are very well suited to investigate the dynamics of RNA molecules on different timescales, depending on the resonance frequency and the experimental time resolution of the method. Additionally, these spectroscopic methods deliver structural information, making them especially attractive for investigations on RNA molecules. NMR and IR spectroscopy probe structural information on a sub-nanometer-length scale, whereas FRET spectroscopy and pulsed electron-electron double resonance (PELDOR) spectroscopy [2] deliver distance information in the 1- to 10-nm range. This is very valuable to determine the global tertiary structure of functionally important RNA motives. Both methods measure the dipole-dipole interaction between two covalently attached marker molecules – for FRET by the energy transfer induced by the coupling of the electric dipole moments of two chromophores, and for PELDOR via the magnetic dipole interaction between two spin labels. Whereas FRET spectroscopy uses two chemically different chromophores (acceptor and donor), PELDOR spectroscopy usually uses two identical nitroxide spin labels attached to the biomolecule. Due to a much higher excitation frequency, FRET has a superior sensitivity (single spins) and time resolution (in principle down to the femtosecond range). Nowadays, for PELDOR spectroscopy at least 2 nmol of molecules is necessary, and the time resolution is in the nanosecond range. One advantage of PELDOR spectroscopy is that the distances between two nitroxide spin labels can be determined model free with high precision (less than the N-O bond length, where the unpaired electron spin is localized). Due to the small size of the attached nitroxide spin labels, this leads to high-quality nanometer-range distance restrains for structure determinations.

A number of different EPR methods have been used to investigate RNA molecules, schematically depicted in Fig. 11.1. Continuous-wave EPR (CW-EPR) has been applied extensively to analyze dynamics of nucleic acids at ambient temperature [3,4]. This allows not only the determination of the overall rotational motion of the nucleic acid molecule [5], but also local changes of dynamic introduced by ligand binding [6]. Hyperfine spectroscopy methods, like electron nuclear double resonance, electron spin echo envelope modulation (ESEEM), or hyperfine sublevel correlation spectroscopy have been used to investigate the local structure of paramagnetic metal-binding sites of RNA molecules [7,8]. More recently, PELDOR spectroscopy has been added as a further method to determine nanometer-range distances in nucleic acid molecules [9–12].

In this article, we will exclusively concentrate on applications of the PELDOR method to nucleic acid molecules, explaining and illustrating specific features of the method for this class of systems and the necessary requirements to perform and analyze the experiments. For all the other applications of EPR on RNA molecules, we refer to existing review and book articles [12,13].

PELDOR spectroscopy was invented in 1983 by the Novosibirsk group [2]. They used the two-pulse Hahn-echo sequence, explored in 1950 in the field of solid-state NMR spectroscopy [14] to refocus inhomogeneous line-width contributions in solids. In EPR, this sequence eliminates the effect of the differences in the Larmor resonance frequency of the excited electron spin (also called A-spin) arising from the

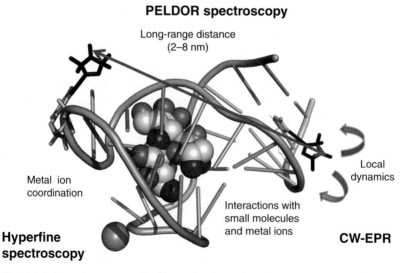

Fig. 11.1 Schematic representation illustrating the application areas of the different EPR techniques on RNA molecules.

anisotropies of the g- and hyperfine tensors at the time of the echo signal. Unfortunately, this sequence also refocuses the dipolar interaction to the second nonexcited electron spin (called B-spin). Thus, an inversion pump pulse at a second microwave frequency was introduced, which flips the B-spin within the echo sequence. This pump pulse breaks the symmetry for the time evolution of the dipolar coupling and therefore leads to a partially defocused echo signal, which is only caused by the dipolar coupling itself. This is analog to the Spin-echo Double Resonance (SEDOR) experiment, again invented in the field of solid-state NMR spectroscopy [15]. The application of this method for nanometer structural investigations of macromolecules was triggered by the development of site-directed spin labeling of proteins [16,17]. From that time on, the method has also been called double electron-electron resonance (DEER) [18]. To avoid signal distortions, when pump and probe pulses overlap temporally, a four-pulse sequence was invented [19] based on the remote echo detection scheme [20]. In this pulse sequence, which is used almost exclusively nowadays, the time zero of the PELDOR time trace is given, when the pump pulse is applied at the center of the first Hahn-echo signal at the time $2\tau 1$. Therefore, this experiment can detect the PELDOR time trace dead-time free.

It should also be mentioned that a number of other methods exist to detect the dipolar coupling between two unpaired electron spins: the 2+1 sequence [21], where pump and probe frequency are identical; the out-of-phase Hahn-echo sequence for radical pairs created by photoexcitation out of a singlet state [22]; and a number of methods, where the dipolar relaxation induced by the B-spin is observed [23–25]. The most elegant experiment by far is the double-quantum EPR experiment [26], which observes the double-quantum coherence between the two electron spins. Unfortunately, the excitation field strengths of most existing pulse EPR spectrometers are too low to perform this experiment efficiently for nitroxide spin labels. Thus, all experiments performed on RNA molecules so far rely on the four-pulse DEER sequence. The recorded PELDOR time trace has to be Fourier transformed to obtain the dipolar Pake pattern. Instead of that, Tikhonov regularization is often used to convert the time trace directly into a distance distribution function $P(R)$. This gives direct access to the average distance between both spin labels attached to the biomolecule and additionally to the distance distribution, which reflects the conformational flexibility of the spin label and the biomolecule, especially interesting for nucleic acid oligomers. Excellent software exists, which allows user friendly analysis of experimental PELDOR time traces [27].

Momentarily, PELDOR is the most applied pulse EPR method, with important applications to proteins [28–32], polymers, and material sciences [33,34]. Most of the applications are performed with nitroxide spin labels, but natural paramagnetic cofactors in proteins have also been used, like amino acid radicals [35] and metal centers [36–39]. For rigid spin labels, additional information on the mutual orientation between two paramagnetic centers can be deduced [35,38,40,41], giving valuable additional information. Several reviews on this method and its applications to biological and material sciences exist [9,10,42]. As mentioned before, the focus of this article will be a description of all necessary steps to successfully perform PELDOR experiments on nucleic acid molecules, starting from the spin-labeling procedure and sample preparation to PELDOR data acquisition and analysis, which consist of the determination of the average distance, the distance distribution, and the mutual orientation between two spin labels. For all these steps, remarks about quality control and potential pitfalls will be made and illustrated by examples.

11.2 Pulse EPR spectroscopy on RNA

11.2.1 Spin labeling of nucleic acids

Nucleic acids are diamagnetic molecules that require the incorporation of paramagnetic centers, so-called spin labels, in order to be studied by EPR spectroscopy. Although some nucleic acids contain coordinated diamagnetic Mg^{2+} that can be easily substituted with paramagnetic Mn^{2+}, for distance measurements by PELDOR the most commonly used paramagnetic markers are stable nitroxide radicals. They can be covalently incorporated into nucleic acids without significant disturbance of the oligonucleotide secondary structure. Nitroxides can be attached at selected nucleotides either during chemical synthesis of nucleic acids or postsynthetically using one of three available approaches:

1. Labeling the backbone phosphate group [43]
2. Labeling the sugar moiety [44]
3. Labeling the nucleobases [45]

Several reviews and monographs [10,46–48] are available providing the reader with more detailed descriptions regarding site-specific spin labeling of nucleic acids. In this chapter, we will focus on two nitroxide spin labels, namely, ethynyl-substituted pyrroline type of nitroxide (TPA) and isoindol moiety fused to cytosine by an oxazine connection (Ç) (Fig. 11.2), extensively used in our laboratory to study structure and conformational flexibility of nucleic acids.

The RNA-labeling technique based on a Pd-catalyzed Sonogashira cross-coupling reaction of TPA (Fig. 11.2A) and iodinated nucleotides has been introduced by the group of Engels (Goethe University Frankfurt) [49]. Such labeled RNA molecules have melting temperatures less than 3.5 °C lower than the unmodified oligonucleotides, indicating that the spin labels do not considerably disturb the secondary structure of the RNA. A series of double-labeled double-stranded (ds) DNAs and RNAs carrying this spin label was used to benchmark the precision of the distance measurements with PELDOR on nucleic acid molecules [49,50].

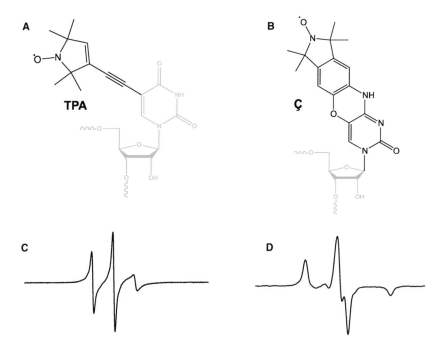

Fig. 11.2 Spin labels TPA and Ç used in our work (A, B), and CW-EPR spectra of corresponding radicals attached to double-stranded nucleic acid molecules recorded at room temperature (C, D).

The extremely rigid spin label Ç (Fig. 11.2B) was covalently incorporated into DNA and RNA by Sigurdsson and coworkers [51]. This nitroxide moiety was integrated into a polycyclic fused ring system of a cytidine analogue that forms a base pair with guanine and stacks on the neighboring bases within a stem. The synthesis starts with the preparation of an isoindoline derivative, 5-hydroxy-6-amino-1,1,3,3-tetramethylisoindoline. It is, then, merged into a nucleoside through reaction with activated 3',5'-diacetyl-5-bromo-2'-deoxyuridine, followed by closing of the phenoxazine ring. Attachment of Ç to a 14-bp DNA duplex causes insignificant effects on duplex stability and conformation, as revealed by circular dichroism, thermal denaturation, and PELDOR studies. This spin label cannot rotate around the bonds in the linker and therefore possesses a negligible conformational flexibility, thus reducing the width of the distance distribution function introduced by the local dynamics of the spin label itself. Moreover, orientation-selective PELDOR measurements can disclose the fixed relative orientation of the two spin labels, and, since the spin labels are rigidly integrated into oligonucleotide, the method can reveal the relative orientation between two helical structural elements.

Efficiency of the chemical incorporation of spin labels into target nucleic acid molecules can easily be accessed by CW-EPR spectroscopy. The EPR spectrum of nitroxide spin labels, such as depicted in Fig. 11.2, is determined by the Zeeman interactions of the electron spin $S = 1/2$ with the external magnetic field vector $\mathbf{B_0}$ and the hyperfine interaction of the electron spin with the nitrogen nuclear spin ^{14}N ($I = 1$). This leads to the three hyperfine lines observed in liquid CW-EPR spectra. The detailed shape of the CW-EPR spectrum depends on dynamic processes (e.g. molecular tumbling or the spin label's flexibility). This is due to the anisotropic, orientation-dependent nature of the hyperfine and the Zeeman interaction, described by the hyperfine tensor A and g-tensor, respectively. At typical X-band frequencies (9.5 GHz, corresponding to a magnetic field B_0 of 0.3 T), the anisotropy of the hyperfine interaction is the dominant contribution, whereas at G-band frequencies (180 GHz/6.4 T), the spectrum is dominated by the anisotropy of the g-tensor. If the tumbling rates are on the order of these interaction strengths (or slower), they significantly contribute to the spectral shape.

Fig. 11.2C shows the narrow three-line EPR spectrum of TPA attached to a DNA duplex. Due to the fast rotation of the spin label around the tether, the anisotropic hyperfine interaction is almost averaged to zero and only contributes to the different line widths of the three hyperfine lines. In contrast, the EPR spectrum of the rigid spin label Ç incorporated into a DNA duplex is much broader, resembling more the spectrum of a frozen solution (so-called powder spectrum). This is due to the stiffness of the spin label and the very slow overall rotation of the DNA duplex molecule, leading to a slow-motion EPR spectrum with partially resolved anisotropic hyperfine interactions (Fig. 11.2D). Thus, it can be used to obtain information on the anisotropic tumbling of the DNA duplex molecule itself [52]. In both cases, such spectra can be also used to quantify the spin-labeling efficiency for the synthesized spin-labeled oligonucleotide. Due to the magnetic field modulation technique with lock-in detection, the CW-EPR signal is recorded in the form of a first derivative. Therefore, double integration is required to quantify the number of radicals. Using an external nitroxide spin standard measured under the same experimental conditions and comparing the double-integral values, the number of spins in the sample can be rather precisely determined ($\pm 10\%$). The ratio between the calculated spin concentration and the known nucleic acid concentration then gives the labeling efficiency, a very important parameter for reliable interpretation of experimental PELDOR data.

11.2.2 Theoretical description of the PELDOR experiment

The unpaired electron spin ($S = 1/2$) of a nitroxide spin label is delocalized between the nitrogen and oxygen atoms. If a spin label is placed into an external magnetic field $\mathbf{B_0}$, the electron spin state is split by the Zeeman interaction into two electron spin states ($m_s = \pm 1/2$), and by the hyperfine interaction of the electron spin with the nitrogen nuclear spin ($I = 1$), each electron spin level is further split into three hyperfine spin states ($m_I = -1, 0, +1$). For the magnetic field strengths commonly applied (> 0.1 T), the electron Zeeman interaction, which determines the frequency band of the applied microwave, is much stronger (> 3 GHz) than the hyperfine interaction (< 100 MHz). Thus, the hyperfine interaction leads to a splitting of the EPR line into three hyperfine lines in liquid solutions and to a broadening and asymmetric line shape in disordered solids. In a double-labeled RNA molecule, the two unpaired electron spins also interact with each other via the magnetic dipole-dipole interaction. The spin Hamiltonian of this system is given by

$$\hat{H} = \hat{H}_Z + \hat{H}_{hf} + \hat{H}_D \tag{11.1}$$

where

$$\hat{H}_Z = \mu_B \mathbf{B}_0 \ddot{g}_1 \hat{\mathbf{S}}_1 + \mu_B \mathbf{B}_0 \ddot{g}_2 \hat{\mathbf{S}}_2 - \mu_N g_N \mathbf{B}_0 \hat{\mathbf{I}}_1 - \mu_N g_N \mathbf{B}_0 \hat{\mathbf{I}}_2 \tag{11.2}$$

describes the Zeeman interactions of both electron spins and nitrogen nuclei with the external magnetic field $\mathbf{B}_0$. μ_B and μ_N are the Bohr and nuclear magneton, respectively. $\hat{\mathbf{S}}_i$, $\hat{\mathbf{I}}_i$ with $i = 1$ or 2 are the vector spin operators of the ith electron and nucleus, respectively. $\ddot{g}_i$ is the orientation-dependent electron spin g-tensor, reflecting the anisotropy of the Zeeman interaction caused by the spin orbit coupling. The contribution of the hyperfine coupling to the total spin Hamiltonian is given by the term

$$\hat{H}_{hf} = \hat{\mathbf{S}}_1 \ddot{A}_1 \hat{\mathbf{I}}_1 + \hat{\mathbf{S}}_2 \ddot{A}_2 \hat{\mathbf{I}}_2 \tag{11.3}$$

Similar to the g-tensor, the hyperfine interaction tensor $\ddot{A}_i$ reflects the anisotropy of the electron-nuclear spin interaction. The last term in equation 11.1 corresponds to the magnetic dipolar interaction between the two unpaired electrons:

$$\hat{H}_D = \hat{\mathbf{S}}_1 \ddot{D} \hat{\mathbf{S}}_2 \tag{11.4}$$

Here $\ddot{D}$ is the distance (R)–dependent magnetic dipolar interaction tensor, which in the principle axes system is given by $\ddot{D} = \dfrac{D}{R^3} diag\,(1,1,-2)$. D is the dipolar interaction constant determined via fundamental physical constants, resulting in 2π·52.16 MHz·nm3 for two nitroxide radicals. For the first term in the dipolar interaction alphabet, $D_{zz}S_{1z}S_{2z}$, the interaction strength is given by:

$$\omega_d = \frac{D}{R^3}(1 - 3\cos^2\theta) \tag{11.5}$$

determined solely via the distance R and the angle θ between the distance vector $\mathbf{R}$ and the magnetic field $\mathbf{B}_0$.

In a frozen sample with immobilized molecules, all tensors in this Hamiltonian are time independent. Calculating the eigenvalues and transition moments for the

Hamiltonian (equation 11.1) allows simulating the CW-EPR spectrum of the biradical. For two electron spins separated by a 2-nm distance, the dipolar interaction frequency is about 6.5 MHz. This is much smaller compared to the line width of the CW-EPR spectrum, dominated by the hyperfine coupling to the nitrogen nuclei and to close-by proton spins of the solvent. Hence, it is rather difficult to deconvolute the distance information from a CW-EPR spectrum when it is longer than 2 nm, as demonstrated in Fig. 11.3.

To determine such nanometer-range distances between the spin labels, the PELDOR experiment is typically used, as explained previously. PELDOR is a double frequency pulsed EPR experiment, as shown in Fig. 11.4. The three probe pulses at the frequency v_A form the refocused Hahn echo, which is observable. The excitation bandwidth of the microwave probe pulses is chosen narrower than the width of the nitroxide EPR powder spectrum, which is about 7 mT at X-band frequencies. Therefore, the probe pulses excite only a small part of the spin ensemble (A-spins) characterized by specific radical orientations with respect to the external magnetic field. The pump pulse at the frequency v_B, which is applied at time T after the first echo between the second and the third probe pulses, inverts the spins resonant with this frequency (B-spins). The B-spin inversion changes the local magnetic field at the positions of the A-spins and therefore their precession frequencies, if A- and B-spins are close to each other. Thus, the phase of the transversal magnetization at the time of the remote echo signal is shifted by an angle $+\omega_d T$ or $-\omega_d T$ in comparison to the situation when the B-spins do not flip.

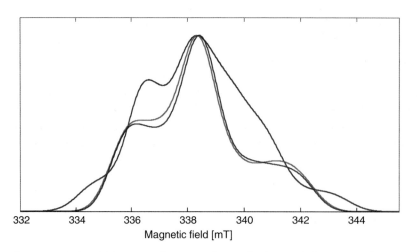

Fig. 11.3 Calculated CW-EPR powder spectra for two dipolar-coupled nitroxide radicals separated by different distances (blue, 1.2 nm; green, 2 nm; and red, 3 nm). As can be seen, if the distance between the electron spins is longer than 2 nm, almost no changes introduced by the electron magnetic dipolar interaction are observable.

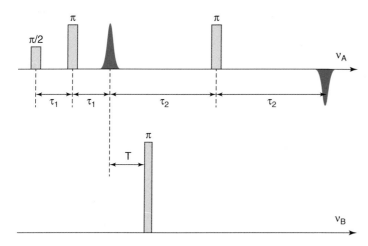

Fig. 11.4 Dead-time free four-pulse PELDOR sequence [19].

The sum of the two magnetization components with the phases $+\omega_d T$ and $-\omega_d T$ results in an oscillating echo intensity $(\exp[i\omega_d T] + \exp[-i\omega_d T])/2$ depending on the time T, the placement of the pump pulse. Also the pump pulse in PELDOR experiments excites only a fraction of the spin ensemble. Therefore, not every double-labeled molecule that has an A-spin at the probe frequency will also have a B-spin at the corresponding pump frequency, depending on the mutual orientation of the two spin labels on that molecule. If the probability of flipping the spin by the pump pulse is equal to p, then the spin echo magnetization consist of two parts. One is constant and proportional to $1-p$, and the second is an oscillating function $p \cdot \cos(\omega_d T)$.

The probability of affecting spins by microwave pulses depends on their resonance frequencies, which are determined by the spatial orientation of nitroxides with respect to the external magnetic field. In order to contribute to the oscillating PELDOR signal, the molecule has to have one nitroxide oriented in such a way that the resonance frequency matches with the frequency v_A, whereas the orientation of the second nitroxide has to be so that its resonance frequency matches with the frequency v_B. We introduce the function $\lambda(\theta)$, which accounts for this weighted contribution of molecules. The total signal of an ensemble of double-labeled randomly oriented molecules is then given by the expression:

$$S(T) = e^{-\gamma T}\left(1 + \int_0^{\pi/2} \lambda(\theta)(\cos(\omega_d T)-1)\sin\theta d\theta\right) \tag{11.6}$$

The exponential factor $e^{-\gamma T}$, with $\gamma = 8\pi^2 Dpn/(9\sqrt{3})$, accounts for the intermolecular dipolar signal relaxation, depending on the overall spin concentration n. The term in parentheses is responsible for the signal modulation caused by the intramolecular interaction. For molecules with flexible spin labels, like TPA, the modulation depth λ is constant and does not depend on the angle θ. In this case, the expression for the signal simplifies to

$$S(T) = e^{-\gamma T}\left(1 - \lambda + \lambda \int_0^\infty P(R)K(R,T)dR\right) \tag{11.7}$$

where $P(R)$ is the distance distribution function between the unpaired electron spins and

$$K(R,T) = \int_0^{\pi/2} \cos(\omega_d T)\sin\theta\, d\theta \tag{11.8}$$

is the integral kernel function.

11.2.3 Practical aspects of the PELDOR experiment

To study RNA by PELDOR, it is important to first consider some properties of the sample necessary to perform such experiments successfully. A first important property is the spin concentration, which depends on the number of spin labels the RNA contains, the spin-labeling efficiency, and the distance that has to be measured. For optimal sensitivity and distances in the 2- to 4-nm range, the RNA concentration should be about 0.1 mM of double-labeled RNA, which might be lowered for longer distances (5 nm and more) and can be increased if the labeling efficiency is lower than 100%. The labeling efficiency should be as high as possible; a labeling efficiency of 50% would amount to only 25% of double-labeled RNA molecules, leading to a factor 4 decreased modulation depth and therefore very low sensitivity. This cannot be compensated for by higher RNA concentration; thus, a good sample should at least have 80% spin-labeling efficiency, as can be checked by CW-EPR (see Section 11.2.1). At X-band frequencies, the sample volume is about 20–100 μL (depending on the used cavity), resulting in a total amount of about 2 to 10 nmol of spin-labeled RNA molecules. At higher RNA concentrations (> 0.5 mM), the RNA molecules come so close that intermolecular dipolar effects dominate the PELDOR time trace, leading to a fast monotonic decay of the signal.

PELDOR experiments with nitroxide radicals have to be performed in frozen solution due to the fast relaxation times induced by the fast tumbling of the nitroxide radials in liquids. Cryoprotection like ethylene glycol has to be used for RNA in water-based buffer solutions to prevent microcrystallization, which leads to inhomogeneous spatial distributions of the RNA molecules. Usually, about 20% ethylene glycol is enough to build glass-like structures at low temperatures. This is important to achieve a long transversal relaxation time T_2. Without cryoprotection, the T_2 could be as short as 1 μs, and typical values in a properly prepared sample are 3 μs or more. Of course, whether the RNA molecule is perturbed by the cryoprotectant (e.g. by Circular Dichroism (CD) spectroscopy) has to be checked. Long transversal relaxation times T_2 are crucial to determining distances between the two spins above 4 nm. One other contribution to the spin-spin relaxation is the hyperfine coupling to nuclear spins; in protonated solvents, the proton nuclear spins significantly shorten the transversal relaxation time. One way to circumvent this is to use deuterated water and deuterated cryoprotectant, which can easily prolong the T_2 to 4 μs or longer, which is important to reaching distances of 5 nm or more.

In PELDOR experiments, pump and detection pulses should not interfere with each other, meaning that they should not excite the same spins, to avoid ESEEM effects between the electron spin and nearby nuclei spins and strongly reduced detection intensity. Therefore the pump and detection frequency should be separated by more than 60 MHz at typical X-band frequencies with pulse lengths of 12 ns for the pump pulse and 32 ns for the probe pulses. Due to the excitation profile of the rectangular pulses, a separation of 65 MHz to 70 MHz is optimal, giving the smallest spectral overlap. The ESEEM effect still observed results mainly from weakly coupled "matrix" protons or deuterons, if present. These could lead to a modulation of the PELDOR time trace with a frequency of 14 MHz or 2.2 MHz for protons or deuterons, respectively. The proton modulation frequency would induce an artificial distance of 1.5 nm, which is out of the PELDOR accessible distance range but would be visible on the PELDOR time trace. Deuterium on the other side would induce an artificial distance of about 2.8 nm. This ESEEM effect can be circumvented by not using deuterated solvent or by performing PELDOR measurements at higher magnetic field values and respective frequencies. Already at Q-band frequencies (33.5 GHz/1.2 T) the ESEEM effect is strongly reduced in intensity because it involves forbidden transitions. Additionally, the deuterium frequency would be 7.8 MHz resulting in an artificial distance of about 1.9 nm.

PELDOR experiments with nitroxide spin labels are typically performed in frozen solution, at a temperature of 40–50 K. For biological samples, the signal-to-noise ratio is always an issue; therefore, the experiments are averaged typically between several hours up to 1 day. Commercial spectrometers operating at X-band (9.5 GHz), Q-band (33.5 GHz), or W-band (95 GHz) frequencies are nowadays stable enough to perform such measurements without difficulty. Some home-built spectrometers have increased sensitivity due to higher pump and probe excitation bandwidth [53,54] leading to reduced experiment times or allowing measurements with lower spin concentrations.

11.2.4 PELDOR experiments with rigid spin labels

The determination of both distance and orientation has been successfully applied to protein dimers with intrinsic paramagnetic centers [35,55] and to nucleic acids

[41,56,57]. Gaining the additional orientation information is possible with two rigid spin labels that are incorporated into a biomolecule with a fixed relative orientations. Due to the g-tensor and ^{14}N-hyperfine tensor anisotropy, the frozen solution EPR spectra contain orientation information. The PELDOR technique can encode this by soft pulses, which excite only parts of the spectrum, corresponding to molecules with a specific orientation with respect to the external magnetic field $B0$.

The rigid nitroxide spin label Ç (Fig. 11.3B) allows the preparation of samples with a specific, fixed orientation of spin labels relative to the framework of the nucleic acid. Consequentially, both a well-defined distance between such a pair of spin labels and their mutual orientations can be extracted from an orientation-selective PELDOR experiment, as we have recently shown on dsDNA molecules [58]. Moreover, this spin label allows studying the mobility of nucleic acids directly without the need to consider the conformational flexibility of the spin label itself.

In Section 11.2.2, it was described that the PELDOR time traces oscillate with the frequency ω_d of the dipole-dipole interaction between two electron spin centers (equation 11.5). By choosing the probe frequency, a well-defined subensemble of nitroxide orientations with respect to the external magnetic field B_0 can be selected. This also implies a specific selection of θ angles because the two spin labels are rigidly incorporated into the nucleic acid molecule. Therefore, the θ dependence of the dipolar frequency allows deducing the relative orientation between the two nitroxides from a set of PELDOR time traces with different probe frequencies [58]. The PELDOR modulation depth and oscillation frequency depend on the mutual orientation of the two nitroxide spin labels in a nontrivial way. In simple cases, like molecules with inversion symmetry, the angular and distance information can be extracted by a mathematical procedure [57]. In other cases, simulation of the PELDOR signals measured at different probe/pump frequencies and at different magnetic field values is required to disentangle distance and orientation information [59,60].

The X-band nitroxide spectrum is dominated by the anisotropic 14N-hyperfine tensor, which has a large A_{zz} component in the molecular frame as shown in the inset of Fig. 11.5. Therefore, the PELDOR signal depends on the angle between the out-of-plane vector Azz and the interspin distance vector R in addition to the distance R. In contrast to X-band frequencies, PELDOR experiments performed at high magnetic fields (e.g. G-band: 180 GHz/6.4 T) are sensitive to all three Euler angles, which describe the relative orientation between the two spin labels and with respect to the distance vector R due to the spectrally resolved g-tensor (Fig. 11.5). As explained in Section 11.2.3, for RNAs labeled with flexible nitroxides, like TPA, the PELDOR experiments are carried out with a pump pulse frequency at the center of the spectrum to achieve maximum PELDOR modulation depth λ and the probe frequency at the edge of the spectrum (typically leading to a difference in pump and probe frequency $\Delta\nu$ of 70 MHz at X-band) providing the best separation between pump and probe excitation profiles and therefore the best performance.

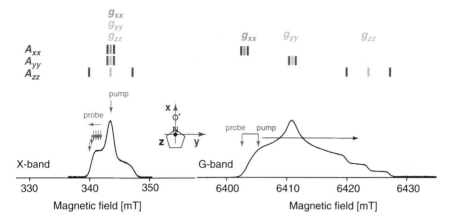

Fig. 11.5 Echo-detected field swept EPR spectra of a doubly Ç-spin-labeled bent DNA recorded at 40 K and corresponding 14N stick diagram for X-band and G-band, respectively. Arrows indicate pump and probe frequencies with varied detection positions Δν ranging from 40 to 90 MHz at X-band. At G-band, Δν is usually kept constant, whereas the magnetic field is stepped over the spectral range.

In order to obtain unambiguous information regarding the distances and the relative orientations of two rigidly incorporated paramagnetic centers, orientation-selective PELDOR measurements at different magnetic field strengths and at different microwave frequencies have to be performed. Orientation-selective PELDOR experiments at X-band frequency on biomacromolecules labeled with rigid nitroxides are usually performed with a fixed pump frequency (located at the maximum of the anisotropic powder spectra) and a series of probe frequencies with a frequency offset Δν relative to the pump frequency (Fig. 11.5). At high magnetic fields, the EPR spectrum is much broader than the microwave resonator bandwidth. In this case, usually the magnetic field B_0 is stepped over the full spectrum, whereas the frequency separation Δν between pump and probe pulse is kept constant (e.g. Δν = 70 MHz).

Oscillation frequency, modulation depth, and damping of the PELDOR signal change as a function of the probe frequency offset Δν. The strong probe pulse frequency (and therefore orientation) dependence of the experimentally observed PELDOR time traces is a clear signature of strong angular correlations, as expected for the rigid spin label Ç. A deep modulation depth requires that pump and probe pulses excite individually one radical of the pair attached to the biomolecule. At X-band frequency, the pump pulse excites all molecular orientations at the center of the nitroxide spectrum, whereas the probe pulse is more selective. Thus, at higher frequency offsets (around 90 MHz), only nitroxide radicals with the plane normal almost parallel to the external magnetic field and in the ^{14}N nuclear $m_I = +1$ spin states are excited. For this offset frequency, the angle between A_{zz} and R can be directly extracted from the PELDOR time trace after the distance R is deduced. To estimate the average distance R, all time traces with different frequency offsets $\Delta\nu$ can be summed up to reduce the orientation selection effects in the individual time traces [61]. Then, this averaged time trace can be analyzed, for example, by Tikhonov regularization, as described in Section 11.2.2. Offset-dependent measurements and quantitative simulations thus allow us to obtain information about the angle between the nitroxide normal vectors with respect to R and with respect to each other.

At high magnetic field and accordingly high microwave frequencies, the spectral shape is dominated by the anisotropy of the nitroxide g-tensor. Therefore, different spectral positions correspond to different orientations of the nitroxide molecule with respect to the external magnetic field and can be distinguished. In contrast to PELDOR experiments at X-band frequency, here in principle, all Euler angles defining the orientation between the two nitroxides and with respect to the dipolar vector R can be determined by systematically varying pump and probe pulses over the whole spectrum and a quantitative simulation of all PELDOR time traces simultaneously. Unfortunately, because of restricted pulse excitation width, the modulation depth is rather small in most cases, and because of limited microwave resonator bandwidth, not all combinations of pump and probe frequencies are experimentally accessible. Therefore, usually pump and probe frequencies are kept at a constant separation $\Delta\nu$, and PELDOR time traces are recorded by only varying the magnetic field over the spectral range.

Orientation-selective PELDOR experiments can be performed as fast and effectively as PELDOR experiments with flexible spin labels at X-band frequencies. However, this frequency band is only sensitive to three angles: the angle between the respective ç out-of-plane normal A_{zz} and the distance vector R for each nitroxide and the angle between the two A_{zz} vectors. Therefore, it can be difficult to find a unique solution with only these data sets. This can be compensated by additional restraints obtained from samples labeled at different positions or by PELDOR measurements performed at higher magnetic field values, where the g-tensor anisotropy is fully resolved. Thus, orientation-selective PELDOR experiments at different magnetic field values are very important to improve the uniqueness of the solution for the overall structure of biomolecules from a few double-labeled mutants only. This is especially important for oligonucleotides, which exhibit a large degree of conformational flexibility.

11.2.5 Data analysis and interpretation

Finally, the experimental PELDOR time traces have to be analyzed. A first and crucial step is the removal of the intermolecular interactions. This is done by fitting a (stretched) exponential to the end of the PELDOR time trace and dividing the experimental time trace by this function. In a second step, a Tikhonov regularization is usually performed to obtain the distance distribution function P(R) from the experimental PELDOR time trace, if it does not contain orientation selection. Easy-to-use free software is available to perform these steps [27].

Equation 11.7 is usually used to determine distances between spin labels attached to biomolecules. In order to obtain the distance distribution function, the integral equation has to be solved with respect to the function P(R) after extraction of the intermolecular contribution from the experimental time traces. This is an ill-posed problem, meaning that a small amount of noise in the experimental data set can lead to large differences in P(R). Tikhonov regularization is employed to find a smooth solution in P(R), which still satisfies equation 11.7 quite well. Another possibility to determine the distance distribution function from equation 11.7 is to assume a certain form for it (e.g. a Gaussian form) and to vary the fitting parameters until S(T) calculated from equation 11.7 provides the best fit to the experimental data.

The determination of the distance distribution function can be difficult when the estimation of the factor $e^{-\gamma T}$ in equation 11.7 that arises from the intermolecular interactions is uncertain. This is the case when the period of the observed PELDOR oscillations is comparable to the length of the PELDOR time trace. Fig. 11.6 demonstrates how the shape of the function P(R) depends on the intermolecular background correction. As can be seen, too much background correction yields an artificially narrowed distance distribution, whereas too little correction leads to a broader distance distribution function.

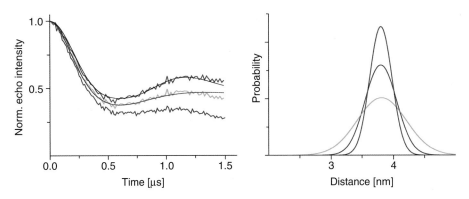

Fig. 11.6 Determination of distance distribution function subject to the estimation of the intermolecular background function $e^{-\gamma T}$. The blue time trace (left) has been simulated according to equation 11.7 for the distance distribution function shown on the right sight side in blue and an intermolecular relaxation rate $\gamma = 0.5\ \mu s^{-1}$. The green and red time traces on the right side are obtained by a background correction with an underestimated value for $\gamma = 0.4\ \mu s^{-1}$ or an overestimated value for $\gamma = 0.6\ \mu s^{-1}$, respectively. As shown in the right panel, an incorrect assumption of the background function leads to either a too broad (green) or a too narrow (red) distance distribution function.

Finally, it should be mentioned that equation 11.7 is only valid for distances longer than 2 nm. For shorter distances, the dipolar splitting is too large to be efficiently excited by the pump pulse, leading to decreased PELDOR modulation amplitude.

In contrast to distance determination with flexible attached nitroxides, disentangling the orientation/angular information from the distance information of rigid spin labels is more cumbersome and not so straightforward. Tikhonov regularization cannot be used anymore, as is the case for more than two spin labels attached to a biomolecule. The separation and determination of the distance and orientation contributions can only be achieved by a quantitatively simultaneous simulation of PELDOR time traces taken at different probe frequencies and at different magnetic fields. Two major strategies can be distinguished: on the one hand, only the structure of the spin labels is considered for the simulations [60,62], and on the other hand, the biomolecule itself as well as the spin labels are taken into account (molecular modeling) [58].

The first approach aims to detect a set of spin label orientations and distances that can be responsible for the observed experimental PELDOR signal set. PELDOR traces for all possible orientations and distances for pairs of spin labels are computed and stored in a database. An iterative fitting algorithm finds a combination of these presimulated PELDOR time traces that reconstruct the experimental data. This is a very fast way to obtain simulations that are in quantitative agreement with the experimental data sets [60]. Unfortunately, it is not clear yet if obtained solutions are unique, especially for biological systems such as RNAs, which have a large conformational flexibility. This approach considers two spin labels in free space without any knowledge of the structure of a biomolecule between them. However, steric restraints in biological systems restrict the number of possible angles and distances of the spin labels. Therefore, a more restricted solution can be achieved by modeling the biomolecular structure between the pair of spin labels. Different molecular modeling approaches and additional experimental restraints from other methods can be considered to take the biomolecular structure into account. The uniqueness of the solution of the structure, and especially the conformational flexibility of oligonucleotides, is an important issue. One of the advantages with oligonucleotides is the possibility of spin labeling the molecule at almost every position. Restricted are positions in the ligand-binding pockets or involved in *interactions* that stabilize RNA *tertiary structure*.

11.3 Application examples

Within the past decade, PELDOR has demonstrated its applicability and reliability to provide structural information on nucleic acids. In the following, we will use some of our applications to highlight the potential of this method on RNA and DNA molecules to map the global structure and determine conformational flexibility with high accuracy.

11.3.1 Applications on dsRNA and DNA

Application of PELDOR spectroscopy on a series of helical RNAs and DNAs carrying two TPA spin labels has been used to measure distances in the range from 2 to 6 nm with high accuracy (±0.25 nm) [49,50]. Fig. 11.7 depicts the sequences of three selected dsRNAs with various numbers of nucleotides between the labeled positions, the corresponding PELDOR time traces after division of the background function, and the distance distributions obtained by Tikhonov regularization. Although there are variations in modulation depths observable in the dipolar evolution functions (Fig. 11.7B) originating from differences in labeling efficiency (75%–100%), good signal-to-noise ratios and visible dipolar oscillations permitted reliable data analysis. The high precision of the PELDOR experiment due to very small distance variations within the spin label rotamer space for the TPA spin label allowed distinguishing A-form RNA from B-form DNA duplex structures. Molecular Dynamics (MD) simulations using an all-atom force field with explicit water were in good accordance with the measured distances, indicating that the double-stranded nucleic acids with the two spin labels attached retain their structure in frozen aqueous solution.

11.3.2 Application on RNA with more complex structure

Mapping the global structure of the neomycin-responsive riboswitch was the first application of PELDOR spectroscopy on a tertiary-folded RNA molecule [63]. This 27-mer oligonucleotide is an engineered riboswitch developed by combination of in vitro selection and in vivo screening [64]. Upon insertion into 5'-untranslated region of messenger RNA and binding the cognate ligand, it inhibits translational initiation in yeast. The nucleotide positions labeled with TPA marker were chosen outside of the binding pocket not to interfere with ligand binding. Efficient ligand binding was

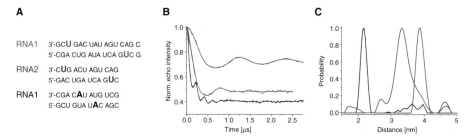

Fig. 11.7 (A) Three examples of dsRNAs containing various numbers of base pairs between labeled nucleotides used for PELDOR measurements. (B) Corresponding PELDOR time traces after background correction. (C) Distance distributions derived by the Tikhonov regularization of the PELDOR data.

proved by alteration of the CW-EPR spectra and by thermal stabilization of 20.3 ± 3.3°C upon addition of neomycin. The distances observed on the neomycin-bound riboswitch (Fig. 11.8A) are in very good agreement with the structure determined by NMR spectroscopy [65]. In contrast to NMR measurements performed at room temperature, the same distances were also observed for the riboswitch without neomycin. This led to the conclusion that the neomycin-bound tertiary structure is already preformed in absence of the ligand. At room temperature, an entopically favored high conformational flexibility exists, collapsing to the neomycin-bound structure at the freezing temperature, which therefore is energetically favored. In this special case, the conformational space observed in PELDOR experiments is not the same as observed at physiological temperatures.

The double-labeled cUUCGg tetraloop hairpin RNA was studied in view of the NMR results [66] where the order parameter (S^2) indicates a higher mobility of U7 (second U of the cUUCGg tetraloop) compared to U6; these data were recorded near room temperature. Comparing the double-labeled U7/U11 with U6/U11 PELDOR experiments showed in both cases a narrow distance distribution with comparable width (Fig. 11.9). The differences in the observed mobility of the NMR data could have several reasons: First of all, the mobilities observed in NMR are in the sub-nanometer range and might be hard to observe if the spin label has internal flexibility. Second, the spin label attached to the nucleotide will change its dynamics. Finally, freezing could again lead to a locked conformation. MD studies, room-temperature CW-EPR studies, and rigid spin labels can be used to distinguish these options.

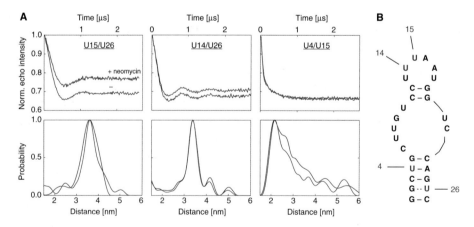

Fig. 11.8 (A) Upper panel: Background-corrected PELDOR time traces of double-labeled neomycin-responsive riboswitch samples in presence (red) and in absence (blue) of neomycin. Lower panel: Distance distribution derived by Tikhonov regularization of the PELDOR data. (B) The secondary structure of the neomycin-responsive riboswitch with marked spin-labeled positions.

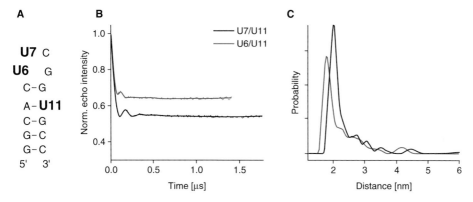

Fig. 11.9 Double-labeled cUUCGg tetraloop hairpin RNA with TPA labels. (A) Structure. (B) PELDOR time trace. (C) Distance distribution obtained by Tikhonov regularization.

11.3.3 Applications on DNA with rigid spin labels

As was shown in the previous section, PELDOR data are sensitive to the interspin distance and relative orientation of the nitroxide radicals. If the spin labels are rigidly attached to biomolecules, then orientation-selective PELDOR experiments allow determining the relative orientation between compact fragments of the studied molecule. This was demonstrated on rigidly labeled dsDNA molecules, where two cytosine bases are substituted with the analog Ç at specific positions.

Since the Ç spin label has a planar form, the planes of the nitroxide rings are parallel to the base-pair planes of the DNA molecule. As shown in Fig. 11.10, we can define the angle φ between the nitroxide normal vector and the vector R connecting both unpaired electrons. The X-band PELDOR data are sensitive to this angle, and it is possible to determine it by a quantitative analysis of the full experimental data set [56,57]. For the two samples shown in Fig. 11.10, values of 0° and 45° are obtained for φ, in good agreement with the known geometry.

In addition to the determination of the averaged distances and angles between the two rigid spin labels, the PELDOR data can also be used to unravel conformations of the molecule that appear during its motion and which are trapped by freezing. Therefore PELDOR spectroscopy can detect the molecular flexibility of the DNA molecule at the freezing temperature indirectly via a frozen-in snapshot. For dsDNA molecules, there are several different models for cooperative dynamics discussed in the literature [67,68]. Predicted PELDOR time traces for each of these models have been compared with two-dimensional PELDOR data sets collected at X-band and G-band frequencies.

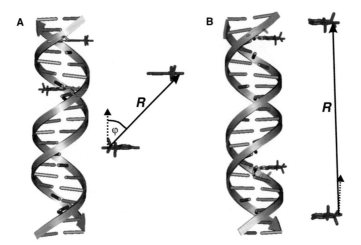

Fig. 11.10 Double-stranded DNA molecule with two rigid spin labels Ç incorporated.

The only model consistent with all PELDOR data from a series of dsDNA molecules, labeled at different positions, is a correlated twist-stretch motion where the elongation of a helix is accompanied by the reduction of its radius [41]. Thus, PELDOR experiments with quantitative simulations allowed us to detect very precise details of nucleic acid dynamics (Fig. 11.11).

11.4 Outlook and summary

As shown previously, PELDOR is a very powerful and precise technique to determine tertiary structural restraints in the nanometer range on nucleic acid molecules. Moreover, with rigid spin labels, a very detailed description of the conformational flexibility of the biomolecule at the freezing temperature can be obtained (Fig. 11.11). Different from proteins, a variety of spin-labeling strategies exists for nucleic acids, which allows modifying the oligomer almost at every position. Nevertheless, PELDOR restraints alone cannot determine the full structure of the oligonucleotide. Therefore, combination with other methods, such as NMR or MD, is important. If the tertiary structure of the nucleic acid molecule is already known (e.g. from X-ray or NMR studies, see Chapters 12 and 13), PELDOR restraints obtained in frozen solution or in-cell applications can probe the presence of these structures in different environments [69].

Also, changes of the tertiary structure upon ligand or protein binding can be followed by PELDOR measurements. Metal ions are crucial for the tertiary structure and the function of RNA molecules such as ribozymes. The extension of the PELDOR method to such metal ions is underway in several laboratories. That this is feasible

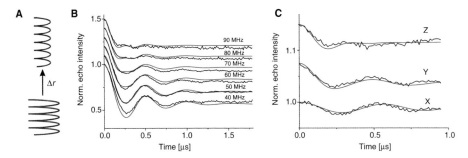

Fig. 11.11 Comparison of the X-band (B) and G-band (C) experimental PELDOR data with the simulations based on the twist-stretch dynamic model (A) of dsDNA. The PELDOR experiments have been performed at the spectral positions described in Section 11.2.4 (see Fig. 11.5).

at high magnetic fields has already been shown for high-spin metal ions such as Gd^{2+} and Mn^{2+} [36,37]. Because of the higher sensitivity compared to NMR and the possibility of background-free measurements, in-cell applications with spin-labeled nucleic acid molecules seem very attractive [69]. Unfortunately, the nitroxide spin labels used so far exhibit rather fast reduction in cell media (within 30 minutes). Therefore, new classes of protected nitroxide- or carbon-based spin labels have to be developed to monitor nucleic acid reactions or interactions with proteins in cells by PELDOR. So far, all PELDOR experiments have to be performed in frozen solutions. The reason is the short relaxation time of the spin label in liquid solution, caused by the fast rotational tumbling of the spin label itself. Extension of the method to liquid state has been demonstrated recently on an immobilized lysozyme protein with a carbon-based trityl radical attached as spin label [70]. More synthetic organic work in this direction is necessary to find spin labels with long relaxation times, long-term stability, and easy synthetic ways to attach them to nucleic acids. Finally, the sensitivity of the method is still an issue of continuous research. High-field EPR and improved pulsing schemes might increase the sensitivity of the method further. This will open up new avenues in the future to investigate tertiary structure, conformational dynamics, and interactions under physiological conditions by PELDOR spectroscopy.

11.5 Acknowledgments

Our own EPR work relied on the synthesis of spin-labeled RNA and DNA molecules performed by Nelly Piton and Olga Romainczyk from the group of Joachim W. Engels (Institute of Organic Chemistry and Chemical Biology, Goethe University) and by Pavol Cekan from the group of Snorri Th. Sigurdsson (University of Iceland) for the rigid spin labels on DNA. Vasyl Denysenkov is thanked for high-field G-band PELDOR experiments on DNA samples. Olav Schiemann (University of St. Andrews) is thanked for his major impact in the initial phase of this work as Habilitand in Frankfurt. Robert Hänsel from the group of Volker Dötsch (Institute of Biophysical Chemistry, Goethe University) is thanked for the insertion of spin-labeled nucleic acids into cells. Funding from the German Research Society (DFG) within the Collaborative Research Center 579 *RNA-Ligand Interaction* and the Collaborative Research Center 902 *Molecular Principles of RNA-based Regulation* is gratefully acknowledged as well as support from the Center of Biomolecular Magnetic Resonance (BMRZ) and the Center of Excellence Frankfurt *Macromolecular Complexes* (CEF).

References

1. Gesteland RF, Cech T, Atkins JF. The RNA World: The Nature of Modern RNA Suggests a Prebiotic RNA World. Cold Spring Harbor, NY: Cold Spring Harbor Laboratory Press; 2006.
2. Milov AD, Ponomarev AB, Tsvetkov YD. Electron-electron double resonance in electron spin echo: model biradical systems and the sensitized photolysis of decalin. Chem Phys Lett. 1984;110:67–72.
3. Zhang X, Cekan P, Sigurdsson ST, Qin PZ. Studying RNA using site-directed spin-labeling and continuous-wave electron paramagnetic resonance spectroscopy. Methods Enzymol. 2009:303–28.
4. Okonogi TM, Alley SC, Reese AW, Hopkins PB, Robinson BH. Sequence-dependent dynamics in duplex DNA. Biophys J. 2000;78:2560–71.
5. Liang ZC, Freed JH, Keyes RS, Bobst AM. An electron spin resonance study of DNA dynamics using the slowly relaxing local structure model. J Phys Chem B. 2000;104:5372–81.
6. Edwards TE, Sigurdsson ST. Electron paramagnetic resonance dynamic signatures of TAR RNA-small molecule complexes provide insight into RNA structure and recognition. Biochemistry. 2002;41:14843–7.
7. Schiemann O, Fritscher J, Kisseleva N, Sigurdsson ST, Prisner TF. Structural investigation of a high-affinity Mn(II) binding site in the hammerhead ribozyme by EPR spectroscopy and DFT calculations. Effects of neomycin B on metal-ion binding. ChemBioDhem. 2003;4:1057–65.
8. Morrissey SR, Horton TE, Grant CV, Hoogstraten CG, Britt RD, DeRose VJ. Mn^{2+}-nitrogen interactions in RNA probed by electron spin-echo envelope modulation spectroscopy: application to the hammerhead ribozyme. J Am Chem Soc. 1999;121:9215–18.
9. Schiemann O, Prisner TF. Long-range distance determinations in biomacromolecules by EPR spectroscopy. Q Rev Biophys. 2007;40:1–53.
10. Krstić I, Endeward B, Margraf D, Marko A, Prisner T. Structure and dynamics of nucleic acids. Top Curr Chem. 2012;321:159–98.
11. Sicoli G, Wachowius F, Bennati M, Hobartner C. Probing secondary structures of spin-labeled RNA by pulsed EPR spectroscopy. Angew Chem Int Ed. 2010;49:6443–7.
12. Sowa GZ, Qin PZ. Site-directed spin labeling studies on nucleic acid structure and dynamics. In: Conn PM, ed. Progress in Nucleic Acid Research and Molecular Biology. Vol. 82. San Diego, CA: Elsevier Academic Press; 2008:147–97.
13. Hunsicker-Wang L, Vogt M, Derose VJ. EPR methods to study specific metal-ion binding sites in RNA. Methods Enzymol. 2009;468:335–67.
14. Hahn EL. Spin echoes. Phys Rev. 1950;80:580–94.
15. Thomann H, Bernardo M. Indirect detection of internuclear dipolar couplings in paramagnetic solids. J Am Chem Soc. 1996;118:5806–7.
16. Caron M, Dugas H. Specific spin-labeling of transfer ribonucleic-acid molecules. Nucleic Acids Res. 1976;3:19–34.

17. Ramos A, Varani G. A new method to detect long-range protein-RNA contacts: NMR detection of electron-proton relaxation induced by nitroxide spin-labeled RNA. J Am Chem Soc. 1998;120:10992–3.
18. Larsen RG, Singel DJ. Double electron-electron resonance spin-echo modulation: spectroscopic measurement of electron spin pair separations in orientationally disordered solids. J Chem Phys. 1993;98:5134–46.
19. Pannier M, Veit S, Godt A, Jeschke G, Spiess HW. Dead-time free measurement of dipole-dipole interactions between electron spins. J Magn Reson. 2000;142:331–40.
20. Cho H, Pfenninger S, Gemperle C, Schweiger A, Ernst RR. Zero deadtime pulsed ESR by remote echo detection. Chem Phys Lett. 1989;160:391–5.
21. Kurshev VV, Raitsimring AM, Tsvetkov YD. Selection of dipolar interaction by the "2 + 1" pulse train ESE. J Magn Reson. 1989;81:441–54.
22. Timmel CR, Fursman CE, Hoff AJ, Hore PJ. Spin-correlated radical pairs: microwave pulse effects on lifetimes, electron spin echo envelope modulations, and optimum conditions for detection by electron spin echo spectroscopy. Chem Phys. 1998;226:271–83.
23. Lyubenova S, Siddiqui MK, de Vries MJMP, Ludwig B, Prisner TF. Protein-protein interactions studied by EPR relaxation measurements: cytochrome c and cytochrome c oxidase. J Phys Chem B. 2007;111:3839–46.
24. Jäger H, Koch A, Maus V, Spiess HW, Jeschke G. Relaxation-based distance measurements between a nitroxide and a lanthanide spin label. J Magn Reson. 2008;194:254–63.
25. MacArthur R, Brudvig GW. Location of EPR-active spins buried in proteins from the simulation of the spin-lattice relaxation enhancement caused by Dy(III) complexes. J Phys Chem B. 2004;108:9390–6.
26. Borbat PP, Freed JH. Multiple-quantum ESR and distance measurements. Chem Phys Lett. 1999;313:145–54.
27. Jeschke G, Chechik V, Ionita P, et al. DeerAnalysis2006 – a comprehensive software package for analyzing pulsed ELDOR data. Appl Magn Reson. 2006;30:473–98.
28. Hellmich UA, Lyubenova S, Kaltenborn E, et al. Probing the ATP hydrolysis cycle of the ABC multidrug transporter LmrA by pulsed EPR spectroscopy. J Am Chem Soc. 2012;134:5857–62.
29. Altenbach C, Kusnetzow AK, Ernst OP, Hofmann KP, Hubbell WL. High-resolution distance mapping in rhodopsin reveals the pattern of helix movement due to activation. Proc Natl Acad Sci U S A. 2008;105:7439–44.
30. Borbat PP, Surendhran K, Bortolus M, Zou P, Freed JH, Mchaourab HS. Conformational motion of the ABC transporter MsbA induced by ATP hydrolysis. PLoS Biol. 2007;5:2211–19.
31. Zou P, Bortolus M, Mchaourab HS. Conformational cycle of the ABC transporter MsbA in liposomes: detailed analysis using double electron-electron resonance spectroscopy. J Mol Biol. 2009;393:586–97.
32. Joseph B, Jeschke G, Goetz BA, Locher KP, Bordignon E. Transmembrane gate movements in the type II ATP-binding cassette (ABC) importer BtuCD-F during nucleotide cycle. J Biol Chem. 2011;286:41008–17.
33. Jeschke G. Determination of the nanostructure of polymer materials by electron paramagnetic resonance spectroscopy. Macromol Rapid Commun. 2002;23:227–46.
34. Poppl A, Rudolf T, Manikandan P, Goldfarb D. W- and X-band pulsed electron nuclear double-resonance study of a sodium-nitric oxide adsorption complex in NaA zeolites. J Am Chem Soc. 2000;122:10194–200.
35. Denysenkov VP, Prisner TF, Stubbe J, Bennati M. High-field pulsed electron-electron double resonance spectroscopy to determine the orientation of the tyrosyl radicals in ribonucleotide reductase. Proc Natl Acad Sci U S A. 2006;103:13386–90.
36. Banerjee D, Yagi H, Huber T, Otting G, Goldfarb D. Nanometer-range distance measurement in a protein using Mn^{2+} tags. J Phys Chem Lett. 2012;3:157–60.

37. Yagi H, Banerjee D, Graham B, Huber T, Goldfarb D, Otting G. Gadolinium tagging for high-precision measurements of 6 nm distances in protein assemblies by EPR. J Am Chem Soc. 2011;133:10418–21.
38. Roessler MM, King MS, Robinson AJ, Armstrong FA, Harmer J, Hirst J. Direct assignment of EPR spectra to structurally defined iron-sulfur clusters in complex I by double electron-electron resonance. Proc Natl Acad Sci U S A. 2010;107:1930–5.
39. Elsasser C, Brecht M, Bittl R. Pulsed electron-electron double resonance on multinuclear metal clusters: assignment of spin projection factors based on the dipolar interaction. J Am Chem Soc. 2002;124:12606–11.
40. Marko A, Margraf D, Yu H, Mu Y, Stock G, Prisner T. Molecular orientation studies by pulsed electron-electron double resonance experiments. J Chem Phys. 2009;130:064102.
41. Marko A, Denysenkov V, Margraf D, et al. Conformational flexibility of DNA. J Am Chem Soc. 2011;133:13375–9.
42. Jeschke G, Polyhach Y. Distance measurements on spin-labelled biomacromolecules by pulsed electron paramagnetic resonance. Phys Chem Chem Phys. 2007;9:1895–910.
43. Qin PZ, Haworth IS, Cai Q, et al. Measuring nanometer distances in nucleic acids using a sequence-independent nitroxide probe. Nat Protoc. 2007;2:2354–65.
44. Edwards TE, Sigurdsson ST. Site-specific incorporation of nitroxide spin-labels into 2[prime]-positions of nucleic acids. Nat Protoc. 2007;2:1954–62.
45. Schiemann O, Piton N, Plackmeyer J, Bode BE, Prisner TF, Engels JW. Spin labeling of oligonucleotides with the nitroxide TPA and use of PELDOR, a pulse EPR method, to measure intramolecular distances. Nat Protoc. 2007;2:904–23.
46. Shelke SA, Sigurdsson ST. Site-directed spin labelling of nucleic acids. Eur J Org Chem. 2012;12:2291–301.
47. Wachowius F, Hobartner C. Chemical RNA modifications for studies of RNA structure and dynamics. ChemBioChem. 2010;11:469–80.
48. Berliner LJ. Spin Labeling: The Next Millennium. Biological Magnetic Resonance. New York: Plenum Press; 1998.
49. Schiemann O, Piton N, Mu Y, Stock G, Engels JW, Prisner TF. A PELDOR-based nanometer distance ruler for oligonucleotides. J Am Chem Soc. 2004;126:5722–9.
50. Piton N, Mu YG, Stock G, Prisner TF, Schiemann O, Engels JW. Base-specific spin-labeling of RNA for structure determination. Nucleic Acids Res. 2007;35:3128–43.
51. Barhate N, Cekan P, Massey AP, Sigurdsson ST. A nucleoside that contains a rigid nitroxide spin label: a fluorophore in disguise. Angew Chem Int Ed Engl. 2007;46:2655–8.
52. Sezer D, Sigurdsson ST. Simulating electron spin resonance spectra of macromolecules labeled with two dipolar-coupled nitroxide spin labels from trajectories. Phys Chem Chem Phys. 2011;13:12785–97.
53. Polyhach Y, Bordignon E, Tschaggelar R, Gandra S, Godt A, Jeschke G. High sensitivity and versatility of the DEER experiment on nitroxide radical pairs at Q-band frequencies. Phys Chem Chem Phys. 2012;14:10762–73.
54. Cruickshank PAS, Bolton DR, Robertson DA, Hunter RI, Wylde RJ, Smith GM. A kilowatt pulsed 94 GHz electron paramagnetic resonance spectrometer with high concentration sensitivity, high instantaneous bandwidth, and low dead time. Rev Sci Instrum. 2009;80:103102–15.
55. Denysenkov VP, Biglino D, Lubitz W, Prisner TF, Bennati M. Structure of the tyrosyl biradical in mouse R2 ribonucleotide reductase from high-field PELDOR. Angew Chem Int Ed. 2008;47:1224–7.
56. Schiemann O, Cekan P, Margraf D, Prisner TF, Sigurdsson ST. Relative orientation of rigid nitroxides by PELDOR: beyond distance measurements in nucleic acids. Angew Chem Int Ed Engl. 2009;48:3292–5.

57. Marko A, Margraf D, Cekan P, Sigurdsson ST, Schiemann O, Prisner TF. Analytical method to determine the orientation of rigid spin labels in DNA. Phys Rev E. 2010;81:021911.
58. Schiemann O, Cekan P, Margraf D, Prisner TF, Sigurdsson ST. Relative orientation of rigid nitroxides by PELDOR: beyond distance measurements in nucleic acids. Angew Chem Int Ed. 2009;48:3292–5.
59. Margraf D, Bode BE, Marko A, Schiemann O, Prisner TF. Conformational flexibility of nitroxide biradicals determined by X-band PELDOR experiments. Mol Phys. 2007;105:2153–60.
60. Marko A, Prisner TF. An algorithm to analyze PELDOR data of rigid spin label pairs. Phys Chem Chem Phys. 2013;15:619–27.
61. Godt A, Schulte M, Zimmermann H, Jeschke G. How flexible are poly(para-phenyleneethynylene)s? Angew Chem Int Ed. 2006;45:7560–4.
62. Abe C, Klose D, Dietrich F, et al. Orientation selective DEER measurements on vinculin tail at X-band frequencies reveal spin label orientations. J Magn Reson. 2012;216:53–61.
63. Krstic I, Frolow O, Sezer D, et al. PELDOR spectroscopy reveals preorganization of the neomycin-responsive riboswitch tertiary structure. J Am Chem Soc. 2010;132:1454–5.
64. Weigand JE, Sanchez M, Gunnesch EB, Zeiher S, Schroeder R, Suess B. Screening for engineered neomycin riboswitches that control translation initiation. RNA. 2008;14:89–97.
65. Duchardt-Ferner E, Weigand JE, Ohlenschlager O, Schtnidtke SR, Suess B, Wohnert J. Highly modular structure and ligand binding by conformational capture in a minimalistic riboswitch. Angew Chem Int Ed. 2010;49:6216–19.
66. Nozinovic S, Furtig B, Jonker HR, Richter C, Schwalbe H. High-resolution NMR structure of an RNA model system: the 14-mer cUUCGg tetraloop hairpin RNA. Nucleic Acids Res. 2010;38:683–94.
67. Cluzel P, Lebrun A, Heller C, et al. DNA: an extensible molecule. Science. 1996;271:792–4.
68. Mathew-Fenn RS, Das R, Harbury PAB. Remeasuring the double helix. Science. 2008;322:446–9.
69. Krstic I, Hansel R, Romainczyk O, Engels JW, Dotsch V, Prisner TF. Long-range distance measurements on nucleic acids in cells by pulsed EPR spectroscopy. Angew Chem Int Ed Engl. 2011;50:5070–4.
70. Yang ZY, Liu YP, Borbat P, Zweier JL, Freed JH, Hubbell WL. Pulsed ESR dipolar spectroscopy for distance measurements in immobilized spin labeled proteins in liquid solution. J Am Chem Soc. 2012;134:9950–2.

Christina R. Mozes and Mirko Hennig
12 NMR-based characterization of RNA structure and dynamics

12.1 Introduction

Genome-wide RNA-sequencing studies have revealed unexpected diversity and complexity of RNA species. It is estimated that less than 2% of the human genome encodes for proteins, yet the majority of DNA is transcribed into RNA. The biological motivation for this energetically expensive task is not fully understood, but we now know that noncoding RNAs have widespread roles including catalysis and gene regulation. Transcribed as single-stranded molecules, RNA can pair to itself, other RNA molecules, or DNA, and it can bind a host of biological macromolecules and cellular metabolites. The varied and specific three-dimensional (3-D) shapes that confer biological function on RNA are surprising, given the restricted set of building blocks: four chemically similar nucleotides. Like the protein-folding problem, the RNA-folding problem considers the hierarchy of secondary and tertiary structures in deriving a compact fold from a primary sequence. While the base pairing that defines RNA secondary structure can be fairly accurately predicted from nearest-neighbor thermodynamic models (see Chapter 14), tertiary interactions are often weak enough that they can be disrupted by mild changes in solution conditions and temperatures and have mostly eluded predictive models. As a result of transient tertiary interactions, RNA molecules are dynamic, often displaying conformational heterogeneity that defies the designation of a single native-state structure. While multiple conformers and inherent dynamics can present serious challenges for structure determination, these are the very properties that allow RNA to sense external cues and transduce these cellular signals into functional responses. Solution nuclear magnetic resonance (NMR), which reports on dynamic ensembles of structures, has a powerful advantage in studying RNAs and their intermolecular interactions in physiologically relevant solutions. As methods in sample preparation, pulse schemes, restraint collection, and structure calculation progress, solution NMR continues to address fundamental questions of RNA folding and dynamics in increasingly complex systems. In this review, we briefly summarize modern methods of sample preparation, cover current NMR methodology in RNA structure and dynamics determination, and present a number of applications to biological systems found in the literature.

12.2 Part I: RNA structure

12.2.1 Primary structure

12.2.1.1 RNA sequence determinants on structure

Polymeric RNA structure begins with its monomeric chemical composition, that is, its primary sequence. The four standard RNA nucleotides – adenylic, cytidylic, guanylic, and uridylic acid – compose simple sequences that achieve combinatorial complexity [1]. Contained in the purine- and pyrimidine-derived nucleobases is the chemical prescription for the π-stacking interactions that drive RNA folding and sufficiently separate the highly repulsive phosphodiester linkages when intertwined as a helix.

12.2.1.2 Unusual nucleotides

The standard nucleotides are, however, only 4 out of more than 100 naturally occurring RNA nucleotides. Unusual nucleotides are the product of posttranslational enzymatic modifications, and they are far more varied in RNA than in DNA. The most common modified RNA nucleotide is pseudouridine (Ψ), an isomer of uridine, which tends to rigidify the sugar-phosphate backbone by coordinating a structural water and enhancing base stacking [2]. Other RNA modifications include methylation of various positions of the nucleobase or the 2'-hydroxyl, sulfur-for-oxygen substitutions in the nucleobase, and a number of other chemical substitutions [3]. Deamination of adenosine to form inosine (A-to-I editing) is common especially in transfer RNAs (tRNAs) where it alters coding since inosine can base pair with U, C, or A [4]. Because of their low abundance and physiochemical similarity to standard nucleotides, the presence and position of modified nucleotides is difficult to detect by most methods. NMR, with its atomic resolution, is suited to the study of modified nucleotides and their impact on structure, once the hurdle of incorporating such a modification is overcome. (Isolation from tissue has been successful for a number of posttranscriptionally modified tRNAs [5].)

12.2.1.3 Torsion angles in the polynucleotide sequence

The conformational space of nucleic acids is far greater than that of proteins because the torsion angles along the polynucleotide backbone outnumber those of poly-peptide backbone. Along the phosphate backbone, there are six rotatable bonds, denoted with torsion angles α through ζ (Fig. 12.1A). Within the ribose sugar poly, five single bonds account for torsion angles v_0 through v_4 that can be represented by the pseudorotation phase angle [6], defined as

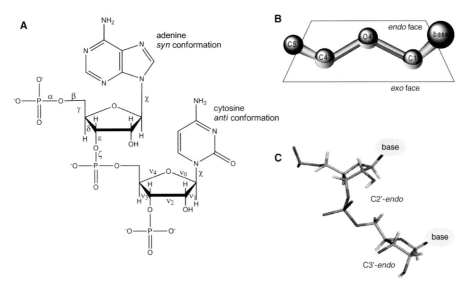

Fig. 12.1 Torsional degrees of freedom in RNA. (A) The sugar-phosphate backbone torsion angles α through ζ, ribose torsions v_0 through v_4, and the glycosidic torsion χ are illustrated for the dinucleotide 5'-pApCp-3'. (B) *Endo* and *exo faces* of the pentose ring are used to reference the sugar conformation of nucleic acids. (C) The predominantly populated C2'-*endo* and C3'-*endo* puckers are demonstrated for a dinucleotide.

$$\psi = \tan^{-1}\{[(v_4 + v_1) - (v_3 + v_0)]/[2v_2(\sin 36° + \sin 72°)]\}. \tag{12.1}$$

The sugar pucker, which describes the nonplanar conformation of the five-member ring, is a function of ψ, with the two dominant conformations being C3'-*endo* (0° ≤ ψ ≤ 36°) and C2'-*endo* (144° ≤ ψ ≤ 190°) (Fig. 12.1B,C).

While DNA sugar puckers dynamically interconvert, RNA typically adopts the C3'-*endo* conformation, due to the restrictions imposed by the 2'-hydroxyl group. Exceptions can be found in regions that do not form regular helical structures due to the influence of local sterics and electrostatics. Finally, there is the rotatable glycosidic bond joining the sugar and base, for which the torsion angle χ defines the *anti-* or *syn*-arrangement of the nucleotide (Fig. 12.1A). While the *anti*-conformation is preferred, the *syn*-arrangement is tolerated by purines where the smaller five-member ring of the base lies directly over the sugar, instead of the bulky six-member ring as is the case for pyrimidines.

12.2.2 Secondary structure: base pairing and helices

12.2.2.1 Regular structure and base pairing

In spite of the six-dimensional space along the polynucleotide backbone, the conformational space of RNA is limited. If backbone dimensions are reduced to two pseudorotation angles, η and θ (anchored at phosphate and C4' atoms, where η is the torsion of C4'$_{i-1}$, P$_i$, C4'$_i$, and P$_{i+1}$, and θ is the torsion of P$_i$, C4'$_i$, P$_{i+1}$, and C4'$_{i+1}$), the correlation of these angles among known RNA structures reveals a conformational map with distinct structural preferences [7–9], reminiscent of the φ/ψ space for proteins.

Regular RNA structures are helical and are generally formed from two antiparallel strands held together by hydrogen bonds between complementary bases. In single-stranded RNA, this complementarity is satisfied by intramolecular pairing of bases sometimes very distant in the primary sequence. The canonical Watson-Crick base pairs (A paired to U, and G paired to C) display an extraordinary design feature in their inherent complementarity and isostericity, with any of the four combinations able to substitute for another pair without distortion of the A-form helix. However, in structured RNAs, only about 60% of the bases on average engage in Watson-Crick pairing. The remaining bases may reside in unpaired regions or engage in a number of non-canonical hydrogen-bonding interactions. In these "edge-to-edge" interactions, each nucleotide interacts through one of three edges: the Watson-Crick edge, the Hoogsteen edge (also called the "C-H edge" in pyrimidines), or the sugar edge (also termed the shallow-groove edge) (Fig. 12.2).

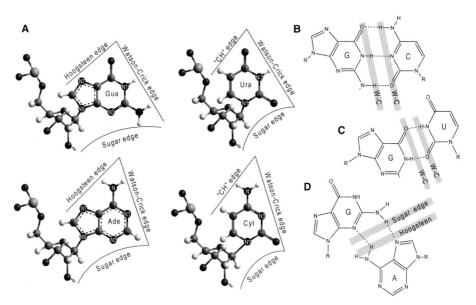

Fig. 12.2 Edge-to-edge pairing interactions. (A) The three edges available for hydrogen-bonding interactions are shown for Gua, Ade, Ura, and Cyt. Standard atom numbering is indicated for the nucleobases. The interaction of two Watson-Crick faces is shown for a (B) canonical G-C base pair and a (C) semicanonical G·U wobble pair. (D) The commonly termed sheared G-A base pair is formed from the interaction of the sugar edge of Gua with the Hoogsteen edge of Ade.

In total, there are 12 possible edge-to-edge base-pair geometries, 11 of which have been observed in high-resolution RNA structures. (Leontis and Westhof have proposed a systematic nomenclature for RNA base pairs to simplify the vernacular of base-pair geometries and the shorthand representation of them [10].) The inclusion of noncanonical base pairs lends structural and dynamic properties to RNAs that convey specific function, many examples of which are discussed in Section 12.4. Base triples are also possible, arising when a nucleotide makes hydrogen bonds along two edges, pairing with two other nucleotides. Base triples can tie together distant domains to stabilize a tertiary fold.

12.2.2.2 Helical secondary structure

As previously noted, RNA has a strong preference for the C3'-*endo* sugar pucker because of the 2'-hydroxyl (-OH) group, with steric restrictions and hydrogen bonding to neighboring phosphodiester bonds both playing a role. This conformation disposes RNA to the right-handed A-form helix, whose topology is broader and shorter than its B-form counterpart and features a hollow core along the central axis. While the major groove is the site of most protein-DNA interactions, the deep and narrow major groove of A-RNA excludes most base-specific tertiary and quaternary contacts, such that the shallow and wide minor groove plays more prominently in many protein-RNA interactions. The parameters of regular A-form helices are well defined (right-handed, 11 base pairs/turn, h = 0.24 nm/bp) [1], but breaks in form are abundant in structural RNAs and are the basis for their unique architecture and biological specificity.

12.3 Part II: NMR studies of RNA

12.3.1 NMR sample preparation and labeling

12.3.1.1 Sample preparation

Typically, a few milligrams of a RNA is needed to achieve a sample that is at least submillimolar in concentration when dissolved in volumes ranging from 250 to 500 µL. Isotopic labeling of the sample becomes necessary for the assignment and study of larger RNAs, and additional labeling schemes may be desirable for a particular strategy. The standard method for large-scale production of labeled or unlabeled RNA is *in vitro* transcription of a DNA template, using standard, labeled, or modified nucleotides as desired. Bacteriophage T7 RNA polymerase is widely used to catalyze *in vitro* transcription because it has been characterized more thoroughly than other common DNA-directed RNA polymerases (T3 and SP6) [11,12].

12.3.1.2 Labeling schemes

Nucleotides enriched in ^{15}N and ^{13}C are required for the production of isotopically labeled RNA molecules for multidimensional heteronuclear NMR experiments [13–16]. The incorporation of NMR-inactive deuterons partially alleviates the relaxation problems that plague NMR studies of larger molecules. Deuterated nucleotides [17–21] can be used with and without protonated counterparts to reduce signal overlap and line broadening or as the starting point for site-specific protonation schemes. One strategy for simplifying spectra of large RNAs is the use of protonated nucleotides in combination with perdeuterated nucleotides to achieve base-specific signals in separate samples. The obvious disadvantage here is the need to prepare multiple samples, coupled with the loss of some sequential connectivity information. For nucleotides adjacent to purines, sequential information can be recovered by site-specific protonation the C8 position in a deuterium background [22]. This is achieved by subjecting perdeuterated purines to base-catalyzed exchange in an H_2O buffer. An analogous method does not currently exist for site-specific protonation of H5 or H6 in pyrimidines, though developments are ongoing in this area.

Aside from traditional isotopic labeling, modification of nucleotides may be desired for a number of other experimental strategies. Incorporation of the 100% naturally abundant isotope ^{19}F in RNA nucleobases (5F-U, 5F-C [23–25], or 2F-A [26]) or in the 2'-position of the ribose [27], for example, offers a sensitive, minimally perturbing NMR-active probe at specified positions throughout the molecule of study. The ^{19}F chemical shift has a wide chemical shift distribution (ca. 50-fold larger than that of ^{1}H) and is intensely sensitive to changes in its microenvironment due to the anisotropic distribution of electrons in the fluorine 2-p orbitals. These features allow for simplified, one-dimensional ^{19}F spectra that can report on aspects of conformation, molecular interactions, and mobility that may be obscured in ^{1}H-NMR spectroscopy.

Segmental isotopic labeling is the labeling of a subdomain within the context of a larger, intact RNA, and it offers another strategy for simplifying spectra, compatible with any of the previously discussed labeling schemes [28–30]. Ligation of labeled and unlabeled fragments of RNA can be catalyzed by T4 DNA ligase [31], T4 RNA ligase [32], or deoxyribozymes [33]. Each of these enzymes has requirements for the chemical identity of the ends to be joined (e.g. T4 DNA ligase requires a 3'-OH and 5'-monophosphate) and other requirements for efficient yields.

12.3.1.3 RNA purification

Efficient purification of *in vitro* transcribed RNA is most commonly achieved by anion exchange fast-performance liquid chromatography [34,35]. Preparative quantities of transcript can be purified in this manner under native conditions, a distinct advantage of this method over earlier denaturing polyacrylamide gel electrophoresis separations [36]. Where a native anion exchange column is not sufficient to separate the product, high-performance liquid chromatography anion exchange in combination with heat denaturation during purification can offer up to single-nucleotide resolution [37]. Alternatively, general affinity-tag purification methods under native conditions have also been developed [38,39].

12.3.2 NMR parameters to characterize RNA structure

12.3.2.1 Sequence-specific assignment of NMR resonances

Assignment of NMR resonances is prerequisite to analysis aimed at high-resolution structural detail of RNAs. The unambiguous assignment of RNA molecules is hampered by the severe spectral overlap in the proton spectrum and broad line widths of

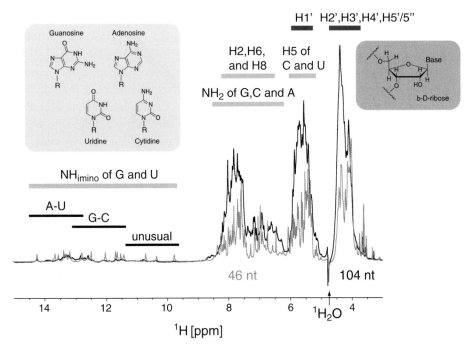

Fig. 12.3 Typical ^{1}H chemical shift ranges and line widths in RNA. 1D-excitation sculpting ^{1}H spectra of pri-mir16–1 (104 nt, black line) and the lower half stem of the same molecule (46 nt, gray line). Proton chemical shift ranges of nucleobase signals are indicated using light gray bars, whereas dark gray bars highlight chemical shift ranges of ribose resonances. Insets show the four different nucleobases (light gray box) and common β-D-ribose sugar (dark gray box) building blocks. The arrow indicates the residual H$_2$O resonance.

fast-relaxing, larger RNAs (>15 kDa, ca. 45 nucleotides). Fig. 12.3 gives the standard ^{1}H chemical shift ranges for RNA resonances and shows the typical line broadening observed for large RNA molecules.

Detection of ^{31}P signals along the sugar-phosphate backbone does not require isotopic labeling and in small RNAs (<5 kDa, ca. 15 nucleotides), phosphate resonances can be sequentially assigned by through-bond connectivities from ^{1}H-^{31}P heteronuclear correlation experiments that connect H3'$_i$-P$_i$-H5'/H5''$_{i+1}$ systems [40,41], with the possibility of appending homonuclear ^{1}H-^{1}H nuclear Overhauser effect spectroscopy (NOESY) or total correlation spectroscopy (TOCSY) [42] experiments to correlate these poorly resolved resonances to the more distinct H1' or aromatic H8/H6 resonances [43]. Typically, sequential assignment of even medium-size RNA molecules is not readily achieved by through-bond connectivities. Instead, sequential information is gained from through-space transfer.

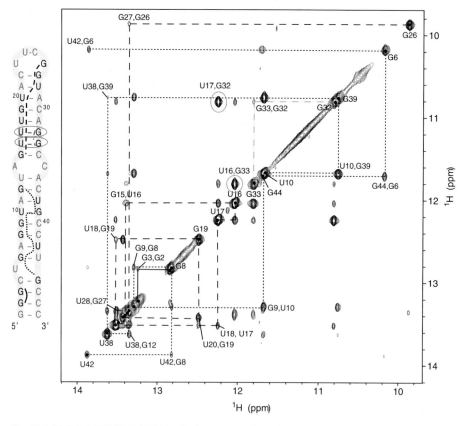

Fig. 12.4 ^{1}H-^{1}H WATERGATE-NOESY for the lower stem of pre-let-7g. Tie lines highlight the NOE correlations between G and U imino protons. Four separate sequential "walks" are indicated by solid black, dotted black, dashed black, and dashed gray lines in the spectrum and the schematic representation of the pre-let-7g hairpin. Cross peaks between G and U of two wobble pairs (U16-G33, U17-G32) are circled. Tie lines between the dashed black and dashed gray walks are omitted for clarity.

Sequential assignments and secondary structure (i.e. the base-pairing pattern) of an unlabeled RNA can be determined from ^{1}H-^{1}H nuclear Overhauser effect (NOE) measurements, typically from a WATERGATE –2-D NOESY [44], and a reasonably accurate starting point – the knowledge of the thermodynamically predicted secondary structure. Guanine and uracil imino protons are observable when protected from solvent exchange because of hydrogen bonding, and these signals conveniently occupy the otherwise empty region of the proton spectrum between 15 and 10 ppm. For each Watson-Crick pair, one such imino proton is observed, and two signals are observed in a G·U wobble pair. The so-called NOESY walk, demonstrated in Fig. 12.4 for the lower stem of pre-let-7g, identifies sequential, through-space connections between these exchangeable iminos, building up a picture of the paired stems of the RNA. Guanine versus uracil identity is made possible by NOE signal to other base protons, notably the AH2 proton in the case of uracil paired to adenine, and the CH41/2 amino protons in the case of guanine paired to cytosine.

The "NOESY walk" for imino assignments is combined with sequential assignment of base H8/H6 to ribose H1' NOE patterns seen in helical regions of the molecule [45]. Assignment of the ribose protons is difficult due to the limited dispersion of the ribose H2', H3', H4', H5', and H5" spin system (ca. 1 ppm) and inefficient through-bond magnetization transfer from the better resolved H1' resonances (i.e. $^3J_{H1'-H2'}$ ~ 1 Hz in C3'-*endo* puckers) but is undertaken with correlation spectroscopy (COSY) and TOCSY spectra. The proton of the 2'-hydroxyl, typically unobserved at standard temperature due to exchange with solvent, can be assigned and analyzed in low-temperature TOCSY and NOESY experiments [46–50].

Correlation of carbon shifts in a ^{1}H,^{13}C–heteronuclear single quantum correlation (HSQC) spectrum facilitates site-specific assignment, and even natural abundance ^{13}C signal (1.1%) can aid this effort and is highly desirable for RNAs larger than ca. 15 nucleotides. The C1', C4', C5', Ade C2, Cyt C5, and Ura C5 carbon chemical shift ranges are all identifying.

The most common measure taken to increase spectral resolution is isotopic labeling of the RNA sample, which allows protons to be correlated to the chemical shifts of ^{15}N and ^{13}C signals. The chemical shift ranges of these heteronuclei, shown in Fig. 12.5, are again restricted due to the limited chemical diversity in RNA, but the ability to resolve peaks in multiple dimensions removes much of the degeneracy. The usefulness of ^{31}P chemical shifts remains limited, and through-bond assignment using HCP-like experiments [51,52] is not feasible for larger RNA molecules (>15 kDa, ca. 45 nucleotides) due to notoriously overlapped resonances and modest sensitivity (Fig. 12.5).

A large suite of experiments has been developed for resonance assignment of RNAs, with different experiments tailored to the size and labeling scheme of the RNA sample. A number of excellent reviews of RNA assignment strategies are also available [53–59]. Briefly, uniform ^{15}N, ^{13}C-labeling allows for isotopic editing in experiments to reduce the ambiguities inherent in NOESY-based assignment approaches. The HCN [60,61] and HCNCH [61] experiments identify intranucleotide H6/H8-H1' correlations,

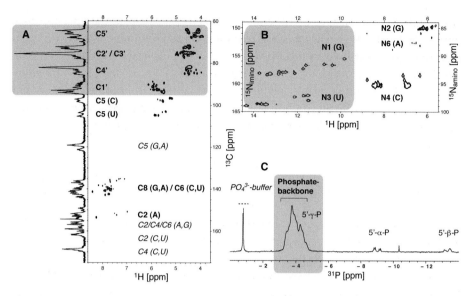

Fig. 12.5 Typical ^{13}C, ^{15}N, and ^{31}P-chemical shift ranges in RNA. (A) Carbon-13 chemical shift ranges in the 30-nt HIV-2 TAR RNA recorded at 850 MHz and 298 K. A ^{1}H,^{15}N-decoupled ^{13}C-1D spectrum is shown along the left axis of an HSQC spectrum. ^{1}H-^{13}C ribose resonances (C1'-C5') are boxed in gray. Protonated carbon resonances are labeled using bold letters; quaternary carbons are italic. (B) Protonated nitrogen-15 chemical shift ranges in the 46-nt pri-mir16–1 RNA recorded at 800 MHz and 288 K. Imino GH1-N1 and UH3-N3 resonate between 150 and 165 ppm (gray box). Amino GH2-N2, AH6-N6, and CH4-N4 resonate between 85 and 100 ppm. (C) Phosphorous-31 chemical shift ranges in the 30 nt HIV-2 TAR RNA recorded at 600 MHz and 298 K. RNA phosphate backbone resonances cluster around −4 ppm (gray box); the terminal 5'-pppG (5'-α,β,γ-P) and the buffer PO$_4^{3-}$ signals are labeled.

distinguishing them from the corresponding internucleotide correlations. Versions of the through-bond HCCH-COSY and HCCH-TOCSY experiments [63,64] provide unambiguous assignment of ribose as well as nucleobase H5-H6 pyrimidine and H2-H8 adenosine spin systems [65,66]. Additional through-bond correlation experiments facilitate the assignments of exchangeable imino and amino proton resonances linked to nonexchangeable base H6 and H8 protons [67–72]. As the size of the molecule of study increases, problems with fast transverse relaxation hamper the resolution and sensitivity of NMR experiments. Many of the standard experiments mentioned have transverse relaxation-optimized spectroscopy (TROSY) counterparts that attempt to mitigate these effects [73–75]. Even so, for RNA molecules exceeding ca. 40 nt, dwindling T_2 values (i.e. fast transverse relaxation) render the through-bond experiments too insensitive to achieve assignments in a conventional manner. A variety of deuteration and site-specific labeling schemes have been developed to meet this challenge (see Section 12.3.1.2), and the experimental strategies are largely determined case by case (see the references for excellent recent reviews [17,76–78]). In large systems where full assignment is not possible, identifying particular resonances may nonetheless provide important information about a biological system. For example, structural elements of the 356-nt 5' leader sequence of human immunodeficiency virus type-1 (HIV-1) were studied by long-range probing by adenosine interaction detection (lr-AID) [79]. This method involved the substitution of three consecutive native base pairs with three U-A pairs in regions of interest, resulting in an easily identified, upfield-shifted adenosine signal. The substitutions did not affect the overall structure or stability of the stems where they were introduced, and they revealed the alternate base pairing of a hairpin sequence, which promotes the molecule's dimerization and viral packaging.

12.3.2.2 NMR measurements for torsion angle restraints

NMR methods offer three types of measurements that are each useful in defining a subset of RNA backbone torsion angles: chemical shifts, scalar coupling constants [80], and cross-correlated relaxation rates [81,82].

Chemical shift has been computed and correlated with a number of backbone torsions. The ^{13}C chemical shifts for ribose carbons C1'–C5' correlate with **sugar pucker**, as determined from *ab initio* calculations [83], as well as an evaluation of database RNA structures [84]. ^{13}C-detected experiments can be employed to collect chemical shifts of unprotonated carbons of RNA bases, for which correlations with base pairing and base stacking can be made, indirectly restraining torsions during structure calculation [85,86]. Finally, the **α and ζ torsion angles** are major determinants of the ^{31}P isotropic chemical shift [87].

NMR scalar coupling constants can be analyzed quantitatively according to empirical Karplus relationships to extract a number of backbone torsion angles [80]. The **sugar pucker** can be qualitatively determined from the ^{3}J(H1',H2') coupling, which is large (>8 Hz) for the C2'-*endo* conformation and small (ca. 1 Hz) for the C3'-*endo* conformation. The opposite trend is observed for the ^{3}J(H3',H4') coupling, for which larger values are observed for the C3'-*endo* pucker [88–91]. The **glycosidic torsion χ** impacts the heteronuclear vicinal ^{3}J(H1',C4/8) in purines and ^{3}J(H1',C2/C6) in pyrimidines [89]. Additionally, in a fluorinated sample (5F-Ura and 5F-Cyt), the five-bond scalar ^{5}J(H1',F) coupling can unambiguously distinguish the glycosidic torsion χ [92]. The **β and ε torsion angles** are accessible from a number of ^{13}C, ^{31}P, and ^{1}H,^{31}P [41,93] scalar couplings. A lack of dispersion among proton and phosphorous signals generally requires that accurate measurements be obtained from phosphorous-fitting of doublets from singlets [89], or spin-echo difference experiments [94–96]. J-HMBC techniques can be applied to determine ^{3}J(H,P) couplings [97], and a quantitative version of the HCP experiment allows for the measurement of ^{3}J(C4',P) [98,99]. Finally, the **γ torsion** is measured from the two-bond ^{2}J(C4',H5'/H5") and the vicinal ^{3}J(H4',H5'/H5"), though this can be difficult in practice due to the requirement for stereospecific assignments for H5' and H5" resonances [100]. The α and ζ torsion angles cannot be obtained from scalar coupling measurements because the involved ^{16}O nuclei have no magnetic moment.

Cross-correlated relaxation between a ribose ^{13}C-^{1}H dipole and the ^{31}P chemical shift anisotropy (CSA) can be used to determine **α and ζ torsion angles** [101,102]. Similarly, the **glycosidic torsion χ** can be determined from cross-correlated relaxation between two ^{13}C-^{1}H dipoles or a dipole and the glycosidic ^{15}N CSA [103,104]. The cross-correlated relaxation rates of neighboring ^{13}C-^{1}H dipoles within the ribose ring may also be exploited in an HCCH-TOCSY for quantitative determination of the **sugar pucker** [105,106].

12.3.2.3 NMR measurements for distance restraints

Traditionally, local RNA conformation has been defined by ^{1}H-^{1}H NOE measurements, which provide through-space distance restraints between hydrogen pairs, effective over a short range of about 5 Å. (Increasingly, residual dipolar couplings [RDCs] are used to refine the backbone conformation, relying on the global fitting of "dipolar waves," the periodic variation in the magnitude of dipolar couplings observed for repeating geometries [107]. We consider RDCs further in Section 12.3.2.5.) Many variants of the NOESY experiment exist, but in all cases, the intensities I of cross peaks between proton pairs are evaluated according to the proportionality

$$I \sim 1/r_{ij}^6 \tag{12.2}$$

where r_{ij} is the distance between protons i and j, assumed to be an isolated spin system. NOE-derived distance restraints are often determined semiquantitatively by binning cross-peak intensities in four categories, from strongest to weakest, and assigning upper and lower distance bounds to these categories. Negative constraints may also be set, where the absence of a particular ^{1}H-^{1}H NOE cross peak can be interpreted as a lower bound distance of 4 Å between the two protons involved. Negative restraints are introduced only if the protons in question show observable NOE to at least one other proton. The intensity-distance NOE correlation can break down when spin-diffusion is encountered in experiments where long NOE mixing times (>100 ms) are used. In these instances, proton pairs no longer approximate isolated spin systems because relaxation pathways to nearby spins become very efficient, and the NOE signal is quickly dissipated among neighboring protons. While interfering with distance restraint collection, spin diffusion can be helpful in the early stages of resonance assignment where it can identify extended spin systems. NOESY spectra acquired at multiple mixing times (between 50 and 200 ms) can be useful for this purpose.

12.3.2.4 Scalar couplings across hydrogen bonds

Within a ^{15}N-labeled RNA, secondary structure can also be determined via experiments directly detecting hydrogen bonding. This ability for direct observation contrasts with crystallographic studies where hydrogen bonding can only be inferred and is a powerful tool for confirming secondary structure. In the HNN-COSY [108,109], the donor and acceptor nitrogen atoms are correlated across N-H · · · N hydrogen bonds involving G or U iminos. The magnetization is transferred via the two-bond scalar coupling ($^{h2}J(N,N)$), and the sensitivity of the experiment is high such that the measurement is feasible even for large RNAs. Because of the ^{15}N chemical shift dispersion, which distinguishes N1, N3, N7, and N9 atoms, the HNN-COSY experiment can detect noncanonical base pairs [110–112] and can also be useful in differentiating alternate secondary structures. Additionally, if imino protons are involved in hydrogen bonds to an external ligand, as is the case for adenine and guanine riboswitches, the coordination of the binding site can be determined [113].

The HN(N)-TOCSY [114] and H(N)CO [115] experiments can be used for direct detection of N-H · · · O hydrogen bonds involving imino proton donors and carbonyl oxygen acceptors. Magnetization is transferred via $^{h4}J(N,N)$ and $^{h3}J(N,CO)$ couplings, respectively. The smaller values of these couplings limits the sensitivity of these experiments, restricting their use for large RNAs. The 2'-OH of RNA can also mediate important tertiary interactions by serving as a hydrogen bond donor to nitrogen acceptors. Normally unobserved due to fast chemical exchange with solvent at room temperature, the 2'-OH can be observed in cases of H-bond protection. The $^{h1}J(2'OH,N)$ coupling can be measured from quantitative spin-echo difference experiments [116].

12.3.2.5 Residual dipolar couplings

The additive errors in a structure derived from short-range NOE distance restraints can result in large uncertainty in the placement of secondary structural elements within an overall fold. Long-range structural information can instead be gained from RDCs, which report on the orientation of internuclear vectors relative to a static, global frame [117,118].

Dipolar couplings, which are ordinarily averaged to zero by the isotropic tumbling of molecules in solution, can be recovered by partial alignment of the sample [119]. RDCs depend on the average value of an orientational function, $½(3\cos^2θ – 1)$, and the inverse cube of the distance between the coupled nuclei. When the distance is known, as it is for bonded ^1H-^{15}N and ^1H-^{13}C groups, the axial orientation $θ$ of the bond vector with respect to the principal axis system of the molecular alignment tensor can be determined. In RNA studies, popular external alignment media include Pf1 bacteriophage solutions [120,121], polyacrylamide gels [122], and dilute liquid crystalline bicelles [123], all of which induce a slight orientational preference among the codissolved RNA molecules. Partial or weak alignment is key for measuring RDCs, as strong coupling induces excessive line broadening. RDCs are calculated from the difference in coupling values measured under isotropic and anisotropic (aligned) conditions, and it is desirable to obtain two or more independently aligned data sets to resolve degenerate solutions for the orientation of two or more secondary structural elements with respect to each other. Linearly independent alignment tensors are obtained by using different media. In practice, this can be difficult to achieve because polyanionic RNA exhibits nearly identical alignment due to electrostatic repulsion and steric mechanisms in diverse media conditions [124]. Self-alignment with the magnetic field can alleviate this problem, particularly for nucleic acids because of the large magnetic susceptibility of the double helix structure [125,126]. In this case, the requirement for multiple aligned data sets must be satisfied by making measurements at multiple fields, as the alignment scales with the square of the magnetic field. Magnetic alignment is the least perturbing method of RDC measurement, and its popularity may increase as high-field instruments become more widely available.

The ^{1}H-^{15}N one-bond RDCs (^{1}D(H,N)), which are critical in protein structure determination, are less helpful in RNA samples where H-N vectors are sparse and redundantly oriented in regular A-form helix, and those belonging to unpaired G and U imino groups are unobserved due to rapid exchange with the solvent. ^{1}D(H,N) values are derived either from a J-modulated ^{1}H-^{15}N-HSQC [127] or a gradient-enhanced, interleaved inphase-antiphase (IPAP) HSQC [117,128]. More useful RDCs in RNA molecules, however, are derived from ^{13}C-^{1}H one-bond vectors and are calculated from the difference in coupling values measured in J-modulated or IPAP ^{1}H,^{13}C-HSQC [129–131] experiments recorded for the sample under isotropic and anisotropic conditions. Better resolved ^{1}H-^{13}C one-bond RDCs typically correspond to A C2-H2, A/G C8-H8, C/U C5-H5, C/U C6-H6, and C1'-H1'. Because of the planar geometry of bases, just three linearly independent RDCs are needed to predict all other RDCs in the base ring. Additional RDCs, however, will improve accuracy, and a number of additional experiments can be used to measure one-bond ^{1}D(C,C/N) and two-bond ^{2}D(H,C) and ^{2}D(H,N) RDCs in RNA bases (including ^{1}D(N1/N9,C1')) [129,132,133].

RDCs defining the sugar-phosphate backbone are more difficult to obtain largely because of NMR-inactive, bridging oxygen nuclei and spectral overlap (which also hampers the collection of useful NOEs for restraining the backbone). Nevertheless, routines employing two-dimensional and 3-D experiments have been developed for the measurement of one-, two-, and three-bond RDCs in the ribose sugar (^{1}D(C5',H5'/H5"), ^{2}D(H5',H5"), ^{1}D(C2',H2'), ^{1}D(C3',H3'), ^{1}D(C2',H2'), ^{1}D(C1',H1'), ^{2}D(C1',H2'), ^{2}D(C2',H1'), ^{3}D(H1',H2')) [134], and three-bond RDCs (^{3}D(C2'/C4',P), ^{3}D(H3',P)) [99,135], as well as more distant P-Hbase and P-H2' couplings defining the phosphate linkages [41,135].

Dipolar coupled nuclei interact through space, not through bond, and the $1/r^3$ distance dependence means that dipolar couplings can observe longer distances than is possible through NOEs ($1/r^6$ dependence). Longer range (~5–9 Å) ^{1}H-^{1}H RDCs can be obtained [136], but generally only where methods exist to band-selectively decouple alternative ^{1}H-^{1}H RDC pathways leading to prohibitive line broadening [137,138].

12.3.2.6 NMR-based structure calculation

Structure calculations take as input both experimental restraints and nonexperimental constraints such as the planarity constraints for bases and conformational database potentials of mean force [139]. The latter potentials are derived from RNA crystal structures and will bias base-pair geometry, torsion angles, and base-base interactions toward conformations represented in known structures. Therefore, their inclusion requires careful monitoring and validation with the experimental data. Other experimental evidence can also be used to gauge the validity of a structure calculation. The low-resolution structural information provided by small angle X-ray scattering (SAXS; see Chapter 9), for example, can report on overall dimensions, radius of gyration, and shape of a biomolecule. Use of SAXS data to reduce the degrees of freedom in a structural calculation has been employed in a number of NMR-based structure determinations [140–142].

Either distance geometry methods or simulated annealing protocols, both utilizing the experimental NMR restraints, can be used to generate ensembles of structures for refinement and validation. The final converged ensemble is assessed according to the root mean square deviation statistic and validated by back-calculating a subset of the restraints (normally a subset of RDCs) [143] intentionally removed from a round of structure calculation. Agreement between the omitted experimental data and the values back-calculated from structures determined in their absence is indication of a valid structural ensemble. Similarly, the structures can be cross-validated by a comparison of observed and calculated ^{1}H chemical shifts [144].

12.3.3 NMR parameters to characterize RNA dynamics

The 3-D structure of an RNA makes possible its quaternary interactions with other biomacromolecules and metabolites. Equally fundamental to its functional roles, however, is the dynamic nature of RNA.

The free-energy landscape of RNA is often characterized by deep wells, each containing a number of similar conformations with degenerate base pairing, and separated from alternate secondary structures by high free-energy barriers. Equilibrium fluctuations may not be sufficient to easily convert between conformations on a biological timescale, resulting in the possibility of kinetically trapped, nonfunctional RNA conformations. Nature has used this feature to its advantage, modeling RNA "binary switches" from alternate RNA secondary structures that only interconvert upon sensing the proper cellular cue. Rather than thermodynamically stabilizing the functional conformation, these cellular triggers – which may be other RNAs, target proteins, small molecule metabolites, or conditions like temperature or pH – often effect the transition by lowering the energy barriers of separation. That is, the presence of a ligand can induce RNA mobility such that equilibrium fluctuations occur more rapidly, without stabilizing any of the preexisting conformations. Helicases, for example, catalyze duplex melting, lowering the energy barrier between alternate base-pairing patterns. Other "chaperone" proteins can achieve the same effect. The nucleocapsid protein destabilizes helices in the 5' leader sequence of the HIV-1 genome to disrupt its monomeric transcription/translation-competent fold and promote a long-distance interaction that ultimately facilitates dimerization and packaging of the RNA genome in late-phase infection [145–147].

While a protein chaperone or helicase will, on average, contact a region of the RNA spanning 4–10 bp, small molecule metabolites are limited in contact surface. They cannot effect large enough changes in the RNA energy landscape to facilitate the melting of stable helices. Typically, metabolites influence the RNA fold cotranscriptionally—before and during the generation of stable secondary structure—to guide a particular folding pathway. RNA structures that respond to the presence of metabolite concentrations include the growing class of riboswitches, genetic elements embedded in the 5' untranslated region of some prokaryotic messenger RNAs (mRNA) that control gene expression by regulating transcription elongation or translation initiation. The riboswitch architecture includes a 5' aptamer domain, which binds a target metabolite with high affinity, followed by an expression platform that responds allosterically to the conformational change of the aptamer. The expression platform may pair to sequester a downstream element that positively or negatively affects expression. For example, an expression platform may adopt a fold that precludes the formation of an intrinsic terminator stem, allowing transcription to proceed unchecked. In other cases, expression platforms are known to pair with the Shine-Delgarno sequence to inhibit translation initiation. This is the case for the S-adenosyl-methionine (SAM)–II riboswitch, which blocks translation of the *metX* gene to regulate sulfur metabolism upon SAM binding to its aptamer domain. Resistant to crystallization efforts, the free SAM-II riboswitch was studied by NMR titration studies and SAXS [148]. A free state dynamically sampling multiple conformations was concluded, with varied levels of compaction induced upon Mg^{2+} and SAM ligand binding.

RNA conformational switches do not necessarily rely on alternate secondary structures and may instead be facilitated by tertiary interactions, often in the context of ligand binding. The binding of a cognate protein, nucleic acid, or metabolite can alter local motifs or elicit large orientational changes in helical domains.

12.3.3.1 NMR measurements for RNA dynamics

One of the powerful advantages of NMR as a structure determination technique is its ability to detect dynamics. This makes NMR highly amenable to studying equilibrium-state RNA dynamics, as well as the structural and dynamic response of RNAs upon binding a cofactor [149–151]. Qualitatively, slow conformational dynamics in an RNA molecule can be appreciated from peak line widths and intensities. For example, base-pair fraying in a pseudoknot of the mouse mammory tumor virus was identified based on an analysis of imino proton exchange (as monitored by line widths) and heterogeneous H5-H6 cross-peak intensities in COSY and NOESY spectra [152]. Dynamics due to base-pair instability can also be interpreted from base- or temperature-catalyzed imino exchange rates [153,154]. The most quantitative analyses of RNA dynamics, spanning picosecond to millisecond motions, are taken from measurements of NMR relaxation parameters and residual dipolar couplings.

12.3.3.2 Dynamics probed by relaxation parameters

Heteronuclear spin relaxation parameters (T_1, T_2, heteronuclear NOE) [155] are sensitive to the fast librational motions that allow an RNA to sample similar conformations separated by low-energy barriers. The slower (microsecond to millisecond) motions that facilitate events such as base-pair opening, base flipping, sugar pucker flips, or large-scale domain reorientation are more rigorously analyzed by the power dependence of $T_{1\rho}$, where $T_{1\rho}$ is the rotating-frame transverse relaxation time constant, maintained by a spin-lock field that continuously refocuses transverse magnetization. Measuring this parameter at varying spin-lock strength gives information on the chemical exchange lifetimes between interconverting states. The utility of this method was first shown for the lead ribozyme, where the power dependence analysis of ^{13}C $T_{1\rho}$ at an adenine C2 position revealed the motions that mediate conversion between the catalytically active and inactive forms [156].

Intermediate timescale motions that are not adequately captured by solution NMR relaxation can be probed by solid-state deuterium NMR, where nanosecond to microsecond motions modulate the quadrupolar line shape. In conjunction with explicit motional models, deuterium line-shape measurements have been employed to characterize the free and bound (to HIV-1 Tat derived peptides) states of the transactivation response element (TAR) RNA [157].

12.3.3.3 Dynamics probed by residual dipolar couplings

The dynamics information of RDCs (picosecond to millisecond) is encoded in the alignment tensor magnitude [158]. For a fully rigid molecule, the constituents of the alignment tensor – the generalized degree of order S_{RDC} and the asymmetry parameter η – will be the same across all domains. Discrepancies among domains indicate that the domains orient independently of one another in the media. The amplitude of motion can be analyzed but requires some knowledge of an appropriate motional model to apply (e.g. hinge, cone, etc.). RDCs were used to refine the structure and characterize the dynamics of the wild-type human telomerase (TER) pseudoknot, in comparison to its catalytic-deficient mutant U177. While the structures between wild-type and mutant show little variation, changes in base-pair stability in the region of the bulge U177 implicate the role of dynamics for catalysis [159].

Both NMR relaxation parameters and RDCs rely on the anisotropic interactions between a nuclear spin and the magnetic field. The fast local motions of individual bond vectors can ordinarily be decoupled from the slower overall molecular reorientation. However, in flexible RNAs undergoing large amplitude motions, this approximation breaks down because (1) the timescale of conformational dynamics can approach that of the molecular correlation time and (2) the conformational rearrangement can significantly change the shape of the overall molecule, affecting its molecular tumbling. The method of domain elongation has been developed to decouple internal and overall motional contributions to spin relaxation measurements and RDCs [160]. The ligation of a long, stable helix to a terminal helix in an RNA of interest ensures that the correlation time of the molecule will be slow and dominated by the rod-shaped helix. This method has been used to further characterize the correlated twisting and swinging of two stems of the HIV-1 TAR RNA [160].

12.4 Part III: examples of RNA tertiary structure

RNA tertiary structure, or suprahelical structure, refers to the contacts between regular helices, between helices and unpaired strands, or between unpaired regions. Tertiary interactions are driven by the tendency of an RNA molecule to maximize base stacking, similar to the hydrophobic effect in protein folding. (The energetic contribution of stacking to RNA folding is classically displayed in tRNA. While only 41 out of 76 bases are paired in a helical structure, 72 bases in total are involved in stacking interactions [161].) In this section, we describe common tertiary motifs and illustrate their utility in stabilizing 3-D RNA architectures (see also Chapter 15) within the context of NMR-resolved structures.

12.4.1 Helix-helix interactions

12.4.1.1 Coaxial stacking

The most utilized motif for stabilization of a tertiary RNA fold is coaxial stacking, by which two helical domains abut end to end to form a quasi-continuous helix with uninterrupted π-stacking along one or both strands. Coaxial stacking is often observed in conjunction with a divalent metal-binding site, particularly at higher-order junctions of helices. In the NMR and SAXS-refined structure of the yeast U2/U6 small nuclear RNA complex, a three-way junction of helices displays coaxial stacking between two of the helices and a Mg^{2+}-binding site. The continuous stacking observed may be responsible for the helical twist of this "Y"-shaped complex that aligns elements required for spliceosome activity along a single face [142].

12.4.1.2 A-platform and A-C platform

An adenosine platform refers to the noncanonical pairing (Hoogsteen/sugar edge) of two sequential adenosines. The analogous A-C platform is composed of adenosine followed by cytidine, forming an A-C pair. In both cases, the approximately planar platform stacks upon a G-U wobble or Hoogsteen A-U pair immediately below it. The opposite face of the platform ordinarily stacks with another helix to mediate a long-range interaction. The platform motif may be more general in large RNAs. The pseudopairing of sequential U-A bases near the center of stem-loop 2 of the HIV-1 psi-site forms an A-U-A "base triple platform," stabilized by a hydrogen-bonding network. This NMR study was limited to stem-loop 2, however, and could not identify long-range stacking interactions that may be made to other domains of the psi-site RNA [162].

12.4.2 Helix-strand interactions

12.4.2.1 Base triples and A-minor motifs

Base triples, either in isolation or in succession to form a short triplex, involve the noncanonical pairing of a loop that lies typically in the minor groove of a duplex. Most often, the third nucleotide is adenosine ("A-minor" interaction), owing to the complementarity of its sugar edge to the shape of the minor groove. Examples of major groove triples also exist, however, and commonly involve the Hoogsteen edge of a purine already involved in a Watson-Crick pair of the duplex. An analysis of the large ribosomal subunit crystal structure revealed that A-minor interactions are more common than long-range Watson-Crick pairings [163], and they may represent a universal mode of RNA packing. Other classified structural motifs, such as pseudoknots, frequently make extensive use of base triple interactions. In the Sugarcane Yellow Leaf Virus (ScYLV) pseudoknot, four consecutive base triples (three of them with adenosine as the third base) position loop 2 (L2) of the pseudoknot in the minor groove of stem 1 (S1). Those four stacked L2-S1 minor groove base triples are further stacked on the L1-S2 C8$^+$·(G12-C28) major groove base triple (Fig. 12.6). The hydrogen bonding and stacking of the terminal nucleotide of the triplex was determined to modulate the programmed frameshift efficiency of this mRNA [164,165].

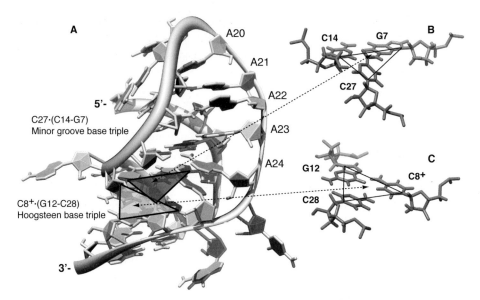

Fig. 12.6 Stacked Hoogsteen and minor groove base triples stabilize the ScYLV pseudoknot. (A) Global architecture of the ScYLV mRNA pseudoknot (PDB identifier 1YG4) determined by NMR. The helical junction region reveal stacked C27·(C14-G7) minor groove (dark gray triangle) and C8+·(G12-C28) Hoogsteen (light gray triangle) base triples. Nearly continuous stacking of A20 through A24 can also be observed. (B) C27·(C14-G7) minor groove loop 2–stem 1 base triple displayed in the same orientation as in (A). (C) C8+·(G12-C28) Hoogsteen loop 1–stem 2 base triple displayed in the same orientation as in (A).

12.4.2.2 Tetraloops

Tetraloops are four nucleotide sequences that frequently form the terminal loop of hairpins in ribosomal and other structural RNAs. Phylogenetic analyses identified the prevalence of three tetraloop consensus sequences – UNCG, GNRA, and CUUG [166,167] – that were subsequently shown to be thermodynamically favored over other four nucleotide sequences [168,169]. Contrary to their name, these sequences are not unstructured loops but display highly stable hydrogen bond networks and, in many cases, provide a docking platform for tertiary contacts (the "tetraloop-receptor'" motif). A GAAA tetraloop, for example, docks in the minor groove of its receptor in the group I intron of *Tetrahymena thermophila*, making use of an adenosine platform and a base triple and base quadruple. NOE-restrained structures of the GNRA family of tetraloops have applied an additional electrostatic energy term in calculation refinements to examine the hydrogen-bonding network [170]. The loop is observed to be asymmetric, where the first G stacks on 5' side of the stem, followed by a sharp turn in the phosphate backbone, and the remaining NRA sequence stacks on the 3' side. A sheared G-A base pair (sugar edge/Hoogsteen) between the first and last nucleotides further stabilizes the motif. The chemical shift patterns of tetraloops are distinctive owing to the ring current effects produced by unusual twists angles, and the considerable number of solved tetraloop structures provides a resource in structure determination efforts of other tetraloop-containing targets. An NMR structure of the GAAA tetraloop-receptor complex revealed the spectroscopic fingerprint of the adenosine platform involved in this motif, and confirmed the hydrogen bonds stabilizing this complex [171].

12.4.3 Loop-loop interactions

Tertiary contacts between unpaired regions, or loop-loop interactions, have demonstrably important biological roles. We consider two such interactions here, the kissing loop and variants of the pseudoknot.

12.4.3.1 Kissing loop

The kissing loop motif is formed when single-stranded loops of two hairpins interact to form a third duplex. It is utilized commonly in retroviral replication [172–175]. Due to coaxial stacking, a single composite helix is formed, with continuous stacking from the 3' side of one stem helix through the loop-loop helix to the other stem helix. The retroviral nucleocapsid proteins can chaperone conversion of the noncovalently linked kissing loop to a more thermodynamically stable extended duplex-mediated dimer linkage in a structural rearrangement suggested to be associated with viral particle maturation [147]. The ColE1 plasmid of *Escherichia coli* also uses this motif, formed between its sense and antisense transcripts, to regulate its replication. The solution structure of the complex revealed phosphate clusters – unique phosphodiester bonds that bridge and narrow the major groove – which stabilize the kissing loop helix [176]. A similar kissing loop complex with coaxial stacking is formed by the HIV-2 TAR hairpin and its complement. The NOE-restrained structure reveals alternating helical twists between adjacent base pairs in the loop-loop helix and other helix-distorting features that are likely important for recognition by Rom protein, which stabilizes the complex [177].

12.4.3.2 Pseudoknot

Similar to the kissing loop motif, the simplest pseudoknot, known as the hairpin (H)–type pseudoknot, involves a hairpin loop pairing with a complementary strand elsewhere in the sequence. The resulting helix coaxially stacks with the hairpin stem. Unlike the kissing loop, the pseudoknot can retain long stretches of unpaired bases, termed loops L1 and L2, which are free for further tertiary interactions such base triples that stabilize the structure (see Section 12.4.2.1). Pseudoknots have been implicated in a number of functional roles, often forming essential components of ribozymes, as in the TER [159]. NMR studies have also given structural insight into the pseudoknot's role in programmed frameshifts in the translation of several coronavirus genomes to produce viral fusion proteins [178,179].

12.5 Conclusion

The inherently dynamic nature of RNA in combination with its polyanionic character poses serious obstacles to crystallization for structure determination. NMR is capable of elucidating RNA structure, but also the dynamics integral to its function. Our understanding of RNA dynamics, especially in large biological RNAs, is in its infancy. With continuing methodological developments, NMR promises to detail the biological activities of RNA molecules and complexes of increasing size.

12.6 Acknowledgments

The authors acknowledge past and present members of the Hennig laboratory for helpful discussions and Braden Roth for help in preparing ^{15}N,^{13}C-labeled RNA samples. This work was supported by funding from the National Institutes of Health (AI081640 and RR024442 to M.H. and GM095128 to C.R.M.) and by the National Science Foundation (NSF 0845512 to M.H.).

References

1. Saenger W. Principles of Nucleic Acid Structure. New York: Springer; 1984.
2. Charette M, Gray MW. Pseudouridine in RNA: what, where, how, and why. IUBMB Life. 2000;49:341–51.
3. Kellner S, Burhenne J, Helm M. Detection of RNA modifications. RNA Biol. 2010;7:237–47.
4. Curran JF. Decoding with the A:I wobble pair is inefficient. Nucleic Acids Res. 1995;23:683–8.
5. Puglisi EV, Puglisi JD. Secondary structure of the HIV reverse transcription initiation complex by NMR. J Mol Biol. 2011;410:863–74.
6. Altona C, Sundaralingam M. Conformational analysis of the sugar ring in nucleosides and nucleotides. A new description using the concept of pseudorotation. J Am Chem Soc. 1972;94:8205–12.
7. Duarte CM, Pyle AM. Stepping through an RNA structure: a novel approach to conformational analysis. J Mol Biol. 1998;284:1465–78.
8. Malathi R, Yathindra N. A novel virtual bond scheme to probe ordered and random coil conformations of nucleic acids: configurational stastistics of polynucleotide chains. Curr Sci. 1980;49:5.
9. Olson WK, Flory PJ. Spatial configurations of polynucleotide chains. I. Steric interactions in polyribonucleotides: a virtual bond model. Biopolymers. 1972;11:1–23.
10. Leontis NB, Westhof E. Geometric nomenclature and classification of RNA base pairs. RNA. 2001;7:499–512.

11. Milligan JF, Groebe DR, Witherell GW, Uhlenbeck OC. Oligoribonucleotide synthesis using T7 RNA polymerase and synthetic DNA templates. Nucleic Acids Res. 1987;15:8783–98.
12. Milligan JF, Uhlenbeck OC. Synthesis of small RNAs using T7 RNA polymerase. Methods Enzymol. 1989;180:51–62.
13. Batey RT, Battiste JL, Williamson JR. Preparation of isotopically enriched RNAs for heteronuclear NMR. Methods Enzymol. 1995;261:300–22.
14. Batey RT, Cloutier N, Mao H, Williamson JR. Improved large scale culture of *Methylophilus methylotrophus* for $^{13}C/^{15}N$ labeling and random fractional deuteration of ribonucleotides. Nucleic Acids Res. 1996;24:4836–7.
15. Batey RT, Inada M, Kujawinski E, Puglisi JD, Williamson JR. Preparation of isotopically labeled ribonucleotides for multidimensional NMR spectroscopy of RNA. Nucleic Acids Res. 1992;20:4515–23.
16. Nikonowicz EP, Sirr A, Legault P, Jucker FM, Baer LM, Pardi A. Preparation of 13C and 15N labelled RNAs for heteronuclear multi-dimensional NMR studies. Nucleic Acids Res. 1992;20:4507–13.
17. Scott LG, Tolbert TJ, Williamson JR. Preparation of specifically 2H- and 13C-labeled ribonucleotides. Methods Enzymol. 2000;317:18–38.
18. Tolbert TJ, Williamson JR. Preparation of specifically deuterated RNA for NMR studies using a combination of chemical and enzymatic synthesis. J Am Chem Soc. 1996;118:7929–40.
19. Tolbert TJ, Williamson JR. Preparation of specifically deuterated and 13C-labeled RNA for NMR studies using enzymatic synthesis. J Am Chem Soc. 1997;119:12100–8.
20. Nikonowicz EP, Kalurachchi K, DeJong E. Comparison of H5 and H8 relaxation rates of a 2H/13C/15N labeled RNA oligonucleotide with selective protonation at C5 and C8. FEBS Lett. 1997;415:109–13.
21. Nikonowicz EP, Michnicka M, Kalurachchi K, DeJong E. Preparation and characterization of a uniformly 2 H/ 15 N-labeled RNA oligonucleotide for NMR studies. Nucleic Acids Res. 1997;25:1390–6.
22. Huang X, Yu P, LeProust E, Gao X. An efficient and economic site-specific deuteration strategy for NMR studies of homologous oligonucleotide repeat sequences. Nucleic Acids Res. 1997;25:4758–63.
23. Hennig M, Scott LG, Sperling E, Bermel W, Williamson JR. Synthesis of 5-fluoropyrimidine nucleotides as sensitive NMR probes of RNA structure. J Am Chem Soc. 2007;129:14911–21.
24. Kreutz C, Kahlig H, Konrat R, Micura R. A general approach for the identification of site-specific RNA binders by 19F NMR spectroscopy: proof of concept. Angew Chem Int Ed Engl. 2006;45:3450–3.
25. Puffer B, Kreutz C, Rieder U, Ebert MO, Konrat R, Micura R. 5-Fluoro pyrimidines: labels to probe DNA and RNA secondary structures by 1D 19F NMR spectroscopy. Nucleic Acids Res. 2009;37:7728–40.
26. Scott LG, Geierstanger BH, Williamson JR, Hennig M. Enzymatic synthesis and 19F NMR studies of 2-fluoroadenine-substituted RNA. J Am Chem Soc. 2004;126:11776–7.
27. Kreutz C, Kahlig H, Konrat R, Micura R. Ribose 2'-F labeling: a simple tool for the characterization of RNA secondary structure equilibria by 19F NMR spectroscopy. J Am Chem Soc. 2005;127:11558–9.
28. Kim I, Lukavsky PJ, Puglisi JD. NMR study of 100 kDa HCV IRES RNA using segmental isotope labeling. J Am Chem Soc. 2002;124:9338–9.
29. Nelissen FH, van Gammeren AJ, Tessari M, Girard FC, Heus HA, Wijmenga SS. Multiple segmental and selective isotope labeling of large RNA for NMR structural studies. Nucleic Acids Res. 2008;36:e89.
30. Tzakos AG, Easton LE, Lukavsky PJ. Complementary segmental labeling of large RNAs: economic preparation and simplified NMR spectra for measurement of more RDCs. J Am Chem Soc. 2006;128:13344–5.

31. Moore MJ, Query CC. Joining of RNAs by splinted ligation. Methods Enzymol. 2000;317:109–23.
32. England TE, Uhlenbeck OC. Enzymatic oligoribonucleotide synthesis with T4 RNA ligase. Biochemistry. 1978;17:2069–76.
33. Purtha WE, Coppins RL, Smalley MK, Silverman SK. General deoxyribozyme-catalyzed synthesis of native 3'-5' RNA linkages. J Am Chem Soc. 2005;127:13124–5.
34. Anderson AC, Scaringe SA, Earp BE, Frederick CA. HPLC purification of RNA for crystallography and NMR. RNA. 1996;2:110–7.
35. Wincott F, DiRenzo A, Shaffer C, et al. Synthesis, deprotection, analysis and purification of RNA and ribozymes. Nucleic Acids Res. 1995;23:2677–84.
36. Wyatt JR, Chastain M, Puglisi JD. Synthesis and purification of large amounts of RNA oligonucleotides. Biotechniques. 1991;11:764–9.
37. Shields TP, Mollova E, Ste Marie L, Hansen MR, Pardi A. High-performance liquid chromatography purification of homogenous-length RNA produced by trans cleavage with a hammerhead ribozyme. RNA. 1999;5:1259–67.
38. Batey RT, Kieft JS. Improved native affinity purification of RNA. RNA. 2007;13:1384–9.
39. Kieft JS, Batey RT. A general method for rapid and nondenaturing purification of RNAs. RNA. 2004;10:988–95.
40. Sklenar V, Miyashiro H, Zon G, Miles HT, Bax A. Assignment of the 31P and 1H resonances in oligonucleotides by two-dimensional NMR spectroscopy. FEBS Lett. 1986;208:94–8.
41. Carlomagno T, Hennig M, Williamson JR. A novel PH-cT-COSY methodology for measuring JPH coupling constants in unlabeled nucleic acids. Application to HIV-2 TAR RNA. J Biomol NMR. 2002;22:65–81.
42. Kellogg GW. Proton-detected hetero-TOCSY experiments with application to nucleic acids. J Magn Reson. 1992;98:176–82.
43. Kellogg GW, Szewczak AA, Moore PB. Two-dimensional hetero-TOCSY-NOESY. Correlation of phosphorus-31 resonances with anomeric and aromatic proton resonances in RNA. J Am Chem Soc. 1992;114:2727–8.
44. Lippens G, Dhalluin C, Wieruszeski JM. Use of a water flip-back pulse in the homonuclear NOESY experiment. J Biomol NMR. 1995;5:327–31.
45. Wuthrich K. NMR of Proteins and Nucleic Acids. New York: Wiley; 1986.
46. Hennig M, Fohrer J, Carlomagno T. Assignment and NOE analysis of 2'-hydroxyl protons in RNA: implications for stabilization of RNA A-form duplexes. J Am Chem Soc. 2005;127:2028–9.
47. Fohrer J, Hennig M, Carlomagno T. Influence of the 2'-hydroxyl group conformation on the stability of A-form helices in RNA. J Mol Biol. 2006;356:280–7.
48. Fohrer J, Reinscheid U, Hennig M, Carlomagno T. Calculation of the dependence of homo- and heteronuclear 3J and 2J scalar couplings for the determination of the 2'-hydroxy conformation in RNA. Angew Chem Int Ed Engl. 2006;45:7033–6.
49. Ying J, Bax A. 2'-Hydroxyl proton positions in helical RNA from simultaneously measured heteronuclear scalar couplings and NOEs. J Am Chem Soc. 2006;128:8372–3.
50. Nozinovic S, Gupta P, Furtig B, et al. Determination of the conformation of the 2'OH group in RNA by NMR spectroscopy and DFT calculations. Angew Chem Int Ed Engl. 2011;50:5397–400.
51. Heus HA, Wijmenga SS, van de Ven FJM, Hilbers CW. Sequential backbone assignment in 13C-labeled RNA via through-bond coherence transfer using three-dimensional triple resonance spectroscopy (1H, 13C, 31P) and two-dimensional hetero TOCSY. J Am Chem Soc. 1994;116:4983–4.
52. Marino JP, Schwalbe H, Anklin C, Bermel W, Crothers DM, Griesinger C. Three-dimensional triple-resonance 1H, 13C, 31P experiment: sequential through-bond correlation of ribose protons and intervening phosphorus along the RNA oligonucleotide backbone. J Am Chem Soc. 1994;116:6472–3.

53. Furtig B, Richter C, Wohnert J, Schwalbe H. NMR spectroscopy of RNA. ChemBioChem. 2003;4:936–62.
54. Wijmenga SS, van Buuren BNM. The use of NMR methods for conformational studies of nucleic acids. Prog Nucl Magn Reson Spectrosc. 1998;32:287–387.
55. Latham MP, Brown DJ, McCallum SA, Pardi A. NMR methods for studying the structure and dynamics of RNA. ChemBioChem. 2005;6:1492–505.
56. Pardi A. Multidimensional heteronuclear NMR experiments for structure determination of isotopically labeled RNA. Methods Enzymol. 1995;261:350–80.
57. Scott LG, Hennig M. RNA structure determination by NMR. Methods Mol Biol. 2008;452:29–61.
58. Roth BM, Hennig M. RNA Structure Determination by NMR: Combining Labeling and Pulse Techniques. Washington, DC: IOS Press; 2011.
59. Zidek L, Stefl R, Sklenar V. NMR methodology for the study of nucleic acids. Curr Opin Struct Biol. 2001;11:275–81.
60. Sklenar V, Dieckmann T, Butcher SE, Feigon J. Optimization of triple-resonance HCN experiments for application to larger RNA oligonucleotides. J Magn Reson. 1998;130:119–24.
61. Sklenar V, Peterson RD, Rejante MR, Feigon J. Two- and three-dimensional HCN experiments for correlating base and sugar resonances in 15N,13C-labeled RNA oligonucleotides. J Biomol NMR. 1993;3:721–7.
62. Sklenar V, Rejante MR, Peterson RD, Wang E, Feigon J. Two-dimensional triple-resonance HCNCH experiment for direct correlation of ribose H1' and base H8, H6 protons in 13C,15N-labeled RNA oligonucleotides. J Am Chem Soc. 1993;115:12181–2.
63. Hu W, Kakalis LT, Jiang L, Jiang F, Ye X, Majumdar A. 3D HCCH-COSY-TOCSY experiment for the assignment of ribose and amino acid side chains in 13C labeled RNA and protein. J Biomol NMR. 1998;12:559–64.
64. Nikonowicz EP, Pardi A. An efficient procedure for assignment of the proton, carbon and nitrogen resonances in 13C/15N labeled nucleic acids. J Mol Biol. 1993;232:1141–56.
65. Marino JP, Prestegard JH, Crothers DM. Correlation of adenine H2/H8 resonances in uniformly ^{13}C labeled RNAs by 2D HCCH-TOCSY: a new tool for ^{1}H assignment. J Am Chem Soc. 1994;116:2205–6.
66. Simon B, Zanier K, Sattler M. A TROSY relayed HCCH-COSY experiment for correlating adenine H2/H8 resonances in uniformly 13C-labeled RNA molecules. J Biomol NMR. 2001;20:173–6.
67. Fiala R, Jiang F, Patel DJ. Direct correlation of exchangeable and nonexchangeable protons on purine bases in 13C,15N-labeled RNA using a HCCNH-TOCSY experiment. J Am Chem Soc. 1996;118:689–90.
68. Simorre JP, Zimmermann GR, Mueller L, Pardi A. Triple-resonance experiments for assignment of adenine base resonances in C-13/N-15-labeled RNA. J Am Chem Soc. 1996;118:5316–7.
69. Simorre JP, Zimmermann GR, Pardi A, Farmer BT II, Mueller L. Triple resonance HNCCCH experiments for correlating exchangeable and nonexchangeable cytidine and uridine base protons in RNA. J Biomol NMR. 1995;6:427–32.
70. Wohnert J, Gorlach M, Schwalbe H. Triple resonance experiments for the simultaneous correlation of H6/H5 and exchangeable protons of pyrimidine nucleotides in C-13, N-15-labeled RNA applicable to larger RNA molecules. J Biomol NMR. 2003;26:79–83.
71. Wohnert J, Ramachandran R, Gorlach M, Brown LR. Triple-resonance experiments for correlation of H5 and exchangeable pyrimidine base hydrogens in (13)C,(15)N-labeled RNA. J Magn Reson. 1999;139:430–3.
72. Simorre JP, Zimmermann GR, Mueller L, Pardi A. Correlation of the guanosine exchangeable and nonexchangeable base protons in 13C-/15N-labeled RNA with an HNC-TOCSY-CH experiment. J Biomol NMR. 1996;7:153–6.

73. Fiala R, Czernek J, Sklenar V. Transverse relaxation optimized triple-resonance NMR experiments for nucleic acids. J Biomol NMR. 2000;16:291–302.
74. Riek R, Pervushin K, Fernandez C, Kainosho M, Wuthrich K. [(13)C,(13)C]- and [(13)C,(1)H]-TROSY in a triple resonance experiment for ribose-base and intrabase correlations in nucleic acids. J Am Chem Soc. 2001;123:658–64.
75. Van Melckebeke H, Pardi A, Boisbouvier J, Simorre JP, Brutscher B. Resolution-enhanced base-type-edited HCN experiment for RNA. J Biomol NMR. 2005;32:263–71.
76. Dayie KT. Key labeling technologies to tackle sizeable problems in RNA structural biology. Int J Mol Sci. 2008;9:1214–40.
77. Lu K, Miyazaki Y, Summers MF. Isotope labeling strategies for NMR studies of RNA. J Biomol NMR. 2010;46:113–25.
78. Nikonowicz EP. Preparation and use of 2H-labeled RNA oligonucleotides in nuclear magnetic resonance studies. Methods Enzymol. 2001;338:320–41.
79. Lu K, Heng X, Garyu L, et al. NMR detection of structures in the HIV-1 5'-leader RNA that regulate genome packaging. Science. 2011;334:242–5.
80. Marino JP, Schwalbe H, Griesinger C. J-coupling restraints in RNA structure determination. Acc Chem Res. 1999;32:614–23.
81. Reif B, Hennig M, Griesinger C. Direct measurement of angles between bond vectors in high-resolution NMR. Science. 1997;276:1230–3.
82. Schwalbe H, Carlomagno T, Hennig M, et al. Cross-correlated relaxation for measurement of angles between tensorial interactions. Methods Enzymol. 2001;338:35–81.
83. Rossi P, Harbison GS. Calculation of 13C chemical shifts in RNA nucleosides: structure-13C chemical shift relationships. J Magn Reson. 2001;151:1–8.
84. Ohlenschlager O, Haumann S, Ramachandran R, Gorlach M. Conformational signatures of 13C chemical shifts in RNA ribose. J Biomol NMR. 2008;42:139–42.
85. Fares C, Amata I, Carlomagno T. 13C-detection in RNA bases: revealing structure-chemical shift relationships. J Am Chem Soc. 2007;129:15814–23.
86. Fiala R, Munzarova ML, Sklenar V. Experiments for correlating quaternary carbons in RNA bases. J Biomol NMR. 2004;29:477–90.
87. Precechtelova J, Padrta P, Munzarova ML, Sklenar V. ^{31}P chemical shift tensors for canonical and non-canonical conformations of nucleic acids: a DFT study and NMR implications. J Phys Chem B. 2008;112:3470–8.
88. Duchardt E, Richter C, Reif B, et al. Measurement of 2J(H,C)- and 3J(H,C)-coupling constants by alpha/beta selective HC(C)H-TOCSY. J Biomol NMR. 2001;21:117–26.
89. Schwalbe H, Marino JP, King GC, Wechselberger R, Bermel W, Griesinger C. Determination of a complete set of coupling constants in 13C-labeled oligonucleotides. J Biomol NMR. 1994;4:631–44.
90. Glaser SJ, Schwalbe H, Marino JP, Griesinger C. Directed TOCSY, a method for selection of directed correlations by optimal combinations of isotropic and longitudinal mixing. J Magn Reson B. 1996;112:160–80.
91. Schwalbe H, Marino JP, Glaser SJ, Griesinger C. Measurement of H,H-coupling constants associated with v_1, v_2, and v_3 in uniformly ^{13}C-labeled RNA by HCC-TOCSY-CCH-E.COSY. J Am Chem Soc. 1995;117:7251–2.
92. Hennig M, Munzarova ML, Bermel W, Scott LG, Sklenar V, Williamson JR. Measurement of long-range ^1H-^{19}F scalar coupling constants and their glycosidic torsion dependence in 5-fluoropyrimidine-substituted RNA. J Am Chem Soc. 2006;128:5851–8.
93. Clore GM, Murphy EC, Gronenborn AM, Bax A. Determination of three-bond 1H3'-31P couplings in nucleic acids and protein-nucleic acid complexes by quantitative J correlation spectroscopy. J Magn Reson. 1998;134:164–7.

94. Hoogstraten CG, Pardi A. Measurement of carbon-phosphorus J coupling constants in RNA using spin-echo difference constant-time HCCH-COSY. J Magn Reson. 1998;133:236–40.
95. Hu W, Bouaziz S, Skripkin E, Kettani A. Determination of 3J(H3i, Pi+1) and 3J(H5i/5i, Pi) coupling constants in 13C-labeled nucleic acids using constant-time HMQC. J Magn Reson. 1999;139:181–5.
96. Szyperski T, Fernandez C, Ono A, Wuthrich K, Kainosho M. The 2D [31P] spin-echo-difference constant-time [13C, 1H]-HMQC experiment for simultaneous determination of 3J(H3'P) and 3J(C4'P) in 13C-labeled nucleic acids and their protein complexes. J Magn Reson. 1999;140:491–4.
97. Gotfredsen CH, Meissner A, Duus JO, Sorensen OW. New methods for measuring 1H-31P coupling constants in nucleic acids. Magn Reson Chem. 2000;38:692–5.
98. Richter C, Reif B, Worner K, et al. A new experiment for the measurement of nJ(C,P) coupling constants including 3J(C4'i,Pi) and 3J(C4'i,Pi+1) in oligonucleotides. J Biomol NMR. 1998;12:223–30.
99. O'Neil-Cabello E, Wu Z, Bryce DL, Nikonowicz EP, Bax A. Enhanced spectral resolution in RNA HCP spectra for measurement of (3)J(C2'P) and (3)J(C4'P) couplings and (31)P chemical shift changes upon weak alignment. J Biomol NMR. 2004;30:61–70.
100. Marino JP, Schwalbe H, Glaser SJ, Griesinger C. Determination of γ and stereospecific assignment of H5' protons by measurement of ^{2}J and ^{3}J coupling constants in uniformly ^{13}C labeled RNA. J Am Chem Soc. 1996;118:4388–95.
101. Nozinovic S, Richter C, Rinnenthal J, et al. Quantitative 2D and 3D Gamma-HCP experiments for the determination of the angles alpha and zeta in the phosphodiester backbone of oligonucleotides. J Am Chem Soc. 2010;132:10318–29.
102. Richter C, Reif B, Griesinger C, Schwalbe H. NMR spectroscopic determination of angles α and ζ in RNA from CH-dipolar coupling, P-CSA cross-correlated relaxation. J Am Chem Soc. 2000;122:12728–31.
103. Duchardt E, Richter C, Ohlenschlager O, Gorlach M, Wohnert J, Schwalbe H. Determination of the glycosidic bond angle chi in RNA from cross-correlated relaxation of CH dipolar coupling and N chemical shift anisotropy. J Am Chem Soc. 2004;126:1962–70.
104. Rinnenthal J, Richter C, Ferner J, Duchardt E, Schwalbe H. Quantitative gamma-HCNCH: determination of the glycosidic torsion angle chi in RNA oligonucleotides from the analysis of CH dipolar cross-correlated relaxation by solution NMR spectroscopy. J Biomol NMR. 2007;39:17–29.
105. Boisbouvier J, Brutscher B, Pardi A, Marion D, Simorre JP. NMR determination of sugar puckers in nucleic acids from CSA-dipolar cross-correlated relaxation. J Am Chem Soc. 2000;122:6779–80.
106. Richter C, Griesinger C, Felli I, Cole PT, Varani G, Schwalbe H. Determination of sugar conformation in large RNA oligonucleotides from analysis of dipole-dipole cross correlated relaxation by solution NMR spectroscopy. J Biomol NMR. 1999;15:241–50.
107. Walsh JD, Cabello-Villegas J, Wang YX. Periodicity in residual dipolar couplings and nucleic acid structures. J Am Chem Soc. 2004;126:1938–9.
108. Dingley AJ, Grzesiek S. Direct observation of hydrogen bonds in nucleic acid base pairs by internucleotide (2)J(NN) couplings. J Am Chem Soc. 1998;120:8293–7.
109. Dingley AJ, Nisius L, Cordier F, Grzesiek S. Direct detection of N-H[...]N hydrogen bonds in biomolecules by NMR spectroscopy. Nat Protoc. 2008;3:242–8.
110. Cornish PV, Giedroc DP, Hennig M. Dissecting non-canonical interactions in frameshift-stimulating mRNA pseudoknots. J Biomol NMR. 2006;35:209–23.
111. Hennig M, Williamson JR. Detection of N-H...N hydrogen bonding in RNA via scalar couplings in the absence of observable imino proton resonances. Nucleic Acids Res. 2000;28:1585–93.

112. Wohnert J, Dingley AJ, Stoldt M, Gorlach M, Grzesiek S, Brown LR. Direct identification of NH...N hydrogen bonds in non-canonical base pairs of RNA by NMR spectroscopy. Nucleic Acids Res. 1999;27:3104–10.
113. Noeske J, Richter C, Grundl MA, Nasiri HR, Schwalbe H, Wohnert J. An intermolecular base triple as the basis of ligand specificity and affinity in the guanine- and adenine-sensing riboswitch RNAs. Proc Natl Acad Sci U S A. 2005;102:1372–7.
114. Liu AZ, Majumdar A, Hu WD, Kettani A, Skripkin E, Patel DJ. NMR detection of N-H center dot center dot center dot O=C hydrogen bonds in C-13,N-15-labeled nucleic acids. J Am Chem Soc. 2000;122:3206–10.
115. Cordier F, Nisius L, Dingley AJ, Grzesiek S. Direct detection of N-H[...]O=C hydrogen bonds in biomolecules by NMR spectroscopy. Nat Protoc. 2008;3:235–41.
116. Giedroc DP, Cornish PV, Hennig M. Detection of scalar couplings involving 2'-hydroxyl protons across hydrogen bonds in a frameshifting mRNA pseudoknot. J Am Chem Soc. 2003;125:4676–7.
117. Bax A, Kontaxis G, Tjandra N. Dipolar couplings in macromolecular structure determination. Methods Enzymol. 2001;339:127–74.
118. Zhou H, Vermeulen A, Jucker FM, Pardi A. Incorporating residual dipolar couplings into the NMR solution structure determination of nucleic acids. Biopolymers. 1999;52:168–80.
119. Tjandra N, Bax A. Direct measurement of distances and angles in biomolecules by NMR in a dilute liquid crystalline medium. Science. 1997;278:1111–4.
120. Hansen MR, Hanson P, Pardi A. Filamentous bacteriophage for aligning RNA, DNA, and proteins for measurement of nuclear magnetic resonance dipolar coupling interactions. Methods Enzymol. 2000;317:220–40.
121. Hansen MR, Mueller L, Pardi A. Tunable alignment of macromolecules by filamentous phage yields dipolar coupling interactions. Nat Struct Biol. 1998;5:1065–74.
122. Sass HJ, Musco G, Stahl SJ, Wingfield PT, Grzesiek S. Solution NMR of proteins within polyacrylamide gels: diffusional properties and residual alignment by mechanical stress or embedding of oriented purple membranes. J Biomol NMR. 2000;18:303–9.
123. Ruckert M, Otting G. Alignment of biological macromolecules in novel nonionic liquid crystalline media for NMR experiments. J Am Chem Soc. 2000;122:7793–7.
124. Latham MP, Hanson P, Brown DJ, Pardi A. Comparison of alignment tensors generated for native tRNA(Val) using magnetic fields and liquid crystalline media. J Biomol NMR. 2008;40:83–94.
125. Zhang Q, Throolin R, Pitt SW, Serganov A, Al-Hashimi HM. Probing motions between equivalent RNA domains using magnetic field induced residual dipolar couplings: accounting for correlations between motions and alignment. J Am Chem Soc. 2003;125:10530–1.
126. Ying J, Grishaev A, Latham MP, Pardi A, Bax A. Magnetic field induced residual dipolar couplings of imino groups in nucleic acids from measurements at a single magnetic field. J Biomol NMR. 2007;39:91–6.
127. Tjandra N, Grzesiek S, Bax A. Magnetic field dependence of nitrogen-proton J splittings in N-15-enriched human ubiquitin resulting from relaxation interference and residual dipolar coupling. J Am Chem Soc. 1996;118:6264–72.
128. Yao L, Ying J, Bax A. Improved accuracy of 15N-1H scalar and residual dipolar couplings from gradient-enhanced IPAP-HSQC experiments on protonated proteins. J Biomol NMR. 2009;43:161–70.
129. Boisbouvier J, Bryce DL, O'Neil-Cabello E, Nikonowicz EP, Bax A. Resolution-optimized NMR measurement of (1)D(CH), (1)D(CC) and (2)D(CH) residual dipolar couplings in nucleic acid bases. J Biomol NMR. 2004;30:287–301.
130. Tjandra N, Bax A. Measurement of dipolar contributions to 1JCH splittings from magnetic-field dependence of J modulation in two-dimensional NMR spectra. J Magn Reson. 1997;124:512–5.
131. Andersson P, Weigelt J, Otting G. Spin-state selection filters for the measurement of hetero-nuclear one-bond coupling constants. J Biomol NMR. 1998;12:435–41.

132. Yan J, Corpora T, Pradhan P, Bushweller JH. MQ-HCN-based pulse sequences for the measurement of ^{13}C1'-^{1}H1', ^{13}C1'-^{15}N, ^{1}H1'-^{15}N, ^{13}C1'-^{13}C2', ^{1}H1'-^{13}C2', ^{13}C6/8-^{1}H6/8, ^{13}C6/8-^{15}N, ^{1}H6/8-^{15}N, ^{13}C6-^{13}C5, ^{1}H6-^{13}C5 dipolar couplings in ^{13}C, ^{15}N-labeled DNA (and RNA). J Biomol NMR. 2002;22:9–20.
133. Zidek L, Wu H, Feigon J, Sklenar V. Measurement of small scalar and dipolar couplings in purine and pyrimidine bases. J Biomol NMR. 2001;21:153–60.
134. Miclet E, O'Neil-Cabello E, Nikonowicz EP, Live D, Bax A. 1H-1H dipolar couplings provide a unique probe of RNA backbone structure. J Am Chem Soc. 2003;125:15740–1.
135. Wu Z, Tjandra N, Bax A. Measurement of 1H3'-31P dipolar couplings in a DNA oligonucleotide by constant-time NOESY difference spectroscopy. J Biomol NMR. 2001;19:367–70.
136. Latham MP, Pardi A. Measurement of imino 1H-1H residual dipolar couplings in RNA. J Biomol NMR. 2009;43:121–9.
137. Boisbouvier J, Delaglio F, Bax A. Direct observation of dipolar couplings between distant protons in weakly aligned nucleic acids. Proc Natl Acad Sci U S A. 2003;100:11333–8.
138. Hennig M, Carlomagno T, Williamson JR. Residual dipolar coupling TOCSY for direct through space correlations of base protons and phosphorus nuclei in RNA. J Am Chem Soc. 2001;123:3395–6.
139. Clore GM, Kuszewski J. Improving the accuracy of NMR structures of RNA by means of conformational database potentials of mean force as assessed by complete dipolar coupling cross-validation. J Am Chem Soc. 2003;125:1518–25.
140. Zuo X, Wang J, Foster TR, et al. Global molecular structure and interfaces: refining an RNA:RNA complex structure using solution X-ray scattering data. J Am Chem Soc. 2008;130:3292–3.
141. Grishaev A, Ying J, Canny MD, Pardi A, Bax A. Solution structure of tRNAVal from refinement of homology model against residual dipolar coupling and SAXS data. J Biomol NMR. 2008;42:99–109.
142. Burke JE, Sashital DG, Zuo X, Wang YX, Butcher SE. Structure of the yeast U2/U6 snRNA complex. RNA. 2012;18:673–83.
143. Clore GM, Garrett DS. R-factor, free R, and complete cross-validation for dipolar coupling refinement of NMR structures. J Am Chem Soc. 1999;121:9008–12.
144. Cromsigt J, van Buuren B, Schleucher J, Wijmenga S. Resonance assignment and structure determination for RNA. Methods Enzymol. 2001;338:371–99.
145. Feng YX, Copeland TD, Henderson LE, et al. HIV-1 nucleocapsid protein induces "maturation" of dimeric retroviral RNA in vitro. Proc Natl Acad Sci U S A. 1996;93:7577–81.
146. Abbink TE, Berkhout B. A novel long distance base-pairing interaction in human immunodeficiency virus type 1 RNA occludes the Gag start codon. J Biol Chem. 2003;278:11601–11.
147. Paillart JC, Shehu-Xhilaga M, Marquet R, Mak J. Dimerization of retroviral RNA genomes: an inseparable pair. Nat Rev Microbiol. 2004;2:461–72.
148. Chen B, Zuo X, Wang YX, Dayie TK. Multiple conformations of SAM-II riboswitch detected with SAXS and NMR spectroscopy. Nucleic Acids Res. 2012;40:3117–30.
149. Rinnenthal J, Buck J, Ferner J, Wacker A, Furtig B, Schwalbe H. Mapping the landscape of RNA dynamics with NMR spectroscopy. Acc Chem Res. 2011;44:1292–301.
150. Bothe JR, Nikolova EN, Eichhorn CD, Chugh J, Hansen AL, Al-Hashimi HM. Characterizing RNA dynamics at atomic resolution using solution-state NMR spectroscopy. Nat Methods. 2011;8:919–31.
151. Dethoff EA, Chugh J, Mustoe AM, Al-Hashimi HM. Functional complexity and regulation through RNA dynamics. Nature. 2012;482:322–30.
152. Shen LX, Tinoco I Jr. The structure of an RNA pseudoknot that causes efficient frameshifting in mouse mammary tumor virus. J Mol Biol. 1995;247:963–78.
153. Snoussi K, Leroy JL. Imino proton exchange and base-pair kinetics in RNA duplexes. Biochemistry. 2001;40:8898–904.

154. Rinnenthal J, Klinkert B, Narberhaus F, Schwalbe H. Direct observation of the temperature-induced melting process of the *Salmonella* fourU RNA thermometer at base-pair resolution. Nucleic Acids Res. 2010;38:3834–47.
155. Abragam A. The Principles of Nuclear Magnetism. Oxford: Clarendon Press; 1961.
156. Hoogstraten CG, Wank JR, Pardi A. Active site dynamics in the lead-dependent ribozyme. Biochemistry. 2000;39:9951–8.
157. Olsen GL, Echodu DC, Shajani Z, Bardaro MF Jr, Varani G, Drobny GP. Solid-state deuterium NMR studies reveal micros-ns motions in the HIV-1 transactivation response RNA recognition site. J Am Chem Soc. 2008;130:2896–7.
158. Tolman JR, Ruan K. NMR residual dipolar couplings as probes of biomolecular dynamics. Chem Rev. 2006;106:1720–36.
159. Kim NK, Zhang Q, Zhou J, Theimer CA, Peterson RD, Feigon J. Solution structure and dynamics of the wild-type pseudoknot of human telomerase RNA. J Mol Biol. 2008;384:1249–61.
160. Zhang Q, Al-Hashimi HM. Domain-elongation NMR spectroscopy yields new insights into RNA dynamics and adaptive recognition. RNA. 2009;15:1941–8.
161. Sussman JL, Kim S. Three-dimensional structure of a transfer RNA in two crystal forms. Science. 1976;192:853–8.
162. Amarasinghe GK, De Guzman RN, Turner RB, Summers MF. NMR structure of stem-loop SL2 of the HIV-1 psi RNA packaging signal reveals a novel A-U-A base-triple platform. J Mol Biol. 2000;299:145–56.
163. Nissen P, Ippolito JA, Ban N, Moore PB, Steitz TA. RNA tertiary interactions in the large ribosomal subunit: the A-minor motif. Proc Natl Acad Sci U S A. 2001;98:4899–903.
164. Cornish PV, Hennig M, Giedroc DP. A loop 2 cytidine-stem 1 minor groove interaction as a positive determinant for pseudoknot-stimulated -1 ribosomal frameshifting. Proc Natl Acad Sci U S A. 2005;102:12694–9.
165. Cornish PV, Stammler SN, Giedroc DP. The global structures of a wild-type and poorly functional plant luteoviral mRNA pseudoknot are essentially identical. RNA. 2006;12:1959–69.
166. Gutell RR, Weiser B, Woese CR, Noller HF. Comparative anatomy of 16-S-like ribosomal RNA. Prog Nucleic Acid Res Mol Biol. 1985;32:155–216.
167. Woese CR, Winker S, Gutell RR. Architecture of ribosomal RNA: constraints on the sequence of "tetra-loops." Proc Natl Acad Sci U S A. 1990;87:8467–71.
168. Antao VP, Lai SY, Tinoco I Jr. A thermodynamic study of unusually stable RNA and DNA hairpins. Nucleic Acids Res. 1991;19:5901–5.
169. Antao VP, Tinoco I Jr. Thermodynamic parameters for loop formation in RNA and DNA hairpin tetraloops. Nucleic Acids Res. 1992;20:819–24.
170. Jucker FM, Heus HA, Yip PF, Moors EH, Pardi A. A network of heterogeneous hydrogen bonds in GNRA tetraloops. J Mol Biol. 1996;264:968–80.
171. Davis JH, Tonelli M, Scott LG, Jaeger L, Williamson JR, Butcher SE. RNA helical packing in solution: NMR structure of a 30 kDa GAAA tetraloop-receptor complex. J Mol Biol. 2005;351:371–82.
172. Clever JL, Wong ML, Parslow TG. Requirements for kissing-loop-mediated dimerization of human immunodeficiency virus RNA. J Virol. 1996;70:5902–8.
173. Haddrick M, Lear AL, Cann AJ, Heaphy S. Evidence that a kissing loop structure facilitates genomic RNA dimerisation in HIV-1. J Mol Biol. 1996;259:58–68.
174. Laughrea M, Jette L. A 19-nucleotide sequence upstream of the 5' major splice donor is part of the dimerization domain of human immunodeficiency virus 1 genomic RNA. Biochemistry. 1994;33:13464–74.
175. Skripkin E, Paillart JC, Marquet R, Ehresmann B, Ehresmann C. Identification of the primary site of the human immunodeficiency virus type 1 RNA dimerization in vitro. Proc Natl Acad Sci U S A. 1994;91:4945–9.

176. Lee AJ, Crothers DM. The solution structure of an RNA loop-loop complex: the ColE1 inverted loop sequence. Structure. 1998;6:993–1005.
177. Chang KY, Tinoco I Jr. The structure of an RNA "kissing" hairpin complex of the HIV TAR hairpin loop and its complement. J Mol Biol. 1997;269:52–66.
178. Plant EP, Perez-Alvarado GC, Jacobs JL, Mukhopadhyay B, Hennig M, Dinman JD. A three-stemmed mRNA pseudoknot in the SARS coronavirus frameshift signal. PLoS Biol. 2005;3:e172.
179. Michiels PJ, Versleijen AA, Verlaan PW, Pleij CW, Hilbers CW, Heus HA. Solution structure of the pseudoknot of SRV-1 RNA, involved in ribosomal frameshifting. J Mol Biol. 2001; 310:1109–23.

Adrian R. Ferré-D'Amaré
13 Crystallization of RNA for structure determination by X-ray crystallography

13.1 Introduction

X-ray crystallography is the most powerful technique for elucidating the three-dimensional structure of biological macromolecules. Successful structure determination of an RNA or a ribonucleoprotein (RNP) complex requires, at a minimum, the preparation of well-ordered crystals, diffraction data collection, the experimental or computational solution of the phase problem, model building, crystallographic refinement, and structure validation. The underlying crystallographic theory and methodology is the same for RNAs or RNPs as for proteins and has been covered elsewhere (e.g. [1–3]). In this review, I will touch on specific approaches particularly relevant to RNA and RNP crystallization, with emphasis on recent methodological developments.

13.2 General strategy for crystallization

Typically, the rate-limiting step in the structure determination of RNA by X-ray crystallography is the preparation of well-ordered crystals suitable for diffraction data collection. General strategies and approaches for macromolecular crystallization [4], as well as RNA crystallization [5], have been previously reviewed. Success is achieved through an iterative optimization process that starts with appropriate construct design, followed by synthesis (chemical or enzymatic), purification, crystallization trials, evaluation of resulting crystals (if any), evaluation of postcrystallization treatments, and redesign of constructs as needed for the next cycle of optimization. In general, the most efficient strategy is to carry out this iterative optimization on many related RNA or RNP constructs, in parallel.

The design of RNA constructs suitable for crystallization is based on four criteria. First, the construct must embody the chemical or biological process whose elucidation is the ultimate goal of the structure determination. Thus, a ribozyme construct should be catalytically active, an aptamer or riboswitch construct should bind its cognate ligand specifically and with affinity comparable to that of wild-type, and so forth. Second, the construct can be designed to incorporate features that might facilitate intermolecular crystal contacts that will stabilize crystals or increase their degree of order. Third, to the extent possible, the RNA or RNP should be covalently and conformationally homogeneous. Fourth, the construct can incorporate features that will facilitate structure determination.

From the earliest days of protein X-ray crystallography, it has been recognized that a powerful strategy for obtaining high-quality crystals is to attempt crystallization with orthologous proteins from different sources. Although such proteins will have sequence differences resulting from genetic drift, conservation of function requires that the three-dimensional structure also be conserved. For globular proteins, this means that amino acids that differ between orthologs will be concentrated on the molecular surface, rather than in the core of the protein. Since it is these surface residues that largely determine the solution properties of a protein and that might also participate in forming crystal contacts, different orthologs can be expected to differ in their ability to form useful crystals.

Advances in the sequencing of genomes and bioinformatics has resulted in the characterization of many orthologous RNA families [6]. Like proteins, members of these families differ in functionally and structurally less important regions, which are often at the molecular periphery. In addition to providing crystallization candidates with different properties, the variation within a family of orthologous RNAs provides useful information for crystallization construct design. Watson-Crick covariation is strongly indicative of conserved helical elements, and if putative loops capping these helices are poorly conserved, they might constitute good candidates for creation of artificial variants that differ in loop length and composition, and in having the loop entirely removed and replaced by a duplex terminus, either blunt or with overhangs. In the case of artificial or engineered molecules, in vitro evolution techniques can also provide artificial phylogenies [7] and fitness landscapes [8] on which to base molecular engineering efforts.

13.2.1 Oligonucleotides and duplex termini

The principal source of stability of RNA structures is the stacking of nucleotides and base pairs, which offsets repulsion of the negatively charged backbone [9, 10]. This has three important consequences for crystallization. First, base stacking renders nucleic acid helices much more stable than protein helices. Therefore, it is often possible to engineer RNA and DNA helices (e.g. by changing their lengths or inserting loops at their termini) without drastically destabilizing or altering the structure of the whole molecule. Second, many crystalline RNAs and DNAs make intermolecular contacts, that is, crystal packing interactions, through the stacking of exposed helical termini. This is the basis for the canonical strategy for crystallizing oligonucleotide duplexes and protein-DNA complexes [11, 12], in which the sequence element of interest is embedded into a duplex whose ends are systematically varied (in length and in the nature of overhanging nucleotides, if any). Typically, crystals of such molecules or molecular complexes will contain pseudoinfinite helices that traverse their length, formed by (noncovalent) association of the oligonucleotide duplexes through stacking. It is because of this that overhanging nucleotides are often introduced (and varied) at helical termini. When such duplexes stack, and depending on the composition of the overhangs, these may function as "sticky ends" to further increase the likelihood of crystallization. This strategy has been taken further by creating the binding site for one protein through association of two duplexes with sticky ends, for instance, to facilitate chemical access to longer DNA sequences and combinatorially increase construct numbers [12], as well as to introduce asymmetry in the crystallization construct [13].

A third, usually undesirable, consequence of the stabilization of DNA and RNA duplexes through stacking often arises when simple stem-loop structures are subjected to crystallization. Because the stem portion of a stem-loop structure is, by definition, self-complementary, such molecules may exist in an equilibrium between a unimolecular stem loop and a bimolecular duplex, in which the loop nucleotides are either unpaired and bulged or forming noncanonical base pairs. If stacking of duplex ends drives crystallization, the duplex form will be depleted from the solution in which crystals are growing (i.e. the "mother liquor"), and, by mass action, the majority of stem loops will ultimately be converted into crystalline duplexes. If the structure of the loop is what is of interest, such duplexes will provide little or no useful information. This tendency of stem-loop structures to dimerize can sometimes be overcome by introducing very stable loops or by having a protein bind to the loop. For instance, the pseudouridine synthase TruB was cocrystallized with a stem loop "minihelix" that mimics the T stem loop of transfer RNA [14]. In the crystal, the protein binds to the T-loop portion, presumably stabilizing it. In addition, the exposed duplex corresponding to the distal end of the T-stem packs against the equivalent portion of a symmetry-related molecule in the crystal. Thus, crystallization and crystal quality varied depending on the specific length and termini of the end of the T-stem present in different RNA crystallization candidates. Even protein stabilization of a specific stem loop might not overcome the strong driving force represented by helical stacking. In the case of the archaeal K-loop-binding protein L7Ae, very well-ordered crystals were obtained by subjecting the complex between the protein and RNA stem loops derived from a box H/ACA RNP [15] to crystallization. When the structure was determined, it was found that the RNA had duplexed, and that the internal bulge formed by the noncomplementary nucleotides was such that each half could be recognized by one L7Ae protein. Although this provided high-resolution information on the minimal L7Ae-RNA interface, the structure of a native box H/ACA K-loop was only visualized eventually as part of a much larger RNA-protein complex (reviewed by Hamma and Ferré-D'Amaré [16]).

Some RNA structural motifs interrupt helices and therefore might be amenable to structural characterization by embedding them into a duplex. This was successfully carried out, (e.g. for the loop-E motif, originally characterized as part of 5S RNA). An RNA construct comprising approximately half of 5S RNA produced crystals that diffracted X-rays to only 3.0 Å resolution, which was sufficient to determine global conformation, but not to visualize structural detail. The loop-E motif, present in this larger construct, was then embedded in a short oligonucleotide duplex. A dodecamer produced very high-quality crystals that diffracted X-rays to 1.5 Å resolution. Structure determination unambiguously revealed the nature of the base pairing and stacking characterizing this motif, as well as details of metal ion binding and hydration [17]. In contrast to this success story, attempts to obtain high-quality crystals of the K-turn as part of short oligonucleotides have not thus far succeeded. The K-turn was identified as a motif in studies of the large ribosomal subunit [18]. The Lilley laboratory has taken advantage of the relatively easy crystallization of a riboswitch that has a K-turn in its wild-type form for crystallographic structure characterization K-turn sequence variants [19]. This riboswitch has also been employed to determine the structure of a bacterial K-turn-binding protein by cocrystallization [20].

13.2.2 Loop engineering and RNP formation and topological permutation

Complex RNA architecture results from association of RNA helices and helical stacks that are constrained by their connectivity, the structures of junctions, bound metal ions, and other long-range (tertiary) interactions [21, 22]. As a result, helical ends (capped by various loops) often project away from the core of the folded RNA. Some of these loops provide the opportunity to engineer the RNA to facilitate crystallization.

Many loops participate in the formation of intramolecular interactions. For instance, GNRA tetraloops (N is any nucleotide, and R is either purine) are employed by large RNAs to pack against a helical segment that has a tetraloop receptor sequence [23]. GAAA tetraloops appear to be particularly adept at making these contacts. Thus, if a crystallization target RNA contains a loop capping a helix that does not participate in intramolecular interactions and is functionally dispensable, converting it into a GAAA tetraloop may result in favorable crystal contacts. This strategy can be taken further, by introducing both a GAAA tetraloop and a tetraloop receptor into functionally dispensable and solvent exposed parts of the target RNA [24]. Another common RNA tertiary interaction is the kissing loop motif, in which loop nucleotides that are self-complementary drive association of two loops [25]. A kissing loop motif was employed successfully by the Nagai laboratory, for instance, to crystallize the very large U1 small nuclear RNP (snRNP) complex, where it replaced the binding site (in a terminal loop) for the peripheral U1A protein [26].

As in the case of the U1A protein, the cognate sites for many sequence-specific RNA-binding proteins occur in the terminal loops of stem-loop structures [27]. Thus, as an extension of the "loop engineering" approaches discussed previously, it is possible to introduce a protein-binding site to RNAs of interest, form a complex between the RNA and the protein, and subject this RNP rather than the "naked" RNA to crystallization [28]. Although any RNA-binding protein could in principle be used, the U1A RRM-1 domain has been particularly successful in this regard (reviewed by Ferré-D'Amaré [29]). The usefulness of this protein probably arises from several factors.

First, the U1A protein binds to its cognate site with very high affinity and specificity, and it does so in a variety of solution conditions commonly employed for crystallization. Second, like many RNA-binding proteins, U1A has a net positive charge, and its molecular surface properties contrast with those of RNA. Thus, it can facilitate formation of a crystal by helping RNA molecules pack. Third, it is a relatively small domain, of molecular mass ~11 kDa. Thus, while it can form crystal contacts, it is less likely to dominate the molecular structure of the RNA into which its binding site was grafted than a larger protein. Therefore, the introduction of the U1A RNP "crystallization module" is less likely to distort the structure of the target RNA. Fourth, as discussed subsequently, the U1A protein can be employed to facilitate solving the phase problem.

Recent work from the Piccirilli laboratory has brought a technique from the protein crystallization field to extend the RNP crystallization module approach described previously. Fab (Fragment, antigen-binding) fragments of antibodies raised against proteins have been employed previously to facilitate crystallization of difficult targets. These workers were able to employ phage display techniques to select anti-RNA Fabs starting with artificial libraries of a composition that was skewed to facilitate recognition of nucleic acids [30]. One of the Fabs they produced was found to bind to an RNA stem loop of a particular sequence, independent of structural context. Thus, this Fab and its epitope-containing RNA stem-loop sequence can together be employed as a crystallization module, as demonstrated in the structure determination of the class I RNA ligase ribozyme [31].

In the successful crystallization of many RNAs with complex architecture, functionally dispensable loops are often omitted. There are three common motivations for this. First, omission of a loop produces a new pair of 5' and 3' termini, thereby allowing the RNA to be assembled from two covalently disconnected (and shorter) strands, if the RNA folds correctly when this is done. The shorter strands allow for chemical synthesis (rather than transcription) of either or both of them, which in turn can allow the introduction of site-specific chemical modifications. The latter have been useful, for instance, to trap ribozymes in various states [32], to introduce heavy atom derivatives [33], to covalently link a substrate to a ribozyme [34], and so forth. Second, elimination of a loop results in a new exposed helix terminus (blunt or with overhangs), which might facilitate formation of stacking interactions with helix termini of other molecules in the crystal (see Section 13.2.1). Third, omission of a loop (and thereby creation of new 5' and 3' termini) makes it possible to circularly permute an RNA, if a new loop (for instance, a GAAA tetraloop) is introduced in place of the original termini.

13.2.3 An example of success through construct engineering

How the various strategies described previously to generate molecular diversity can be employed in concert is illustrated by the successful crystallization of the tetracycline aptamer (Fig. 13.1). This is an in vitro selected RNA that binds the antibiotic specifically and with a sub-nanomolar dissociation constant. Biochemical characterization prior to the start of crystallization efforts had delineated the secondary structure of the RNA, which consists of three helices (P1, P2, P3) connected by a junction,

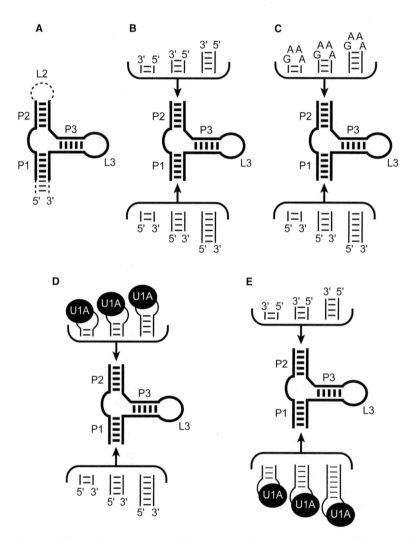

Fig. 13.1 Schematic representation of the construct design strategy employed for crystallization of the tetracycline aptamer. (A) The biochemically characterized minimal construct is composed of three paired regions or helices (P1, P2, P3) that come together at a junction and two loops (L2 and L3). L2, as well as the distal portion of P1, is functionally dispensable. (B) A series of constructs in which L2 has been deleted. (C) A series of constructs in which L2 has been replaced with a GAAA tetraloop. (D) A series of constructs in which L2 has been replaced with a U1A-binding site. (E) A series of constructs in which L2 has been deleted and the distal end of P1 has been closed with a U1A-binding site. This results in circular permutation of the RNA.

and two terminal loops (Fig. 13.1A). Of the latter, L2 is functionally dispensable, but L3 was critical for ligand recognition. Moreover, it was known that the distal portion (close to the 5' and 3' termini) of helix P1 was dispensable for function [35, 36]. Based on this, a first set of constructs was prepared in which L2 was omitted. Therefore, the RNA could be assembled from two synthetic oligonucleotides. A series of oligonucleotides was prepared that differed in the length and composition of the termini. These could be "mixed and matched" to generate a large number of RNA constructs differing in the length and end composition of P1 and P2 (Fig. 13.1B). In a second set of constructs, L2 was replaced with a GAAA tetraloop. Molecular diversity was created by generating a series of constructs in which the GAAA tetraloop was separated from the body of P2 by variable numbers of base pairs. Simultaneously, the length and composition of the distal end of P1 was also varied (Fig. 13.1C). In a third set of constructs, L2 was replaced with a U1A-binding site. A series of constructs was generated that differ in the number of spacer base pairs between the U1A-binding loop and the distal end of P2. Simultaneously, the length and composition of the distal end of P1 was also varied (Fig. 13.1D). Finally, a series of circularly permuted constructs was generated in which L2 was opened, and a U1A-binding site was introduced at the distal end of P1 A series of constructs was generated that differ in the number of spacer base pairs between the U1A-binding loop and the distal end of P1. Simultaneously, the length and composition of the distal end of P2 was also varied (Fig. 13.1E). A construct from this series, in complex with the RNA-binding domain of U1A, produced crystals that ultimately allowed its structure to be solved at 2.2 Å resolution [37].

13.3 Purity and monodispersity

For RNA crystallization, it is particularly important to consider three sources of heterogeneity. First, chemical synthesis and transcription both often yield failure products that differ from the desired crystallization target in the length and composition of terminal nucleotides. Because such variant molecules are otherwise very similar to the authentic crystallization target, they will easily poison crystal growth by formation of defects. Second, RNAs of complex structure need to be properly folded into a unique aggregation state and conformation (i.e. be monodisperse). Third, ribonucleases are ubiquitous and need to be excluded from crystallization samples. This can be particularly problematic when recombinant proteins are employed, either as crystallization targets themselves (as part of an RNP) or as crystallization modules.

Because no chemical step goes to completion during stepwise synthesis, solid-phase synthesized oligonucleotides are contaminated by failure products that are closely related to the desired RNA but differ in being shorter by one or more nucleotides. Run off in vitro transcription with bacteriophage polymerases, in contrast, results in addition of several nontemplated nucleotides at the 3' terminus of the RNA. In addition, transcription from poor promoters can also result in 5' heterogeneity. Synthetic oligonucleotides of moderate length can be purified to homogeneity through various chromatographic and electrophoretic techniques, some of which take advantage of hydrophobic protecting groups. For transcript RNAs, which can often be several hundred nucleotides long, achieving single-nucleotide resolution on a preparative (multi-milligram) scale can be difficult. One commonly employed strategy that reduces 3' heterogeneity is to transcribe RNA from duplex DNA templates (produced through the polymerase chain reaction) that contain two 2'-OMe nucleotides at the 5' end of the nontemplate strand. Transcription under those conditions typically yields only N+1 and N+2 species, and these are much less abundant than when RNA is transcribed from templates lacking the 2'-OMe modifications [38]. Another strategy is to cleave the RNA at a particular location near the 3' end. This can be achieved using ribozymes (either in *cis* or in *trans*) [39, 40], catalytic DNAs [41], end-filling using polymerases [42, 43], or site-directed cleavage with ribonuclease (RNase) H [44], among others. Suitable choice of the cleavage site results in RNA products that differ sufficiently in length so that they can be easily resolved by electrophoresis or chromatography.

In protein crystallization, conformational homogeneity is often examined by dynamic light scattering [45]. Although this and other hydrodynamic methods are useful for examining RNA, a simpler alternative that often is indicative of serious polydisperisty is native gel electrophoresis [46]. In addition, small-angle X-ray scattering (SAXS, see Chapter 9) is particularly powerful for RNA because of the large scattering cross section of the phosphodiester backbone [47, 48]. The key is to employ a method that allows rapid sampling of multiple solution conditions as well as refolding, or ligand-binding conditions, if relevant. For some RNAs, it might be useful to examine the impact of nondenaturing versus denaturing purification [49, 50]. Although nondenaturing purification schemes might result in better folded RNA, there is no warranty that transcription employing phage polymerases results in a folded RNA population similar to that which would result from transcription by the cognate polymerase under native conditions.

Since RNA crystallization typically takes from days to months, often at room temperature or above, it is important that the preparation be scrupulously free of nucleases. This can be a considerable challenge when recombinant proteins are present in the crystallization solution. Even a few molecules of RNase can degrade all the RNA in a crystallization experiment over the course of days. Nonetheless, some examples of serendipitous crystallization due to (limited) nuclease action have been noted in the literature (e.g. [51]).

13.4 Postcrystallization treatments

Successful growth of RNA or RNP crystals of useful size (typically, larger than 50 μm per Cartesian dimension), does not, unfortunately, represent the end point of a crystallization project. The vast majority of RNA crystals prove not sufficiently ordered to provide useful information when examined using bright, well-collimated X-ray sources. In general, resolution of the order of ~3.5 to 2.5 Å (what is colloquially known as "medium resolution") is needed for a crystal structure to yield biochemical insight, and better than 1.8 Å ("high resolution") is desirable if mechanistic information is being sought. Chemical information only becomes available at better than 1.2 Å resolution, which is properly called "atomic" resolution, since individual atoms become resolved only when diffraction data extending that far are available. Many RNA crystals diffract only to resolutions comparable to the level of detail that can be obtained by techniques such as cryoelectron microscopy or SAXS (10–20 Å nominal resolution) and thus do not justify the efforts needed to solve the crystallographic phase problem (see Section 13.5). However, macromolecular crystals are rarely employed for diffraction experiments directly under their growth conditions. Postcrystallization treatment of crystals (reviewed by Heras and Martin [52]) can have dramatic impact on diffraction properties and should be explored before abandoning a crystal form.

The primary event in X-ray damage of macromolecules is the ejection of a photoelectron. Since the latter is a very strong reducing agent, it is the more oxidized functional groups in macromolecules that suffer radiation damage first. Thus, in proteins, damage to disulfide bonds precedes damage to carbonyl groups [53]. Nucleic acids are overall more oxidized than proteins. Therefore, nucleic acid crystals in general are more susceptible to radiation damage than protein crystals. Radiation damage cannot be avoided insofar as X-rays are being used. Its effects can be mitigated, however, by cooling crystals to liquid nitrogen temperature and collecting diffraction data on crystals held near 100 K [54]. Although cryocrystallography has been carried out using crystals held in solutions that remain liquid at liquid nitrogen temperatures [55], the majority of contemporary macromolecular crystallography is performed using crystals that have been flash cooled (typically by plunging into liquid nitrogen), such that they are encased in vitrified aqueous solutions [56]. Given typical crystal sizes, the cooling rates achievable by plunging into liquid nitrogen are not sufficient to yield vitrified water [57]. (The water needs to be vitreous, rather than crystalline, since the latter will diffract strongly. In addition, freezing, rather than vitrification, of water will result in expansion. Most macromolecular crystals contain 40%–70% water by volume, and freezing of this internal water will destroy crystalline order.) Therefore, the composition of the mother liquor of most crystals needs to be adjusted, or the mother liquor exchanged with a "cryoprotecting" solution that will vitrify readily [58]. Many different cryoprotectants have been employed, but the most common ones in RNA crystallography are either low-molecular-weight alcohols (including glycerol and sugars), polyethylene glycol (PEG) or PEG-monomethyl ether (PEG-MME) of moderate average molecular mass (typically less than 2,000 Da), and lithium salts. Typically, ~ 30% (v/v) of a small molecule or PEG, or at least a 1 M concentration of lithium salts [59] is needed to achieve success. Representative examples are listed in Tab. 13.1 for crystals of RNAs and RNPs ranging from a dodecamer duplex to the U1 snRNP. The choice of suitable cryoprotectant is unknown a priori and has to be explored through an additional iterative optimization process for which many crystals may be needed.

Many of the "cryoprotectant" solutions to which crystals are subjected have lower water activity than the mother liquors in which the crystals grew (this is most obviously the case when the concentration of PEG or a lithium salt is increased to achieve a concentration sufficient for efficient vitrification). As a consequence, crystals subjected to cryoprotection may shrink. This often results in irreversible crystal cracking, which precludes data collection. However, shrinkage of crystals can also result in increased crystalline order. Therefore, it is important to examine the consequences of cryoprotection beyond just successful vitrification. A dramatic example from our laboratory is the *glmS* ribozyme-riboswitch. Crystals that had been treated with solutions containing just enough lithium salts and PEG to achieve efficient vitrification when plunged into liquid nitrogen diffracted X-rays to 2.8–3.2 Å resolution. When the concentration of the cryoprotecting agents was increased, cracks appeared on the crystals. When some apparently intact fragments of the crystals that cracked were examined at the synchrotron, it was found that they diffracted X-rays to 1.7 Å resolution. Structure determination showed that the unit cell of the crystals had contracted ~9%. Presumably, this contraction resulted in tighter packing of the RNA within the crystal, giving rise to improved order [51].

13.5 Construct design and structure determination

With well-diffracting crystals in hand, the structure determination progress moves to crystallography. At typical resolutions, macromolecular crystal structures are determined either through experimental phasing techniques such as multiple isomorphous replacement (MIR) and multiwavelength anomalous dispersion (MAD) or computationally by molecular replacement (MR). The theory and practice of these techniques is reviewed elsewhere [1–3].

For the application of MIR or MAD, it is necessary that the crystals contain heavy atoms (i.e. atoms considerably more electron-rich than the hydrogen, carbon, nitrogen, oxygen, phosphorus, and sulfur, "light" atoms of which the bulk of proteins and nucleic acids are composed) or atoms that absorb X-rays in wavelengths that are outside of the range where the light atoms absorb X-rays. These heavy or "anomalous" atoms need to be bound at specific sites, ordered, and bound at high occupancy. Construct engineering can incorporate features that facilitate structure determination. For instance, binding sites for metal hexammines [mimics of hydrated magnesium, such as iridium (III) hexammine and cobalt (III) hexammine] can be introduced into helices [60], synthetic oligonucleotides can contain halogen (Br or I) or selenium atoms [61], and protein components of crystallization modules can contain cysteine

Tab. 13.1 Representative cryoprotectants used for RNA and RNP crystals.

Structure, PDB ID*, reference	Cryoprotectant	Mother liquor components that may have aided in cryoprotection
Loop-E dodecamer, 354D [17]	2-Methyl-2,4-pentanediol, 25%	2-Methyl-2,4-pentanediol, 5%–25%
preQ$_1$ riboswitch, 3FU2 [63]	PEG 200, 50%	None
Flexizyme, 3CUL [64]	LiCl, 1.8 M	PEG 3000, 14.5%–15.0%
HDV ribozyme, 1DRZ [65]	(2R,3R)-(–)-2,3-butanediol, 3%–4%	PEG-MME 2000, 25%
SAM-I riboswitch–YbxF complex, 3V7E [20]	None	PEG 400, 25%
glmS ribozyme, 2NZ4 [66]	PEG 8000, 35%; 1,6-Hexanediol, 1.5 M	PEG 8000, 11%; DMS, 9%
FMN riboswitch, 3F2Q [67]	Glycerol, 25%	PEG 4000, 10%
Group II intron, 3EOG [49]	Ethylene glycol, 33%	PEG 3350, 7.8%
RNase P holoenzyme-tRNA complex, 3OKB [68]	Xylitol, 15%	1.8 M LiSO$_4$
U1 snRNP, 3CW1 [26]	None	2-Methyl-2,4-pentanediol, 38%–42%

*Protein Data Bank structure identifier

residues that can be derivatized with mercury compounds, or their methionine residues can be replaced with selenomethionine.

Success in MR is a function of the availability of an accurate three-dimensional model of the RNA or RNP that has been crystallized. Traditionally, a model that comprised most of the crystalline molecule was thought necessary. Advances in crystallographic theory and computing have resulted in the ability to carry out molecular replacement with partial models (i.e. fragments of known structure of what might otherwise be an unknown structure). This is particularly true for complex RNAs because the structure of their basic building block, the A-form double helix, is well known. In favorable cases, it has been possible to carry out iterative molecular replacement using small duplexes of arbitrary sequence as search models to, ultimately, build and refine a complete crystallographic model for an RNA of previously unknown structure [62]. It should be clear that the presence of crystallization modules of known structure can greatly facilitate phase determination by MR.

13.6 Conclusion

Crystallization and structure determination of RNA remains one of the most challenging arenas in structural biology. This partly stems from the nature of RNA folds, which are stabilized by relatively weak tertiary interactions. This means that RNA is prone to misfold, since many local energy minima that do not differ greatly from the global minimum exist. The resulting heterogeneous population of molecules is difficult to crystallize. In addition, throughout most of its history, structural biology has targeted proteins, and therefore the techniques and strategies that developed are better suited to that class of macromolecules. As we have summarized in this review, strategies specifically aimed at RNA and RNP crystallization have gradually emerged, and as these are applied to new and different targets, it can be expected that the pace of methodological progress will accelerate.

13.7 Acknowledgments

This work was supported by the Intramural Program of the National Heart, Lung and Blood Institute, NIH.

References

1. Drenth J. Principles of Protein X-ray Crystallography. New York: Springer-Verlag; 1994.
2. Blow D. Outline of Crystallography for Biologists. Oxford: Oxford University Press; 2002.
3. Rupp R. Biomolecular Crystallography. New York: Garland Science; 2010.
4. McPherson A. Crystallization of Biological Macromolecules. Cold Spring Harbor: Cold Spring Harbor Laboratory Press; 1999.
5. Ferré-D'Amaré AR, Doudna A. Methods to crystallize RNA. In: Beaucage SL, Bergstrom DE, Glick GD, Jones RA, eds. Current Protocols in Nucleic Acid Chemistry. New York: John Wiley & Sons; 2000:7.6.1–10.
6. Gardner P, Daub J, Tate J, et al. Rfam: updates to the RNA families database. Nucleic Acids Res. 2009;37:D136–40.
7. Ellington AD, Szostak JW. In vitro selection of RNA molecules that bind specific ligands. Nature. 1990;346:818–22.
8. Pitt JN, Ferré-D'Amaré AR. Rapid construction of empirical RNA fitness landscapes. Science. 2010;330:376–9.
9. Saenger W. Principles of Nucleic Acid Structure. New York: Springer-Verlag; 1984.
10. Richert C, Roughton AL, Benner SA. Nonionic analogs of RNA with dimethylene sulfone bridges. J Am Chem Soc. 1996;118:4518–31.
11. Aggarwal AK. Crystallization of DNA binding proteins with oligodeoxynucleotides. Methods. 1990;1:83–90.
12. Schultz SC, Shields GC, Steitz TA. Crystallization of *Escherichia coli* catabolite gene activator protein with its DNA binding site the use of modular DNA. J Mol Biol. 1990;213:159–66.
13. Párraga A, Bellsolell L, Ferré-D'Amaré AR, Burley SK. Co-crystal structure of sterol regulatory element binding protein 1a at 2.3 Å resolution. Structure. 1998;6:661–72.
14. Hoang C, Ferré-D'Amaré AR. Cocrystal structure of a tRNA Ψ55 pseudouridine synthase: nucleotide flipping by an RNA-modifying enzyme. Cell. 2001;107:929–39.
15. Hamma T, Ferré-D'Amaré AR. Structure of protein L7Ae bound to a K-turn derived from an archaeal box H/ACA sRNA at 1.8 Å resolution. Structure. 2004;12:893–903.

16. Hamma T, Ferré-D'Amaré AR. The box H/ACA ribonucleoprotein complex: interplay of RNA and protein structures in post-transcriptional RNA modification. J Biol Chem. 2010;285:805–9.
17. Correll CC, Freeborn B, Moore PB, Steitz TA. Metals, motifs, and recognition in the crystal structure of a 5S rRNA domain. Cell. 1997;91:705–12.
18. Klein DJ., Schmeing TM, Moore PB, Steitz TA. The kink-turn: a new RNA secondary structure motif. EMBO. J 2001;20:4214–21.
19. Schroeder KT, Daldrop P, Lilley DMJ. RNA tertiary interactions in a riboswitch stabilize the structure of a kink turn. Structure. 2011;19:1233–40.
20. Baird NJ, Zhang J, Hamma T, Ferré-D'Amaré AR. YbxF and YlxQ are bacterial homologs of L7Ae and bind K-turns but not K-loops. RNA. 2012;18:759–70.
21. Draper DE. Strategies for RNA folding. Trends Biochem Sci. 1996;21:145–9.
22. Ferré-D'Amaré AR, Doudna JA. RNA folds: insights from recent crystal structures. Annu Rev Biophys Biomol Struct. 1999;28:57–73.
23. Costa M, Michel F. Frequent use of the same tertiary motif by self-folding RNAs. EMBO J. 1995;14:1276–85.
24. Ferré-D'Amaré AR, Zhou K, Doudna JA. A general module for RNA crystallization. J Mol Biol. 1998;279:621–31.
25. Marino JP, Gregorian RS, Csankovski G, Crothers DM. Bent helix formation between RNA hairpins with complementary loops. Science. 1995;268:1448–54.
26. Pomeranz Krummel DA, Oubridge C, Leung AKW, Li J, Nagai K. Crystal structure of human spliceosomal U1 snRNP at 5.5 Å resolution. Nature. 2009;458:475–80.
27. Varani G, Nagai K. RNA recognition by RNP proteins during RNA processing. Annu Rev Biophys Biomol Struct. 1998;27:407–45.
28. Ferré-D'Amaré AR, Doudna JA. Crystallization and structure determination of a hepatitis delta virus ribozyme: use of the RNA-binding protein U1A as a crystallization module. J Mol Biol. 2000;295:541–56.
29. Ferré-D'Amaré AR. Use of the spliceosomal protein U1A to facilitate crystallization and structure determination of complex RNAs. Methods. 2010;52:159–67.
30. Ye JD, Tereshko V, Frederiksen JK, et al. Synthetic antibodies for specific recognition and crystallization of structured RNA. Proc Natl Acad Sci U S A. 2008;105:82–7.
31. Koldobskaya Y, Duguid EM, Shechner DM, et al. A portable RNA sequence whose recognition by a synthetic antibody facilitates structural determination. Nat Struct Mol Biol. 2011;18:100–6.
32. Rupert PB, Massey AP, Sigurdsson ST, Ferré-D'Amaré AR. Transition state stabilization by a catalytic RNA. Science. 2002;298:1421–4.
33. Klein DJ, Ferré-D'Amaré AR. Structural basis of *glmS* ribozyme activation by glucosamine-6-phosphate. Science. 2006;313:1752–6.
34. Serganov A, Keiper S, Malinina L, et al. Structural basis for Diels-Alder ribozyme catalyzed carbon-carbon bond formation. Nature Struct Mol Biol. 2005;12:218–24.
35. Berens C, Thain A, Schroeder R. A tetracycline-binding RNA aptamer. Bioorg Med Chem. 2001;9:2549–56.
36. Müller M, Weigand JE, Weichenrieder O, Suess B. Thermodynamic characterization of an engineered tetracycline-binding riboswitch. Nucleic Acids Res. 2006;34:2607–17.
37. Xiao H, Edwards TE, Ferré-D'Amaré AR. Structural basis for specific, high-affinity tetracycline binding by an *in vitro* evolved aptamer and artificial riboswitch. Chem Biol. 2008;15:1125–37.
38. Kao C, Zheng M, Rüdisser S. A simple and efficient method to reduce nontemplated nucleotide addition at the 3' terminus of RNAs transcribed by T7 RNA polymerase. RNA. 1999;5:1268–72.

39. Price SR, Ito N, Oubridge C, Avis JM, Nagai K. Crystallization of RNA-protein complexes I. Methods for the large-scale preparation of RNA suitable for crystallographic studies. J Mol Biol. 1995;249:398–408.
40. Ferré-D'Amaré AR, Doudna JA. Use of cis- and trans-ribozymes to remove 5' and 3' heterogeneities from milligrams of in vitro transcribed RNA. Nucleic Acids Res. 1996;24:977–8.
41. Santoro SW, Joyce GF. A general purpose RNA-cleaving DNA enzyme. Proc Natl Acad Sci U S A. 1997;94:4262–6.
42. Huang Z, Szostak JW. A simple method for 3'-labeling of RNA. Nucleic Acids Res. 1996;24:4360–1.
43. Ledoux S, Uhlenbeck OC. [3'-32P]-labeling tRNA with nucleotidyltransferase for assaying aminoacylation and peptide bond formation. Methods. 2008;44:74–80.
44. Xu J, Lapham J, Crothers DM. Determining RNA solution structure by segmental isotopic labeling and NMR: application to *Caenorhabditis elegans* spliced leader RNA 1. Proc Natl Acad Sci U S A. 1996;93:44–8.
45. Ferré-D'Amaré AR, Burley SK. Dynamic light scattering in evaluating crystallizability of macromolecules. Methods Enzymol. 1997;276:157–66.
46. Ferré-D'Amaré AR, Doudna JA. Establishing suitability of RNA preparations for crystallization. Determination of polydispersity. Methods Mol Biol. 1997;74:371–8.
47. Lipfert J, Doniach S. Small-angle X-ray scattering from RNA, proteins, and protein complexes. Annu Rev Biophys Biomol Struct. 2007;36:307–27.
48. Baird NJ, Ferré-D'Amaré AR. Idiosyncratically tuned switching behavior of riboswitch aptamer domains revealed by comparative small-angle X-ray scattering analysis. RNA. 2010;16:598–609.
49. Toor N, Rajashankar K, Keating KS, Pyle AM. Structural basis for exon recognition by a group II intron. Nat Struct Mol Biol. 2008;15:1221–2.
50. Batey RT, Kieft JS. Improved native affinity purification of RNA. RNA. 2007;13:1384–9.
51. Klein DJ, Ferré-D'Amaré AR. Crystallization of the *glmS* ribozyme-riboswitch. Methods Mol Biol. 2009;540:129–39.
52. Heras B, Martin JL. Post-crystallization treatments for improving diffraction quality of protein crystals. Acta Crystallogr D. 2005;61:1173–80.
53. Weik M, Ravelli RBG, Kryger G, et al. Specific chemical and structural damage to proteins produced by synchrotron radiation. Proc Natl Acad Sci U S A. 2000;97:623–8.
54. Garman EF, Owen RL. Cryocooling and radiation damage in macromolecular crystallography. Acta Crystallogr D. 2005;62:32–47.
55. Rasmussen BF, Stock AM, Ringe D, Petsko GA. Crystalline ribonuclease A loses function below the dynamical transition at 220K. Nature. 1992;357:423–4.
56. Rodgers DW. Practical cryocrystallography. Methods Enzymol. 1997;276:183–203.
57. Dubochet J, Adrian M, Chang JJ, et al. Cryo-electron microscopy of vitrefied specimens. Q Rev Biophys. 1988;21:129–228.
58. Alcorn T, and Juers D. Progress in rational methods of cryoprotection in macromolecular crystallography. Acta Crystallogr D. 2010;66:366–73.
59. Rubinson KA, Ladner JE, Tordova M, Gilliland GL. Cryosalts: suppression of ice formation in macromolecular crystallography. Acta Crystallogr D. 2000;56:996–1001.
60. Keel A, Rambo RP, Batey RT, Kieft J. A general strategy to solve the phase problem in RNA crystallography. Structure. 2007;15:761–72.
61. Olieric V, Rieder U, Lang K, et al. A fast selenium derivatization strategy for crystallization and phasing of RNA structures. RNA. 2009;15:707–15.
62. Robertson MP, Chi YI, Scott WG. Solving novel RNA structures using only secondary structural fragments. Methods. 2010;52:168–72.

63. Klein D, Edwards T, Ferré-D'Amaré A. Cocrystal structure of a class I preQ$_1$ riboswitch reveals a pseudoknot recognizing an essential hypermodified nucleobase. Nat Struct Mol Biol. 2009;16:343–4.
64. Xiao H, Murakami H, Suga H, Ferré-D'Amaré AR. Structural basis of specific tRNA aminoacylation by a small *in vitro* selected ribozyme. Nature. 2008;454:358–61.
65. Ferré-D'Amaré AR, Zhou K, Doudna JA. Crystal structure of a hepatitis delta virus ribozyme. Nature. 1998;395:567–74.
66. Cochrane JC, Lipchock SV, Strobel SA. Structural investigation of the GlmS ribozyme bound to its catalytic cofactor. Chem Biol. 2007;14:97–105.
67. Serganov A, Huang L, Patel DJ. Coenzyme recognition and gene regulation by a flavin mononucleotide riboswitch. Nature. 2009;458:233–7.
68. Reiter NJ, Osterman A, Torres-Larios A, Swinger KK, Pan T, Mondragón A. Structure of a bacterial ribonuclease P holoenzyme in complex with tRNA. Nature. 2010;468:784–9.

Gerhard Steger and Robert Giegerich
14 RNA structure prediction

The secondary and tertiary folding of RNA as well as the kinetics of RNA folding is of critical importance for RNA's biological function. To help guide biologists in experimental structure determination of RNA, quite a large number of computational programs have been developed in the past 30 years that are able to predict various aspects of RNA structure folding. In the following, we will try to give an overview on which programs might be of use given a certain question.

Structure formation is driven by a gain in energy, and most programs implement such an energy model, which we will summarize in Section 14.1.

Most programs are only able to predict secondary structure(s) of a single RNA molecule. In secondary structure, positions i and j of a base pair are restricted by $j \geq 4 + i$, which gives the minimum size of a hairpin loop, and the order of two base pairs $i:j$ and $k:l$ has to satisfy

$$i < k < l < j \quad \text{or} \quad i < j < k < l \tag{14.1}$$

which allows for nested pairs and neighboring helices and disallows any triple pair $i:k:l$ with $i = k$ and $j \neq l$, while "crossing" interactions leading to pseudoknots or kissing loops in the form

$$i < k < l < j \tag{14.2}$$

are disregarded.

Given the thermodynamic model (see Section 14.1), one might ask for the secondary structure of minimum free energy (MFE structure, for short; see Section 14.2).

The MFE structure is only a single structure out of the huge structural space possible for a certain sequence. Even if the thermodynamic model was perfect and the (in vivo or in vitro) solvent conditions would fit to the model parameters, the MFE structure might be present only as a tiny fraction of all possible structures. To get an impression of all possible structures and their relative concentrations, one should look for calculation of the partition function and its intuitive representation showing the probability of all possible base pairs (see Section 14.3).

One might search through possible suboptimal structures that fulfill experimental criteria; these structures may be thermodynamically less stable (have less negative free energy) than the MFE structure. Again, several computational alternatives are available: one might ask for suboptimal structures in a certain energy range or those differing "somehow" from the MFE structure (see Section 14.3.1), "representative" structures (see Section 14.3.2), or families of structures that share the same "abstract shape" (see Sections 14.1.2 and 14.3.3).

If several, homologous RNA sequences are available, one might ask for a consensus structure common to this RNA family (see Section 14.4). We will summarize a few methods that combine sequence and structure alignment.

Finally, we will give some hints on questions that are beyond secondary structure prediction. One might ask for structures containing pseudoknots or kissing loop interactions (see Section 14.5.1) or for hybridization of two different RNAs, as in the case of micro-RNA (miRNA)/target (see Section 14.5.2). For prediction and modeling of three-dimensional structures, we refer to Chapter 15.

RNA bioinformatics has produced a large body of programs that help with the analysis of RNA structure and is continuing to do so. We cannot strive for completeness here and apologize to all authors whose methods and tools we have not mentioned in this review for lack of space.

14.1 The thermodynamic model of RNA folding

14.1.1 Free energy and partition function

Structure formation of a single-stranded nucleic acid sequence x—from an unfolded, random coil structure C into the folded structure S—is a standard equilibrium reaction with temperature-dependent free energy ΔG_T^0, equilibrium constant K_T, and temperature-independent enthalpy ΔH^0 and entropy ΔS^0:

$$C \rightleftharpoons S$$

$$K_T = \frac{[S]}{[C]} \quad (14.3)$$

$$\Delta G_T^0 = -RT \ln K_T = \Delta H^0 - T \cdot \Delta S^0$$

At the denaturation temperature (melting temperature or midpoint of transition)

$$T_m = \Delta H^0 / \Delta S^0 \quad (14.4)$$

the folded structure S has the same concentration as the unfolded structure ($K_{T_m} = 1; \Delta G_{T_m}^0 = 0$). This is only true if the structure S denatures in an all-or-none transition. In most cases, however, temperature-dependent structural rearrangements and/or partial denaturations take place prior to the complete denaturation.

For hybridization of two different sequences x and y—from unfolded, random coil structures C_x and C_y into the folded structure S—similar equations do hold:

$$C_x + C_y \rightleftharpoons S$$

$$K_T = \frac{[S]}{[C_x] \cdot [C_y]} \quad (14.5)$$

The denaturation temperature

$$T_m = \Delta H^0 / (\Delta S^0 + R \ln c_t / 4) \tag{14.6}$$

is dependent on the total molar strand concentration $c_t = [C_x] + [C_y]$ if $[C_x] = [C_y]$. For more complex reactions, see the study by Schmitz and Tinoco [1].

The number of possible secondary structures of a single sequence (i.e. the folding space $F(x)$ of x) grows exponentially ($\sim 1.8^N$) [2,3] with the sequence length N. The possible structures S_i of a single sequence coexist in solution with concentrations dependent on their free energies $\Delta G_T^0(S_i)$; that is, each structure is present as a fraction p_{S_i} according to its Boltzmann probability

$$p_{S_i} = \exp\left[-\frac{\Delta G_T^0(S_i)}{RT}\right] / Q \tag{14.7}$$

given by its molar Boltzmann weight $\exp(-\Delta G_T^0(S_i)/(RT))$ and the partition function Q for the ensemble of all possible structures

$$Q = \sum_{\text{all structures } S_i} \exp\left[-\frac{\Delta G_T^0(S_i)}{RT}\right] \tag{14.8}$$

The structure S_0 of lowest free energy $\Delta G_T^0(S_0) = \Delta G_{T,\text{MFE}}^0$ is called the optimal structure or structure of MFE. By equation 14.7, the MFE structure is also the structure with highest probability, but this individual probability is typically marginal. Quite different structures with similar or even identical energies might exist for a single sequence. This is of special biological relevance for RNA switches [4]. Thus, one should not assume that an RNA folds into a single, static structure.

The relative concentration of the MFE structure may be very low when there are many structural alternatives (see equation 14.7). Furthermore, the free energy of the ensemble is lower (more negative) than that of the MFE structure due to the entropy caused by the many different structures:

$$\Delta G_{T,\text{ensemble}}^0 = -RT \ln Q$$

$$= \Delta G_T^0(S_0) - RT \ln\left[\sum_{\text{all structures } S_i} \exp\left(-\frac{\Delta G_T^0(S_i) - \Delta G_T^0(S_0)}{RT}\right)\right] \tag{14.9}$$

$$\leq \Delta G_T^0(S_0)$$

14.1.2 Abstract shapes

Many of the possible structures S_i of a sequence x differ from each other by only tiny structural rearrangements like addition or removal of a base pair, or a slight shift in position of a small bulge loop. Thus it is very helpful to pool structures according to their abstract shape. Generally, an abstract shape gives information about the arrangement of structural elements such as helices, but no concrete base pairs [5,6]. The MFE structure within each shape class is called "shrep," which is short for shape representative structure. The partition function Q_p for the ensemble of all structures of shape p is

$$Q_p = \sum_{\text{all structures } S_i \in p} \exp\left(-\frac{\Delta G_T^0(S_i)}{RT}\right) \qquad (14.10)$$

Of course, the structures from all shape classes sum up to the ensemble of all structures:

$$Q = \sum_{\text{all shape classes } p} Q_p \qquad (14.11)$$

and the probability of shape p is

$$\text{Prob}(p) = Q_p / Q \qquad (14.12)$$

The advantage of probabilistic information over pure energy minimization is the following: assume we have an MFE prediction of structure S_0 with energy $\Delta G_{T,\text{MFE}}^0$. As discussed previously, this does not exclude the existence of different structures with energies similar to $\Delta G_{T,\text{MFE}}^0$. However, if the shape class of S_0 also achieves (say) 80% probability in the Boltzmann example, this practically rules out the existence of significantly different, competing structures from a different shape class.

14.1.3 Free-energy computation of an RNA structure

The free energy ΔG_T^0 of a structure S is the sum of the free-energy contributions of all structural elements of S:

$$\Delta G_T^0 = \sum(\Delta H_{\text{stack}}^0 - T \cdot \Delta S_{\text{stack}}^0) + \sum(\Delta H_{\text{loop}}^0 - T \cdot \Delta S_{\text{loop}}^0) \qquad (14.13)$$

Energetic contributions of base pairs are favorable ($\Delta G^0 < 0$) due to stacking of the base pairs, thus forming regular helices. Formation of loops is often, but not necessarily, unfavorable ($\Delta G^0 > 0$); exact values depend on loop type, nucleotides adjacent to the loop-closing base pair(s) as well as on the exact sequence of the loop, and whether the loop nucleotides form a stable, structured motif. Loop types are classified according to the number of loop-closing base pair(s): a single base pair closes hairpin loops, while bulge loops (with no unpaired nucleotides in the other strand) and interior loops (with symmetric or asymmetric numbers of nucleotides in both strands) are closed by two base pairs, and multiloops (bifurcations, junctions) connect more than two helices. Note that for a bulge or interior loop of n nucleotides total length, there are (including the possibilities for the closing base pairs) up to $6*6*4^n$ different sequence combinations with different energetic contributions.

A major set of these parameters has been measured by the group of D. Turner. The parameters implemented by most programs are known as the Turner-1999 set [7]; this parameter set is used for the examples in this review. The Turner-2004 [8] set is slightly refined and is not implemented in programs with all its subtleties. A parameter set based on these thermodynamically measured parameters but computationally enhanced to maximize prediction accuracy is known as the Andronescu set [9,10]. Parameters are known only within certain error limits; because these errors are smallest near $T = 37°C$, mostly $\Delta G^0_{37°C}$ values are reported.

A loop should not be thought of as a floppy structural element; in many cases, loop nucleotides have fixed orientations due to stacking and/or non-Watson-Crick (non-WC) interactions with other loop nucleotides. Famous examples are the loop E of eukaryotic 5S rRNA and the multiloop of transfer RNA (tRNA). The loop E (see Fig. 14.1) is an internal loop of five and four bases in its strands; all nucleotides are involved in non-WC interactions including one triple-strand interaction. In tRNA, the stacking of multiloop-closing base pairs across the multiloop is a major energetic contribution to the stability of the cloverleaf and is critical for formation of tRNA's tertiary structure.

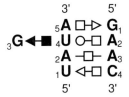

Fig. 14.1 Structure of a loop E or sarcin/ricin motif [86,87]. The loop has five and four bases in its strands and is usually closed by neighboring WC base pairs (not shown). The symbols connecting the bases indicate the interacting edges for each non-WC base pair [88]; circles indicate WC edges, squares Hoogsten edges, and triangles sugar edges; open and closed symbols indicate *trans* and *cis* base pairs, respectively.

14.1.4 Influence of solvent

Compensation of the negatively charged phosphate backbone of nucleic acids by positively charged counterions M^+ leads to stabilization of structural elements according to

$$C + n \cdot M^+ \rightleftharpoons S$$

$$K = \frac{[S]}{[C] \cdot [M^+]^n} \tag{14.14}$$

From this, a logarithmic dependence between denaturation temperature T_m (see equation 14.4) and salt concentration (ionic strength) follows:

$$\frac{dT_m}{d\ln[M^+]} = -n\frac{RT_m^2}{\Delta H^o} \tag{14.15}$$

All thermodynamic parameters for RNA structure formation were determined in 1 M NaCl. In terms of nucleic acid backbone neutralization, this is not far from the ionic conditions in cells except when specific interactions with divalent cations play a role [11,12]. If necessary, however, values for the ionic strength dependence of a structure or a structural element may be found in the literature, including functions accounting for the G : C content of the RNA or depending on various types of buffers (e.g. Tris/borate) and cosolvents like formamide or urea [13–20].

14.2 MFE structure

The Vienna RNA, MFOLD/UNAFOLD, and RNASTRUCTURE packages include, for example, programs for prediction of the MFE (see Section 14.1.1) structure of a single RNA sequence; for availability of these packages, see Tab. 14.1.

A single RNA secondary structure is conveniently represented as a dot-bracket string—such as shown in the third line of Fig. 14.2A—where matched parentheses indicate a base pair and dots indicate unpaired bases. Alternatives to the dot-bracket representation are dot plot (see lower triangle in Fig. 14.2C), sketchy (see Fig. 14.2D), or mountain plot representations (see black line in Fig. 14.2F); these representations are more space consuming but do not add any informative value in comparison to the dot-bracket representation. In a dot plot, base pairs present in the MFE structure are marked by squares. A mountain representation of the MFE structure shows each base pair $i : j$ with $\Delta y = 1$ and $\Delta y = -1$ at positions $x = i$ and $x = j$, respectively, and $\Delta y = 0$ for loops; thus helices are slopes, loops are plateaus, and hairpin loops are peaks.

Tab. 14.1 Tools for prediction of RNA secondary structure.

Package name	Address[a]	Reference
Vienna RNA	C: www.tbi.univie.ac.at/~ivo/RNA/[bc]	[92]
(RNAFOLD)	W: rna.tbi.univie.ac.at/	[93]
MFOLD[d]	C: mfold.bioinfo.rpi.edu/download/[be]	[24]
	W: mfold.bioinfo.rpi.edu/cgi-bin/rna-form1.cgi	[94]
UNAFOLD	C: dinamelt.bioinfo.rpi.edu/download.php[bce]	[95]
	W: dinamelt.bioinfo.rpi.edu/	[84]
RNASTRUCTURE	C: rna.urmc.rochester.edu/RNAstructure.html[bce]	[96]
RNASHAPES	C: bibiserv.techfak.uni-bielefeld.de/rnashapes/[bce]	[97]
	W: bibiserv.techfak.uni-bielefeld.de/rnashapes/	[5]

[a]C, source code and binary are available at given address; W, address of web service. [b]Unix, Linux. [c]Windows. [d]MFOLD is replaced by UNAFOLD. [e]Mac OS X.

Note that the MFE structure is usually present only as a very tiny fraction of all possible structures; in the example of Fig. 14.2, its fraction is unusually high with 12% (see last line of Fig. 14.2A), which points to an evolutionary highly selected sequence. The free energy of the MFE structure is given at the end of the dot-bracket string $\Delta G^0_{37°C,MFE} = -33.5$ kcal/mol $= -140.3$ kJ/mol.

Although all these programs are based on the same thermodynamic model, they implement it with a different degree of detail. This is not by mistake but is motivated by concerns about computational effort. Hence, their predictions do not always agree, and even when the predicted structures are the same, their assigned energies may come out different. In rare cases, the true MFE structure may even be missed. Janssen et al. [21] have analyzed these differences quantitatively.

14.3 Partition folding

The Vienna RNA, UNAFOLD, and RNASTRUCTURE packages allow for prediction of the partition function (see Section 14.1.1) of a single RNA sequence.

RNAFOLD with options "-p -d3 --noLP --MEA -P rna_turner1999.par" was used to produce the output shown in Fig. 14.2A,C,D,F: "-p" forces calculation of the partition function and drawing of the base-pairing probability matrix (dot plot shown in Fig. 14.2C); "-d3" allows for coaxial stacking of adjacent helices in multiloops (only in MFE calculation); "--noLP" tries to avoid lonely base pairs (helices of length 1); "--MEA" forces calculation of a "maximum expected accuracy" structure; and with

A
```
> human_gut_metag_1
GAAACGGAGCGCCACCUCUUUUAACCCUUGAAGUCACUGCCGUUUCGAGAGUUCUCAACUCGAAUAACUAAAGCCAACGUGAACUUUGCGAUCUCCAGGAUCCGCU
(((((((.((((.((((((((.((((((....))))....)))))))))))..(((((.(....))))).((((((......))))))...)))))))))))).  (-33.50)
(((((((.((((.(((((((((((((....,,,.)))))))))))))..........((((......))))...............)))))))))))).       [-34.80]
(((((((.((((.((((((((.((((((....))))....)))))))))))..(((((......))))).((((((......))))))...)))))))))))).  [-33.00 d=8.28]
(((((((.((((.((((((((.((((((....))))....)))))))))))..(((((......))))).((((((......))))))...)))))))))))).  (-33.00 MEA=96.13)
frequency of mfe structure in ensemble 0.120724; ensemble diversity 13.13
```

B
```
> human_gut_metag_1
GAAACGGAGCGCCACCUCUUUUAACCCUUGAAGUCACUGCCGUUUCGAGAGUUCUCAACUCGAAUAACUAAAGCCAACGUGAACUUUUGCGAUCCAGGAUCCGCU
-33.50  (((((((.((((.((((((((.((((((....))))....)))))))))))..(((((......))))).((((((......))))))...))))))))))).  [][][][]
-31.80  (((((((.(((((((((((...((((....))))...))))))))))).((((......)))).(((((......)))))....))))))).              [][][][]
-29.50  (((((((.(((((((((((....))))))))))).((((......)))).(((((......))))).....)))))))                            [][[]][]
-28.70  (((((((.((((.((.((((....))))...)).)))).....((((......))))...((((......))))....))))))).                    [][[[]][]
```

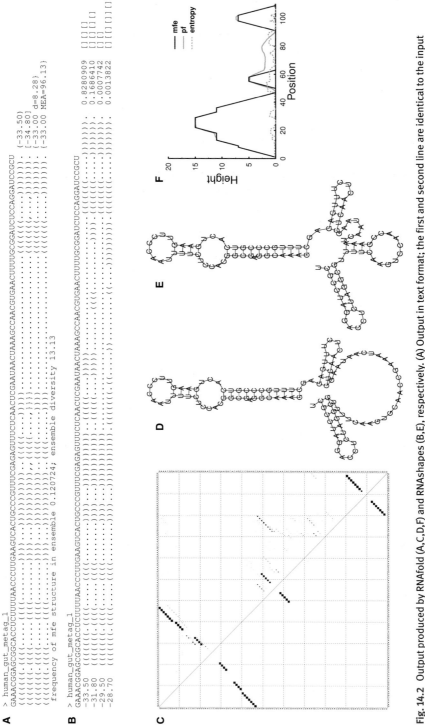

Fig. 14.2 Output produced by RNAfold (A,C,D,F) and RNAshapes (B,E), respectively. (A) Output in text format; the first and second line are identical to the input sequence (*E. coli* OxyS); the following lines are the optimal (MFE) structure in dot-bracket notation, the centroid structure, and the "maximum expected accuracy" structure, respectively; the command to produce this output was "RNAfold -p -d3 --noLP --MEA < seq.vie." (B) Representation of the four best shape classes and their optimal representatives; the command to produce this output was "RNAshapes -P 4 -e 6 -f seq.vie." (C) Dot plot; the part below the diagonal presents the MFE structure, the part above illustrates the partition function; (D) MFE structure; (E) MFE structure of second best shape class; (F) mountain plot representation of MFE structure (black), partition function (gray), and pairing entropy (dotted gray).

"-P," the parameter set is selected. For further options and details, see the main page of RNAFOLD.

A convenient representation of the partition function is a dot plot (see part above the diagonal in Fig. 14.2C) in which base pairs $i : j$ present in any possible structure are marked by squares with area proportional to the probability of pairing ($0 \leq p_{i:j} \leq 1$); usually the MFE structure is presented in the part below the diagonal (with $p_{i:j} = 1$).

In a mountain plot representation of the partition function (see gray line in Fig. 14.2F), the Δy_i values correspond to the pairing probability of base i.

The partition function is also approximated by a dot-bracket string (see fourth line of Fig. 14.2A) in which the characters ". , | { } ( )" denote bases that are essentially unpaired, weakly paired, strongly paired without preference, weakly upstream (downstream) paired, or strongly up-/(down)stream paired bases, respectively.

The free energy of the structure ensemble is usually more negative than that of the MFE structure (see equation 14.9, and compare free-energy values given at the end of lines 3 and 4 in Fig. 14.2A).

At the end of RNAFOLD's last output line (see Fig. 14.2A) is given the ensemble diversity [22]

$$\langle d \rangle = \sum_{i,j} p_{i:j}(1-p_{i:j}) \tag{14.16}$$

which is the average base-pair distance between all structures in the Boltzmann ensemble; $p_{i:j}$ is the probability of forming the base pair $i : j$. $\langle d \rangle$ is low if the sequence has a unique structure.

A further measure of reliability is the entropy of base i given by

$$S = -\sum_{j \neq i} p_{i:j} \log p_{i:j} - (1-p_{i:j}) \log(1-p_{i:j}) \tag{14.17}$$

which is zero if base i is either paired to a unique base or unpaired in all structures (see dotted gray line in Fig. 14.2F).

The advantage of the partition function and its representation as a dot plot is the overview on any pairing alternatives not present in the MFE structure or, in other words, its sound thermodynamic basis on the equilibrium between all possible structures of a sequence (see equations 14.7 and 14.8); a (slight) disadvantage is the lack of concrete, individual structures. This lack should be compensated for by the methods described in Sections 14.3.1 to 14.3.3.

14.3.1 Suboptimal structures

In principle, programs like RNASUBOPT [23] from the Vienna package are able to calculate all suboptimal structures (structures with energies above that of the MFE structure). The drawbacks are that the number of structures increases exponentially with sequence length (see Section 14.1.1) and that many of these structures are quite similar to each other; both make such an approach quite boring. An alternative is to find a restricted set of suboptimal structures that are optimal for each possible base pair $i:j$, which is implemented in RNASUBOPT by option "-z." A further restriction [24] is used by MFOLD: the optimal structures for each possible base pair $i:j$ have to be in a defined energy range below the MFE structure and to differ from each other by a defined number of base pairs. In Fig. 14.2E is shown one of two suboptimal structures with identical $\Delta G^0_{37°C} = -31.8\,\text{kcal/mol} = -113.4\,\text{kJ/mol}$, which differs by an additional helix from the MFE structure shown in Fig. 14.2D.

The concentration ratio of the structures shown in Fig. 14.2D,E is

$$K_{37°C} = \frac{[S_{14.2E}]}{[S_{14.2D}]} = \exp\left(-\frac{\Delta G^0_{37°C,14.2E} - \Delta G^0_{37°C,14.2D}}{RT}\right) \approx 0.063 \tag{14.18}$$

that is, the concentration of the MFE structure in Fig. 14.2D is higher by a factor of about 15.8 than the concentration of the structure in Fig. 14.2E.

14.3.2 Mean and sampled structures

The centroid structure S_c has the minimal average distance d to all other structures [25–27]:

$$\langle d(S_c)\rangle = \sum_{i,j\in S_c}(1-p_{i:j}) + \sum_{i,j\notin S_c} p_{i:j} \tag{14.19}$$

Thus, the centroid is simply the structure containing all pairs with $p_{i:j} > 0.5$. A high similarity between centroid and MFE structure indicates a reliable prediction. The centroid structure is given in the fifth line of RNAFOLD's output as a dot-bracket string followed by the free energy of this structure and its distance (see Fig. 14.2A).

The structure S_{MEA} of maximum expected accuracy (MEA) [28,29] is calculated by

$$A(S_{\text{MEA}}) = \sum_{i,j\in S_{\text{MEA}}} 2\gamma p_{i:j} + \sum_{i\notin S_{\text{MEA}}} p_i^u; \quad p_i^u = 1 - \sum_{i\neq j} p_{i:j} \tag{14.20}$$

That is, the each base pair $i:j$ is scored by weight $0 \leq \gamma \leq 1$ times its pairing probability, and each unpaired base is scored by its probability to be unpaired. For small values of γ, the MEA structure will contain only pairs with very high probability; its default value is $\gamma = 1$. The MEA structure is given in the sixth line of RNAFOLD's output as a dot-bracket string followed by the free energy of this structure and its accuracy (see Fig. 14.2A).

While the two previous single structures might be more trustworthy than the single MFE structure, it might also be of interest to extract "random" structures from the partition function according to their Boltzmann weight. This is performed by SFOLD [26,30] and RNASUBOPT with option "-p"' by stochastic sampling (backtracking).

14.3.3 Shape representative structures and shape probabilities

The program RNASHAPES (see Tab. 14.1) provides shape abstraction functions $\pi_1, \ldots, \pi_5$ that implement different levels of abstraction, with π_5 being the most abstract. Shapes can be represented as strings, similar to structure representations, where a single pair of square brackets marks a helix (of any length), and an underscore marks a stretch of unpaired bases, also of any length. Levels of abstraction differ in the amount of information they retain about unpaired regions. The MFE structure shown in the third line of Fig. 14.2A is mapped to shape strings on abstraction levels 2 and 5 as follows:

$$\begin{aligned} \pi_2: &\quad [\_[\_[]\_]][][] \\ \pi_5: &\quad [][][] \end{aligned} \tag{14.21}$$

Both shapes indicate that the structure consists of three stem loops. The less abstract level 2 shape indicates, in addition to the more abstract level 5, that the left-most stem is interrupted after the first helix by a bulge loop on the 5' side and by an internal loop between the second and third helix, whereas the second and third stem loops do not contain bulges or internal loops. For a detailed definition of shape abstraction levels, see Janssen et al. [31].

RNASHAPES was used to calculate in abstraction level 5, which is the default, the four best shapes (by option -P 4) in an energy range up to 6 kcal/mol (about 25 kJ/mol) above the energy of the MFE structure (by option -e 6); the output is shown in Fig. 14.2B. The first and second lines are identical to the input sequence. Each following line shows the result for a certain shape: free energy of the shrep (MFE structure in this shape class), its structure in dot-bracket format, the probability of the shape class, and the shape string. The shrep of the optimal shape class has to be the "true" MFE structure (as in Fig. 14.2A, third line), but the probability of all structures in this shape class sums up to 82.8%, whereas the second best shape class (with three stem loops as shown in Fig. 14.2E) has a probability of 16.9%.

14.4 Structure prediction and multiple alignment

The accuracy of (MFE) secondary structure prediction for a single RNA sequence is relatively low due to several factors including simplifications in the underlying model (e.g. only nearest-neighbor interactions), uncertainties of the energy parameters (especially with stacking in larger loops and junctions), ignorance of kinetic factors (which are of increasing importance with increasing sequence length), and disregard of energy contributions of tertiary interactions. Mean values of accuracy for predicting the correct base pairs range from as low as 45% up to 100% mostly depending on the tested sequence families [7,32–34]. A formidable improvement in prediction accuracy can be achieved, however, by using the additional information from sufficiently diverged homologous sequences: the secondary and tertiary structure of a noncoding RNA changes more slowly than its sequence during evolution. Mutations in base-paired regions are mainly compensated by further mutations that retain the pairing scheme. Due to the isostericity of all WC pairs (and other groups of non-WC pairs [35,36]), the structure common to homologous RNAs can easily be conserved while their sequences might differ from each other to a large extent.

The common structure for a set of homologous sequences is called the consensus structure. To find it, one would like to perform simultaneously a sequence and structure alignment, which has a prohibitive computational cost of $O(N^{3m})$ for m sequences of length N [37]. Hence, several simplifying and more pragmatic approaches for consensus structure prediction have been developed (see Tab. 14.2) that can be classified as follows [38]:

1. Align the sequences first, then predict the structure common to the aligned sequences [34,39,40].
   For the primary alignment step, pure sequence alignment programs or one of the sequence + structure alignment programs (see class 3) can be used. Dynamic programming (for secondary structure prediction) [34,40] or "maximum weighted matching" (for secondary structure prediction including pseudoknots or base triples) [34,41] might be used in the structure prediction step given the fixed alignment. Several RNA sequence + structure editors are available [34,42–46] that allow a user to refine the initial alignment.
2. Predict structures for all single sequences, then align these structures [47–51].
3. Align and predict structures at the same time, using heuristics, and/or restrict the alignment to two sequences to lower the computing cost of Sankoff's algorithm [52–63].

This separation of approaches should not to be taken too strictly; for example, several of the Sankoff-like approaches restrict first the sequence + structure search space by taking into account a sequence alignment and partition functions for the individual sequences. Other approaches do also exist: for example, RNACAST (RNASHAPES with option -c) predicts an abstract shape common to all sequences [48], where each shape of an RNA molecule comprises a class of similar structures and has a representative MFE structure within the class. That is, RNACAST predicts a consensus structure for the sequence but leaves their alignment to a structure alignment program such as RNAFORESTER [47].

In general, all methods for consensus structure prediction outperform the single-sequence methods, but several prerequisites have to be met:

Tab. 14.2 Tools for prediction of RNA consensus secondary structure.

Name	Address[a]	Reference
CARNAC	C/W: http://bioinfo.lifl.fr/RNA/carnac/carnac.php	[52,98]
CONSTRUCT	C: http://biophys.uni-duesseldorf.de/construct3/	[34]
FOLDALIGNM	C: http://foldalign.ku.dk/software/index.html	[57]
KNetFold	C/W: http://knetfold.abcc.ncifcrf.gov/	[39]
LARA	C: https://www.mi.fu-berlin.de/w/LiSA/Lara	[59]
LOCARNA	C/W: http://www.bioinf.uni-freiburg.de/Software/LocARNA/	[58]
MAFFT-(Q|X)INSi	C/W: http://align.bmr.kyushu-u.ac.jp/mafft/online/server/	[63]
MURLET	W: http://murlet.ncrna.org/murlet/murlet.html	[61]
MXSCARNA	C/W: http://mxscarna.ncrna.org/mxscarna/mxscarna.html	[99]
R-COFFEE	C/W: http://www.tcoffee.org/	[51,100]
RNASAMPLER	C: http://ural.wustl.edu/~xingxu/RNASampler/index.html	[50]
RNASHAPES (RNAcast)	C/W: http://bibiserv.techfak.uni-bielefeld.de/rnashapes/	[48]
RNASTRUCTURE/ DYNALIGN	C: http://rna.urmc.rochester.edu/dynalign.html	[56]
RNASTRUCTURE/ PARTS	C: http://rna.urmc.rochester.edu/	[62]
STEMLOC	C: http://biowiki.org/StemLoc	[54]
STRAL	C/W: http://www.biophys.uni-duesseldorf.de/stral/	[49]
Vienna/RNAALIFOLD	C/W: http://rna.tbi.univie.ac.at/cgi-bin/RNAalifold.cgi	[40,101]
WAR	W: http://genome.ku.dk/resources/war/	[102]

[a]C, source code is available at given address; W, address of web service.

- The performance of most (iterative) programs improves with an increasing number of input sequences and decreasing identities of sequences. Optimal values might be five sequences with an average pairwise sequence identity (APSI) of 55% to 70%.
- Only the structure alignment programs (approach 3 or RNAcast) might give reasonable results for a sequence set with an APSI below 55%, but most of these programs are very demanding in computer resources.
- While even a single compensating base-pair change might hint to a certain structure, a pure statistical analysis (e.g. via information theory [34,64] or other methods [65]) needs more than 10 sequences and still does not reach the accuracy of thermodynamics-based approaches.

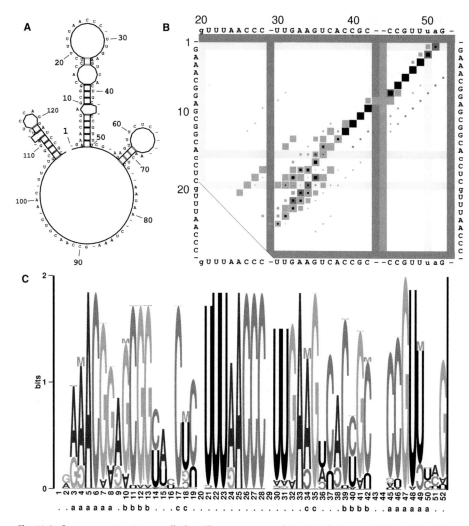

Fig. 14.3 Consensus structure prediction. The sequence and structural alignment is based on 59 nonredundant OxyS sequences (RF00035) from RFam 10.1 [89]. A primary aligned calculated by MAFFT-QINSI [63] was slightly improved using CONSTRUCT [34]. All numberings correspond to the structural alignment (including gaps). (A) Sketch of consensus structure with *E. coli* OxyS sequence as in Fig. 14.2; consensus base-pairing probability is coded from white to dark gray. (B) Section of dot plot encompassing the 5' stem loop; black, dark gray and light gray dots depict probability of consensus base pairings, base pairings of *E. coli* OxyS, and base pairings of all other sequences, respectively. The horizontal and vertical bars depict gaps in grayscale corresponding to number of sequences with this gap; the consensus sequence surrounds the dot plot. (C) Structure logo [66] encompassing the 5' stem loop; the height of base symbols is in proportion to the fraction of their observed frequency; bases that appear less than expected are displayed with upside-down symbols.

Fig. 14.3 summarizes a simple example on consensus structure prediction: the sequences of 59 nonredundant OxyS sequences were aligned by the fast, RNA-structure aware alignment program MAFFT-Q-INSI [63]. The predicted consensus structure (annotated with sequence of an *Escherichia coli* OxyS) is shown in Fig. 14.3A. A (partial) dot plot containing the base-pairing probabilities of all sequences in range of the 5' stem loop, depicted by CONSTRUCT [34], is shown in Fig. 14.3B. Obviously, most sequences share the first two helices (note the high consensus probability of helices at positions 3–8/45–50 and 10–13/39–42), while different helices are probable close to the hairpin loop. Fig. 14.3C shows a "structure logo" [66] of the 5' range, where the information content I_i at column i [67] of the alignment is

$$I_i = \sum_{k \in \{A,U,G,C\}} f_{i,k} \log_2 \frac{f_{i,k}}{p_k}; \quad 0 \leq I_i \leq 2[\text{bit}] \tag{14.22}$$

with the fraction f of base k and the probability $p = 0.25$ to find the base k in a random sequence, and the mutual information H [64] at columns i and j is

$$H_{ij} = \sum_{k \in \{A,U,G,C\}} \sum_{l \in \{A,U,G,C\}} f_{ij,kl} \log_2 \frac{f_{ij,kl}}{p_{i,k} p_{j,l}}; \quad 0 \leq H_{ij} \leq 2[\text{bit}] \tag{14.23}$$

where $f_{ij,kl}$ denotes the fraction of the joint occurrence of base types k and l in columns i and j. That is, the base most common in a column i is depicted largest according to I_i. Columns are marked by an "M" in proportion to the information $H_{ij} - I_i - I_j$ given by the base composition in column i on the composition of another column j (and vice versa). In the example of Fig. 14.3C, many positions show bases with a high information content, which fits to the high average pairwise sequence identity of all sequences (APSI ≈ 88%); consequently, there is only a low number of base pair changes and only low statistical support for the consensus structure (only small "M"s).

14.5 Beyond secondary structure prediction

14.5.1 Pseudoknots

A pseudoknot is an RNA structure characterized by WC base pairing between nucleotides in a loop with complementary residues outside the loop. A classical or H-type (hairpin-type) pseudoknot consists of two helical regions named S1 and S2 (or H1 and H2) and three loop regions L1, L2, and L3 (see Fig. 14.4). In sequence, the serial arrangement of these elements is S1, L1, S2, L2, S1' (complement of S1), L3, and S2' (complement of S2). The crossing order S1 < S2 < S1' < S2' of the interacting bases is allowed as an element of tertiary structure. In many cases, the loop region L2 is absent

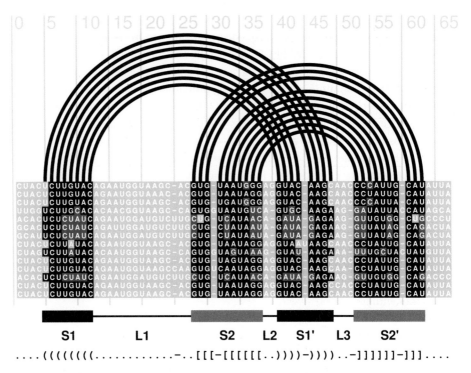

Fig. 14.4 Visualization of an H-type pseudoknot in the 3' untranslated region (UTR) of coronavirus (+) strands. The alignment and structure (both from Rfam-10; RF00165) was depicted by R-CHIE [90]. Bases in paired regions are shown on black background; base pairs showing covariation are on gray background; non-WC pairs in paired regions, bases in loop regions, and gaps in the alignment are on light gray background; base pairs are connected by black arcs. Below the alignment is given the annotation of H-type pseudoknots: two helical regions S1/S1' and S2/S2' are separated by loops L1, L2, and L3.

and the two helices coaxially stack. Such or more complex pseudoknots are tertiary structural motifs that occur widely in RNA.

Drawing pseudoknots as we do with secondary structures leads to crossing arcs; writing them as dot-bracket string s requires different types of brackets to tell the helices apart. In Fig. 14.4, two types of brackets suffice, but for more complex pseudoknots, even more bracket types are required.

In analogy to abstract shape notation, the topology of pseudoknots can be characterized by the nesting pattern of helices, where each helix is indicated by a single pair of brackets. A complete topological classification of pseudoknots has been worked out by Reidys et al. [68]. Subsequently, we show the four topologies that make up the

Abstract shape	Classification [68]	Colloquial name
([ ])	H	Simple H-type
([ ])([ ])	K	Kissing hairpin
([{ }])	L	--
([{ }({ })])	M	--
([[{ }]){ }])	--	HDV ribozyme [112] (PDB entry 1SJ3)

simplest class of pseudoknots. Note that nested between helix parts, any structure of the same type is allowed.

The HDV 1SJ3 pseudoknot shown in the last entry does not fit in the classes H to M. At the time of this writing, it appears to be the topologically most complicated pseudoknot with a resolved structure.

Let us now turn to the prediction of structures potentially containing pseudoknots. None of the tools mentioned in Sections 14.2 to 14.4 are capable of predicting pseudoknots or any form of tertiary interactions, due to the restrictions in their dynamic programming algorithms. In contrast to plain secondary structure prediction, computation of pseudoknots is a largely unsolved problem. This is so for two independent reasons.

The first one is the lack of data: energy parameters for pseudoknot formation have not been determined, and probably never will, as their experimental determination appears infeasible (for further details see the review by Aigner et al. [69]). To estimate the energy parameters from data, we do not have enough experimentally resolved structures that contain pseudoknots of various topologies and sizes.

The second reason is computational complexity. Expanding structure prediction algorithms to general pseudoknot prediction is computationally expensive. Actually, Lyngsø and Pedersen [70] have proved that the general problem of predicting RNA secondary structures containing pseudoknots is non-polynomial time (NP) complete for a large class of reasonable models of pseudoknots. Thus, numerous heuristic approaches were developed—only a subset thereof is listed in Tab. 14.3.

Which program to use in a particular application context is a difficult question. Programs fold different classes of pseudoknots, and since all programs use heuristics in one way or another, even on the simpler classes they barely agree in the same optimal structure. The more general a program, the less suitable it is for use in a large-scale study, while it may still be of use in a specific application, focusing on a small number of related sequences.

Tab. 14.3 Tools for prediction of structures including pseudoknots.

Name	Address[a]	Effort[b]	Reference
DotKnot	W: http://dotknot.csse.uwa.edu.au		[103]
GFold	C: http://www.combinatorics.cn/cbpc/gfold.tar.gz	$\mathcal{O}(N^6)$	[68]
HotKnots	C/W: http://www.cs.ubc.ca/labs/beta/Software/HotKnots		[71]
HPknotter	W: http://bioalgorithm.life.nctu.edu.tw/HPKNOTTER/		[104]
ILM	C/W: http://www.cse.wustl.edu/~zhang/projects/rna/ilm/	$\mathcal{O}(N^3) - \mathcal{O}(N^4)$	[105]
KnetFold	W: http://knetfold.abcc.ncifcrf.gov/	$\mathcal{O}(n^2)$[c]	[39]
DotKnot	C/W: http://dotknot.csse.uwa.edu.au/		[103,106]
NUPACK	C/W: http://nupack.org/	$\mathcal{O}(n^5)$	[107]
pKiss	C/W: http://bibiserv.techfak.uni-bielefeld.de/pkiss/	$\mathcal{O}(N^4)$, $\mathcal{O}(N^5)$[d]	[74]
pknots	C: ftp://selab.janelia.org/pub/software/pknots/	$\geq \mathcal{O}(N^6)$	[108]
pknotsRG	C/W: http://bibiserv.techfak.uni-bielefeld.de/pknotsrg	$\mathcal{O}(N^4)$	[72,73]
ProbKnot	C: http://rna.urmc.rochester.edu/RNAstructure.html	$\mathcal{O}(N^3)$	[75]
PSTAG	C: http://phmmts.dna.bio.keio.ac.jp/pstag/	$\mathcal{O}(on^4 + mn^5)$[e]	[109]
vsfold5	W: http://www.rna.it-chiba.ac.jp/~vsfold/vsfold5	$\mathcal{O}(N^{4.7})$	[110]

[a]W, address of web service; C, code is available at given address. [b]Computing effort; N, length of sequence. [c]n, length of alignment. [d]Depending on the strategy chosen. [e]n, length of unfolded pair of sequences; m, o, nodes on structure tree.

We include a diagram from a recent evaluation of four of the simpler programs, which also demonstrates their lack of agreement:

– HotKnots [71] is a heuristic program that does not restrict the class of pseudoknots and constructs them in a greedy fashion, starting from thermodynamically favorable structures.
– pknotsRG [72,73] is a program that optimally folds structures including the class of canonical simple (H-type) recursive pseudoknots. Canonization means that helices extend maximally toward their inside. This considerably reduces computation

time and space requirements. Each pseudoknot has a canonical representative, but (the representative of) the optimal pseudoknot is found only if it is optimal among the canonical pseudoknots.

- pKiss [74] is a recent extension of PKNOTSRG to kissing hairpin structures. It is interesting to observe how such an extension by just one structural feature affects the predictions.
- ProbKnot [75] is a heuristic program that starts from base-pair probabilities of unknotted structures. Iteratively, the most likely base pairs are added to a structure, allowing for crossing interactions. In the end, short helices are eliminated. This procedure makes the program very fast, but no assertions about (energetic) optimality can be made.

Consider Fig. 14.5. The first column shows that all programs achieve a similar proportion of structures similar to the true structure, with a slight advantage of pKiss. But this does not mean that the programs produce similar results: The other entries in the upper triangle show that all programs do not agree very well in their predictions. The lower triangle shows that all programs perform a bit better on the unknotted structures. Here, they are also more similar to each other, probably because the different classes of pseusdoknots considered make no difference when no pseudoknot is actually predicted.

14.5.2 RNA-RNA hybridization

Many noncoding RNAs exert their biological function via base pairing to their target RNAs; examples are miRNA/messenger (mRNA) interactions leading to inhibition of mRNA translation or mRNA degradation, or small interfering RNAs (siRNAs) inducing RNA interference. In such cases, a correct prediction of the secondary structure of the long target RNA, which would be very error prone, is not necessary: it is sufficient to find potential targets for a given miRNA or to predict an optimal target position for hybridization to the synthetic siRNA.

Several programs have been developed to predict such interactions. Note that none of the programs mentioned in the following (see Tab. 14.4) allows for intramolecular base pairings that lead to a tertiary interaction (e.g. hybridization of an oligonucleotide to a hairpin loop of another RNA) as defined in the introduction (see equations 14.1 and 14.2).

RNAHYBRID [76–78] predicts multiple potential binding sites of miRNAs in large target RNAs by finding the energetically most favorable hybridization sites of a small RNA in a large RNA. The time complexity of the algorithm is linear in the target length, allowing many long targets to be analyzed. To reflect the "seed region" of miRNA interactions, RNAHYBRID allows one to force formation of certain base pairs and to restrict loop sizes.

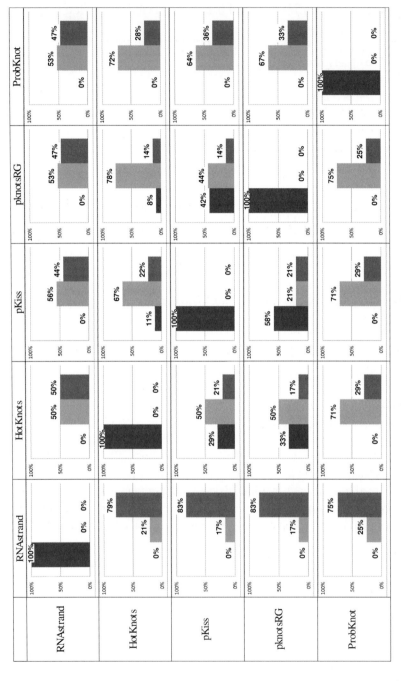

Fig. 14.5 Percentages of identical (dark gray), similar (light gray), or different (gray) predictions, compared between validated structured from the RNA Strand database [91], and predictions from the four programs. The upper triangle uses a set of 277 structures known to contain pseudoknots; the lower triangle uses a data set of the same size known *not* to contain pseudoknots.

Tab. 14.4 Tools for prediction of RNA-RNA interaction.

Package name/ program	Address[a]	Reference
RNAhybrid	C/W: http://bibiserv.techfak.uni-bielefeld.de/rnahybrid/	[76,77]
RNAstructure/ OligoWalk	C/W: http://rna.urmc.rochester.edu/servers/oligowalk	[79]
Vienna/	C: http://www.tbi.univie.ac.at/~ivo/RNA/	
RNAcofold	W: http://rna.tbi.univie.ac.at/cgi-bin/RNAcofold.cgi	[85]
RNAup	W: http://rna.tbi.univie.ac.at/cgi-bin/RNAup.cgi	[111]
RNAxs	C: http://www.bioinf.uni-leipzig.de/~htafer/RNAxs/RNAxs.html	[82]
	W: http://rna.tbi.univie.ac.at/cgi-bin/RNAxs	
UNAfold/Hybrid	C: http://dinamelt.bioinfo.rpi.edu/?q=DINAMelt/software	[95]
	W: http://dinamelt.bioinfo.rpi.edu/?q=DINAMelt/Hybrid2	[84]

[a]C, source code is available at given address; W, address of web service.

OligoWalk [79,80] adds to the algorithm of RNAhybrid partition function calculations of target as well as oligonucleotide: self-structure of the siRNA sequence or the target RNA at the binding site may prevent a favorable interaction.

RNAplfold [81] from the Vienna package (see Tab. 14.1) is able to scan long sequences for short, stable secondary structures. For example, a user can search for structural elements of length L bases in a sequence of total length N with a computing effort proportional to $O(NL^2)$ [in contrast to the usual $O(N^3)$], for example, to find non-coding RNA structures in a genome. We mention it here because RNAplfold and the web service RNAxs, respectively, can be used to find base stretches that have a high probability to be unpaired; for example, this is useful to find optimal binding sites for a designed, synthetic siRNA [82].

DINAMelt ("di-nucleic acid hybridization and melting prediction") [83,84] and similarly RNAcofold [85] predicts the hybridization of two (identical or different) RNA sequences A and B including possible monomer and homodimer interactions:

$$\begin{aligned} A_u &\rightleftharpoons A_f \\ B_u &\rightleftharpoons B_f \\ 2A_u &\rightleftharpoons A \cdot A \\ 2B_u &\rightleftharpoons B \cdot B \\ A_u + B_u &\rightleftharpoons A \cdot B \end{aligned} \quad (14.24)$$

In the example shown in Fig. 14.6, the structure ensemble of two different sequences is predicted. The heterodimer AB, observable only at temperatures below 40°C, contains an asymmetric internal loop (see Fig. 14.6C, top), which leads to only a low stability of the

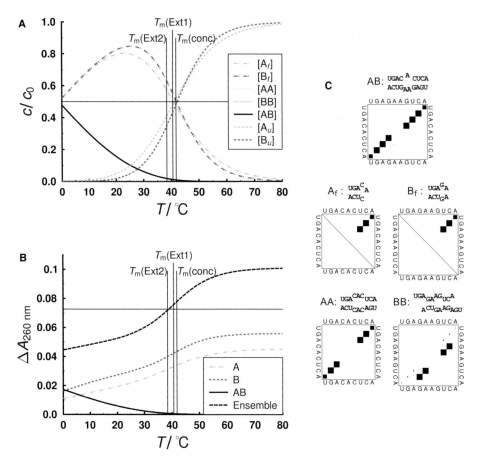

Fig. 14.6 Temperature-dependent hybridization of two different sequences A and B calculated by unafold/DINAMelt [84]. The sequences were A: 5'UGACACUCA 3' and B: 5'UGAGAAGUCA 3' at total concentrations $[A_0] = [B_0] = 1 \times 10^{-5}$ M. (A) Plot of concentrations; the hybrid AB is present only at temperatures below 50°C; the homodimers AA and AB have negligible concentrations; the monomeric structures, A_f and B_f, are present at highest concentrations at about 30°C. (B) Simulated optical melting curve; the experimentally measurable ensemble curve (dotted black) is dominated by the denaturation of the monomeric hairpin structures, while the denaturation of the hybrid AB makes a marginal effect at low temperatures. For a description of the T_m values given in (A) and (B), see text. (C) MFE structures and dot plots of the different structure distributions at 0°C.

hybrid (black curves in Fig. 14.6A,B). The homodimers AA and BB (see Fig. 14.6C, bottom) are only marginally stable, while the monomeric structures A_f and B_f are the dominant structures in the range from 0 to 40°C. Consequently, the main transition, as observable in a UV melting experiment (Fig. 14.6B, see Chapter 1), is the denaturation of these two hairpins at about 40°C. In Fig. 14.6A,B, three versions of melting temperature (T_m) are given: T_m(conc) is defined as the temperature where the concentration of folded molecules is halved; T_m (Ext1) is at the inflection point of the ensemble absorbance curve; and T_m (Ext2) is at halfway between minimum and maximum of the ensemble absorbance curve. Note that values of T_m (conc) and T_m (Ext2) depend on the chosen T_{min} and T_{max} of the calculation. The denaturation of the heterodimer AB is much lower with denaturation temperature (see equation 14.6)

$$T_m = \frac{\Delta H^0}{\Delta S^0 + R \ln \frac{[A_0]+[B_0]}{4}} - 273.15 = 7.1°C \qquad (14.25)$$

and $\Delta H^0 = -286.4$ kJ/mol and $\Delta S^0 = -920.3$ J/(K mol) as calculated by DINAMELT.

14.6 Acknowledgments

The authors thank Sonja Klingberg for performing the evaluation of four pseudoknot folding programs and providing the graphics in Fig. 14.5.

References

1. Schmitz M, Tinoco I Jr. Solution structure and metal-ion binding of the P4 element from bacterial RNase P RNA. RNA. 2000;6:1212–25.
2. Waterman M. Introduction to Computational Biology. Maps, Sequences and Genomes. London: Chapman & Hall; 1995.
3. Nebel M, Scheid A. On quantitative effects of RNA shape abstraction. Theory Biosci. 2009;128:211–25.
4. Nagel J, Pleij C. Self-induced structural switches in RNA. Biochimie. 2002;84:913–23.
5. Giegerich R, Voss B, Rehmsmeier M. Abstract shapes of RNA. Nucleic Acids Res. 2004;32:4843–51.
6. Voss B, Giegerich R, Rehmsmeier M. Complete probabilistic analysis of RNA shapes. BMC Biol. 2006;4:5.
7. Mathews D, Sabina J, Zuker M, Turner D. Expanded sequence dependence of thermodynamic parameters improves prediction of RNA secondary structure. J Mol Biol. 1999;288:911–40.
8. Mathews D, Disney M, Childs J, Schroeder S, Zuker M, Turner D. Incorporating chemical modification constraints into a dynamic programming algorithm for prediction of RNA secondary structure. Proc Nat Acad Sci U S A. 2004;101:7287–92.
9. Andronescu M, Pop C, Condon A. Improved free energy parameters for RNA pseudoknotted secondary structure prediction. RNA. 2009;16:26–42.
10. Andronescu M, Condon A, Hoos H, Mathews D, Murphy K. Computational approaches for RNA energy parameter estimation. RNA. 2010;16:2304–18.
11. Draper D. RNA folding: thermodynamic and molecular descriptions of the roles of ions. Biophys J. 2008;95:5489–95.
12. Ramesh A, Winkler W. Magnesium-sensing riboswitches in bacteria. RNA Biol. 2010;7:77–83.
13. McConaughy B, Laird C, McCarthy B. Nucleic acid reassociation in formamide. Biochemistry. 1969;8:3289–95.

14. Klump H. Thermodynamic values of the helix-coil transition of DNA in the presence of quaternary ammonium salt. Biochim Biophys Acta. 1977;475:605–10.
15. Steger G, Müller H, Riesner D. Helix-coil transitions in double-stranded viral RNA: fine resolution melting and ionic strength dependence., Biochim Biophys Acta. 1980;606:274–84.
16. Michov B. Specifying the equilibrium constants in Tris-borate buffers. Electrophoresis 1986;7:150–1.
17. Riesner D, Steger G. Viroids and viroid-like RNA. In: Saenger W, ed. Nucleic Acids, Subvolume d, Physical Data II, Theoretical Investigations. Landolt-Bornstein, Group VII Biophysics. Vol. 1. Berlin: Springer-Verlag; 1990:194–243.
18. Record M, Lohmann T. A semiempirical extension of polyelectrolyte theory to the treatment of oligonucleotides: application to oligonucleotide helix-coil transitions. Biopolymers. 1978;17:159–66.
19. Sadhu C, Gedamu L. In vitro synthesis of double stranded RNA and measurement of thermal stability: effect of base composition, formamide and ionic strength. Biochem Int. 1987;14:1015–22.
20. Shelton V, Sosnick T, Pan T. Applicability of urea in the thermodynamic analysis of secondary and tertiary RNA folding. Biochemistry. 1999;38:16831–9.
21. Janssen S, Schudoma C, Steger G, Giegerich R. Lost in folding space? Comparing four variants of the thermodynamic model for RNA secondary structure prediction. BMC Bioinformatics. 2011;12:429.
22. Freyhult E, Gardner P, Moulton V. A comparison of RNA folding measures. BMC Bioinformatics. 2005;6:241.
23. Wuchty S, Fontana W, Hofacker I, Schuster P. Complete suboptimal folding of RNA and the stability of secondary structures. Biopolymers. 1999;49:145–65.
24. Zuker M. On finding all suboptimal foldings of an RNA molecule. Science. 1989;244:48–52.
25. Ding Y, Chan C, Lawrence C. RNA secondary structure prediction by centroids in a Boltzmann weighted ensemble. RNA. 2005;11:1157–66.
26. Chan C, Lawrence C, Ding Y. Structure clustering features on the Sfold Web server. Bioinformatics. 2005;21:3926–8.
27. Hamada M, Kiryu H, Sato K, Mituyama T, Asai K. Prediction of RNA secondary structure using generalized centroid estimators. Bioinformatics. 2009;25:465–73.
28. Lu Z, Gloor J, Mathews D. Improved RNA secondary structure prediction by maximizing expected pair accuracy. RNA. 2009;15:1805–13.
29. Hamada M, Sato K, Asai K. Prediction of RNA secondary structure by maximizing pseudo-expected accuracy. BMC Bioinformatics. 2010;11:586.
30. Ding Y, Lawrence C. A statistical sampling algorithm for RNA secondary structure prediction. Nucleic Acids Res. 2003;31:7280–301.
31. Janssen S, Reeder J, Giegerich R. Shape based indexing for faster search of RNA family databases. BMC Bioinformatics. 2008;9:131.
32. Doshi K, Cannone J, Cobaugh C, Gutell R. Evaluation of the suitability of free-energy minimization using nearest-neighbor energy parameters for RNA secondary structure prediction. BMC Bioinformatics. 2004;5:105.
33. Wilm A, Mainz I, Steger G. An enhanced RNA alignment benchmark for sequence alignment programs. Algorithms Mol Biol. 2006;1:19.
34. Wilm A, Linnenbrink K, Steger G. ConStruct: improved construction of RNA consensus structures. BMC Bioinformatics. 2008;9:219.
35. Leontis N, Stombaugh J, Westhof E. The non-Watson-Crick base pairs and their associated isostericity matrices. Nucleic Acids Res. 2002;30:3497–531.
36. Stombaugh J, Zirbel C, Westhof E, Leontis N. Frequency and isostericity of RNA base pairs. Nucleic Acids Res. 2009;37:2294–312.

37. Sankoff D. Simultaneous solution of the RNA folding, alignment and protosequence problems. SIAM J Appl Math. 1985;45:810–25.
38. Gardner P, Giegerich R. A comprehensive comparison of comparative RNA structure prediction approaches. BMC Bioinformatics. 2004;5:140.
39. Bindewald E, Shapiro B. RNA secondary structure prediction from sequence alignments using a network of k-nearest neighbor classifiers. RNA. 2006;12:342–52.
40. Bernhart S, Hofacker I, Will S, Gruber A, Stadler P. RNAalifold: improved consensus structure prediction for RNA alignments. BMC Bioinformatics. 2008;9:474.
41. Tabaska J, Cary R, Gabow H, Stormo G. An RNA folding method capable of identifying pseudoknots and base triples. Bioinformatics. 1998;14:691–9.
42. Griffiths-Jones S. RALEE – RNA ALignment Editor in Emacs. Bioinformatics. 2005;21:257–9.
43. Jossinet F, Westhof E. Sequence to Structure (S2S): display, manipulate and interconnect RNA data from sequence to structure. Bioinformatics. 2005;21:3320–1.
44. Seibel P, Müller T, Dandekar T, Schultz J, Wolf M. 4SALE–a tool for synchronous RNA sequence and secondary structure alignment and editing. BMC Bioinformatics. 2006;7:498.
45. Andersen ES, Lind-Thomsen A, Knudsen B, et al. Semiautomated improvement of RNA alignments. RNA. 2007;13:1850–9.
46. Stombaugh J, Widmann J, McDonald D, Knight R. Boulder ALignment Editor (ALE): a web-based RNA alignment tool. Bioinformatics. 2011;27:1706–7.
47. Höchsmann M, Voss B, Giegerich R. Pure multiple RNA secondary structure alignments: a progressive profile approach. IEEE/ACM Trans Comput Biol Bioinform. 2004;1:53–62.
48. Reeder J, Giegerich R. Consensus shapes: an alternative to the Sankoff algorithm for RNA consensus structure prediction. Bioinformatics. 2005;21:3516–23.
49. Dalli D, Wilm A, Mainz I, Steger G. StrAl: progressive alignment of non-coding RNA using base pairing probability vectors in quadratic time. Bioinformatics. 2006;22:1593–9.
50. Xu X, Ji Y, Stormo G. RNA Sampler: a new sampling based algorithm for common RNA secondary structure prediction and structural alignment. Bioinformatics. 2007;23:1883–91.
51. Moretti S, Wilm A, Higgins D, Xenarios I, Notredame C. R-Coffee: a web server for accurately aligning noncoding RNA sequences. Nucleic Acids Res. 2008;36:W10–13.
52. Perriquet O, Touzet H, Dauchet M. Finding the common structure shared by two homologous RNAs. Bioinformatics. 2003;19:108–16.
53. Hofacker I, Bernhart S, Stadler P. Alignment of RNA base pairing probability matrices. Bioinformatics. 2004;20:2222–7.
54. Holmes I. Accelerated probabilistic inference of RNA structure evolution. BMC Bioinformatics. 2005;6:73.
55. Yao Z, Weinberg Z, Ruzzo W. CMfinder–a covariance model based RNA motif finding algorithm. Bioinformatics. 2005;22:445–52.
56. Harmanci A, Sharma G, Mathews D. Efficient pairwise RNA structure prediction using probabilistic alignment constraints in Dynalign. BMC Bioinformatics. 2007;8:130.
57. Torarinsson E, Havgaard J, Gorodkin J. Multiple structural alignment and clustering of RNA sequences. Bioinformatics. 2007;23:926–32.
58. Will S, Reiche K, Hofacker I, Stadler P, Backofen R. Inferring noncoding RNA families and classes by means of genome-scale structure-based clustering, PLoS Comput Biol. 2007;3:e65.
59. Bauer M, Klau G, Reinert K. Accurate multiple sequence-structure alignment of RNA sequences using combinatorial optimization. BMC Bioinformatics. 2007;8:271.
60. Lindgreen S, Gardner P, Krogh A. MASTR: multiple alignment and structure prediction of non-coding RNAs using simulated annealing. Bioinformatics. 2007;23:3304–11.
61. Kiryu H, Tabei Y, Kin T, Asai K. Murlet: a practical multiple alignment tool for structural RNA sequences. Bioinformatics. 2007;23:1588–98.

62. Harmanci A, Sharma G, Mathews D. PARTS: probabilistic alignment for RNA joinT secondary structure prediction. Nucleic Acids Res. 2008;36:2406–17.
63. Katoh K, Toh H. Improved accuracy of multiple ncRNA alignment by incorporating structural information into a MAFFT-based framework. BMC Bioinformatics. 2008;9:212.
64. Chiu D, Kolodziejczak T. Inferring consensus structure from nucleic acid sequences. Comput Appl Biosci. 1991;7:347–52.
65. Gruber A, Bernhart S, Hofacker I, Washietl S. Strategies for measuring evolutionary conservation of RNA secondary structures. BMC Bioinformatics. 2008;9:122.
66. Gorodkin J, Heyer L, Brunak S, Stormo G. Displaying the information contents of structural RNA alignments: the structure logos. Comput Appl Biosci. 1997;13:583–6.
67. Schneider T, Stephens R. Sequence logos: a new way to display consensus sequences. Nucleic Acids Res. 1990;18:6097–100.
68. Reidys C, Huang F, Andersen J, Penner R, Stadler P, Nebel M. Topology and prediction of RNA pseudoknots. Bioinformatics. 2011;27:1076–85.
69. Aigner K, Dreßen F, Steger G. Methods for predicting RNA secondary structure. In: Leontis N, Westhof E, eds. RNA 3D Structure Analysis and Prediction. Nucleic Acids and Molecular Biology 27. Berlin: Springer-Verlag; 2012:19–41.
70. Lyngsø R, Pedersen C. RNA pseudoknot prediction in energy-based models. J Comput Biol. 2000;7:409–27.
71. Ren J, Rastegari B, Condon A, Hoos H. HotKnots: heuristic prediction of RNA secondary structures including pseudoknots. RNA. 2005;11:1494–504.
72. Reeder J, Giegerich R. Design, implementation and evaluation of a practical pseudoknot folding algorithm based on thermodynamics. BMC Bioinformatics. 2004;5:104.
73. Reeder J, Steffen P, Giegerich R. pknotsRG: RNA pseudoknot folding including near-optimal structures and sliding windows. Nucleic Acids Res. 2007;35:W320–4.
74. Theis C, Janssen S, Giegerich R. Prediction of RNA secondary structure including kissing hairpin motifs. In: Moulton V, Singh M, eds. Lecture Notes in Bioinformatics. Lecture Notes in Computer Science 6293. Berlin: Springer-Verlag; 2010:52–64.
75. Bellaousov S, Mathews D. ProbKnot: fast prediction of RNA secondary structure including pseudoknots. RNA. 2010;16:1870–80.
76. Rehmsmeier M, Steffen P, Höchsmann M, Giegerich R. Fast and effective prediction of microRNA/target duplexes. RNA. 2004;10:1507–17.
77. Krüger J, Rehmsmeier M. RNAhybrid: microRNA target prediction easy, fast and flexible. Nucleic Acids Res. 2006;34:W451–4.
78. Rehmsmeier M. Prediction of microRNA targets. Methods Mol Biol. 2006;342:87–99.
79. Lu Z, Mathews D. Efficient siRNA selection using hybridization thermodynamics. Nucleic Acids Res. 2007;36:640–7.
80. Mathews D. Using OligoWalk to identify efficient siRNA sequences. Methods Mol. Biol. 2010;629:109–21.
81. Bernhart S, Hofacker I, Stadler P. Local RNA base pairing probabilities in large sequences. Bioinformatics. 2005;22:614–15.
82. Tafer H, Ameres S, Obernosterer G, et al. The impact of target site accessibility on the design of effective siRNAs. Nat. Biotechnol. 2008;26:578–83.
83. Dimitrov R, Zuker M. Prediction of hybridization and melting for double-stranded nucleic acids. Biophys. J. 2004;87:215–26.
84. Markham N, Zuker M. DINAMelt web server for nucleic acid melting prediction. Nucleic Acids Res. 2005;33:W577–81.
85. Bernhart S, Tafer H, Mückstein U, Flamm C, Stadler P, Hofacker I. Partition function and base pairing probabilities of RNA heterodimers. Algorithms Mol. Biol. 2006;1:3.

86. Wimberly B, Varani G, Tinoco I Jr. The conformation of loop E of eukaryotic 5 S ribosomal RNA. Biochemistry. 1993;32:1078–87.
87. Leontis N, Stombaugh J, Westhof E. Motif prediction in ribosomal RNAs: lessons and prospects for automated motif prediction in homologous RNA molecules. Biochimie. 2002;84:961–73.
88. Leontis N, Westhof E. Geometric nomenclature and classification of RNA base pairs. RNA. 2001;7:499–512.
89. Gardner P, Daub J, Tate J, et al. Rfam: Wikipedia, clans and the "decimal" release. Nucleic Acids Res. 2010;39:D141–5.
90. Lai D, Proctor J, Zhu J, Meyer I. R-CHIE: a web server and R package for visualizing RNA secondary structures. Nucleic Acids Res. 2012;40:e95.
91. Andronescu M, Bereg V, Hoos H, Condon A. RNA STRAND: the RNA secondary structure and statistical analysis database. BMC Bioinformatics. 2008;9:340.
92. Lorenz R, Bernhart S, Höner Zu Siederdissen C, et al. ViennaRNA Package 2.0. Algorithms Mol Biol. 2011;6:26.
93. Gruber A, Lorenz R, Bernhart S, Neuböck R, Hofacker I. The Vienna RNA websuite. Nucleic Acids Res. 2008;36:W70–4.
94. Zuker M. Mfold web server for nucleic acid folding and hybridization prediction. ucleic Acids Res. 2003;31:3406–15.
95. Markham N, Zuker M. UNAFold: software for nucleic acid folding and hybridization. In: Keith J, ed. Bioinformatics: Structure, Function and Applications. Methods in Molecular Biology 453. Totowa, NJ: Humana Press; 2008:3–31.
96. Reuter J, Mathews D. RNAstructure: software for RNA secondary structure prediction and analysis. BMC Bioinformatics. 2010;11:129.
97. Steffen P, Voss B, Rehmsmeier M, Reeder J, Giegerich R. RNAshapes: an integrated RNA analysis package based on abstract shapes. Bioinformatics. 2006;22:500–3.
98. Touzet H, Perriquet O. CARNAC: folding families of related RNAs. Nucleic Acids Res. 2004;32:W142–5.
99. Tabei Y, Kiryu H, Kin T, Asai K. A fast structural multiple alignment method for long RNA sequences. BMC Bioinformatics. 2008;9:33.
100. Wilm A, Higgins D, Notredame C. R-Coffee: a method for multiple alignment of non-coding RNA. Nucleic Acids Res. 2008;36:e52.
101. Hofacker I, Fekete M, Stadler P. Secondary structure prediction for aligned RNA sequences. J Mol Biol. 2002;319:1059–66.
102. Torarinsson E, Lindgreen S. WAR: webserver for aligning structural RNAs. ucleic Acids Res. 2008;36:W79–84.
103. Sperschneider J, Datta A. DotKnot: pseudoknot prediction using the probability dot plot under a refined energy model. Nucleic Acids Res. 2010;38:e103.
104. Huang C, Lu C, Chiu H. A heuristic approach for detecting RNA H-type pseudoknots. Bioinformatics. 2005;21:3501–8.
105. Ruan J, Stormo G, Zhang W. An iterated loop matching approach to the prediction of RNA secondary structures with pseudoknots. Bioinformatics. 2004;20:58–66.
106. Sperschneider J, Datta A, Wise M. Heuristic RNA pseudoknot prediction including intramolecular kissing hairpins. RNA. 2010;17:27–38.
107. Dirks R, Pierce N. A partition function algorithm for nucleic acid secondary structure including pseudoknots. J Comput Chem. 2003;24:1664–77.
108. Rivas E, Eddy S. A dynamic programming algorithm for RNA structure prediction including pseudoknots. J Mol Biol. 1999;285:2053–68.
109. Matsui H, Sato K, Sakakibara Y. Pair stochastic tree adjoining grammars for aligning and predicting pseudoknot RNA structures. Bioinformatics. 2005;21:2611–17.

110. Dawson W, Fujiwara K, Kawai G. Prediction of RNA pseudoknots using heuristic modeling with mapping and sequential folding. PLoS One. 2007;2:e905.
111. Mückstein U, Tafer H, Hackermüller J, Bernhart S, Stadler P, Hofacker I. Thermodynamics of RNA-RNA binding. Bioinformatics. 2006;22:1177–82.
112. Ke A, Zhou K, Ding F, Cate J, Doudna J. A conformational switch controls hepatitis delta virus ribozyme catalysis. Nature. 2004;429:201–5.

Anton I. Petrov, Blake A. Sweeney, and Neocles B. Leontis
15 Analyzing, searching, and annotating recurrent RNA three-dimensional motifs

This chapter focuses on modular, recurrent three-dimensional (3D) motifs found in structured RNA molecules. RNA 3D motifs are structurally diverse and play important architectural and molecular recognition roles. Predicting their 3D structures is a crucial step in solving the RNA structure prediction problem. To set the context, we first review the characteristics of structured RNA molecules, including the nature of the RNA polymer backbone, the properties of the nucleotide building blocks/nucleobases, and the different ways nucleotides interact with each other to structure 3D motifs. Then we discuss how to define 3D motifs to enable automated search procedures in RNA 3D structures. We review programs that identify and extract 3D motifs from structures and databases created to annotate, compare, and store them. Finally, we provide examples of 3D motifs that constitute exceptions to the currently implemented definitions and discuss ways that automatic motif searching and classification in recurrent families can be improved.

15.1 Characteristics of structured RNAs

RNA molecules play many different roles in living systems and come in a great variety of shapes and sizes. RNA molecules are especially interesting because they are so similar to DNA at the nucleotide level, and like DNA, they are able to encode genetic information, while at the same time being much more versatile in their functions and diverse in their structures, resembling proteins in this respect. Unlike cellular DNA, RNA molecules are single-stranded polymers, which, like proteins, fold into complex, compact structures. Depending on their biological function, they may be partially or entirely structured, but even highly structured RNA molecules, such as transfer RNAs (tRNAs), are flexible [1,2]. As is the case for proteins, this flexibility is important for biological function [3,4].

15.1.1 RNA molecules are structurally diverse

Like DNA, RNA molecules are composed of just four nucleotide building blocks, linked together in chains that can form regular, antiparallel double helices through Watson-Crick (WC) basepairing. How is it then that RNA can form so many different structures? A partial answer derives from the observation that about one-third of the bases in typical structured RNAs do not form WC basepairs [5]. Consequently, the WC helices in RNA are generally short and punctuated frequently by nominally single-stranded

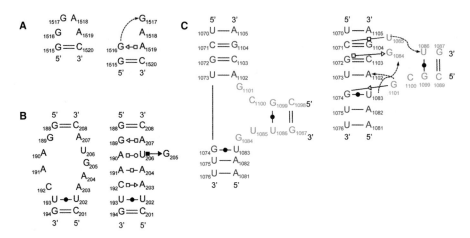

Fig. 15.1 Examples of classical internal, hairpin, and junction loop RNA 3D motifs. (A) GNRA-hairpin from *E. coli* 16S rRNA (PDB 2AW7). (B) Sarcin-ricin internal loop from *E. coli* 23S (PDB 2AW4). (C) Three-way junction from *E. coli* 16S (helices h34-h35-h36, PDB 2AW7). The left side of each panel shows the unstructured "loops" as seen in 2D diagrams. The right of each panel shows the same structures annotated according to the Leontis-Westhof nomenclature are shown. The 3D structures can be interactively explored online at http://rna.bgsu.edu/publications/deGruyter2013/.

regions that appear in secondary (2° or two-dimensional [2D]) structures as "loops," or more rarely as "linker strands" (defined subsequently). However, nucleotides in most of these loop regions are not "looped out" and disorganized, as implied by 2D representations of the structure. Rather, they form structural elements with distinct shapes, dynamic characteristics, and specific stabilities that we refer to as "RNA 3D motifs." Many of these structures are modular or recurrent. Examples are shown in Fig. 15.1.

15.1.2 "Loops" in RNA secondary structures and RNA 3D motifs

There is a rough one-to-one correspondence between RNA 3D motifs and the "loops" apparent in RNA 2D representations. This correspondence is the basis for automated methods designed to identify and extract modular 3D motifs from atomic-resolution structures.

Loops in RNA secondary structures are generally classified into three distinct types, depending on the number of WC helices to which they are attached, as shown in Fig. 15.1. Each hairpin loop is attached to one helix, each internal loop to two helices, and each multihelix junction loop to three or more helices. Three- and four-way junctions are the most common multihelix junctions (see Chapter 5), but large RNAs such as the ribosomal RNAs (rRNAs) have junction loops that, at least formally, comprise many more helices. Except for very large junction loops, RNA 3D motifs are relatively small in size. For example, hairpin loops extracted from atomic resolution RNA structures in the Protein Data Bank (PDB) [6] range in size from 3 to more than 20 nucleotides, including the 2 nucleotides of the flanking WC basepair. Internal loops vary in size from 5 to almost 30 nucleotides, including the flanking basepairs. The distribution of hairpin and internal loop lengths is shown in the histograms in Fig. 15.2. These loop data were extracted from the 4.0 Å Reduced Redundancy data set of RNA-containing PDB files available as of January 19, 2013, maintained by the Bowling Green State University (BGSU) RNA group (http://rna.bgsu.edu/rna3dhub/nrlist). The peak of the hairpin loop distribution occurs at 6 nucleotides, consistent with the large number of tetraloops, hairpins with four nominally unpaired bases, and one flanking WC pair. Internal loops tend to be larger on average because they include four flanking pairs. Junctions are generally more diverse in size and shape [7].

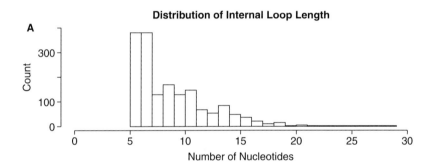

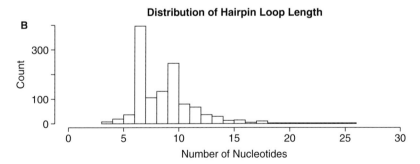

Fig. 15.2 Distribution of hairpin and internal loop sizes. (A,B) The sizes (in nucleotides) of all internal loops and hairpins found by the FR3D software in a 4 Å nonredundant set of RNA 3D structures available as of January 19, 2013. These sizes include four nucleotides belonging to the two flanking Watson-Crick basepairs for internal loops and the two nucleotides of the flanking Watson-Crick basepair for hairpin loops.

15.1.3 The 3D motifs and hierarchical organization of RNA

Large RNAs, like proteins, are organized hierarchically and consist of structural domains and subdomains. In the RNA structural hierarchy, RNA 3D motifs are intermediate in size and complexity between the nucleotides, the basic units of primary structure, and the large discrete domains that define the overall structure. Structural domains are linked to each other covalently by double-stranded helices or by single-stranded linkers, depending on the topology of the RNA, as discussed subsequently. Structural domains of large RNAs also interact with each other noncovalently. RNA 3D motifs mediate most noncovalent interactions, within and between domains. The 3D motifs serve at least three functions: (1) they organize the 3D structures of domains (architectural role); (2) they mediate tertiary interactions within and between domains (organizational role); and (3) they provide functional binding sites for other molecules, including proteins, RNAs, and small molecules.

15.1.4 Linker regions and 3D motifs

In some structured RNAs the 5'- and 3'-terminal ends are basepaired to each other, for example, in the tRNAs and 5S rRNA. In such molecules, each nominally single-strand nucleotide of the RNA formally belongs to some hairpin, internal, or junction loop in the 2D representation, with the possible exception of short overhangs on the 5' or 3' ends such as the CCA-3' sequence of tRNAs. For example, each nucleotide of 5S RNA belongs to one of the five helical regions (numbered I to V), one of the two hairpin loops (loop C or D), one of the two internal loops (loop B or E), or to the central three-way junction (loop A) [8]. In RNAs in which the 5' and 3' ends are not paired to form a terminal end helix, there will be single-stranded linkers that are not formally included in any loop defined by the 2D. For example, the 5' and 3' ends of 16S rRNA are not paired to each other; consequently, major structural domains of this RNA are connected to the rest of the structure by nominally single-stranded linkers. For example, the head is connected to the body by a single strand common to helix 28 (h28) located at the base of the head, and h2, in the central pseudoknot, is located in the upper part of the body of the small subunit. Likewise, the 3'-minor domain (h44 and h45) is connected to the head by a strand common to h44 and h28. The strand topology is apparent in Fig. 15.3A. As discussed subsequently, stacking and non-WC interactions contribute additional interactions that stabilize the 3D motifs formed by the linker strands as shown in Fig. 15.3B. It is important to note that methods of 3D motif extraction that rely on identifying the positions of loops in 2D structures by reference to flanking WC pairs will miss all 3D motifs comprising linker nucleotides.

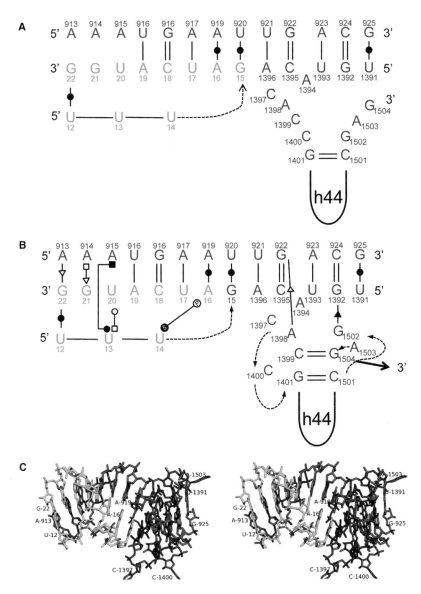

Fig. 15.3 16S linker region. (A) Secondary structure diagram of the linker region connecting the head and the body of the 16S rRNA from *E. coli*. (B) The same fragment annotated with basepair interactions according to the Leontis-Westhof nomenclature. Nucleotides that have interactions not shown in (A) are highlighted with a gray box. (C) The 3D visualization based on PDB 2AW7 created using the PyMOL software (http://www.pymol.org/). The 3D structures can be interactively explored online at http://rna.bgsu.edu/publications/deGruyter2013/.

15.2 Structural diversity of RNA 3D motifs

Although each RNA 3D motif is built from the same set of four nucleotides, there are many ways the nucleotides can be ordered sequentially, so many different motifs are possible. In principle, a loop of n nucleotides could form 4^n different motifs. However, as a general rule, a significant number of different sequences can form essentially the same motif, depending on how motif identity or similarity is defined. Geometrically similar motifs generally have similar functions by virtue of their similar shapes; therefore, identifying the sequence variants of each RNA 3D motif is of great interest in RNA structural analysis and prediction [9]. Prediction of potential sequence variants is facilitated organizing nucleotide interactions, especially basepairs, into recurrent geometric families (see Section 15.3.1). Such classification helps to identify base substitutions that are structurally neutral.

In addition to the sequence, two other factors contribute to the variety of shapes possible in RNA 3D motifs: the flexibility of the RNA backbone and the large variety of interactions possible between two nucleotides.

15.2.1 Contribution of RNA chain flexibility to motif diversity

The RNA chain is very flexible due to the large number of torsion angles in each dinucleotide linkage. There are six single-bond dihedral angles along the backbone, labeled sequentially with Greek letters from α to ζ. Only one of these, δ (delta), is significantly constrained, as it connects the C3' and C4' carbon atoms of the ribose ring. In addition, each nucleotide contains a seventh degree of freedom – the torsion angle labeled χ (Chi), associated with the single bond connecting each base to the C1' atom of the ribose. The large number of torsion angles per nucleotide makes for a very flexible backbone, which allows each pair of covalently linked nucleotides (called a "suite") to form a large number of conformations [10]. Moreover, this flexibility allows the chain to stretch in 3D space over a broad range of distances. The best way to visualize this variation is by measuring successive C1'-C1' distances along the chains of structure RNA. For nucleotides in helices, the distribution of C1'-C1' distances is narrower and peaks at roughly 5.5 Å (see Fig. 15.4A). When all nucleotides from the ribosome (16S and 23S rRNA) are included, the resulting distribution has a long tail extending out to about 11 Å (Fig. 15.4B). As discussed subsequently, this extreme flexibility complicates the analysis of RNA 3D motifs, as it allows bulged bases to interact with nucleotides far away from their local neighbors in the 2D structure.

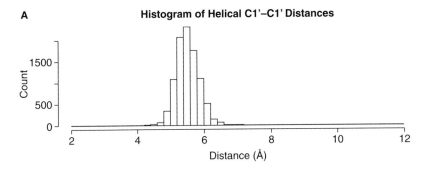

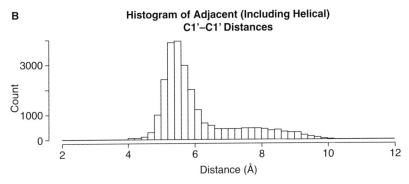

Fig. 15.4 Distributions of C1'-C1' distances. (A) Distances between C1' atoms of adjacent nucleotides in helical regions of structured RNAs. (B) C1'-C1' distances for all adjacent nucleotides including helical regions, "loops," and linker regions. The distributions were computed based on ten ribosomal structures (1S72, 1FJG, 2AW7, 2ZJR, 2QBG, 3V2F, 3U5F, 2XZM, 3U5H, 4A1B) from *E. coli*, *Thermus thermophilus*, *Deinococcus radiodurans*, *Saccharomyces cerevisiae*, and *Tetrahymena thermophila*.

15.2.2 Contribution of internucleotide interactions to motif diversity

Each RNA nucleotide consists of three chemical modules, a central five-member ribose ring, a phosphate group, and a heterocyclic aromatic base. The base is connected to the ribose at its C1' position, and the ribose to its own phosphate at the 5'-hydroxyl oxygen and to the phosphate of the next nucleotide through its 3'-hydroxyl oxygen. Each nucleotide comprises a variety of functional groups capable of forming specific and fairly stable noncovalent interactions. All five ribose oxygens act as H-bond acceptors, and the 2'-OH group also functions as a donor. The phosphate group bears a net negative charge distributed primarily over the two nonbridging oxygens, which serve as strong H-bond acceptors, as well as metal-binding sites [11,12].

The bases themselves comprise a number of H-bond donor and acceptor groups. Including their exocyclic amino and carbonyl groups, RNA bases are shaped approximately like triangles (i.e. oblate triangular prisms in 3D). The edges of the triangles are defined by arrays of H-bonding donor and acceptor groups that interact noncovalently with other bases in a coplanar fashion so as to form basepairs.

15.3 Pairwise nucleotide interactions that stabilize RNA 3D motifs

The diversity of RNA 3D structure results in part from the diversity of 3D motifs. The diversity of 3D motifs, in turn, can be understood in light of the diversity of interactions possible between two nucleotides. This diversity may be surprising, given there are only four different bases (A, C, G, and U) in most RNA structures. Some RNAs (e.g. tRNAs) have significant numbers of modified nucleotides, but most of these are simply methylated versions of the standard ones and tend to be more restricted in their interactions.

15.3.1 Basepairing interactions and 3D motifs

Because each base has three edges, the WC, Hoogsteen, and sugar edges, and because bases can approach each other in the plane in two ways, *cis* or *trans*, for each pair of edges, there are 12 distinct geometric families of basepairs, as described in previous publications [5,13]. Each basepair family is specified according to which edges interact to form the pair and by the mutual orientations of the ribose rings, *cis* or *trans*. The ribose 2'-OH group is considered part of the sugar edge and participates in sugar-edge basepairs. The canonical WC basepairs belong to the *cis* Watson-Crick/Watson-Crick (cWW) family.

The basepair families are abbreviated using "c" for *cis*, "t" for *trans*, and capital letters for the interaction base edges (W, H, and S). Note that the order in which the bases are listed in most non-WC pairs is significant, so that, for example, AC tHS and AC tSH are different basepairs.

15.3.1.1 Occurrence frequencies of basepairs is context dependent

The 12 basepair families vary in their occurrence frequencies; moreover, for most basepair types, the frequency depends on structural context (Tab. 15.1). We distinguish two contexts – "local" and "long range." "Local" means pairing between bases located within the same helix or 3D motif; "long range" means basepairs that form between different 3D motifs or helical elements, as part of tertiary or quaternary interactions. Tab. 15.1 shows frequencies by basepair family for *Escherichia coli* 16S and 23S (PDB files 2AW7 and 2AW4) as computed by the FR3D program [14] and is

Tab. 15.1 Frequencies of occurrences of basepairs in the 16S and 23S rRNA by structural context.

A

	Count	% of all pairs
cWW pairs	1,355	66.9
All local	1,786	88.2
All long range	238	11.8

B

Type	Local	% of all local	Long range	% of all LR	Total count	% of all basepairs
cWW	1,313	**73.5**	42	17.6	1,355	66.9
cWH	18	1.0	10	**4.2**	28	1.4
cWS	16	0.9	14	**5.9**	30	1.5
cHH	1	0.1	0	0.0	1	0.0
cHS	41	**2.3**	0	0.0	41	2.0
cSS	46	2.6	89	**37.4**	135	6.7
tWW	24	1.3	8	**3.4**	32	1.6
tWH	95	**5.3**	4	1.7	99	4.9
tWS	18	1.0	11	**4.6**	29	1.4
tHH	20	1.1	1	0.4	21	1.0
tHS	154	**8.6**	4	1.7	158	7.8
tSS	40	2.2	55	**23.1**		4.7
Total	1,786	100.0	238	100.0	2,024	100.0

"Local" corresponds to basepairs formed inside individual 3D motifs, "long range" (LR) indicates long-range basepairs between 3D motifs. Bold numbers indicate basepair frequencies that are significantly higher in one context. The calculations were based on PDB files 2AW4 and 2AW7.

representative of other structured RNAs [11]. The most common local interaction is, of course, the canonical WC pair. In the bacterial ribosome, cWW pairs constitute 73% of local interactions, but only 18% of long-range interactions. By contrast, more than 50% of long-range interactions, but fewer than 5% of local pairings, are cSS or tSS. The tWH, tHS, and cHS pairings are also more common in local contexts. Pairings that occur more frequently in long-range contexts, in addition to cSS and tSS, are cWH, tWW, and tWS. The cHH pair is the rarest of all and usually requires one base to assume a *syn*-conformation.

15.3.1.2 Basepair isostericity and structure conservation during evolution

The classification of basepairs into geometric families based on edges helps us to predict and to make sense of base substitutions observed at homologous positions of aligned RNA sequences and recurrent RNA 3D motifs alike. To illustrate the basic ideas, we first consider the WC basepairs (i.e. the AU and GC base combinations in the cWW family), which compose the regular helices of RNA molecules. RNA WC helices, regardless of base sequence, are geometrically regular precisely because cWW GC, CG, AU, and UA basepairs can substitute for each other with little or no distortion of the helical geometry. They are said to be isomorphous or isosteric to each other. Basepair isostericity [5] is a relation that holds between two basepairs only if they belong to the same geometric basepair family. Thus, it is essential to specify the geometric family as well as the base combination when discussing basepairs. The same base combination (for example AG) can form many different basepair types, including tHS cWW, cHW, and tWS. However, these are not isosteric with each other. The structural context determines which geometric basepair type forms. Other base combinations can substitute, without changing the structure of the motif, as long as they form the same type of basepair and it is isosteric, or nearly so, with the one it replaces [5].

Basepair geometries are remarkably conserved in RNA evolution, as shown in a recent study comparing the 3D structures of the rRNAs of *E. coli* and *Thermus thermophilus*, two distantly related bacterial species [5]. Not only are basepair families conserved at homologous positions, but in almost all cases where base substitutions occur, isosteric basepairs result. The isostericity relations in each basepair family have been catalogued in previous publications [5,13].

15.3.2 Base-stacking interactions and 3D motifs

Due to their planar geometries, RNA bases exhibit strong propensities to stack. Base stacking provides a major driving force stabilizing RNA 3D motif formation. Almost all bases in highly structured RNAs such as the ribosomal RNAs are stacked on at least one other base. Tertiary interactions often involve stacking as well basepairing interactions. Base stacking is detected and annotated by several RNA structure analysis programs, including FR3D and MC-Annotate [14,15].

15.3.3 Base-phosphate interactions and 3D motifs

RNA is a polyelectrolyte bearing one negative charge per phosphate unit, largely concentrated on the two nonbridging oxygen atoms of the phosphate. For RNA molecules to fold compactly, this negative charge has to be adequately screened or neutralized to overcome intramolecular electrostatic repulsions. Significant electrostatic screening is provided by loosely bound monovalent ions (primarily K^+ in cells) and specifically bound Mg^{2+} ions. K^+ ions are in much higher concentration, but Mg^{2+} ions, with their high charge density, are able to bridge phosphate ions to stabilize highly compact RNA 3D structures [16]. The hydrogen-bonding edges of RNA bases can also screen the negative charge of backbone phosphates through characteristic base-specific "base-phosphate" (BPh) interactions, which were analyzed, classified, and quantified by bioinformatic and quantum chemical methods [11]. The FR3D program suite was modified to locate and annotate these interactions in 3D structures and make them available for motif searching and classification.

Some BPh interactions are intrinsically as stable as WC CG basepairs [11]. Moreover, comparison of *E. coli* and *T. thermophilus* 3D rRNA structures shows these interactions are very conserved in RNA evolution. Bulged Gs frequently stabilize nearby turns and kinks in the backbone though tertiary BPh interactions involving their WC edges, which have two H-bond donor groups and form especially stable BPh interactions. Conserved BPh interactions play crucial roles in stabilizing 3D motifs. An example is the sarcin-ricin motif [17].

15.4 Defining RNA 3D motifs

RNA 3D motifs are local modular elements of structured RNA molecules that are stabilized primarily by interactions among the nucleotides belonging to the motif. RNA 3D motifs roughly correspond to the hairpin, internal, and junction loops evident in correctly drawn secondary structures. The stabilizing interactions include all those discussed in the previous section.

15.4.1 Role of induced fit in RNA motif structure

Autonomous motifs form stable structures in the absence of additional interactions. Other 3D motifs require interactions with a binding partner to achieve a stable, well-defined structure, in a process called "induced fit." The stabilizing binding partner can be another part of the RNA, a protein, or a small molecule, as in allosteric genetic elements called riboswitches [18,19]. Induced fit also plays an important role in protein structure [20]. Obvious examples of induced fit in RNA are motifs in which all nucleotides are looped out and interact extensively with other elements but not with each other. Conformationally flexible motifs can have two or more stable, distinct structures, depending on whether the correct binding partner is present. The internal loop in h44 of 16S rRNA, which is responsible for decoding tRNA–messenger RNA (mRNA) interactions in the A-site of the ribosome, is a salient example. It assumes an extended open conformation in the presence of a cognate tRNA-mRNA interaction in the A-site and a closed conformation in its absence [21].

Generally, it is not possible to predict which motifs are autonomous and which form by induced fit just by examining the sequence or even the 3D structure, because most motifs comprise nucleotides that interact with each other as well as with elements outside of the motif.

15.4.2 Definition of "classic" RNA 3D motifs

The term *motif* is widely applied in biology to designate well-defined, recurrent patterns in space and/or time, at all levels of biological organization. In structural biology, motifs can be defined at the level of the macromolecular sequence (primary structure), of the 2D structure, as well as for the 3D structure [9]. RNA 3D motifs also come in different "flavors," some of which are actually recurrent backbone conformations or submotifs that occur only as parts of different modular motifs but do not themselves constitute autonomous motifs [22,23].

For the purpose of automatically extracting modular motifs corresponding to hairpin, internal, and junction loops, most workers in the field use an operational definition based on the subsequent criteria.

The nucleotides of the motif comprise chain segments that are flanked by cWW basepairs so as to form distinct loops in the 2D representation. Hairpin loops are flanked by one cWW basepair, internal loops by two cWW pairs, and N-fold multihelix junctions by N cWW pairs. In this chapter and in our online databases (http://rna.bgsu.edu/motifs/), the flanking basepairs are considered part of the 3D motif, as they are often constrained in sequence by interactions with the rest of the motif.

Nucleotides of the motif interact with each other through non-WC basepairing, base stacking, and base-backbone interactions. Non-WC pairs include noncanonical cWW pairs such as AG and UU [13]. They may also interact with nucleotides distant in the 2D structure to form stabilizing tertiary (3°) or quaternary (4°) RNA-RNA interactions, but not with nucleotides in the adjacent helices, with the possible exception of flanking basepairs, which in any case are included in the 3D motif.

No canonical cWW basepairs (base combinations AU, UA, GC, CG, GU, or UG) form between any of the nucleotides of the motif, with the exception, by definition, of the flanking basepairs.

Most modular 3D motifs conform to this definition and can be reliably detected and automatically extracted from 3D structures by its application with suitable software. Fig. 15.1 provides examples of "classic" hairpin, internal, and junction loops that conform to all of these criteria. We find that two different programs, one implemented in the FR3D program suite and another by Djelloul et al. (see Section 15.5.4, [24]), consistently identify and extract those 3D motifs that fit this definition and that are well-modeled and accurately annotated (see Section 15.7.3).

There are 3D motifs that, in some respect, do not conform to this definition but which, nonetheless, are modular and recurrent and therefore worth identifying and classifying. These motifs present challenges for automatic motif identification and extraction and are the subject of current research. Some examples are discussed in the last section of this chapter.

15.4.2.1 Definition of modular motifs

RNA 3D motifs are modular when they can be inserted in new contexts and retain their functional properties. In some cases, modular motifs can be removed and replaced by functionally equivalent motifs, without significantly disrupting the global structure and function of the molecule. It is of considerable interest to identify motif families that can functionally substitute for each other. Modular motifs and their interaction rules are of special interest in RNA nanotechonology and nanomedicine [25] and in studies of RNA self-assembly.

15.4.2.2 Conservation of motif sequence and structure

Many RNA 3D motifs display evolutionary conservation at the level of sequence, 3D structure, or function. Generally, the 3D structure is more conserved than the sequence, which can vary, sometimes significantly, while forming essentially the same structure. In any case, for the structure to be conserved, the sequence is constrained to a fairly small number of conserved base substitutions. Most base substitutions are consistent with basepair isostericity, but not all isosteric substitutions are observed in a given motif [5]. Determining the sequence signatures of RNA 3D motifs is an active area of research.

15.4.3 Recurrent RNA 3D motifs

Recurrent motifs occur independently in different RNA molecules and even in distinct, nonhomologous locations of the same large molecule, yet form similar local structures. For example, hairpins loops called T-loops were first observed in the third hairpin loop of tRNAs, and subsequently also found in 16S and 23S ribosomal RNAs [26]. Different T-loop instances differ significantly in sequence but are very similar in structure and function. Motif instances that occur in homologous positions of related molecules, such as the 16S rRNAs from *E. coli* and *T. thermophilus*, usually belong to the same recurrent motif. Although the homologous motifs often vary in sequence, they tend to be highly conserved in 3D structure and function [5]. For example, almost all conserved instances of GNRA hairpin loops, regardless where they occur, mediate tertiary or quaternary RNA-RNA minor-groove interactions, and almost all instances of hairpin T-loops form conserved tertiary interactions by intercalating a bulged base from another motif [26].

Different instances of the same recurrent motif that occur in nonhomologous sites may play similar functional roles, but this is not always the case. Thus, the sarcin-ricin motif, which also occurs widely in the ribosome [27], interacts with ribosomal proteins and translation factors, as well as other RNA elements [28].

15.5 Tools for searching for RNA 3D motifs in atomic-resolution RNA structures

Having defined modular RNA 3D motifs, we discuss individually each of the various tools, both stand-alone and web enabled, that have been developed in recent years to identify motifs in RNA 3D structures. Each method discussed is summarized in

Tab. 15.2, which provides the URL, reference(s), and details regarding search modes implemented and sources of basepair annotations used by the method.

The goal of RNA structure search tools is to find all motif instances in target structures that are geometrically similar to a user-provided query motif or motif description. Two distinct approaches have been implemented for RNA 3D motif search: (1) The first approach involves scanning precomputed structural annotations to find matches to motif descriptors. (2) The second approach involves searching the actual 3D structure space by taking into account the distances between component parts (for example nucleotides, phosphates, bases, or individual atoms) of the motif. We refer to the former approach as "symbolic search" and the latter as "geometric search." All methods described previously are classified as symbolic or geometric in Tab. 15.2.

15.5.1 MC-Search

MC-Search [29,30] is a program that uses text descriptors expressed in a special syntax to specify the sequence and nucleotide interactions within the query motif to find geometrically similar instances in 3D structure files. MC-Search relies on the basepair annotations produced by MC-Annotate, another component of the MC-Tools software suite. A web server implementation of MC-Search is also available, but it has not been specifically described in the literature, to the best of our knowledge.

15.5.2 NASSAM

NASSAM ("Nucleic Acid Search for Substructures and Motifs" [31,32]) represents RNA 3D structures as graphs. Each nucleotide is represented by four pseudoatoms, linked in pairs to form two vectors. The edges of the graph correspond to the distances between the vectors. To carry out searches, the algorithm looks for subgraphs matching the graph description of the query in the graph representation of the target structure. NASSAM was tested by searching for basepairs and GNRA tetraloops.

Recently NASSAM was implemented into a web server [32] that is capable of annotating a user-submitted structure in PDB format with basepairs and base triples. Users can also search a user-submitted PDB file for occurrences of a predefined set of RNA 3D motifs including kink-turns and T-loops.

15.5.3 PRIMOS

PRIMOS ("Probing RNA structures to Identify Motifs and Overall Structural changes" [33]) assigns an ordered sequence of pseudotorsion angles, η and θ, called an "RNA worm," to each continuously bonded segment of nucleotides in an RNA structure, to create a searchable, reduced representation of the 3D structure of the backbone of each

Tab. 15.2 Stand-alone programs and web servers for searching for RNA 3D motifs in 3D structures.

Method	URL	Reference	Search mode	Basepair annotations
Stand-alone programs				
MC-Search*	http://www.major.iric.ca/MajorLabEn/MC-Tools.html	Hoffmann et al.[29]	Symbolic	MC-Annotate
NASSAM*	http://mfrlab.org/grafss/nassam	Harrison et al. [31]	Geometric	Not applicable[a]
PRIMOS	http://pylelab.org/software	Duarte et al. [33]	Symbolic	Not applicable[a]
FR3D*	http://rna.bgsu.edu/FR3D	Sarver et al. [14]	Symbolic, geometric, combined	FR3D
Apostolico et al., 2009	Not available	Apostolico et al. [35]	Geometric	Not applicable[a]
RNAMotifScan	http://genome.ucf.edu/RNAMotifScan	Zhong et al. [36]	Symbolic	RNAView, MC-Annotate
FRMF	http://www.cs.cityu.edu.hk/~yingshen/FRMF.html	Shen et al. [38]	Geometric	FR3D
Web servers				
RNA FRABASE 2.0	http://rnafrabase.ibch.poznan.pl	Popenda et al.[39]	Symbolic	RNAView
FASTR3D	http://genome.cs.nthu.edu.tw/FASTR3D/	Lai et al. [41]	Symbolic, geometric	RNAView
FRASS	http://protein.bio.unipd.it/frass/	Kirillova et al. [42]	Geometric	Not applicable[a]
WebFR3D*	http://rna.bgsu.edu/webfr3d	Petrov et al. [34]	Symbolic, geometric, combined	FR3D
R3D-BLAST	http://genome.cs.nthu.edu.tw/R3D-BLAST/	Liu et al. [43]	Symbolic	Not applicable[b]
NASSAM*	http://mfrlab.org/grafss/nassam	Hamdani et al.[32]	Geometric	Not applicable[a]
MC-Search*	http://www.major.iric.ca/MC-Search	Not applicable[c]	Symbolic	MC-Annotate

*Programs for which both stand-alone and web applications are available.
[a]Not applicable because symbolic search is not supported.
[b]Not applicable because symbolic search uses backbone conformations.
[c]Not applicable because this work is unpublished

chain. RNA worms can be used to compare structures, find recurrent and entirely new motifs, and detect conformational changes between related structures. When tested on S-motifs, kink-turns, and hook-turn motifs, PRIMOS was able to recover known motif instances and identify previously unknown motif variants. However, PRIMOS can only search one strand at a time, so to find an internal loop motif in a 3D structure, one must individually search each of the two strands composing the internal loop and combine the results.

15.5.4 FR3D and WebFR3D

The FR3D ("Find RNA 3D") motif search engine [14] is unique because it enables searching in symbolic, geometric, or mixed symbolic/geometric modes. In symbolic search mode, FR3D scans precomputed annotations of individual nucleotide characteristics and pairwise nucleotide interactions, including basepairing, base stacking, base backbone, and separation in the primary sequence, as specified in the query. In geometric search mode, FR3D exhaustively searches 3D structures to identify combinations of nucleotides, regardless of position in the chain(s), which are geometrically similar to the query motif. FR3D defines a special measure of geometric similarity called geometric discrepancy to select and rank candidate structures. The geometric discrepancy has two terms, one of which resembles root-mean-square deviation (RMSD) and takes into account distances between the corresponding base centers. Note that FR3D uses the locations of base centers and not phosphate centers for structure comparison. The second term takes into account the relative orientations of the planes of the bases of the motifs compared. WebFR3D is the online version of FR3D [34].

15.5.5 Apostolico et al., 2009

The motif search method proposed by Apostolico et al., 2009 [35] is based on comparing shape histograms computed for query motifs and target structures. The histogram computation uses quantized distances of the phosphate backbone atoms from the centroid of the backbone atoms. Nucleobase locations are not taken into account. The method was validated by searching the structure of 23S rRNA from *Haloarcula marismortui* for known GNRA tetraloops, kink-turns, loop E, and several other motifs. Shape histograms were also used for grouping junction loops by overall structure similarity.

15.5.6 RNAMotifScan

RNAMotifScan [36] uses a dynamic programming approach to align the extended secondary structure of the query motif with the target structures and rank the matches. Extended secondary structure includes local non-WC basepairs. RNAMotifScan scores aligned basepairs by isostericity [5], so that structurally compatible basepairs are preferentially aligned. Basepairs are annotated by MC-Annotate, but RNAView annotations can also be used.

The authors benchmarked RNAMotifScan against FR3D [36], and methods proposed by Apostolico et al., 2009 [35] and Djelloul and Denise, 2008 [24] by comparing results from searches for five internal loop motifs (kink-turn, reverse kink-turn, C-loop, sarcin-ricin, E-loop) in PDB file 1S72 [37]. However, it is difficult to interpret the results because the authors used FR3D inconsistently. For example, the authors chose not to apply FR3D to search for C-loop, loop E, and reverse kink-turn motifs, although the program was specifically designed for these kinds of searches. The authors used RNAMotifScan to search the entire PDB for these five motifs and made the search results available online along with the RNAMotifScan source code.

15.5.7 FRMF

FRMF ("Feature-based RNA Motif Filtering" [38]) introduces a new technique for selecting matches from the list of potential candidates generated by FR3D motif searches [14]. For each candidate structure found by FR3D, FRMF computes three feature vectors based on geometric moment invariants for base, ribose, and phosphate components of each nucleotide. Next, these vectors are compared with those of the query motif, and the candidate is accepted if several predefined cutoff values are satisfied. The authors benchmarked FRMF against FR3D and the method proposed by Apostolico et al., 2009 [35] and found that FRMF and FR3D performed comparably and both outperformed the approach reported by Apostolico et al.

15.5.8 RNA FRABASE 2.0

RNA FRABASE 2.0 ("RNA FRAgments search engine and dataBASE" [39]) is a regularly updated online database, containing searchable annotations of RNA 3D structures. Users can query the database to find 3D structural fragments, including internal, hairpin, and junction loops, using the 2D representation expressed in dot-bracket notation. The queries are compared to the secondary structures computed based on RNAView basepair annotations [40]. Users can submit their own structures, which are parsed to identify the secondary structures and the locations of loops for searching.

15.5.9 FASTR3D

FASTR3D, which stands for "Fast and Accurate Search Tool for RNA 3D structures" [41], is a web-based tool for searching the PDB database. The users can input queries in the form of RNA sequence, RNA secondary structure in dot-bracket notation, or a fragment of 3D structure from a PDB file. The server predicts the query secondary structure, if necessary, and compares it to a set of annotations from all RNA 3D structures precomputed using RNAView. Additional screening criteria based on sequence identity and backbone similarity can be applied.

15.5.10 FRASS

FRASS [42] is a web server for global comparisons of RNA 3D structures based on the analysis of backbone conformations. Each RNA structure, represented as a chain of phosphorous atoms, is described by a 30-dimensional vector using Gauss integrals. The server performs global structural comparisons by computing Gauss integrals based on the distance between the query and the target structures. It also has a database scanning mode, in which the user can submit a query structure in PDB format and find matching 3D structures from the PDB.

15.5.11 R3D-BLAST

R3D-BLAST [43] is another tool for searching the PDB database from the authors of FASTR3D (see Section 15.5.9). The method uses the BLAST algorithm to compare the representations of backbone conformations of query and target structures expressed in a 23-character set corresponding to the most common rotamers identified using the reduced η/θ pseudotorsion angles. R3D-BLAST can detect local structural similarities between structures having different lengths and/or sequences. While most other programs construct nucleotide-to-nucleotide structural alignments of the entire query with the matching structure, R3D-BLAST uses partial query coverage, which allows the query to partially match the target.

15.5.12 Comparison of 3D search methods

Symbolic search has the virtue of speed and scalability. It is generally possible to search for much larger motifs using symbolic approaches. However, symbolic search depends critically on the accuracy of structural annotations and is limited by the fact that many RNA 3D structures or parts of structures cannot be annotated adequately, due to low resolution of the underlying data, poor modeling of the electron density, or both.

The best results are obtained by methods that combine geometric and symbolic searching. The use of geometric search allows capturing geometrically similar but poorly modeled motifs, while the judicious use of carefully selected symbolic constraints speeds up the geometric search to make it more practical for larger and more complex motifs. FR3D is unique in combining symbolic and geometric search in an integrated user interface.

To accommodate the interest in different types of RNA 3D motifs, a combination of approaches is fruitful, as different methods may be better suited for different types of motifs. For example, backbone motifs like the π-turn, which lack a distinct base-pairing pattern, are more readily identified by programs like PRIMOS that analyze backbone conformations [33]. Base-centered tools like FR3D are well equipped for accurately distinguishing between similar motifs with complex geometries by fine-tuning the searches using symbolic constraints for specific interactions.

15.6 Classifying RNA 3D motifs

15.6.1 Why classify RNA 3D motifs?

It is useful to group or classify geometrically similar RNA 3D motif instances for several reasons, including the following:

1. To identify sequence variants that can form the same recurrent motif. Alignment of variants provides knowledge of the "sequence signatures" of RNA 3D motifs, which is useful in predicting RNA 3D structure from sequence and in identifying structurally conserved motifs in homologous RNA molecules [9].
2. To identify structural variants of the same RNA sequence to better understand induced fit effects in RNA and to improve prediction of RNA structural dynamics.
3. To characterize the range of structural and functional contexts in which the same motif can occur, so as to improve capabilities for predicting functions of new RNAs identified in genomes.
4. To detect possible errors or inconsistencies in structural modeling of RNA and suggest improvements to experimentally determined structures.

15.6.2 How to classify RNA 3D motifs?

Classifying RNA 3D motifs from experimentally determined structures is complicated by a number of factors: First, there is large variation in the quality of the underlying experimental data of published RNA 3D structures [44]. This variation occurs within individual structures as well as across structures. While the variation among different structures is well appreciated, that within individual structures should not be overlooked, and one should be wary of placing undue confidence in the global refinement statistics reported for any given structure. Use of nucleotide-level statistics [45,46] will allow individual assessment of the confidence to be placed in the geometries of individual motif instances from large structures [47].

Second, there is considerable variation in the quality of the 3D modeling among structures of comparable reported resolution. This is apparent from large variations in the number of interactions, especially basepairs, annotated by programs such as

FR3D or RNAView for different structures of the same RNA molecule. Poorly modeled structures of *E. coli* 16S or 23S rRNA, identified as such by manual inspection, consistently produce fewer annotated basepairs, using automatic annotation programs, than well-modeled structures of the same molecules [44].

Finally, there is real variation due to effects of induced fit and inherent conformational flexibility, which cannot be captured accurately by static models in conventional data formats used in structure databases such as PDB.

Our experience shows that automatic classification of RNA 3D motifs is most successful for 3D motifs that have a high density of internucleotide interactions and is generally more problematic when a large fraction of motif bases are looped out and subject to induced fit through interactions with other structural elements.

15.6.3 Criteria for grouping motif instances in the same recurrent family

The central challenge of 3D RNA motif analysis is to decide which motif instances, extracted from the 3D database using criteria such as those listed previously, should be clustered together in the same recurrent, modular motif families. For the purpose of getting well-defined results, at least for modular motifs that correspond to individual "loops" in the 2D representation, the following criteria can be proposed [48]:

1. Instances of the same recurrent motif should be of the same type, in the sense that hairpin loops belong together and should not be mixed with instances of internal or junction loops, however similar they are otherwise. Similarly, a motif family that contains internal loops should not contain hairpin or junction loops and a family of three-way junction should not contain four-way or higher-order junctions.
2. Instances of the same recurrent motif should be similar in size, but not necessarily identical. Size identity should not be required because motif instances that are geometrically very similar may contain variable numbers of bulged out bases that do not affect the geometry of the rest of the motif. This is possible because of the flexibility of the RNA chain, as discussed previously.
3. The nucleotides of the motif that interact with each other are called the core nucleotides of the motif. The core nucleotides, and the sites at which bulged bases are inserted, are defining characteristics of recurrent RNA 3D motifs. Motifs assigned to the same family should therefore have the same number of core nucleotides. Deciding which nucleotides to include in the core, in a consistent manner, is therefore a crucial step in motif classification. The success of automatic procedures in carrying this out depends critically on the quality of the experimental data and the resulting annotations.
4. Motif instances that belong to the same recurrent family are geometrically similar by some appropriate quantitative measure, like RMSD or geometric discrepancy,

as implemented in FR3D, which takes into account how well the centers of corresponding (i.e. aligned) bases superpose, as well as a measure of the coplanarity of the aligned bases [14]. Measures like geometric discrepancy and RMSD can be calculated only for two motifs that have the same number of core nucleotides.
5. Pairs of corresponding core bases in different instances of the same recurrent motif should form the same types of basepairs. When corresponding bases form different types of basepairs, the motif instances are sufficiently different to be assigned to different families. This criterion underscores the critical role of accurate basepair annotations.

15.6.4 Evaluating 3D motif similarity

RNA 3D motifs can be represented by directed graphs with nucleotides as graph nodes and pairwise interactions as graph edges. This representation is very useful for searching and classification. Its implementation enables rapid text search for RNA 3D motifs in structural annotations [14,15,24]. In general, when the interaction graphs of two motifs are identical and the motifs have a sufficiently high density of pairwise interactions, their 3D structures also superimpose closely, reflecting high geometric similarity.

15.6.5 Application of motif classification criteria

In principle, classification of 3D motifs instances should be straightforward – we simply demand that all motif instances included in the same motif family should share identical interaction graphs. In practice, however, this approach only works consistently for motif instances extracted from the best modeled, highest resolution structures that also have a high density of pairwise interactions. For other motifs, the overwhelming majority, strict application of the criteria listed previously fails to group motifs that belong together and results in creation of large numbers of classes of similar motifs that probably should be grouped together. The following issues therefore arise: what to do with poorly modeled motifs and with the ambiguities resulting when annotating them, and how to deal with structures solved at low to moderate resolution that have many imprecisely modeled and, consequently, inconsistently annotated basepairs.

To deal with the challenge of imprecise structures, we implemented annotation of "near" basepairs in the FR3D program suite [34] This is a unique feature of FR3D annotations. The question then becomes whether to group motif instances where one structure has an easy-to-annotate basepair, another has a near basepair, and a third has no detected basepair at all or a basepair annotated with a different family, for the same pair of corresponding bases. Overly stringent criteria result in too many motif groups, most of which only have one or two members. Loose criteria produce inadequate splitting and a small number of very large and heterogeneous groups.

15.6.6 Automatic classification of RNA 3D motifs

Several approaches for automatic RNA 3D motif classification have been proposed, including COMPADRES [49], methods developed by Huang et al. [50] and Wang et al. [51], Rna3Dmotif [24], and RNAMSC [52]. However, these methods have both theoretical and practical limitations. Most of them were tested only on a limited number of structures, or only on a small set of motifs, and none of the methods implements a pipeline for annotating newly released structures. In addition, many of these methods do not take into account non-WC interactions, which makes them insensitive to the differences in the interaction networks that stabilize the motifs. RNA 3D Motif Atlas, a new online resource developed by our laboratory, aims to address these shortcomings and is discussed subsequently [47].

15.7 RNA 3D motif collections

The existing RNA 3D motif collections can be divided into loop-oriented and motif-oriented collections. Loop-oriented motif collections, which include RNAJunction and RNA CoSSMos, extract secondary structure elements from RNA 3D structures and store them in online searchable databases. Although these databases contain instances of RNA 3D motifs, they do not make an effort to cluster these instances into structurally similar families.

Motif-oriented collections, by contrast, are focused on classifying 3D motif instances into geometrically similar families. These collections are either maintained manually or are collected using a motif searching tool (e.g. RNAMotifScan and FRMF). Tab. 15.3 lists existing loop- and motif-oriented collections described individually in the following sections.

15.7.1 Motif-oriented collections

15.7.1.1 SCOR

The SCOR database [53,54] was the earliest attempt to extract and classify RNA 3D motifs from 3D structures. RNA-containing structures determined using X-ray crystallography or nuclear magnetic resonance were examined manually to create a structural classification of internal and hairpin loops into distinct 3D motifs, as defined previously (for example, kink-turn or GNRA tetraloop), and broader structural groups,

Tab. 15.3 Loop-oriented and motif-oriented RNA 3D motif collections.

Method	URL	Reference	Basepair annotations
Loop-oriented collections			
RNAJunction	http://rnajunction.abcc.ncifcrf.gov	Bindewald et al. [58]	RNAView
RNA STRAND	http://www.rnasoft.ca/strand/	Andronescu et al. [59]	RNAView
RLooM	http://rloom.mpimp-golm.mpg.de	Schudoma et al. [63,64]	MC-Annotate
RNA CoSSMos	http://cossmos.slu.edu	Vanegas et al. [65]	MC-Annotate

Method	URL	Reference	Motifs reported
Motif-oriented collections			
SCOR, SCOR 2.0	http://scor.berkeley.edu/	Tamura et al. [53]	Not Available
Comparative RNA Web (CRW) Site	http://www.rna.icmb.utexas.edu	Cannone et al. [55]	12
K-turn database	http://www.dundee.ac.uk/biocentre/nasg/kturn	Schroeder et al. [56]	1
RNAMotifScan	http://genome.ucf.edu/RNAMotifScan/	Zhong et al. [36]	5
FRMF	http://bioinformatics.oxfordjournals.org/content/suppl/2011/08/27/btr495.DC1/motifs.xls	Shen et al. [38]	8
RNA 3D Motif Atlas	http://rna.bgsu.edu/motifs	Petrov et al. [47]	~250 internal loop and ~250 hairpin loop motifs

denoted by descriptions such as "loops with stacked interdigitated bases" or "loops with dinucleotide platforms." The database also contained functional classifications at molecular and motif levels and a classification of tertiary interactions.

15.7.1.2 Comparative RNA Web Site

The Comparative RNA Web (CRW) Site [55] provides a wealth of sequence and structural information, including a compilation of several motifs and their instances as described in the RNA literature (see http://www.rna.icmb.utexas.edu/SIM/4A/ CRWStructure and http://www.rna.icmb.utexas.edu/SIM/4B/) and an interactive tool called RNA2DMap (http://www.rna.icmb.utexas.edu/SAE/2A/RNA2DMap/index.php), which maps motif instances on to ribosomal secondary structure diagrams.

15.7.1.3 K-turn database

The K-turn database [56] is a manually constructed database dedicated exclusively to kink-turn motifs, both known and predicted on the basis of sequence similarity. The motifs are grouped by the type of RNA molecule in which the motif occurs. The structures can be visualized and superposed using the Jmol applet (http://jmol.org) [57]. Secondary structure diagrams and weblogo diagrams showing sequence variability for each kink-turn are also provided.

15.7.1.4 RNAMotifScan

A compilation of search results obtained by the RNAMotifScan motif searching program [36], described previously, is provided online at http://genome.ucf.edu/ RNAMotifScan/. The list contains instances of kink-turn, reverse kink-turn, C-loop, sarcin-ricin, and E-loop motifs in all 1,445 RNA 3D structures available through the PDB at that time. This is an example of a database limited with respect to the types of motifs included.

15.7.1.5 FRMF

The supplementary information file from the FRMF publication [38] contains GNRA tetraloops, sarcin-ricin, kink-turn, π-turn, A-minor, and ribose zipper motifs from 50S *Haloarcula marismortui* rRNA. These motif instances were collected from the literature to validate the methodology.

15.7.1.6 RNA 3D Motif Atlas

RNA 3D Motif Atlas is the first attempt to create a fully automated and regularly updated collection of internal and hairpin loop RNA 3D motifs found in a nonredundant set of RNA 3D structures from PDB (http://rna.bgsu.edu/rna3dhub/nrlist). The Motif Atlas is constructed by exhaustively comparing a representative set of motif instances in an all-against-all fashion using the FR3D geometric search capabilities [14] and clustering the resulting similarity matrix into coherent groups using maximum cliques. The data are made available online in a database featuring an advanced user interface allowing the users to interactively explore the similarity within and across motif groups. RNA 3D Motif Atlas provides unique and stable identifiers for all motifs and motif instances and supports versioning. Preliminary versions of the RNA 3D Motif Atlas were described in [48], and the first major release is currently under active development [47] (http://rna.bgsu.edu/motifs).

15.7.2 Loop-oriented collections

15.7.2.1 RNAJunction

The RNAJunction database [58] uses basepair annotations produced by RNAView [40] to extract junctions, internal loops (regarded as two-way junctions), and kissing hairpin loops, a type of tertiary motif. The website provides online visualization of 2D and 3D structures of all catalogued loops as well as the raw output of JunctionScanner, the program used for loop extraction. It also contains energy-minimized structures of all loops. The website has versatile search capacity, including the ability to search a nonredundant set of PDB files.

Loops with the same sequence were automatically superposed and clustered using a single linkage algorithm with 3.0Å RMSD cutoff. For each cluster a representative is selected.

15.7.2.2 RNA STRAND

RNA STRAND [59] is an online database containing validated secondary structures collected from multiple, publicly available sequence databases, including RFAM [60], RNAse P Database [61], and tmRNA database [62]. It also provides secondary structures computed using the basepair annotations of 3D structures from PDB and Nucleic Acids Database (NDB) as determined by RNAView [40]. New entries are added to the database after manual curation, and all secondary structures undergo several quality assurance procedures designed to ensure that the structures are valid. In a valid secondary structure, a base can participate only in one canonical basepair. Although extracting RNA loops is not the primary goal of RNA STRAND, the resource can be used to determine the positions of loop regions by parsing the output generated by the RNA Secondary Structure Analyzer program for each database entry.

15.7.2.3 RLooM

The RLooM [63,64] is an online database of RNA loops developed for homology-based loop modeling. The loops are extracted based on MC-Annotate basepair annotations and are organized in a database. Users can search loops by sequence or by using MC-Search query descriptors. Loops are also clustered based on length, sequence identity and structure similarity defined by pairwise RMSD comparisons between backbone atoms of loops with the same length. For each cluster a representative structure is selected. The RLooM web application allows for insertion of the best-fitting loops into a user-submitted structure in PDB format, which can be used in conjunction with RNA 3D structure prediction software.

15.7.2.4 RNA CoSSMos

The RNA CoSSMos ("RNA Characterization of Secondary Structure Motifs") database [65] is an automatically updated online database that provides access to internal and hairpin loops extracted from all RNA-containing PDB files. MC-Search [29,30] and input descriptors similar to the one described by Davis et al. [66] are used to extract the loops. Not unlike RNA FRABASE 2.0, RNA CoSSMos allows user to search loops by size, sequence, PDB file of origin, and several other characteristics.

15.7.3 Comparing RNA 3D motif collections

To compare motif collections one needs to answer the following questions: (1) Is the motif instance present in the collection? (2) Is it grouped with the same set of other instances? These seemingly easy tasks, however, are complicated by numerous obstacles because for some collections even accessing the data can be challenging. For example, motif instances for each PDB entry in RNA CoSSMos are grouped by size, and as a result, to retrieve all motif instances for PDB 1S72 one has to download 25 separate files.

Furthermore, the databases present the data in different formats usually without providing sufficient documentation or parsing software. The situation gets worse for PDB files with multiple models, or with nucleotides with alternate locations, or any other features, which present additional challenges and are treated inconsistently by different methods. A universal naming scheme for nucleotides and nucleotide fragments is fundamental for resolving many of these issues.

The same loop may be extracted inconsistently by different methods. For example, the flanking bases can be determined incorrectly resulting in overlapping or "nested" loops. As discussed subsequently, some methods extract parts of multihelix junctions or pseudoknots as hairpins or internal loops. In some cases differences between methods are due to inconsistencies in basepair annotations produced by the three major annotation programs (RNAView, MC-Annotate, FR3D), but this also reflects the lack of common definitions and clear descriptions of the algorithms employed.

We compared motif instances extracted from 23S rRNA from *Haloarcula marismotrui* (PDB 1S72) and compiled in four motif collections (SCOR, RLooM, RNAJunction, CoSSMos) with those obtained by two motif search methods (FR3D, Rna3Dmotif). The results can be viewed online at http://rna.bgsu.edu/motif-extraction-benchmark. The code used for constructing the benchmark is available at https://github.com/BGSU-RNA/loop-extraction-benchmark, along with the description of the employed procedures. Although more work needs to be done to fully analyze the discrepancies between the results produced by different algorithms, it is clear that there is no clear consensus across these six methods. Nevertheless, it must be noted that FR3D and Rna3Dmotif have a high degree of agreement because they use the same basepair annotations and adopt the same definition of RNA 3D motifs.

Another way of comparing motif collections is to validate them against a "gold standard" set of motifs. The problem with this approach is that, although many motifs have common names, and these are widely used in the literature, some of these names have synonyms; for example, sarcin-ricin, G-bulge, S-motif, and loop E can refer to essentially the same structural motif. Some motifs are named according to their provenance; for example, kink-turn 23, also referred to as kt-23, comes from h23 of 16S rRNA. Other motifs, like α-loop or π-turn [49], may be hard to find in the literature due to non-Latin characters in their names. Differences in spelling, ambiguous usage of terms and multiple synonyms make the preparation of a comprehensive list of motif instances a daunting task.

None of the existing motif collections implements a regularly updated pipeline for annotating all PDB files as they become available (although RNA CoSSMos is regularly updated, it does not organize motifs by geometric similarity). Also, the majority of methods to extract and classify motifs have only been tested on restricted sets of PDB files or only focus on a limited set of 3D motifs.

In conclusion, programmatic data discovery and data access, consistent naming schemes for nucleotides and nucleotide fragments, adoption of unique and stable identifiers for motifs and motif instances, and development of robust automated motif classification pipelines are needed in order to make the accumulated body of knowledge regarding RNA 3D motifs useful for a wide community of scientists not necessarily specializing in RNA structural bioinformatics. We are attempting to address these challenges with the upcoming release of RNA 3D Motif Atlas [47] with the preliminary version available at http://rna.bgsu.edu/motifs.

15.8 RNA 3D motifs that "break the rules"

By and large, most hairpin, internal, and junction loops correspond to one 3D motif, and most RNA 3D motifs correspond to a single loop. However, there are significant numbers of exceptions, and it is important to identify, analyze, and classify the exceptions as these are the cases that will be overlooked or incorrectly extracted and classified by automatic procedures based on the criteria and assumptions summarized previously. Only by identifying and examining each exception and analyzing its characteristic features can we hope to modify the automated procedures appropriately to capture more instances and improve classification into recurrent families. The reader is referred to Section 15.4.1 for the definition of "classical" RNA 3D motifs. The motifs that we discuss in this section constitute exceptions to this definition because they violate one or more criterion and therefore do not correspond neatly in a one-to-one manner to a single loop in the secondary structure.

15.8.1 The 3D motifs that contain isolated cWW basepairs

We start with motifs that comprise "isolated" WC pairs. These motifs violate the rule that no two nucleotides of the motif, except for the flanking basepairs, form canonical WC basepairs, defined to include cWW AU and GC pairs as well as "wobble" GU pairs. An isolated WC pair is one that cannot be considered as flanking the motif because the nucleotides forming it constitute an integral part of the 3D motif as they interact with the rest of the motif. It is unlikely that the motif can form correctly without these nucleotides and they should therefore be included in the description of the motif. Frequently, these WC pairs, embedded as they are inside of a 2D "loop," are not detected by comparative sequence analysis and only become apparent after the 3D structure is solved. They are not detected when they are highly conserved, most likely due to the high density of specific interactions they participate in. Consequently, they do not give a detectable covariation signal when multiple sequences are aligned and compared. However, automatic procedures based on detection of flanking WC pairs in 3D structures invariably separate the original motif into two independent motifs.

As an example, we show the highly conserved h20-h21-h22 three-way junction (3WJ) from 16S rRNA in Fig. 15.5, in the standard 2D representation as a "loop" (Fig. 15.5A), as an annotated schematic that includes symbols representing non-WC basepairs and BPh interactions (Fig. 15.5B), and in 3D, using the same coloring scheme as in Fig. 15.5A,B. Bases that are stacked are shown orthogonally adjacent to each other. Nucleotides are numbered using the *E. coli* numbering, which is standard in the bacterial ribosome field. The "isolated" WC pair comprises G654/C754 but is generally not drawn in 16S secondary structures for reasons discussed in the previous paragraph. It is shown in Fig. 15.5B, which captures the interactions found in 3D. This pair only became apparent once the 3D structure was determined; remarkably, it still is not drawn in most 2D representations of 16S rRNA. However, the isolated G654/C754 cWW pair formally separates the 3WJ into an internal loop, comprising nucleotides 654–655 and 751–754 and flanked by cWW G654/C754 and A655/U751. This loop is identified and extracted by 3D analysis programs, leaving a smaller 3WJ loop comprising only 586–588, 651–654, and 754–755, or just 9 nucleotides including flanking pairs.

The sequence conservation of the G654/C754 pair appears to be due to the formation of the CGG cWW/tHS base triple [67] involving tHS pair between G654 and G752, in addition to the G654/C754 cWW pair, as annotated in Fig. 15.5. U652 and A753 are also highly conserved, perhaps because they form a crucial tWH basepair in the core of the 3WJ structure. G587 forms a 3BPh interaction with phosphate 754 and sometimes becomes U in other organisms which can also form this interaction [11]. All the bases in the 3WJ loop region are conserved between the *T. thermophilus* and *E. coli* 3D structure except for the bulged out base at position 653, which is an A in *T. thermophilus* and U in *E. coli*. G752 lies in the same plane as G654 and C754 to form the base triple, but C754 is sandwiched between U751, which forms the flanking basepair of

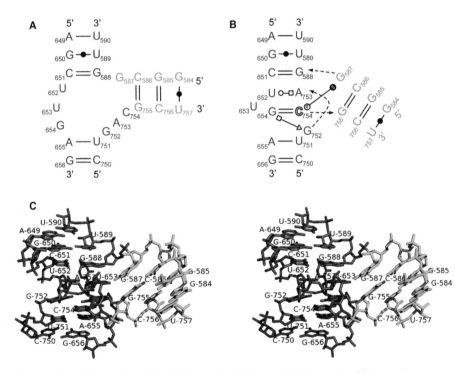

Fig. 15.5 Example of a 3D motif containing isolated Watson-Crick basepairs. (A) Secondary structure diagram showing a three-way junction from 16S *E. coli* rRNA that is treated as a smaller three-way junction and an internal loop by FR3D [14]. (B) The same fragment annotated with basepair interactions according to the Leontis-Westhof nomenclature. (C) Stereo view based on PDB 2AW7. In all panels, h20 is green, h21 is red, and h22 is blue. The 3D structure can be interactively explored online at http://rna.bgsu.edu/publications/deGruyter2013/. The isolated cWW pair is shown to be an integral part of the 3WJ.

h22 with A655, and A753. This arrangement requires a tight turn in the RNA backbone, and this appears to be stabilized by the BPh interaction with G587. Including the flanking pairs, the intact 3WJ comprises 13 nucleotides.

In summary, this detailed analysis of the interactions in the motif demonstrates why in this case the "isolated" G654/C754 cWW pair should not be treated as a flanking basepair.

The difficulty of implementing an automated procedure to identify isolated WC basepairs such as this one is that there are other cases in RNA structures where two distinct motifs share a flanking basepair and where this pair should be treated as a flanking basepair for both motifs. An example is the single CG WC basepair separating the GNRA loop and the sarcin-ricin motif in the very conserved "sarcin-ricin loop" in Domain VI of 23S rRNA. Distinguishing these two cases will be challenging.

15.8.2 Composite 3D motifs: 3D motifs composed of more than one loop

Loops that are close to each other in the 2D structure sometimes interact with each other. In these cases, it is likely that the interaction plays a role in the formation of the observed 3D structure and that without the interaction the loops will form different structures. The 3WJ of 5S rRNA provides an example. Interaction between bases in the 3WJ with a single-base bulge internal loop separated from the 3WJ by two WC basepairs are probably crucial for formation of the 3D structure of the junction, with continuous stacking between helices II and IV.

Helix h6 in *E. coli* 16S rRNA presents a striking example of a composite internal loop 3D motif [68]. This helix extends out from the 30S ribosome to form the "spur" feature, first noted in early electron microscopy work on the ribosome. At the 2D level, this helix is parsed into four relatively simple internal loops, as shown in the 2D representation in Fig. 15.6A. There is two-base bulged loop (G64 and A65) separated by two WC pairs from a three-base asymmetric loop (G68, G100, and A101), followed by two more WC pairs and another two-base bulged loop (A71 and A72). Finally, there are three more WC pairs and a one-base bulged loop (G94). However, in the 3D structure it is apparent that all these internal loops interact with each other to form a single 3D motif stabilized by an integrated network of non-WC basepairs and extended base stacking. This motif mediates RNA-RNA interactions with the HL of h15 and the IL of h8 to structure the lower part of the body domain of 30S.

The specific interactions are shown in the schematic in Fig. 15.6B, which summarizes the 3D structure in Fig. 15.6C. G64 extends across the major groove and forms a cWH pair with G68, forming a triple, as G68 is also cWW paired with A101. A65 is bulged out completely and intercalates in the HL of h15. The G64-A65 bulge increases the twist of the h6 helix, unstacking A66 from C63, which may be necessary to properly orient the distal end of the helix to interact with h8. The unstacking of A66 allows for stacking of U173 on the ribose of A66, as part of the interaction with h8. G100 is not completely

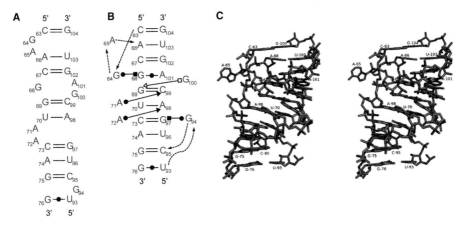

Fig. 15.6 Example of a 3D motif composed of multiple internal loops. (A) Secondary structure diagram of h6 from *E. coli* 16S rRNA. Internal loops are highlighted with gray boxes. (B) The same diagram showing basepairs annotated according to the Leontis-Westhof nomenclature. (C) The 3D visualization of the fragment in stereo (PDB 2AW7). The two RNA strands are consistently colored in red and blue. The four internal loops are combined into a single structural unit stabilized by an intricate interaction network. The 3D structure can be interactively explored online at http://rna.bgsu.edu/publications/deGruyter2013/

bulged out, like A65, but rather stacks between A101 and A71 and forms a tHS pair with G69 in the minor groove of the flanking WC basepair G69/C99. As is typical of adjacent bulged As, A71 and A72 stack on each other and form cWS pairs with the neighboring WC basepairs, A71 with C99 and A72 with A98. This creates a platform for G100 to stack upon. A101 in turn stacks on G100 rather than C99, resulting in a further increase in the twist of the helix. C73 stacks on bulged A72 rather than U70, while U70 stacks on G94. G94 forms a cWH pair with G97, which is also WC paired to c73. There is a dyadic symmetry in the disposition of bulged G64 and G94, which are on opposite strands. Each of these bulged Gs extend into the major groove to form cWH basepairs with WC paired Gs that are located three basepairs away in the 3' direction, G64 with G68 and G94 with G97, resulting in base triples that enhance base stacking in the motif. Thus the motif consists of a series of base triples that change the normal stacking in the helix so as to produce a consistent increase in helical twist and provide docking sites for RNA-RNA interactions. Remarkably, only A64 is completely bulged out from the motif. These observations argue that motifs such as the composite motif in h6 should be treated as single modules. New methods will need to be developed to identify such motifs automatically as current methods separate them into isolated internal loop motifs.

15.8.3 Motifs comprising linker strands

Fig. 15.3 shows that the linker region connecting the head domain to the body of the 16S rRNA in the 30S ribosome. As explained previously, because of the strand topology of 16S, these two major domains are formally linked by a single strand rather than a helix. Analysis of the 3D structure, however, shows that most of the nucleotides shown in 2D as single stranded in the linker region between the pseudoknot (helix 2), helix 28, and helix 44 are involved in multiple pairing, stacking, or base-backbone interactions. This extends the pairing of these helices and results in a continuous helical stacking between h2 and h28 creating multiple reinforcing tertiary interactions involving all these elements. As a result, this motif cannot be neatly classified or extracted by the currently used automatic methods.

15.8.4 Motifs interacting with adjacent helices

The α-loop motif is a recurrent asymmetric internal loop with six bulged nucleotides in one strand and none in the other strand [49]. Fig. 15.7 shows a variation of the α-loop first described by Wadley and Pyle [49]. The flanking WC basepairs of the motif

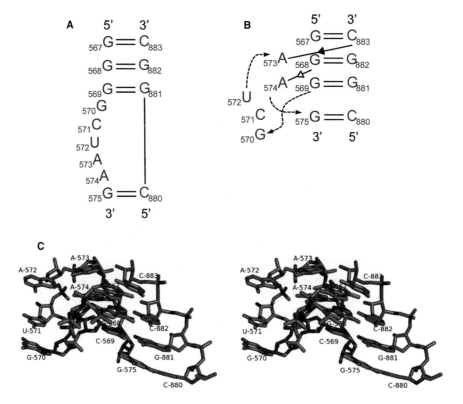

Fig. 15.7 Example of a 3D motif interacting with an adjacent helix. (A) The 2D diagram of a variant of α-loop from 16S *E. coli* rRNA. (B) The same fragment with annotated basepair interactions according to the Leontis-Westhof nomenclature. (C) Stereo view of the fragment from PDB 2AW7. The 3D structure can be interactively explored online at http://rna.bgsu.edu/publications/deGruyter2013/. The loop is shown to make important, conserved interactions with bases outside the loop.

stack on each other with very little distortion of the helical geometry. The bulged nucleotides loop around in the shape of the Greek letter α (alpha), giving the motif its name. None of the nucleotides in the loop interact with the flanking pairs, although there is stacking between bases in the loop.

The two bases on the 3' end of the loop are almost always As. They stack on each other and form cSS and tSS "A-minor" interactions with basepairs in the 5' adjacent helix to the motif. These basepairs are one and two basepairs distant from the flanking pair located 5' to the bulge and should be considered part of the motif, but the current motif searching algorithms ignore these interactions because they occur outside the boundary defined by the flanking pairs. Therefore, current algorithms do not detect the minor groove interactions of the conserved As or the stacking of other bases in the loop that form pseudoknots. In 2D representations of the 16S, the nucleotides of this motif are drawn as part of a complex multihelix junction.

15.9 Conclusions

In writing the chapter, we identified several ways in which automatic RNA 3D motif searching, annotation, and classification can be improved. We conclude this contribution by summarizing these. In some cases, the solutions are obvious and just need to be applied systematically. In other cases, research is needed to identify the best way to accomplish the goal.

1. First, it is important for all workers in the field to explain clearly how they define, search, and extract motif instances from 3D structures, so that consistent results can be obtained.
2. Automated annotation pipelines for motifs need to be developed.
3. Unique identifiers for motifs, motif instances, and their constituent nucleotides are needed for efficient communication and comparison of results and comparison of classification methods.
4. New methods are needed to automatically identify linker regions in 3D structures and identify recurrent motifs found in linker regions.
5. New methods are needed to identify "isolated" WC pairs inside complex motifs to avoid fragmenting one large motif into several smaller nonautonomous submotifs.
6. New methods are needed to identify composite motifs and treat them as integral modules.

15.10 Acknowledgments

This work was funded by NIH Grant 5R01GM085328–03 to N.B.L. The authors thank Prof. Craig L. Zirbel for help with data analysis and production of figures for this chapter.

References

1. Frank J, Gonzalez RL. Structure and dynamics of a processive Brownian motor: the translating ribosome. Annu Rev Biochem. 2010;79:381–412.
2. Valle M, Zavialov A, Li W, et al. Incorporation of aminoacyl-tRNA into the ribosome as seen by cryo-electron microscopy. Nat Struct Biol. 2003;10:899–906.
3. Brown CJ, Johnson AK, Dunker AK, Daughdrill GW. Evolution and disorder. Curr Opin Struct Biol. 2011;21:441–6.
4. Flores S, Echols N, Milburn D, et al. The Database of Macromolecular Motions: new features added at the decade mark. Nucleic Acids Res. 2006;34:D296–301.
5. Stombaugh J, Zirbel CL, Westhof E, Leontis N. Frequency and isostericity of RNA basepairs. Nucleic Acids Res. 2009;37:2294–312.
6. Berman H, Westbrook J, Feng Z, et al. The Protein Data Bank. Nucleic Acids Res. 2000;28:235–42.
7. Laing C, Schlick T. Analysis of four-way junctions in RNA structures. J Mol Biol. 2009;390:547–59.
8. Leontis NB, Ghosh P, Moore PB. Effect of magnesium ion on the structure of the 5S RNA from *Escherichia coli*. An imino proton magnetic resonance study of the helix I, IV, and V regions of the molecule. Biochemistry. 1986;25:7386–92.
9. Nasalean L, Stombaugh J, Zirbel CL, et al. RNA 3D structural motifs: definition, identification, annotation, and database searching. In: Walter NG, Woodson, SA, Batey RT, eds. Non-Protein Coding RNAs. Vol. 13. Berlin: Springer-Verlag; 2009:1–26.
10. Murray LJW, Arendall WB, Richardson DC, Richardson JS. RNA backbone is rotameric. Proc Natl Acad Sci U S A. 2003;100:13904–9.
11. Zirbel CL, Sponer JE, Sponer J, Stombaugh J, Leontis NB. Classification and energetics of the base-phosphate interactions in RNA. Nucleic Acids Res. 2009;37:4898–918.
12. Draper DE. A guide to ions and RNA structure. RNA. 2004;10:335–43.
13. Leontis N, Stombaugh J, Westhof E. The non-Watson-Crick basepairs and their associated isostericity matrices. Nucleic Acids Res. 2002;30:3497–531.
14. Sarver M, Zirbel CL, Stombaugh J, Mokdad A, Leontis NB. FR3D: finding local and composite recurrent structural motifs in RNA 3D structures. J Math Biol. 2008;56:215–52.
15. Lemieux S, Major F. RNA canonical and non-canonical basepairing types: a recognition method and complete repertoire. Nucleic Acids Res. 2002;30:4250–63.
16. Petrov AS, Bowman JC, Harvey SC, Williams LD. Bidentate RNA-magnesium clamps: on the origin of the special role of magnesium in RNA folding. RNA. 2011;17:291–7.
17. Khisamutdinov EK, Sweeney BA, Alenko AL, Leontis N. The role of non-Watson Crick basepairs in stabilizing a recurrent 3D motif. Submitted.
18. Strobel S, Cochrane J. RNA catalysis: ribozymes, ribosomes, and riboswitches. Curr Opin Chem Biol. 2007;11:636–43.
19. Breaker RR. Prospects for riboswitch discovery and analysis. Mol Cell. 2011;43:867–79.
20. Uversky VN, Dunker AK. The case for intrinsically disordered proteins playing contributory roles in molecular recognition without a stable 3D structure. F1000 Biol Rep. 2013;5:1.
21. Ogle JM, Brodersen DE, Clemons WM Jr, Tarry MJ, Carter AP, Ramakrishnan V. Recognition of cognate transfer RNA by the 30S ribosomal subunit. Science. 2001;292:897–902.
22. Leontis NB, Westhof E. Analysis of RNA motifs. Curr Opin Struct Biol. 2003;13:300–8.
23. Leontis NB, Lescoute A, Westhof E. The building blocks and motifs of RNA architecture. Curr Opin Struct Biol. 2006;16:279–87.
24. Djelloul M, Denise A. Automated motif extraction and classification in RNA tertiary structures. RNA. 2008;14:2489–97.

25. Guo P. The emerging field of RNA nanotechnology. Nat Nanotechnol. 2010;5:833–42.
26. Nagaswamy U, Fox GE. Frequent occurrence of the T-loop RNA folding motif in ribosomal RNAs. RNA. 2002;8:1112–9.
27. Leontis N, Westhof E. A common motif organizes the structure of multi-helix loops in 16 S and 23 S ribosomal RNAs. J Mol Biol. 1998;283:571–83.
28. Leontis NB, Stombaugh J, Westhof E. Motif prediction in ribosomal RNAs Lessons and prospects for automated motif prediction in homologous RNA molecules. Biochimie. 2002;84:961–73.
29. Hoffmann B, Mitchell GT, Gendron P, et al. NMR structure of the active conformation of the Varkud satellite ribozyme cleavage site. Proc Natl Acad Sci U S A. 2003;100:7003–8.
30. Olivier C, Poirier G, Gendron P, Boisgontier A, Major F, Chartrand P. Identification of a conserved RNA motif essential for She2p recognition and mRNA localization to the yeast bud. Mol Cell Biol. 2005;25:4752–66.
31. Harrison A-M, South DR, Willett P, Artymiuk PJ. Representation, searching and discovery of patterns of bases in complex RNA structures. J Comput-Aided Mol Des. 2003;17:537–49.
32. Hamdani HY, Appasamy SD, Willett P, Artymiuk PJ, Firdaus-Raih M. NASSAM: a server to search for and annotate tertiary interactions and motifs in three-dimensional structures of complex RNA molecules. Nucleic Acids Res. 2012;40:W35–41.
33. Duarte CM, Wadley LM, Pyle AM. RNA structure comparison, motif search and discovery using a reduced representation of RNA conformational space Nucleic Acids Res. 2003;31:4755–61.
34. Petrov AI, Zirbel CL, Leontis NB. WebFR3D – a server for finding, aligning and analyzing recurrent RNA 3D motifs. Nucleic Acids Res. 2011;39:W50–5.
35. Apostolico A, Ciriello G, Guerra C, Heitsch CE, Hsiao C, Williams LD. Finding 3D motifs in ribosomal RNA structures. Nucleic Acids Res. 2009;37:e29.
36. Zhong C, Tang H, Zhang S. RNAMotifScan: automatic identification of RNA structural motifs using secondary structural alignment. Nucleic Acids Res. 2010;38:e176.
37. Klein DJ, Moore PB, Steitz TA. The roles of ribosomal proteins in the structure assembly, and evolution of the large ribosomal subunit. J Mol Biol. 2004;340:141–77.
38. Shen Y, Wong H-S, Zhang S, Yu Z. Feature-based 3D motif filtering for ribosomal RNA. Bioinformatics. 2011;27:2828–35.
39. Popenda M, Szachniuk M, Blazewicz M, et al. RNA FRABASE 2.0: an advanced web-accessible database with the capacity to search the three-dimensional fragments within RNA structures. BMC Bioinformatics. 2010;11:231–243.
40. Yang H, Jossinet F, Leontis N, et al. Tools for the automatic identification and classification of RNA basepairs. Nucleic Acids Res. 2003;31:3450–3460.
41. Lai CE, Tsai MY, Liu YC, Wang CW, Chen KT, Lu CL. FASTR3D: a fast and accurate search tool for similar RNA 3D structures. Nucleic Acids Res. 2009;37:W287–95.
42. Kirillova S, Tosatto SC, Carugo O. FRASS: the web-server for RNA structural comparison. BMC Bioinformatics. 2010;11:327–335.
43. Liu YC, Yang CH, Chen KT, et al. R3D-BLAST: a search tool for similar RNA 3D substructures. Nucleic Acids Res. 2011;39:W45–9.
44. Leontis N, Zirbel C. Nonredundant 3D structure datasets for RNA knowledge extraction and benchmarking. In: Leontis N, Westhof E, eds. RNA 3D Structure Analysis and Prediction. Berlin: Springer; 2012:281–98.
45. Kleywegt GJ, Jones Ta. xdlMAPMAN and xdlDATAMAN – programs for reformatting, analysis and manipulation of biomacromolecular electron-density maps and reflection data sets. Acta Crystallogr D Biol Crystallogr. 1996;52:826–8.
46. Vaguine AA, Richelle J, Wodak SJ. SFCHECK: a unified set of procedures for evaluating the quality of macromolecular structure-factor data and their agreement with the atomic model. Acta Crystallogr D Biol Crystallogr. 1999;55:191–205.

47. Petrov A, Zirbel C, Leontis N. RNA 3D motif atlas: a comprehensive and biologically pertinent clustering of internal and hairpin loops found in RNA 3D structures. Submitted.
48. Petrov AI. RNA 3D Motifs: Identification, Clustering, and Analysis. Bowling Green: Bowling Green State University; 2012.
49. Wadley LM, Pyle AM. The identification of novel RNA structural motifs using COMPADRES: an automated approach to structural discovery. Nucleic Acids Res. 2004;32:6650–9.
50. Huang HC, Nagaswamy U, Fox GE. The application of cluster analysis in the intercomparison of loop structures in RNA. RNA. 2005;11:412–23.
51. Wang X, Huan J, Snoeyink JS, Wang W, Hill C. Mining RNA tertiary motifs with structure graphs. In: International Conference on Scientific and Statistical Database Management. 2007.
52. Zhong C, Zhang S. Clustering RNA structural motifs in ribosomal RNAs using secondary structural alignment. Nucleic Acids Res. 2012;40:1307–17.
53. Tamura M, Hendrix DK, Klosterman PS, Schimmelman NR, Brenner SE, Holbrook SR. SCOR: Structural Classification of RNA, version 2.0. Nucleic Acids Res. 2004;32:D182–4.
54. Klosterman PS, Tamura M, Holbrook SR, Brenner SE. SCOR: a structural classification of RNA database. Nucleic Acids Res. 2002;30:392–394.
55. Cannone JJ, Subramanian S, Schnare MN, et al. The comparative RNA web (CRW) site: an online database of comparative sequence and structure information for ribosomal, intron, and other RNAs. BMC Bioinformatics. 2002;3:2–33.
56. Schroeder KT, McPhee SA, Ouellet J, Lilley DM. A structural database for k-turn motifs in RNA. RNA. 2010;16:1463–8.
57. Jmol: an open-source Java viewer for chemical structures in 3D. In Accessed April 10, 2013 at http://www.jmol.org/
58. Bindewald E, Hayes R, Yingling YG, Kasprzak W, Shapiro BA. RNAJunction: a database of RNA junctions and kissing loops for three-dimensional structural analysis and nanodesign. Nucleic Acids Res. 2008;36:D392–7.
59. Andronescu M, Bereg V, Hoos HH, Condon A. RNA STRAND: the RNA secondary structure and statistical analysis database. BMC Bioinformatics. 2008;9:340–350.
60. Griffiths-Jones S, Bateman A, Marshall M, Khanna A, Eddy SR. Rfam: an RNA family database. Nucleic Acids Res. 2003;31:439–41.
61. Brown JW. The Ribonuclease P Database. Nucleic Acids Res. 1999;27:314.
62. Zwieb C, Gorodkin J, Knudsen B, Burks J, Wower J. tmRDB (tmRNA database). Nucleic Acids Res. 2003;31:446–7.
63. Schudoma C, May P, Walther D. Modeling RNA loops using sequence and geometric constraints. Bioinformatics. 2010;26:1671–1672.
64. Schudoma C, May P, Nikiforova V, Walther D. Sequence-structure relationships in RNA loops: establishing the basis for loop homology modeling. Nucleic Acids Res. 2010;38:970–80.
65. Vanegas PL, Hudson GA, Davis AR, Kelly SC, Kirkpatrick CC, Znosko BM. RNA CoSSMos: Characterization of Secondary Structure Motifs – a searchable database of secondary structure motifs in RNA three-dimensional structures. Nucleic Acids Res. 2011;40:439–44.
66. Davis AR, Kirkpatrick CC, Znosko BM. Structural characterization of naturally occurring RNA single mismatches. Nucleic Acids Res. 2010;39:1081–94.
67. Abu Almakarem AS, Petrov AI, Stombaugh J, Zirbel CL, Leontis NB. Comprehensive survey and geometric classification of base triples in RNA structures. Nucleic Acids Res. 2012;40:1407–23.
68. Boutorine YI, Steinberg SV. Twist-joints and double twist-joints in RNA structure. RNA. 2012;18:2287–98.

Index

5-bromouridine 91
5-ethynyluridine 83, 85, 90

Absorption 1, 9, 22, 165, 194, 196, 357
Accessibility 30, 45, 51, 53, 55, 57, 61, 62, 66, 68
AdoMet *see* S-adenosyl methionine
AFM *see* Atomic Force Microscopy
Alignment 299
– sequence-based 346
– structure-based 347–349
Alkyne 87–92
Alternating laser excitation 195
Analytical ultracentrifugation 244, 245
Angle
– backbone torsion 288, 297, 368
– glycosidic torsion 289, 297, 298. 368
– euler angle 272
– inter-helical 13–16, 108
– pseudorotation 288, 290
– scattering 216, 220
Anisotropy
– chemical shift 298
– decay 196
– fluorescence 196
– of spin interaction 266, 267
– optical 13
Aptamer 65, 147, 233, 236, 302, 319, 323, 324
Artificial Phylogenies 320
ASAXS *see* Small Angle X-ray Scattering, Anomalous
Association constant 20
Atomic Force Microscopy 125, 133, 137
– cantilever 133–136
– contact mode 134
– deflection 133–135
– split diode 135
– spring constant 133
– tapping mode 134, 135
– tip 133–135
Azide 76, 77, 79, 81, 83–85, 87, 89, 90, 92
– enzymatic incorporation in RNA 85
– postsynthetic incorporation in RNA 84
– solid-phase synthesis of RNA 83

Backbone 35, 36, 53, 58, 59, 63, 75, 76, 82, 83, 92, 137, 236, 264, 288, 289, 294, 297, 298, 300, 326, 340, 368, 372–374, 376, 378, 380, 381, 388, 392, 394
– angles 288, 297, 368
Base pair 288, 290, 291, 306, 335, 338, 339, 343, 344
– intermolecular 125, 151
– long range 129, 305, 370, 371
– non-canonical 35, 61, 67, 291, 299, 305, 321, 374
– reverse Watson-Crick 137
– Watson-Crick 29, 35, 57, 65, 128, 137, 290, 295, 363, 370
– Hoogsteen 137, 290, 305, 306, 370
Base stacking 8, 35, 288, 297, 304, 320, 372, 374, 393
– consequences for crystallization 320
Base triple 291, 305, 306, 307, 346, 376, 391, 393
Bean Pod Mottle Virus 125, 126, 144
Binding
– constant 18
– site 13, 20, 32, 35–37, 176, 235, 250, 299
Bioinformatics 335, 363
Branch 101, 104, 125, 233
Brome Mosaic Virus 138
Bulge 13, 16, 67, 68, 102, 103, 128, 321, 337, 339, 345, 392, 395
Burst identification 193

Calorimetry 1, 6, 17, 18, 21, 22
Capsid protein 145–148
C/D box snoRNP 88
Chemical interference 30, 31, 45
Chemical shift 292–298, 307
Chromatin remodelling 249, 253
Chromophore 1, 9, 81
Circular dichroism 8–13, 265
Click chemistry 75
– bioorthogonal 75
– CuAAC *see* Copper-catalyzed azide-alkyne cycloaddition
– functionality-transfer reaction 84
– photo reaction 80

Coat protein 126, 127, 129, 133, 147–149, 151
Coaxial stacking 110, 111, 113, 114, 115, 118, 119, 304, 307, 341
Cold denaturation 13
Confocal
– microscopy 92, 190, 192
– volume 190–193
Conformational
– change 16, 30, 32, 36, 37, 181, 192, 203, 208, 209, 235, 250, 252, 253, 256, 302, 304, 378
– flexibility 137, 231, 261, 263, 264, 276, 278, 382
– space 229, 230, 278, 288, 289
– transition 136
Cooperativity 11, 13, 22, 237, 252,
Copper-catalyzed azide-alkyne cycloaddition 76
COSY *see* Nuclear Magnetic Resonance, Correlated Spectroscopy
Cryo-EM *see* Cryo-electron microscopy
Cryo-electron microscopy 243
– angular views 246, 249
– class averages 246
– high-resolution maps 248
– missing-wedge 249
– multi-resolution data integration 256
– multivariate statistical analysis 246, 250
– orthogonal tilt 246
– projection matching 246
– random conical tilt 246
– sample heterogeneity 244, 245, 249, 250
– sub-tomogram averaging 249
– three dimensional reconstruction 246, 249, 252, 256
– validation tools 256
Cryo electron tomography 243, 249
– double-tilt 249
– tilt image series 249
Crystallization 319
– circular permutation 324
– use of fitness landscapes 320
– post-crystallization treatment 326
– RNA 319
– RNP 319
– strategy 319, 323
– tetracycline aptamer 324
– U1A module 322, 323
– use of Fab 323
Crystallography *see* X-ray crystallography

Debye equation 224
Dehydration 137
Depurination 62
Desmodium Yellow Mottle Virus 125
Diels Alder reaction 66
– inverse electron demand 80
Dienophile 80
Differential scanning calorimetry 6, 21
Dipolar coupling 181, 233, 263, 298–300, 303, 328, 372
Dipolar interactions
– intermolecular 269, 274, 275
– intramolecular 181, 270
Dipolarophile 78–80
Dissociation constant 11, 324
Distance 181, 193, 207, 261
– constrained docking 204
– distribution 187, 189, 191, 205
– distribution function 219, 220, 234, 263, 270, 275
– restraint 198, 204, 207, 298,
Divalent cations 59, 61, 67, 140, 340
– coordination site 67, 68
– magnesium 8, 11, 13, 36, 56, 61, 67, 68, 103, 108, 110, 118, 199, 201, 203, 205, 235, 237, 264, 302, 328, 372
DLS *see* Dynamic light scattering
DNA
– binding domain 253
– duplexes 265, 266, 277
– fibers 138
– mechanical properties 158
– protein complex 111, 246, 253
DNA-RNA hybrid 89, 108, 161, 166
DSC *see* Differential scanning calorimetry
Dwell time 162, 170, 191, 206
Dynamic light scattering 217, 244, 326

Elasticity 103, 166, 167, 172, 174, 177
– elastic energy 166
– elastic modulus 167
Electron density 63, 126, 217, 221, 236, 237
– maps 126, 128, 251, 380
– difference maps 144
Electron microscopy 243
Electron Paramagnetic Resonance 261
– continuous-wave 262
– detection pulses 271
– deuterated solvent 271

- dipolar interaction constant 267
- dipolar interaction tensor 267
- distance distribution function 263, 265, 270, 275
- double-quantum coherence 263
- ESEEM see Electron Spin Echo Envelope Modulation
- flexible spin label 270, 274
- G-band frequency 266
- G-tensor 266, 267, 272, 274
- high field 281
- hyperfine interaction 266, 267
- hyperfine tensor 263, 266, 272
- probe pulse 273
- proton modulation frequency 271
- pump pulse 263, 268, 271, 272, 273
- rigid spin label 265, 266, 273
- Tikhonov regularization 263, 275–277
Electron Spin Echo Envelope Modulation 262, 271
Ellipticity 9–11
Energy
- considerations 130
- landscape 161, 172, 198, 236, 301, 302
- minimization 149, 232, 338
Enthalpy 3, 6, 18, 20, 23, 24
- enthalpy-entropy compensation 23, 24
- enthalpic and entropic contributions 205
EPR see Electron Paramagnetic Resonance
Equilibrium constant 3, 5–7, 189, 198, 205, 336
Evanescent field 192
Expressed protein ligation 183
Extinction coefficient 1, 186, 194, 196

Fenton reaction 36, 58, 61
Flash-freezing 244
Fluorescence 8, 111, 181, 183, 185, 203,
- anisotropy 196, 197
- decay 184, 187–189, 192, 196, 197
- fluorescent labeling 57, 184
- fluorophores 8, 75, 85, 183, 184, 196, 197
- lifetime 181, 183, 184, 186–189, 192, 195
- quantum yield 183, 186, 193, 195, 196
- rotational correlation time 197
- sensitized emission 186
- steady-state 184
- time-correlated single photon counting 187
Fluorescence Resonance Energy Transfer 8, 181
- correction parameters 194–196, 207
- crosstalk 193

- detection efficiencies 192, 194
- direct acceptor excitation 185, 194
- dwell time 191, 206
- ensemble FRET 184
- frequency domain 187, 188
- histogram 188, 192, 197
- order parameter 197
- overlap integral 196
- steady-state FRET 184
- three-color-FRET 203
- time domain 187
- time-resolved FRET 186
- time trace 190, 193
FRET efficiency 181, 183, 185, 186, 188, 191, 195
- apparent 185
- corrected 193, 195
Fluorophore 183
- acceptor 181, 183
- donor 181, 183
Focused-ion beam 249
Folding
- entropy 4, 5, 22, 336, 337
- intermediate 63, 198, 201, 206
- pathway 7, 12, 13, 36, 127, 129, 199, 201, 203, 206, 207, 235, 302
- two-state 3, 4, 7
Footprinting 29
- in vitro 32
- in vivo 32
Force measurements 157, 158, 161, 172
- flipping 169
- versus time 169
Förster distance 181, 189, 195, 196, 208
Fourier
- transform 219, 225, 263
- synthesis 126
Free energy
- changes 3, 21, 165–168, 236
- landscape 236, 301
FRET see Fluorescence Resonance Energy Transfer
Functionality-transfer reaction 84

Gaussian distribution 189
Gaussian sphere approximation 224
G-band frequency 266
Gel electrophoresis
- comparative gel electrophoresis 101
- theoretical treatment 101
Gel retardation assay 37, 43

Geometric search 376, 378, 380, 387
Gibbs-Helmholtz-equation 4, 7, 18, 21
GlmS ribozyme-riboswitch 328
Global conformation 16, 111, 198, 201, 207, 321
Goniometer 249
Group I intron 8, 206, 236, 306
Group II intron 206, 329
G-tensor 266, 267, 272, 274
Guinier law 219, 220

Hairpin ribozyme 89, 104, 111–113, 201, 202, 204–206
Hairpin 131, 160, 279, 307, 335, 349, 373
– crystallization of 321
– equilibrium with duplex 321
– unfolding 3, 161–172, 278
Hammerhead ribozyme 13, 17, 21–23, 89, 90, 118, 119, 199–201
HDV ribozyme see Hepatitis Delta Virus ribozyme
Heat 18
– capacity 5, 18, 22, 23
Helical stacking 115, 321, 322, 394
Helical RNA 15, 16, 37, 61, 89, 102, 104, 105, 115, 117–119, 126, 128–130, 160, 164, 167, 174, 199, 203, 232, 233, 251, 253, 288, 290, 291, 300, 304, 305, 307, 322, 323, 325, 329, 344, 345, 350, 351, 364, 366, 370, 392–395
– branchpoint 101, 104, 233
– direction of bend 104
– segments 126, 127, 322
– stems 127, 128, 167
Helical Junction 16, 103, 118, 199, 201, 203, 204
– four-way DNA junction 108
– four-way RNA junction 111
– Holliday junction 104, 108
– IRES junction 114, 115
– junction-resolving enzyme 111
– three-way RNA junction 115
Helicase 29, 157, 206, 208, 302
– DEAD-box helicase 206, 208
Helper virus 147
Hepatitis C virus IRES element 112, 115
Hepatitis Delta virus ribozyme 8, 23, 89, 90, 329, 351
High-throughput
– screening 203
– sequencing 57, 60
Hinge region 60, 253, 255
Holey-carbon grid 244

Holliday junction 104, 108, 118
HSQC see Nuclear Magnetic Resonance, Heteronuclear Single Quantum Coherence
Human immunodeficiency virus 13, 16, 104, 296, 297, 302–304, 307
Hydrodynamic model 15
Hydrogen bond 36, 103, 128, 138, 205, 290, 291, 295, 298, 299, 305–307, 372
Hydroxyl radicals 31, 35, 36, 43, 61
– footprinting 229, 236
– multiplexed cleavage 229
Hypochromicity 1, 7
Hysteresis 164, 165, 168

Icosahedral
– symmetry 126, 129, 144
– virus RNA 136, 141
Initiation
– factor 203, 250–252
– Initiator fMet-tRNA^{fMet} 250
Inosine 64, 288
Intermediate state 12, 63, 104, 129, 151, 164, 174, 198, 201, 206, 207, 236, 250
Isostericity 290, 346, 371, 372, 375, 379
Isothermal titration calorimetry 18
Isotope labeling 292
– ^{19}F 292
– segmental labeling 292
Isotropic scattering pattern 216
ITC see Isothermal Titration Calorimetry

Kinetics 7, 80, 81, 90, 170, 198, 203, 256
– rate constant 7, 170, 181–183, 191, 198, 206
– kinetic intermediates 129
– kinetic trap 129, 151, 206
Kink turn 102–104, 322, 376, 378, 379, 384–386, 388, 389
– database 385, 386

L20 protein 174
Lambert-Beer law 1
Lead ribozyme 303
Leontis-Westhof nomenclature 364, 367, 391, 393
Ligand
– binding 19, 23, 31, 35, 36, 194, 235, 253, 262, 277, 302, 326
– binding domain 253
– natural 253

Ligation 75, 80, 81, 87, 89, 90, 106, 184, 292, 304
Linker region 366, 393
Loop 322, 364, 387, 392
– α-loop 389, 394
– C-loop 379, 386
– hairpin loop 41, 307, 335, 349, 353, 365, 375, 385, 387
– internal 66, 104, 233, 339, 345, 355, 364, 373, 378, 379, 391–394
– kissing 7, 172, 307, 322, 335, 336, 387
– loop-E 321, 329, 378, 379, 389
– sizes 128, 165, 353
– tetraloop 23, 232, 278, 306, 307, 322–325, 384
Luciferase reporter gene 92

Macromolecular complex 215, 224, 243
MAD see X-ray crystallography, Multi-wavelength anomalous dispersion
Magnetic tweezers 157, 209
Major groove 83, 89, 291, 305, 307
Mass spectrometry 53, 244, 256
Molecular dynamics simulations 65, 68, 230, 232, 277
Mechanical unfolding 157
Melting
– curve 5, 7, 8, 205, 356
– temperature 5, 18, 22, 264, 336
Metastable state 43, 63, 129, 130, 148, 149
Methyltransferase 87, 88
Mica 133
Minimum energy state 128, 129
Minor groove 291, 305, 306, 375, 393, 395
MIR see X-ray crystallography, Multiple isomorphous replacement
Mitochondrial Phenylalanyl-tRNA synthetase 231
MFold 160, 167, 340, 341, 344
Modeling 65, 198, 221
– ab initio 221–223, 227
– homology 230
– rigid body 224
Modification
– bioorthogonal 74, 91
– chemical 52, 62, 159, 323
– enzymatic 85, 288
– post-synthetic 81, 83, 88, 264
Modulation depth 270, 272–274, 277

Molecular
– construction 159
– diversity, use in crystallization 323, 325
– dynamics 232, 277
– flexibility 279
– motors 157
– ruler 181, 207
Monodispersity 217, 325
Motif 363
– classification 383, 384
MR see X-ray crystallography, Molecular replacement
Multi-angle laser light scattering 244

NAIM see Nucleotide Analog Interference Mapping
Nanopores 157
Neighborhood correlation 150
Neomycin-responsive riboswitch 277, 278
nitroxide 261, 263–284, 286, 271
– orientations 272, 276
– reduction in cells 281
– spin label 261, 263, 264, 266, 271, 272, 281
NMR see Nuclear Magnetic Resonance
NOE see Nuclear Magnetic Resonance, Nuclear Overhauser Effect
NOESY see Nuclear Magnetic Resonance, Nuclear Overhauser enhancement spectroscopy
Nuclear Magnetic Resonance 215, 261, 287–289, 291–293, 295–308
– correlation spectroscopy 294–296, 299, 303
– cross-correlated relaxation 297, 298
– cross-validation 301
– deuteration 292, 296
– heteronuclear single quantum coherence 295, 296, 300
– Nuclear Overhauser Effect 294, 295, 298–300, 303, 306, 307
– Nuclear Overhauser enhancement spectroscopy 278, 292, 294, 295, 298, 303
– quadrupolar line shape 303
– relaxation 292, 298, 303
– scalar coupling 297–299
– T1ρ 303
– Total Correlation Spectroscopy 294–296, 298, 299
– transversal magnetization 268, 303
– transverse relaxation-optimized spectroscopy 296

– transversal relaxation time 270, 271, 281, 303
– transverse relaxation 296, 303
Nuclear Overhauser Effect 294, 295, 298–300, 303, 306, 307
– enhancement spectroscopy 278, 292, 294, 295, 298, 303
Nuclear receptor 243, 244, 249, 253–255
Nuclease protection 12
Nucleic acid
– modifications 52, 53, 78, 80–82, 89, 159
– sequencing 30, 45, 51, 53, 55–58, 62, 64, 65, 232, 287, 320
Nucleoprotein complexes 222, 244, 245, 249, 251, 253–255
Nucleotide analog interference mapping 31, 52, 64, 66–69

Occam's Razor 133
Optical spectroscopy 1
Optical trap 161, 162
Optical tweezers 157–159, 161, 172, 178
– drift 162
– dwell time 162–164, 169, 170
– stretch/release 162, 168
Orientation factor 182, 195, 196
Out-of-equilibrium theory 169
Oxygen scavenger 193

Π-stacking 288, 304
PAGE *see* Polyacrylamide gel electrophoresis
PELDOR *see* Pulsed Electron-Electron Double Resonance
Pariacoto Virus 125, 144, 146
Persistence length 15, 167
Phasing experiment 104, 112, 117
Phosphorothioate 63, 64, 68
Photolysis 79, 80, 84
Phylogenetic analysis 51, 65, 66, 69, 306
Polarization 8, 9, 14, 197
Poliovirus 136, 138, 139, 140, 145
Polyacrylamide gel electrophoresis 32, 56, 101–103, 105, 109, 110, 113–117, 119, 236, 292
Polydispersity 225
Primer extension 30, 31, 33, 57, 58, 60, 63, 68
Probability 162, 166–179, 195, 197, 198, 219, 269, 275, 277–279, 335, 337, 338, 341, 343–345, 348, 349, 355
– distribution analysis 169, 170, 198, 208

– function 167
Probing 30–32, 35, 40, 43, 45, 51, 53, 56, 57, 59, 63, 65, 66
– chemical probing 31, 32, 54, 55, 58, 61, 149, 235
– enzymatic 43, 66, 67
– in-line probing 37, 58, 59, 61
– site-directed chemical probing 31, 43, 45
– structural probing 41, 45, 51–53, 58, 63, 65, 66
Proto-retrovirus 147
Proximity ratio 183, 186, 193
– histogram 198
Pseudoknot 7, 8, 11, 43, 126, 129, 143, 147, 172, 235, 303, 305–308, 311, 335, 336, 346, 349–354, 366, 389, 394, 395
Pseudosymmetry 126
Pseudouridine 288
Pulsed Electron-Electron Double Resonance 261–282
– average distance 263, 264, 274
– comparison with FRET 261
– cryoprotection 270
– data analysis 274, 277
– database 276
– distance range 261, 262, 268, 270, 271, 277
– four-pulse sequence 263
– Hahn-echo sequence 262, 263
– in-cell 280, 281
– on dsRNA 277
– on hairpin RNA 278, 279
– on neomycin-responsive riboswitch 277, 278
– orientation-selective 265, 272, 273, 274, 279
– pake pattern 263
– pulse lengths 271
– Q-band 271
– temperature 270, 271, 278
– using Gd^{2+} and Mn^{2+} ions 281

Radiation damage 217, 227, 244, 327
Radius of gyration 218, 219, 301
RBS *see* Ribosome binding site
RBP *see* RNA binding protein
RDC *see* Residual dipolar couplings
Refractive index 13, 14, 195
Replication 129–131, 133, 147, 150, 199, 201, 307
Reptation 101
Residual dipolar couplings 223, 298–301, 303, 304

Retrotransposon Ty3 147
Retroviruses 155
Reverse transcription 31, 32, 36, 57, 58
Ribonuclease 31, 41, 43, 45, 52, 59, 137, 138, 194, 325
– III 32, 39, 41
– A 59, 138
– H 326
– P RNA 12, 230
– U2 59, 66, 67
– T1 32, 33, 39, 40, 42, 45, 52, 56, 59, 66, 67
– T2 32, 33, 37, 41–43, 45, 66, 67
– V1 32, 33, 37, 39, 40, 43, 45, 59, 66, 67
Ribosome 16, 41–44, 148, 153, 157, 201, 203, 229, 246, 248–252, 368, 371, 373, 375, 392
– assembly 157, 201, 203
– binding site 39, 41–43, 234, 235
– complexes 243, 244, 253
– ribosomal protein S1 43, 45
– ribosomal protein S15 16, 43, 200, 201, 203
– ribosomal RNA 7, 11, 12, 16, 90, 127, 158, 172, 174, 176, 177, 199–201, 203, 217, 227, 229, 246, 306, 339, 364, 366–368, 371–373, 375, 378, 382, 386, 389–394
– ribosomal tunnel 129
Riboswitch 8, 23, 24, 36, 37, 88, 103, 233, 234, 299, 302, 319, 322, 323
– adenine riboswitch 236, 299
– GlmS ribozyme-riboswitch 328, 329
– glycine riboswitch 235, 236
– guanine riboswitch 299
– neomycin-responsive riboswitch 277, 278
– SAM-II riboswitch 234, 235, 302
– TPP riboswitch 235
Ribozyme 7, 62–66, 88, 90, 103, 104, 112, 117, 157, 280, 308, 319, 323, 326
– Diels-Alderase 66–68
– GlmS ribozyme-riboswitch 328, 329
– group I intron 8, 206, 236, 306
– group II intron 206, 329
– hairpin ribozyme 89, 104, 111–113, 201, 202, 204–206
– hammerhead ribozyme 13, 17, 21–23, 89, 90, 118, 119, 199–201
– Hepatitis Delta Virus ribozyme 8, 23, 89, 90, 329, 351
– lead ribozyme 303
– use for purification 326
– Varkud Satellite ribozyme 89, 116, 117
RMSD *see* Root mean square deviation

RNA
– binding protein 29, 41, 131, 172, 206, 321, 322–325
– chaperone 29, 30, 43, 229, 230, 302, 307
– hairpin 3, 4, 8, 11, 37, 39, 41, 42, 131, 132, 158, 160–172, 176, 278, 279, 294, 297, 306, 307, 335, 339, 340, 349, 351, 353, 356, 364–366, 373–375, 379, 382–390
– helix 15, 16, 37, 61, 89, 102, 104, 105, 115, 117–119, 126, 128–130, 160, 164, 167, 174, 199, 203, 232, 233, 251, 253, 288, 290, 291, 300, 304, 305, 307, 322, 323, 325, 329, 344, 345, 350, 351, 364, 366, 370, 392–395
– interference 91, 92, 353
– junction 106, 111–117, 119, 384, 385, 387
– ligand interaction 18, 21, 22, 45
– localization 79, 90, 91, 256
– modifications 53, 63, 78, 79, 82–92, 288
– polymerase 30, 63, 85, 86, 90, 91, 159, 199–204, 206, 249, 254, 291
– primary structure 126, 288, 366
– protein complex 35, 111, 181,183, 184, 192, 198, 201, 202, 206–209, 222, 231, 232, 243–245, 249, 251, 255, 321
– protein interaction 24, 43, 45, 160
– secondary structure 1, 3, 7,10–12, 15, 33, 35, 36, 39, 40, 42, 45, 51, 56, 65–69, 103, 116, 117, 127–132, 137–146, 149, 151, 160, 165, 167, 173, 206, 229, 232–234, 246, 252, 264, 278, 287, 289, 291, 295, 298, 299, 301, 302, 324, 335–337, 340, 341, 345–347, 349–351, 353, 355, 364, 367, 373, 378, 379, 384, 386–393
– single stranded 3, 6, 9, 32, 43, 58, 59, 66–68, 125, 129, 138, 141, 143, 159, 166, 167, 174, 205, 287, 363
– tertiary structure 1, 7, 8, 10, 11, 12, 22, 35, 42, 51, 59, 65, 133, 138, 139, 149, 201, 204, 205, 229, 234, 235, 261, 276, 278, 280, 287, 304, 339, 346, 249
– transport 90
– turnover 90
– viral 53, 90, 125, 130, 136, 147, 148, 157
RNA structural motif 65, 66, 126, 129, 305, 321, 350, 388, 389
– A-minor motif 103, 305, 386, 394
– adenosine platform 305–307
– atlas 384, 385, 387, 389

– GNRA tetraloop 306, 322, 364, 375, 376, 378, 384, 386, 392
– kissing loop 7, 172, 307, 322, 335, 336, 387
– kink turn 102–104, 322, 376, 378, 379, 384–386, 388, 389
– loop-E 321, 329, 378, 379, 389
– pseudoknot 7, 8, 11, 43, 126, 129, 143, 147, 172, 235, 303, 305–308, 311, 335, 336, 346, 349–354, 366, 389, 394, 395
– Sarcin-ricin loop 339, 364, 373, 379, 386, 389, 392
– tetraloop receptor 306, 307, 322, 323
– T-loop 321, 375, 376
RNase *see* Ribonuclease
RNA III 39–41
Root mean square deviation 301, 378, 382, 383, 387, 388
Rot 39
RpsO 43–45

SAM *see* S-adenosyl methionine
S-adenosyl methionine 87, 88, 234, 235, 302, 329
– analogs 87–88
SANS *see* Small Angle Neutron Scattering
Sarcin-ricin loop 339, 364, 373, 379, 386, 389, 392
SAS *see* Small angle scattering
Satellite RNA viruses 125, 136, 201
SAXS *see* Small angle X-ray scattering
SDCP *see* Probing, site-directed chemical probing
Secondary structure 1, 3, 7,10–12, 15, 33, 35, 36, 39, 40, 42, 45, 51, 56, 65–69, 103, 116, 117, 127–132, 137–146, 149, 151, 160, 165, 167, 173, 206, 229, 232–234, 246, 252, 264, 278, 287, 289, 291, 295, 298, 299, 301, 302, 324, 335–337, 340, 341, 345–347, 349–351, 353, 355, 364, 367, 373, 378, 379, 384, 386–393
– elements 1, 3, 7, 126, 136, 139, 144, 145, 235, 246, 252, 299, 300, 384,
– prediction 65, 67, 69, 128, 336, 345, 346, 349, 351
Selective 2'-hydroxyl acylation analyzed by primer extension 31, 33, 35–38, 45, 58
SAFA software 37, 38
Self-assembly 151, 160, 375

Sequencing 30, 45, 51, 53, 55–58, 62, 64, 65, 232, 287, 320
– Maxam & Gilbert 51, 53, 62, 64
– Sanger 51, 57, 58
SHAPE *see* Selective 2'-hydroxyl acylation analyzed by primer extension
Simulated annealing 222, 225, 301
Single molecule 15, 16, 18, 19, 93, 118, 142, 157, 161, 256
– force measurements 157, 158, 161, 162, 172, 176, 178
– FRET 115, 184, 189–198, 202–204, 206–208
Single stranded DNA phages 147
siRNA 76, 91–93, 157, 353, 355
Sliding window analysis 146, 149
Small angle neutron scattering 223, 224
Small angle X-ray scattering 117, 215–217, 221, 223–227, 229–236, 238, 255, 256, 302, 304, 327
– anomalous SAXS 236
– bead models 221, 222, 227, 228
– CRYSOL 224–226, 231
– DAMAVER 223, 234
– DAMCLUST 223
– DAMMIF 221, 222, 227, 232, 234
– DAMMIN 221, 222, 227, 236
– distance distribution function 219, 220, 234, 263,265, 270, 275
– Dmax 219, 220, 226
– dummy residue models 222, 237
– ensemble fitting approach 229
– ensemble optimization method 225, 229
– experimental environment 215, 216
– exposure time 217
– field-emission gun 244, 248
– form factor 217, 218
– forward scattering 219, 220
– GASBOR 222
– indirect transform methods 219, 220, 226
– MC-Sym 232
– momentum transfer 216, 220, 222
– MONSA 222
– multiphase models 222
– normalized spatial discrepancy 223, 234
– normal mode analysis 230
– SASREF 225
– scattering angle 216, 220
– singular value decomposition 226
– SUPCOMB 223
– time-resolved SAXS 215, 216, 226, 238

snoRNA 88, 103
snRNA
– U1 111, 112, 322
– U2/U6 232, 234, 304
– U4 103
Solid-phase RNA synthesis 79, 82–84, 89, 325
Solvation shell 224
Sonogashira reaction 264
Space group symmetry 126
Spin
– diffusion 298
– Hamiltonian 267
– inversion 268
– labeling efficiency 266, 270, 277
– labels' angular correlations 273
Spin labeling 263, 264, 266, 270, 276, 280
– flexible spin label 270, 274
– rigid spin label 263, 265, 266, 271, 273, 276, 278–280
– site-directed 263
– TPA spin label 264–266, 270, 272, 277, 279
Spliceosome 103, 111, 157, 232, 233, 305
sRNA 29, 30, 32, 35, 39
Statistical
– distribution 161
– physics 178
Staudinger reaction 75, 81, 83, 86
Stem-loop *see* hairpin
Steric hindrance 32
Stoichiometry 18, 21, 24, 207, 232
Structural
– model 11, 16, 65, 149, 198, 199, 204, 207, 208, 221, 232, 236, 381
– motif 65, 66, 126, 129, 305, 321, 350, 389
– sorting of particles 246
Structure factor 217, 218
Structure prediction programs
– COMPADRES 384
– Comparative RNA Web Site 385, 386
– CoSSMos 384, 385, 388, 389
– FASTR3D 377, 379, 380
– FR3D 365, 370, 372–374, 377–383, 387, 389, 391
– FRABASE 377, 379, 388
– FRASS 377, 380
– FRMF 377, 379, 384, 385, 386
– MC-Annotate 372, 376, 377, 379, 385, 388, 389
– MC-Search 376, 377, 388
– MotifScan 377–379, 384–386

– NASSAM 376, 377
– PRIMOS 376–378, 381
– RLooM 385, 388, 389
– RNAMSC 384
– RNAView 377, 379, 380, 382, 385, 387, 389
– R3D-BLAST 377, 380
– SCOR 384, 385, 389
Structure-function relationship 45, 233, 243, 250, 251, 253–255
Sugar edge 290, 305, 306, 339, 370
Sugar pucker 289, 291, 295, 297, 298, 303
Super secondary patterns 130
Surface Plasmon Resonance 43
Symbolic search 376–378, 380
Symmetry 101, 106, 108, 126, 127, 131, 132, 135, 222, 225, 263, 272, 321, 393
– dyad axes 126, 128, 144, 150
Synchrotron radiation 61, 215, 227

$T = 1$ viruses 125, 131, 136, 144
$T = 3$ viruses 125, 129, 136, 137, 139, 144, 145, 147
T7 RNA polymerase 86, 291
Tertiary
– fold 33, 128, 277, 291, 335
– interactions 1, 36, 56, 61, 65, 66, 120, 125, 130, 140, 172, 178, 205, 232, 235, 236, 287, 299, 302, 304, 307, 322, 330, 345, 351, 353, 366, 372, 375, 386, 394
Tetraloop 23, 232, 278, 306, 307, 322–325, 384
– GNRA 307, 322–325, 376, 378, 384, 386
– UUCG 232, 278, 279
– receptor 306, 307, 322, 323
Thermodynamic parameters 4–6, 8, 17–20, 22, 24, 176, 340
Thermodynamic stability 3, 7, 10
Threonyl-tRNA synthetase 41
TIR *see* total internal reflection
TMV *see* Tobacco Mosaic Virus
Tobacco Mosaic Virus 136, 141, 151
TOCSY *see* Nuclear Magnetic Resonance, Total Correlation Spectroscopy
Total internal reflection
– fluorescence microscopy 191, 192, 206–208
– objective-type TIRF 192
– prism-type TIRF 192
Transcription 29, 63, 86, 90, 106, 148, 184, 203, 206, 207, 234, 253, 261, 291, 302, 323, 325, 326
– complex 199, 249, 253, 254

– initiation 203, 204
Transient conformational states 147, 157, 172
Transient electric birefringence 13, 14
Transition
– dipole 181, 182, 196
– rates 170
Translation 29, 39, 41, 45, 115, 129, 147, 148, 157, 233, 235, 250, 252, 277, 302, 353
– complex 249–251
– factor 244, 249, 375
Transmission electron microscopy 246, 248
Trefoil 126
Triangulation 202, 208
Trityl radical 281
tRNA 7, 12, 16, 22, 41–43, 52, 53, 87, 88, 90, 128, 142, 217–220, 224, 227, 230, 231, 250–252, 288, 304, 229, 266, 370
– T-loop 321, 375, 376
TROSY *see* Nuclear Magnetic Resonance, Transverse Relaxation-Optimized Spectroscopy
Turnip Yellow Mosaic Virus 125

Ultracentrifugation 43, 217, 244, 245
Ultramicrotom 249
Unimolecular reaction 3, 5
Unit cell 126, 328
UV melting curve 5, 7, 205
UV spectroscopy 21

Van't Hoff enthalpy 3, 22
Van't Hoff plot 205
Varkud Satellite ribozyme 89, 116, 117
VS ribozyme *see* Varkud Satellite ribozyme

Water molecules 128, 224
Wide-field microscopy 191, 192
Worm-Like Chain model 166

X-band frequency 266, 268, 270–274, 281
X-ray crystallography 22, 29, 68, 118, 125, 128, 131, 149, 215, 243, 253, 319, 320
– cryo-crystallography 327
– crystallization conditions 215
– difference density 144
– high-resolution models 223
– molecular replacement 328, 329
– multiple isomorphous replacement 328
– multi-wavelength anomalous dispersion 252, 328
– of oligonucleotides 319
– of RNA 319
– of RNA protein complexes 319
X-rays 36, 65, 215–217, 227, 321, 322, 327, 328

Zeeman interaction 266, 267